PROPHET'S MANUAL

PROPHET'S MANUAL

(FRACTAL SUPERSYMMETRY OF DOUBLE HELIX)

DANIEL SRSA

iUniverse, Inc.
New York Lincoln Shanghai

PROPHET'S MANUAL
(FRACTAL SUPERSYMMETRY OF DOUBLE HELIX)

iUniverse, Inc.

For information address:
iUniverse, Inc.
2021 Pine Lake Road, Suite 100
Lincoln, NE 68512
www.iuniverse.com

http://www.prophetsmanual.com

ISBN: 0-595-29125-2

Printed in the United States of America

CONTENTS

PREFACE...vii

Chapter 1: FRACTAL SPACE/TIME DOUBLE HELIX..................1

Chapter 2: VARIABLE SIZE/MOTION OF CONSTANT GRAVITY...19

Chapter 3: FRACTAL 3/2 DENSITY (BODE'S PERIODICITY) 39

Chapter 4: GRADUAL/INSTANT (QUANTUM) CHANGE.......69

Chapter 5: DOUBLE/INVERSE INERTION 87

Chapter 6: FRACTAL (QUANTUM) SPACE/TIME SEQUENCING...114

Chapter 7: LINE OF SYMMETRY (EIGENVALUE)..................139

Chapter 8: QUANTUM TIME BOUNDARY.............................168

Chapter 9: RELATIVE/UNCERTAIN (FRACTAL) DIFFERENCE..188

Chapter 10: QUANTUM LEAP (DE BROGLIE)..........................209

Chapter 11: QUANTUM COSMOLOGY 242

Chapter 12: CONTINUITY 265

Chapter 13: TIME COORDINATES 289

Chapter 14: RESONANCE...312

Chapter 15: INFINITE SINGULARITY (RENORMALIZATION) 340
Chapter 16: TIME BUBBLE 367
Chapter 17: INVARIANCE 386
Chapter 18: FRACTAL SUPERPOSITION 408
Chapter 19: DISCLOSURE 432
GLOSSARY 451

PREFACE

"The only thing that interferes with my learning is my education"

—A. Einstein

Knowledge, as any tool, can be used in infinite number of ways. Infinity is word often repeated in this book but in contrast with other books, it deals with it, because infinity is only a number. There can be one, two, three or any other number of infinities. There is infinite number of infinities, which is exponential infinity, and then there is infinite number of exponential infinities, which is twice exponential infinity, then infinite number of twice exponential infinities etc. Infinity of exponential infinities is again only one larger infinity, which again repeats itself infinitely. Repetition itself is infinite. Because any infinity is only a different number, resolution of only one is resolution of them all. There is no real difference between one and infinite number of infinities, because infinity of infinities is multiplication/division of only one, singular infinity, regardless which one. What it means is that there is no difference between singularity and infinity—they are two equal opposites or action/reaction of each other. Singularity turned inside out is infinity and inverted infinity is singularity. Infinity in context of larger or exponential infinity is only its singularity.

Anything in between singularity/infinity is only fractal reality of a whole number, just as singularity/infinity is only fractal of larger singularity/infinity. Singularity is inversion that is smallest unit or quantum of inertial infinity, just as infinity is smallest unit or singularity of larger infinity. Singularity/infinity are double (squared) inertion and inversion of inertial here and now in between. Singularity/infinity are equal and opposite boundaries of fractal space/time. Their reality is no different than any horizon in distance. It depends exclusively on point of observation here and now. It is exclusive here and now that makes it real. Any fractal between singularity and equal and opposite infinity is determined by its boundaries, therefore ultimately by point of observation. Here and now is inertion of such singular/infinite duality and all its fractals. There is no reality without inertion, because it is 3rd direction of any dual action/reaction. Cause/effect or action/reaction is real only in its inertial context. Any duality, taken out of its inertial context is totally relative and/or uncertain. It is not reality that is relative and/or uncertain, it is its observation and meaning taken out of its inertial space/time context that is uncertain/relative. Reality itself is changeless in all infinity of its dual changes, because every action is equal and opposite reaction, which is the only principle of reality. All theories and interpretations are only derivates and/or variables of that singular principle, and theories that err from it are simply not correct or complete.

There is also infinite number of different infinities. Every singular point of physical reality (space/time) is infinite number of different infinities within and without. Here and now is singularity of universe and at the same time it is infinity of its atomic reality, which are again infinite number of different infinities. Infinity/singularity are boundaries, while physical reality in between is its fractal content. Singularity/infinity is equal opposite (reaction) of its physical content and equally vice versa. Singularity is smallest fractal of infinity, which is whole number and equally vice versa, because space and time are equal opposites; therefore spatial infinity is time singularity. Relativity is infinite spatial reality in context of singular time, while quantum uncertainty is time infinity in context of singular space. They are two equal and opposite infinities of singular here and now and/or two equal singularities of infinite here and now. Here and now is 3rd (inertial) direction of any duality and relativity/uncertainty are its equal and opposite infinities/singularities. Here and now is constant of two equal and opposite variables (C=d/t= const.). They are equal and opposite fractal boundaries, which are same infinite distortion from inertial here and now, only in diametrically opposite angular directions.

Two equal opposites of inertial framework are equal and opposite curvatures—one is time curvature of space (relativity), while another one is spatial curvature of time (uncertainty). Those two fractal realities are curvatures in opposite directions (anti-parallel). Because every action produces equal and opposite reaction, and where reaction is action of another reaction, such curvature is infinite double spiraling. It is fractal space/time double helix or Fibonacci curvature, where present sequence adds past sequence, in order to generate future sequence. Future sequence is duality of past/present and equally vice versa—present sequence is duality of past/future sequences and past sequence is duality of present/future sequences. Spatial curvature of time and time curvature of space are two strands of double helix and points where they meet are infinities/singularities of inertial here and now. Singular/infinite points are instant addition of past sequence to the present, which creates new (double) sequence. Infinite/singular points are instant inversion of inertion that creates double inertion—i.e. $C/(1/C)=C^2$, where C-light speed. It is instant creation of double duality or 4x1/4 space/time wave that is inversion of inertial 2x1/2 duality. It is instant eigenvalue or resonance of two—inertial and inverse wave, which create double wave. Inversion of inertial time and/or space is its instant mass/energy, while inversion of inertial mass/energy is instant gravity/antigravity (attractive/repulsive forces). Mass/energy is singular/infinite boundary of space/time and equally vice versa. Gravity/antigravity is boundary of mass/energy—one is its singular quantum, while another is infinite number of inverse quanta.

Space/time—mass/energy—gravity/antigravity are one inertial 3D Fibonacci sequence out of infinite number of possibilities, all equally real. It is one loop of double helix. Infinite double spiraling curvature is infinite number of such sequences. One loop is singular relativity/uncertainty without, while curvature itself is infinity of relative/uncertain curvatures within. Singularity/infinity of one curvature is only fractal boundary between its inverse and/or double (squared) sequence. Infinite relativity is singularity of its uncertain sequence in opposite direction. There is no difference between different sizes or periods of time—every inertial size is inversion of its period of time and every inertial period of time is inversion of its size. Inertion/inversion is fractal 3/2D reality, which is relativity in context of uncertainty and/or uncertainty in context of relativity (two anti-parallel spiraling strands). Universe is relative only in direction without; while in direction within it is quantum uncertainty, just as an atom is not only uncertain quantized reality, but also relative in inverse direction.

Only addition of past sequence to the present creates future sequence. Present sequence in itself, without its equal and opposite past sequence is not real, because it is reaction without equal and opposite action. Exclusivity of only one sequence is cul-de-sac that begins/ends in infinity/singularity, which are its boundaries. Reality itself has no bounds—there is always larger/smaller or past/future reality beyond any relative/uncertain (singular/infinite) horizon. Any beginning/end of space in time or time in space is nothing but horizon determined by inertial here and now, because reality is endless double/inverse fractal curvature. It is endless 2x1/2 multiplication/division of inertial here and now. There is spatial universe in every singular point of time, just as there is inverse or time universe in every singular spatial point. Process is repetitive infinitely many times in infinite number of directions. Which infinity/singularity is accepted as boundary is only relative/uncertain choice. Any limitation of reality to its singular/infinite boundary is only subjective and temporary choice, without any real meaning, because it is always same limitation of unlimited. It is always inversion of inertion in both directions of duality (space and time).

Knowledge is infinite/singular boundary of inertial here and now. 3D is duality of action/reaction in context of its inertion. Inertion determines fractality of action and reaction, because it is both at the same time. Inertion is double/inverse cause of its dual action/reaction. Inertion is synchronicity or overlap between equal opposites—it's a whole (quantum) of two equal and opposite halves of duality (2x1/2=1) and at the same time it's only ½ of larger duality. Inertion is a whole (quantum) of horizontal action/reaction and also only action of yet larger (angular) duality. Inertion is synchronicity or resonance between duality and singularity. It is angular ½ of two equal and opposite horizontal ½'s. 3D is 3x1/2 or fractal 3/2. Infinite/singular bounds of inertial knowledge here and now are exclusive function of point of observation. Any inertial knowledge is self-limiting, because through its inertion it determines its self-imposed bounds. Only non-inertial knowledge has no bounds. Non-inertial knowledge is inversion of inertial knowledge; therefore it is its equal opposite. Equal opposites are affirmation/negation or positive/negative polarity. Non-inertial knowledge, which is limitless knowledge, is no knowledge. Any infinite knowledge is only singularity or no knowledge at all in context of infinitely larger (double/inverse) knowledge. Affirmation of inertial knowledge is negation of inverse/double knowledge. Only both—inertial and inverse (knowledge and its negation) create double (squared) knowledge. Only knowledge and its negation (inversion) is twice larger knowledge.

Knowledge understood as absolute is absolute inertion. Only knowledge understood as totally relative/uncertain is non-inertial. Only total relativity of any present understanding of reality is double/inverse understanding. It means, that in order to advance present knowledge, such knowledge has to be negated and forgotten, because it has reached its singular/infinite boundary. Without inversion (negation) of present understanding, which is its past understanding, there is no future advancement. Only double negation is affirmation. Present knowledge is negation of past knowledge, just as past knowledge is negation of present knowledge. Only incorporation of both negations creates one affirmation. There is no advancement of science, without incorporation of negated and forgotten past knowledge, only foggily remembered in vestiges of mysticism and myths of uncertain origins. Uncertainty/relativity are singular/infinite boundaries of science—they are its present zenith. There is no further advancement, but to forget all that is known, just as once all that was known was also forgotten. Only forgetting present axioms of science reveals past achievements. It is endless ritual of Fibonacci sequencing that repeats itself endlessly in determined intervals. It's been done many times before and it will be done yet again. Such ritual that adds past to the present (double negation) in order to generate future (affirmation) is evolutionary curvature of fractal double helix. Only perspective is relative/uncertain—evolution is absolutely certain.

Reaching zenith, present knowledge attained total inertion, which is complete hindrance of further evolution. There is no more room for any further development. Only inversion of such inertion creates yet again infinite room for growth. Larger truth can hatch only by breaking through established shell, which is whole new world. Consequently, thirst for knowledge and improved awareness leads not only through learning, but also through simultaneous forgetting of anything that has been learnt. Anything learnt, taken as absolute is its own negation, because there is no absolute but only fractal (relative/uncertain) understanding. Any knowledge is only one out of infinite number of sequences or possibilities, all of which are equally real. Evolution explores all possibilities, because every Fibonacci sequence is relatively different from any other (infinity of irrational Π and/or φ or golden mean, which are 2D/1D definitions of fractality respectively).

Canonization of any relative understanding is contrary to its own purpose. It defeats its own ends. Canonization of relativity/uncertainty is neither relative nor uncertain—it is negation of own achievements or limiting unlimited truth contained in its premises. Reality is defined by any fractal theory, but subjectivity of its interpretation limits its application. Reality is duality in inertial context—it's a resonant duality, where one of equal opposites negates another, hence double negation that creates affirmation. Only two parts form a whole. Only two fractal theories that contradict each other (inertion/inversion) are twice as large theory, which is not theory at all, but true meaning of them both. Knowledge has to be negated to be advanced. Only questions produce answers, which are only double/inverse questions. Singular answer is composed of infinity of fractal questions, which need further answers, just as any singular question is composed of infinity of fractal answers. Without questioning there is no answers and without answers, there is no questioning. Canonical approach bars questioning and answering—it is solid wall that impedes growth.

Canonization is domestication of wild beast. Attempt to exercise control over uncontrollable is bound to retaliate with revenge. It is legislation of one subjective sequence of thinking, out of an infinite number, while evolution explores all of them. It is anti-evolutionary and outright dangerous, because it undermines one's own existence. Purpose of science is to increase awareness, not putting it to sleep, because only awareness sees ahead. Sleepwalking is dangerous. Canonization is negation of reality itself. It is only partially truth that adaptation and survival are fundamentals of evolution. There is need only for one adaptation, which is adaptation to truth. Such adaptation is no adaptation at all, but rather inversion from all its accumulated artificial adaptations. Every situation is infinite number of life forms; therefore infinite number of different adaptation and any comparison between them is relative. They are all equal. Reality does not change, only its subjective and fractal parts do. Change is only on fractal surface, not in its depth. Every adaptation adds further fractal difference. Larger difference needs larger and faster adaptation, while no difference needs no adaptation. Adaptation is totally relative term, because any kind of adaptation is equal adaptation. Survival based on relativity of adaptation is equally relative. Adaptation and survival based on notion of physical development cannot be further from truth. If survival and adaptation were motivated exclusively by physical needs, there would be no evolution.

There is infinite number of less evolved life forms, which are infinitely better adapted to physical survival. Evolution is equal opposite of adaptation and survival based on physical needs. Evolution is not physical—it is conscious. Physical evolution is only minute fractal in larger evolution of awareness, because only improved awareness is better adaptation. Physical evolution does not lead—it follows evolution of consciousness. Consciousness is its principal, while physical forms are its subordinates. Anticipation and understanding of forthcoming events is pre-requisite of survival. Physical adaptation does not offer that advantage—only raised consciousness does. Only beforehand acquired knowledge and adaptation ensures survival. Only expecting unexpected is true adaptation. Once the change occurred, possibilities of adaptation are infinitely reduced. No amount of exclusive physical adaptation can anticipate or overcome sudden change. Physicality is only a vehicle and not the driver of evolution—it's only a form and not its content. It's only hardware without software to animate it.

Physicality itself is equally relative term, because physical in spatial context is non-physical in context of time and equally vice versa. Large spatial existence is only singular quantum of existence in time, just as existence in time is only infinitely small spatial existence. Spatial physicality is inversion of time physicality—it is only singular quantum of infinity. It is inversion of inertion, which creates double inertion. Non-physical in space is physical in time in context of inertial here and now, which is no difference between space and time or physical and non-physical. Non-physical evolution of consciousness is only infinitely larger (double/inverse) physical evolution, because only non-physical evolution ensures physical survival, while exclusive physical evolution ensures only non-physical survival. Reality is duality of two equal opposites—inertion and its inversion. Only both of them are double inertion or next sequence of double helix here and now. Only negation of negated reality is its affirmation and any fractal adaptation, without its equal opposite is departure from reality—it is its negation. Physical adaptation without non-physical adaptation is not adaptation at all—it is its opposite.

Only two halves are one whole. There is no physical survival without preceding evolution of consciousness. One without the other is self-imposed extinction. Canonization of relative sequence of thinking as absolute truth inhibits that evolution. It is contra-evolutionary and dangerous, because it condemns entire specie to extinction through non-adaptation. Non-adaptation is non-awareness of reality and the ways and means in which it works. Only familiarity with basic law of nature can predict succession of its sequences and ensure its anticipation and its proper understanding. There cannot be intimacy with nature if its superficial physical reality takes precedence. Canonization of subjective fractal sequence as real is self-defeating. It is quintessential to realize fractal nature of reality in order to survive, physically and non-physically. Increased fractality serves its purpose only till it reaches its infinite ceiling. Without inversion of fractal infinity into singular whole, there is no new (double) sequence. Cause of infinite fractal sequencing of Fibonacci curvature is its equal and opposite singularity. Infinite numbers of different sequences are derivates of its non-differentiated singularity. It is singular Π/φ, which is infinite irrational (relative) sequencing.

Every relative (fractal) sequence is real only in context of its wholeness (singularity). Without it, it is not real. Artificial canonization of subjectivity is denunciation of own veracity and willing acceptance of failure. Only true knowledge is acceptance of fact that any knowledge is only irrational fractal of no real consequence. Infinite knowledge understood as no knowledge at all is true wisdom and perfect means of accelerated adaptation demanded by an accelerated evolution. To know everything is to know nothing and to know nothing is to know everything. Indulgence in self-deception has only one outcome—demise. Only repentance leads to salvation and meaning of repentance is only one—rejection of subjective interpretation of reality and acceptance of truth, because only truth withstands all challenges. It's acceptance of singular framework of duality as only reality there is in all its forms. There is no defense against truth, any attack on truth is only attack on oneself.

Canonization of fractal subjectivity is detrimental, because it breeds intolerance and arrogance, which in its turn leads to brutality, destruction and desolation. Absolute knowledge, where there is none, is only unredeemable sin. It's been so, since there is mankind. Anecdote about fruit from tree of knowledge as primeval sin and source of all consequent tribulations of mankind refers to gravity of such erroneous attitude. Absolute tolerance and inclusivity of all fractal parts of a whole is true purpose of evolution. Reality is in understanding, not arrogance, in acceptance, not denial—it is in forgiveness, not punishment, because all errors are born of ignorance. There is no committed vice, which is not self-inflicted. The only vice is not to be true to dictates of consciousness, which has no weight of subjective thinking. Canonical stand is unforgiving attitude, because it is always wrong, regardless of how right it might be. Truth is that there is no absolute wrong or right—it is determined only by inertion of observer. Truth is also that tree of knowledge is self created illusion, because there is only one tree—tree of life, which is only true tree of knowledge. Tree of knowledge is deception and only way back to paradise is to return such false gift back to deceiving snake, which is an ancient translation of spiraling double helix.

Dual fractality of spiraling space/time is and is not science. If science is in fractal theories, than it is not, because it is not a theory—it is singularity of all infinity of different theories. It is singular inversion of infinite inertion that creates double inertion. If science is truth, than it is science. Theories are partial descriptions, limited by its own boundaries, while supersymmetry of fractal double helix has no bounds. Theory is real only in context of its inertion, while supersymmetry is inertionless. Supersymmetry is theory of all theories, because it does not discriminate between them. Every theory is its partial and limited fractal reflection. It reconciles all different theories, not by opposing, but by identifying them as one. It stands alone, with no opposition or comparison from any quarter. Theory that cannot be questioned (has no weak points) is bad theory; therefore it is not only bad—it is the worst, because it is not theory at all—it is reality. It is not search for truth through infinity of fractal answers—it is singular answer of all fractal questions. It is not only science, but also its negation. Inertion and its inversion is double inertion. Consequently, it is square of linear science—it is science of sciences, which is not only science but also magic.

Fractal supersymmetry is addition of past to the present that creates future sequence. It is advancement by reversing in backward direction. It does not discover new—it rediscovers old truth. It is the manual of prophets of old. Prophecy is unexplained knowledge of forthcoming events. It is not science, because it is not understood. Any prediction is prophecy. Science aims to be prophetic by anticipating results. Any knowledge is prophetic. Prophecy is forgotten science. By virtue of being able to predict what modern science cannot, it is infinitely superior, while superficiality of science today is only its pale reflection. It is misguided self-indulgence, supposing that here and now is apex of scientific thinking and human development. Reality is infinite number of possibilities, all equally real. Comparison of equals is meaningless. Here and now is only one fractal sequence, which is only superficially different from any other. It is only inversion of inertion.

If compare we must, then past will come on top, because knowledge that generated prophetic thinking has additional depth of today's superficiality. Only arrogance, bred by ignorance could dismiss prophetic science as groundless superstition. Only understanding principles from which prophecies arise could give an objective assessment. Without having a clue, passing any judgment is irresponsible. Arrogance of exclusive subjectivity and canonization of particular perspective is blind to the truth out of its own volition. It is only relative thinking that here and now is evolutionary advancement when compared with past. It cannot be evolutionary step forward, if answers known before are still being searched for today. Today, which knows all, knows nothing, because it cannot predict what yesterday could.

Any fractal or partial understanding is only local surface of non-local reality—it's only 2D of 3D or duality without its inertial (singular) context. Duality without inertial context is meaningless perspective—it's not reality but its distortion. Only synchronicity or resonance of both equal opposites, where one overlaps another, is real. Reality either is or is not—it cannot be partial. One side of duality is only ½ and only whole reality (quantum) is real. Time is not real without space, just as space is not real without time—only space/time is real. Light is real only in context of darkness, sound only in background of silence and something only in comparison with nothing. It's simultaneous reality of both parts that resonate within each other—one is not more or less real than its counterpart. A duality that differentiates between equal opposites is distortion and not true reality. The only true reality is inertial context of duality—their resonance, synchronicity or overlap, in which both sides have same meaning. It's not this or that but rather this and that. Reality is not exclusivity of only one polarity but inclusiveness of both at the same time. Hence there is no exclusive but only inclusive definition and there is no absolute but only relative interpretation. There is only one "law" of reality that has different names—equivalence, invariance, conservancy, resonance, inertion, synchronicity, balance, symmetry etc. The truth is not in exclusive yes or no, but in their inertial context, when they mean the same thing.

That's the reason why fractality is confusing—it's because it creates confusion before it clarifies it through a pattern. It's a necessity of dualism, because there is no pattern without confusion and equally vice versa. An initial confusion is not discouragement but path toward its resolution.

Relativity and/or uncertainty are prophetic science when set free of its canonical bonds. There is nothing new about fractal space/time supersymmetry—it only incorporates accepted premises embedded in relative or quantum formulations of space/time. It is an honest and unbiased interpretation. Canonization of one exclusive (fractal) aspect of such all-inclusive theories is their handicap. It is not theories that are contradictory—they are in perfect agreement. It is obstinacy in abandoning preconceived ideas about what should and what should not be real that creates illusionary contradictions. It is an effect of fruit from tree of knowledge, which bestows false confidence in distinction between wrong and right, where there is none. It is inertion of point of observation that is arresting their results. Moreover, same inertion is imposing false readings in insatiable wish to prevail in subjectivity over objectivity. True readings are discarded, not because they are false, but because they are not understood (same as prophecies) and replaced by patchwork of ideas that would temporary hid inability of seamless decoding. Result needs no interpretation—it is what it is, nothing more and nothing less. If a particle disappears and reappears somewhere else, without visible connections, is not spookier than Sun disappearing and reappearing beyond horizon or a person disappearing around the corner, just to reappear around other corner. It is particle disappearing and reappearing from beyond singular/infinite horizon, because observed reality is only fractal surface, whose depth is beyond its boundaries.

Particle/wave duality is not contradiction—it is simply reality of particle in motion, because motion is real only in context of another particle. Motion is space/time relativity. Interference between two particles is their relative motion, which is a wave. There is no contradiction between relative universe and quantum reality of an atom, because relativity is gradual curvature only in spatial direction, while in context of time it is the same quantized reality that is atom in spatial context. Only difference is inertial here and now. An atom is within, while universe is without, which creates equal and opposite fractal curvatures. One is concave, while another one is convex non-Euclidean curvature. One is inversion of inertial another. Time of atom is space of universe, while time of universe is space of an atom. Simplicity is not obvious because preconception that exclusion of inertion of point of observation is paramount in determining reality. It is vice versa—its inclusion is paramount in real determination.

That exclusion is an obsession, which distorts fact that reality is comparative only because of point of observation. Comparative reality is duality and duality is 2D—it is only fractal surface of 3D reality. Only ½ of a sphere or any other 3D object is simultaneous in 2D perspective. Another ½ is its equal opposite, which is space of time and/or time of space. Particle (size) is only ½ reality, while its motion (time) is another ½. Spatial reality is only fractal ½; the other ½ is its time. Both—space and time are also only fractal ½ of gravity that is its equal and opposite ½. Space and time are 2x½ in context of its inertial gravity, which is third ½. 3D is fractal 3/2. Spatial observation without its time context or time without its spatial context is meaningless. Disappearing particle is spatial discontinuity and continuity of time. It is spatial reality of time (relativity), beyond boundary of inertial space here and now. Infinite spatial universe taken out of its singular time context is total distortion of its reality. Physical reality taken out of its inverse (non-physical) context of inertial here and now is endless and meaningless complication of its singular simplicity. Reality is simple, while subjective ideas are complex. Degree of complexity is proportional with fractal distance/difference between such idea and the truth. Complexity is fragmentation of only one reality and it leads in opposite direction. Complexity is curvature, while simplicity is straight line. There is no control in forcing an answer by imposing heavy complexity over complexity—it is only fractal patchwork of short duration, because one patch contradicts another. True answer needs no control because it is effortless.

Subjectivity obscures. It obscures, because it sees only what is expected and overlooks what is not. Illusion of distinction between wrong and right is blinding. It repeats one error infinitely many times, till absurdness of idea is not apparent. Absurdity of such subjectivity is instant inversion of inertion. It is burst of a bubble blown out of proportion of its inertion. Error repeats in infinity of directions. It is not an exclusive error of physical science. It is fundamental error of mankind in its conviction of its own exclusivity and superiority. Superiority is inverted inferiority—they are two sides of same coin. True knowledge is neither superior nor inferior—it simply is. It is one equal out of infinite number of different possibilities. Any exclusion is fractal and total subjectivity. Subjectivity is is core of all misunderstandings and forcing convenient answers at all costs. It creates illusion. It is illusion of superiority of present civilization over one in the past or illusion of superiority of one over another. It is illusion that one materialistic approach is different from any other. It is basic premise of materialism as supreme truth and not as valueless superficial belief. Inertial physical reality here and now is only one out of infinite number of different surfaces of 3D, all of them equally real. Any superficial belief is materialistic, which is another name for idolatry.

3D is infinity of singular 2D surfaces—it is its boundary. Physical reality is resonant space/time duality; therefore only its surface, which is small fractal of larger reality. It lacks 3rd direction, which is inertial here and now and which fuses all fractal surfaces into one whole. 2D is subjective exclusivity, while 3D is objective inclusivity. Veneration of any form of materialistic or superficial reality is detrimental to evolutionary process, because its shallowness defeats its own end. There is need for distinction between wrong and right only in superficial materialistic context, because 3D is non-differentiated and such relative distinction is meaningless. Subjective choice of one physical sequence as right precludes infinity of different possibilities as wrong, and equally vice versa—determination of one form as wrong, automatically gives right to infinity of others. Evolution explores all possibilities and obstinate clinging to any superficial differentiation is perpetuation of tribulative chain of learning and ultimate demise, because it is always an idolatric approach. Any superficial error is infinite repetition till accumulated inertion is equal to its fractal inversion, which is instant (quantum) evolutionary step. Uncertainty is inverted relativity—it is its singular quantum. Infinite repetition of relative choice is its instant quantum inversion. Knowledge of that repetitive pattern of fractal double helix, which is infinite precessional curvature of inertion/inversion =double inertion is essence of prophetic science.

Prophecy is comprehensive understanding of reality. It is magic or superstition only in ignorant perspective. Only what is understood is believed and prophecies are not understood. It is essential for them to be understood to be believed, because time of their fulfillment is now. Only belief changes superficial materialistic attitude and inevitable acceptance of truth. Supersymmetry is not only nexus of science and philosophy; it is umbilical cord between physical and non-physical, which merges science with religion and spirituality. Reality is only one. There is no different physical, spiritual or philosophical truth. They are non-differentiated one. Any difference is only fractal subjectivity of no comprehensive value. Fractal supersymmetry is double/inverse symmetry—it is symmetry between symmetry and asymmetry. It is double perfection of perfection and its equal and opposite imperfection. It is reality that fuses real and unreal into twice as real (or right and wrong into twice as right—or wrong). Boundless double fractal precessional curvature is singular source of all fractal infinities. It is prophetic science that is singular source of all ancient wisdom and magic. It is singular source from which all infinity of forgotten and misunderstood ancient beliefs arise.

It is common denominator of all systems of belief, from shamanistic to monotheistic. Fractal supersymmetry is cradle from which all ancient achievements arise. It is one fundamental of all civilizations of old. Egyptian, Sumerian, Indian, Chinese—all major and minor cultures of past age shared one common knowledge, which is fractal supersymmetry of 3D reality, without its relative physical/non-physical distinction. Awareness of superficial fractality of any relative physical sequence infinitely transcends any modern scientific achievement. Apex of modern science is relative/uncertain or gradual/instant (quantum) duality. Fractal supersymmetry not only fuses them into one whole, it also repeats that one whole in infinite number of different forms. Fractality is inertial context in which there is no real difference between relativity of macro-space and quantum uncertainty of micro-space. It's their resonance or symmetry. Moreover—it's resonance between physical and spiritual because any duality is double/inverse perspective of one whole in between. Today's science is infinite fragmentation, while yesterday's science is infinite multiplication of inertial here and now. It is infinity of relative/uncertain (fractal) curvatures, one within the other, where one precedes another because of their fractal 3D/2D symmetry.

The error of both is the same. It is their exclusivity, which is its ultimate demise. Today's science is fractal exclusivity shared by all, while yesterday's science is multiple exclusivity shared by few. Shrouded in secrecy and exclusivity, in context of advancing materialistic beliefs, past science is gradually reduced to one out of many idolatries. All that remained of its infinite depth is only its shallow surface, which is groundless superstition without understanding of its lost foundations. Superficiality is primal error that repeats in infinite number of relatively different sequences.

Fractal supersymmetry exonerates and restores old standards in modern context. It is resurrection of past in present that gives birth to future. It brings to life dead science of Sumerians, Egyptians, Mesoamericans and Indian cultural centers. In light of fractal supersymmetry, pyramids, written and symbolic Egyptian records speak in unison with Sumerian astrology; Tao and Vedic philosophy, science and religion; Kabala; mysticism; Mayan mathematics and astronomy; Celtic folk tales, shamanistic practices and modern monotheism and physical science. It is singular source of all prophecies, from biblical Book of Daniel and Revelation to Mayan calendrics and its end date. Book of Daniel gives an account of fractal reality, while Mayan division of time is its end result. Book of Daniel is explicit description of fractality of dual space/time curvature, where space is 3/2 fractal of time and equally vice versa. Its accurate description of four winds of haven is 4x1/4 multiplication/division of inertial time, which is divided by 4 inversion points of a wave. It is parallel description of De Broglie eigenvalue, where each ¼ of precessional curvature is an instant quantum change that creates different spectral line, which is different Fibonacci sequence of reality. It is description of 2x1/2 wave in context of its inertial 3rd ½ or 3/2 fractality without beginning or end date of such cycle. Mayan calendrics provide accurate dates of such cycle. End date of present cycle of constant passage of time, which is singular quantum of relative reality is 21.12. 2012. It is not just because Mayan calendrics says so, but also because it is date of gravitational (time) eclipse, which occurs at such predetermined intervals.

End of time as fractal spatial reality with winter sun 2012 is not a possibility or speculation—it is an absolute certainty, which has been communicated down the ages and is confirmed by an honest application of modern scientific principles. It is date when present sequence of reality instantly incorporates past sequence to create future sequence. It is instant quantum leap in evolution, which is not gradual spatial change in context of linear passage of time but instant, infinitely larger spatial change in context of infinitely larger/smaller (accelerated) time, when gravity is not singular (changeless) but linear constant (passage of time). It is an instant resonance when time precedes space and gravity precedes time. It is creation of new sequence of inertial Earth's reality in its dual solar/galactic context. Moreover—same prophecy is base of modern division of time into 24 hours/60 minutes/60 seconds, derived from ancient Sumerian sources. To understand motive behind such division is to understand fractal supersymmetry and reason of prophetic passages in Book of Daniel or Mayan calendrics, because they have same meaning.

Vedic, as well as Tao science is reflection of same fractal geometry of space/time and its meaning. Pyramids, ziggurats and monolithic structures are monumental description of such space/time fractality and frequency of their supersymmetry (eigenvalues). Quantum change of time, which is gravitational change, is focal point of ancient science and religion. Celtic folk tales of time when sky falls are illustrative description of such quantum event. Biblical flood is previous and end of ice age is yet another quantum leap of time. Frequency of such changes is 5 184-year interval, which is fractal 4/5 reality of precessional period that repeats in ever larger leaps in multiples of 4x5=20. 5 184 years is singular quantum of relative reality (constant passage of time), while 20x5 184=103 680-years interval is its double quantum known as Milankovich cycle. Any comparison is absolute only within boundaries of its inertial framework. 5 184 is infinity of singular time here and now—it is time boundary of one spatial fractal sequence. Fractality is exponential difference; therefore any comparison between two quantum periods is liable to exponential (relative) difference of 1/10. Fractal difference between two relative periods is reason of often quoted 10-fold longevity in ancient sources. It is not real but relative (fractal) comparison, because any reality is always same here and now.

Quantum event is not meant to be cataclysm, but an opportunity. Nevertheless it is a catastrophe if not anticipated. It will not be anticipated, unless its reality is accepted, and it will not be accepted, unless understood. Repentance, which is acceptance of truth and change of mind, is absolute pre-requisite of anticipation, because it is closure of complete time curvature or instant self-interference. It is judgment day or time of reflection, in which only judge is oneself and only oneself passes sentence according to causality principle, because total freedom of choice is fundamental axiom of reality. It is cause/effect principle. It is total freedom of choice, because any choice is same choice of no choice. There can be only one choice, out of infinite number of equal choices. One choice is infinity of fractal actions/reactions, which is fractal space/time reality of such choice. Reality is not fractal duality but their resonance—it's a whole (quantum). Only space and/or time of a quantum are double/inverse fractals. Fractality is only its superficial or temporary duration. Only quantum leap, which is instant change of singular choice, is multiple of fractal reality. It is framework in which inertial here and now is real. It is non-physical reality that is double/inverse physical reality.

Aim of this book is to help changing minds in anticipation of upcoming event. Only raised awareness is adaptation as mean of survival. Awareness is survival and existence through anticipation. If flock of birds forgets about upcoming seasonal change, they will seal their own fate. Mankind has forgotten of seasonal changes that occur at long intervals, because of superficiality of materialistic beliefs. It is known that sudden changes do occur, but because of prevalent attitude, its frequencies are forgotten. Sudden changes are portrayed as rater irregular unpredictable freaks of nature, through lack of knowledge. Accidental or coincidental explanation is only disguise of ignorance. There is no room for coincidence or accident in context of fractal curvature of precessional space/time. It is perfect imperfection that is double perfection. Sunset and dawn, equinox and solstice are not freak accidents that occur at irregular intervals and neither is quantum change of time, which follows same fractal pattern. Just as sunset/dawn and solstice/equinox are inversion points between two equal opposites of sinus wave, so is gravitational quantum change of time. It's sunset of old and dawn of new age, and it is equally predictable. This discourse is wake-up call before a new dawn.

CHAPTER 1

FRACTAL SPACE/TIME DOUBLE HELIX

In 3D context, constant is a derivate of two equal and opposite variables. Constancy is changeless pattern between two interfering variables (C=d/t=const). Constancy is singular, while variability is its equal and opposite infinity. Difference between constant and its equal and opposite variability is infinite, because difference is exponential. Exponential difference is fractal (dimensional) difference. One constant is infinity of fractal variables, just as 3D cube is infinity of 2D surfaces. 3D is always double fractal difference—it is two equal and opposite infinities (variabilities) of singularity (constant). Singular surface is two equal and opposite cubes on inertial and inverse side. Infinite variability is only surface reality of singular constant and equally vice versa—singular constant is only surface of equal and opposite infinity. Two equal opposites are inertion and its inversion, which are 1D fractal difference in between. 1D fractal difference in between is overlap of both—it is their 3rd direction, which is inertion/inversion=double inertion.

Every 3D singularity (constant) is double and inverse infinity (variability). Singular here and now is double infinity—infinity of space and time. Present is singular point between past and future infinity. Empty space is singular point between two infinite densities—mass and energy, etc. Constant (singularity) is derivate of exponential reality. It is one of fractal surfaces of exponentially larger/

smaller (faster/slower) reality. It is flattening of round curvature (Euclidean projection of non-Euclidean curvature). Every singular direction is constant of its two equal opposites, which are its inversion within and double inertion without (square/inversion). One direction is multiplication, while another is fragmentation of inertial direction (fractal difference or overlap) in between (relativity/uncertainty). Inertial direction between two surfaces (or inertial point between two linear directions and/or inertial surface between two cubes, etc) is their eigenvalue (synchronicity). It is only singular point in which both infinities are simultaneously real. Inertion between inversion and double inertion is their derivate, where derivate in space is exponent in time and equally vice versa. Singular eigenvalue of two equal and opposite infinities is their common inverse/double derivate, which is equal and opposite fractal difference from each—it is 2D of 3D space and/or 2D of 3D time. Eigenvalue is singular point, where derivate of space in time is equal to derivate of time in space.

While multiplication of a whole number with itself is a square (exponent 2), its fragmentation is singular inversion (exponent-1). There is 1D fractal difference between such equal opposites, because of inertial (constant) direction in between (here and now). 2+1=3 and 2-1=1. Synchronicity of inversion (-1) and double inertion (2) is 3D here and now, which is result of fractal 1D difference between such equal and opposite curvatures. Inertial difference of such duality is 2/1 in one and/or ½ in another direction. It is duality in context of inertial ½ (or ½ in context of inertial duality). Two steps in inverse direction is one step in double direction, which is fractal duality of space/time. Double infinite time is singular space and double space is singular time. In 3D context, every action is two equal and opposite (fractal) reactions, which are reactions in space and time. 3D is fractal symmetry, which is infinite asymmetry on surface (2D) and equally vice versa—it is 2D symmetry that is infinite asymmetry in its depth. Surface (2D) symmetry is equality of only two equal opposites, while fractal 3D symmetry is infinite—it is equality of infinite number of different fractal parts. Fractal symmetry is supersymmetry—it is 3rd direction or inertion of symmetry/asymmetry, because it reconciles such horizontal exclusive opposites in context of its inclusive 3rd direction. Symmetry is surface exclusivity of two equal opposites. Asymmetry is 3D inclusivity of all such surface opposites. In context of fractal 3D/2D, there is no difference between symmetry and asymmetry—they are only 2D and 3D projection of same inertial 1D fractal difference in between. Fractal difference is eigenvalue of fractal space/time. It is supersymmetry, which does not differentiate between surface symmetry and asymmetry, because surface asymmetry is only infinitely larger 3D symmetry.

Double inertion of inertion/inversion is $C/(1/C)=C^2$. It is affirmation that is double negation. Double inertion of surface and 3D is their 4D reality. Two fractal opposites are twice inertial wholes. Symmetry in context of asymmetry is double symmetry, which is supersymmetry that does not differentiate between surface symmetry and asymmetry—they are only its two equal opposites. Double inertion is 1D fractal difference in between fractal opposites. In context of double inertion, fractal perfection in context of equal and opposite imperfection is double perfection. Variable and a constant, or derivate and exponent, are in context of inertial fractal difference only double constant and/or double exponential reality. Singularity and infinity is double infinity, which is also double singularity. Every 2D fractal is simultaneously real in its 3D context (infinity), while 3D is only fractally real in context of any surface fractal (singularity). Only all 2D fractals are one whole 3D. Totality of all fractal asymmetries is one symmetry and totality of all fractal imperfections is one perfection. Surface (dual) asymmetry is infinitely larger symmetry—it is singular/infinite symmetry or symmetry that has no equal opposite. Likewise-surface imperfection is infinitely larger perfection that has no comparable opposite. Duality is comparative, while singularity is incomparable reality. Singularity is inertial context of dual action/reaction, which is twice larger/smaller reality. Singularity is eigenvalue of infinite number of variable dualities. It is synchronicity of all fractal relativities/uncertainties.

Duality is always surface reality. It is two equal opposites that are real only one in context of another. Action is real only in context of equal and opposite reaction. Space is real only in time and time only in space. Space is negation of time and time is negation of space. In dual context, affirmation is real only in context of its negation. Up is real only in context of down, left in context of right, strong in context of weak, etc. Duality is comparison that is no comparison. It is comparison of two equal opposites. Only fractal comparison is real, because total comparison is absolute negation of one in context of another. In other words—comparison is curvature between two equal opposites, and in context of curvature, one is totally relative/uncertain reality of another. It is double uncertainty/relativity or double fractal difference. Inertial context of duality (its 3rd or inertial direction) is comparison itself, which is singular point, where two equal opposites meat (have one meaning). Comparison is simultaneous reality of both—it is singularity between two equal and opposite infinities. It is supersymmetry of all fractal dual symmetries and asymmetries. It is 3D or whole number of fractal surfaces. In context of any fractal surface, such infinitely larger reality is only its singular point. It is non-reality of fractal surface or its equal opposite.

Singular comparison is equal opposite of fractal duality—it is its action/reaction, which is infinitely smaller/larger action/reaction. It is space of time that is infinitely larger/smaller time than fractal 3D time here and now. It is time of space that is infinitely larger/smaller space. Comparison is exponential reality of dual action/reaction—it is constant of all its variables and/or variable of all its constants. Comparison, which is singular point of duality is 3rd equal opposite that is simultaneously twice as large/small duality. Duality of two equal opposites is real only in context of its singular comparison and vice versa—comparison is real only in context of comparable opposites. Symmetry is real only in context of asymmetry and perfection only in context of imperfection. Action/reaction (space/time) is real only in context of its inertion (gravity) that is singular here and now of such fractal space/time. Singular here and now is meaning of comparable space/time reality. Without such inertial point, space/time is totally unreal. Inertial singular point here and now and its infinite space/time are action/reaction of each other. One is confirmation/negation of another. Space/time (physical) reality is inversion/double inertion of inertial here and now. Not only that reality cannot be determined objectively, without inclusion of inertial here and now in its definition—it does not exist at all without such point. Inertial here and now and its infinite universe/micro-world are mirror image of each other—each is only ½ of a whole in context of their 3rd direction, which is inertion of all infinity of different inertial points of space/time. Space and time are 2x1/2 multiplication/division of inertial here and now, while space/time (physical reality) and inertial here and now are 2x1/2 of singular reality that is incomparable as physical reality, because it is beyond its relative/uncertain (fractal) boundaries.

Double inertion and inversion of inertial here and now are same pattern—they are singular points of double infinity. In context of double inertion, singular inertion is its inversion, while its double inertion is quadruple inertion in context of here and now. In context of inversion, singular inertion is its double inertion, while its inversion is double inversion (inverse square) in inertial perspective here and now. Reaction of singular inertial difference here and now is infinite 2x1/2 curvature of multiplication/fragmentation. It is infinite repetition of 1D fractal difference 1.111...which is difference between diagonal of a square and diagonal of a cube. Fractal difference is dimensional difference. 3D is synchronicity of 2 variables in context of constant directions—it is eigenvalue of inertion/inversion in context of double inertion (equally vice versa—any combination is the same). 2x1/2 curvature in one direction is 6x1/6 multiplication/fragmentation or triple repetition of 1D fractal difference (3.333..) in all 3 directions. Cube of 6 different surfaces is infinite boundary of any singular direction. 3D is triple duality (2x1/2) and/or double 3D. Cube is determined by any double set of two equal and opposite surfaces (inertion/inversion), which are 2x(2x1/2)=4x1/4 reality. There are 3 different permutations of such surfaces, all of which describe same cube. Any of fractal 6/4=3/2 realities is same inertial cube in different angular perspectives.

Inversion of inertial direction (1D) is double direction (2D square) that creates 3D cube in their duality. 3D coordinate system is singular set of inertion/inversion/double inertion. Each of such 3 directions is comprised of further 2x1/2 multiplication/division or additional 6 directions, which are 1D fractal difference from first (triple) set (in inertial context of first set). 2x1/2 of such 6 directions is another 3 directions (in inertial context of first/second set as its boundaries), which form inverted coordinate system of fist set. All 3 sets are fractal parts of a cube (see illustration 1). Inverted coordinate system is 2D fractal difference from first set, while 6 directions in between are fractal reality of both (eigenvalue), which is 1D fractal difference in opposite directions (same scalars, different vectors) of 1st (inertial) and 2nd (inverse) coordinate system. Inertial and inverse coordinate systems are double 3D—they are space and time duality (inertion/inversion). 3D space and 3D time is inertial framework (two equal and opposite whole numbers) of fractal 3/2 reality in between, which is synchronized reality of both or 2x3=6 space/time directions.

6 space/time directions are ½ of total number of directions and/or twice number of each of dual inertial (coordinate) system. 6 out of 12 directions (fractal ½) is singular space/time here and now in context of universe/micro-world, which is equal and opposite ½ (therefore ¼ each). 9 out of 12 is fractal ¾, which is total number of exclusive spatial or time directions out of total of 12 inclusive directions (including inertion of such duality, which is 6 fractal directions). 6 out of 9 directions is fractal $^2/_3$ reality, which is number of exclusive spatial or time directions in context of its inertion (also-vice versa). Inertion of space is time and inertion of time is space. Fractal 3/2 is space of time and/or time of space. Likewise—fractal $^2/_3$ and/or its inversion—3/2, is here and now in exclusive spatial or time context, while fractal ¾ and/or 4/3 is space inclusive of inertion (here and now) in context of time or time (inclusive of inertion) in spatial context. ¾ is inversion of inertial $^2/_3$, which generates fractal 4/5 reality that is their eigenvalue. One is exclusive spatial or time difference, while other is inclusive or space/time difference. Fractal ½ in curvature of one direction is fractal $^2/_3$ and fractal $^2/_3$ in curvature of additional direction is fractal ¾. (½)/($^2/_3$)=(¾) is triple fractal reality of space/time, which is double fractal difference between such 3 curvatures. ¾ is fractal relativity, $^2/_3$ is fractal uncertainty, while fractal ½ is relativity/uncertainty = double relativity or uncertainty.

Illustration 1 -Fractal geometry of 3D cube

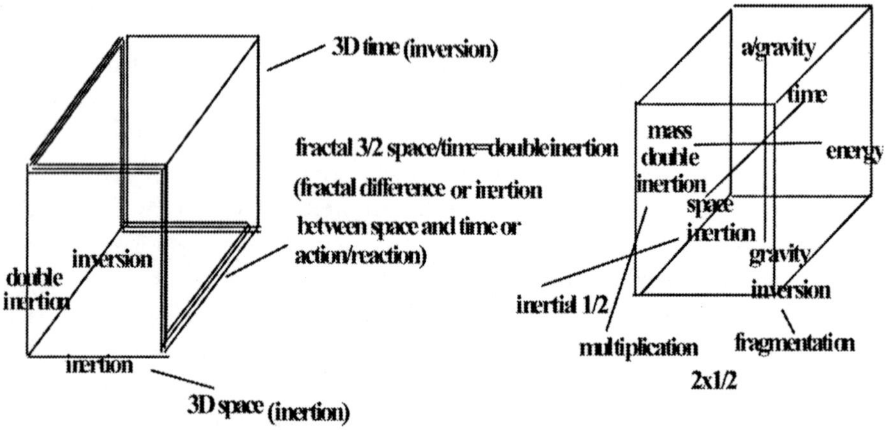

Each singular direction is eigenvalue of two equal and opposite sets of equal and opposite directions—1=2x(2x1/2)=4x1/4. Each singularity is double duality and each double duality is singular 3D. There is no different reality. There is only one reality and its common denominator is 3D cube. 3D is reality of 3 equal and opposite directions and geometry of 3D cube is perfect description of its double fractality. In 2D perspective of 3D, only 3 out of 6 surfaces (½) and 9 out of 12 directions (¾) are simultaneous (eigenvalue), while inertial space/time is $^2/_3$ of its space or time. 3D has 6 surface boundaries, 12 linear and 8 singular boundaries, where linear and surface boundaries are two equal and opposite infinities (1D and 2D fractal difference). Linear infinity is linear (constant) change, while surface infinity is accelerated change of singularity (constant of no change). Two equal and opposite surfaces of a cube are space/time, while two sets of perpendicular surfaces are its mass/energy and gravity/antigravity dualities. Mass/energy is 1D fractal difference of inertial space/time, while gravity/antigravity is 2D fractal difference. Mass/energy and gravity/antigravity are inversion/inertion of double inertion here and now (space/time) and equally vice versa.

Singular and double fractal difference that is mass/energy and gravity/antigravity of space/time (equally vice versa, because reality is triangulation and any fractal difference between space/mass/gravity is relative/uncertain) is singular and double derivate of a constant—i.e. $X^3=3xX^2$ (1D fractal difference) =6X (double fractal difference) =6 (triple fractal difference). In context of illustration 1, $3X^2$

are 3 surfaces of a cube (first coordinate system that is 1D inversion/inertion=double inertion), of which one is inertial space and two are its double/inverse space. Double/inverse spaces of inertial space are two directions of infinite time (i.e. past/present or slow/fast passage of time) of singular space in between. They are its time without and time within, which is universe/micro-world or positive/negative (future/past) time. Universe/micro-world are equal and opposite time curvatures (positive/negative non-Euclidean curvatures), therefore relative/uncertain or +/-1D fractal difference. Relativity is inverted uncertainty, just as uncertainty is inverted relativity. 2 equal and opposite directions of inertial space are relative and uncertain space (uncertain/relative time). Such two different (inert/inverse) space/time curvatures are mass and energy of inertial space (and/or energy/mass of inertial time). Mass/energy are size/motion in context of inertial space and/or—motion/size in context of inertial time. There is 1D fractal difference between mass and energy (size and motion). It means, one is constant, while another one is its variable. It is constant size in variable motion and/or constant motion of variable size (constant wave of variable particle and/or variable wave of constant particle).

In context of fractal space/time duality—there is one fractal space between two whole times and/or one fractal time between two whole spaces. It is infinite space/time/space/time etc fractal curvature; therefore time is singular/infinite boundary of space in between and equally vice versa—space is infinite/singular boundary of time. Each boundary is inversion of inertion that creates double inertion (space of time and/or time of space). Each boundary is exponential increase/decrease of inertial reality within. Each boundary is change from no change that is double change into constant (singular) change and/or vice versa—from constant change into no change/double change, where no change/double change are inversion/double inertion. Every boundary is 2D surface between equal and opposite 3D's and/or 3D between equal and opposites 2D surfaces. There is two equal and opposite 2D times in depth of inertial 3D space and/or two 3D times on each side of inertial 2D space in between. 3D dual space/time reality is one sequence of infinite fractal curvature—it is either space/time/space or time/space/time (3D/2D/3D and/or 2D/3D/2D). It is either double time of singular space or double space of singular time. Space is depth or 4th direction of 3D time and time is depth or 4th direction of 3D space. There is no difference between space and time—time is inverse/double space or infinitely larger/smaller space (1D fractal difference). Likewise, space is double/inverse time. Space and time are same inertial reality as here and now, only in diametrically opposite perspectives (within/without).

Singular space is dual time and singular time is dual space. Space is duality of equal and opposite 1D fractal differences of its dual time, while time is duality of equal and opposite fractal differences of its dual space. It is infinity of spheres or shells, one within the other, where every preceding/succeeding shell of spatial shell is time, and two inward/outward shells of time are spatial spheres that are +/ -1D difference from inertial shell in between (positive in space is negative in time and vice versa). Space/time are 1D fractal difference, while space/space and/or time/time are 2D fractal difference. 3D is duality of 1D and 2D fractal differences, where 1D fractal difference is space of dual time (time of dual space), while 2D fractal difference is space of dual space (time of dual time). 1D fractal difference is space/time and/or time/gravity duality, while 2D fractal difference is space/gravity duality. 1D fractal difference is size/motion duality, while 2D difference is density of such size/motion. 1D/2D fractal differences are singular and double relativity/uncertainty. 1D fractal difference is variation (derivation) of one, while 2D fractal difference is variation (derivation) of two directions. 1D fractal difference is singular non-Euclidean curvature of Euclidean inertion, while 2D fractal difference is double non-Euclidean curvature (3D sphere). Equally vice versa—1D fractal difference is singular flattening of non-Euclidean inertion into Euclidean curvature, while 2D fractal difference is flattening of double non-Euclidean curvature into flat Euclidean reality (or double/inverse flattening). 1D fractal difference is duality (action/reaction or inversion/double inertion) in context of singularity (inertion), while 2D fractal difference is singularity in context of double duality. One is inversion of another (inertion), which generates double inertion (3D here and now).

Inertial difference between two spatial realities (or two times) is 2D, while inertial difference between space and time is only 1D. **Space is fractal ½ reality of time and equally vice versa—time is fractal ½ reality of space.** Space is surface between two equal and opposite 3D times, while time is surface between two equal and opposite 3D spaces (or 3D between two equal and opposite 2D's). Two equal and opposite times of inertial space are relative/uncertain space, just as two equal and opposite spaces of inertial time are its uncertain/relative time. Relative space is uncertain time, while relative time is uncertain space. Uncertainty/ relativity are 1D fractal difference of time within/without space and/or space without/within time. Double relativity/uncertainty (2D fractal difference) is inverse relativity/uncertainty—it is space without/within space and/or time within/without time. Double/inverse space and/or double/inverse time is gravity/ antigravity of inertial space/time, which is double derivate (6X or 6 linear directions of attraction/repulsion).

Mass/energy is 1D fractal difference—it is space of time and/or time of space (size/motion), while gravity/antigravity is 2D fractal difference—it is double/ inverse space and/or time. Mass (spatial size) is positive spatial (non-Euclidean) curvature of time and/or negative time curvature of space. Energy (spatial motion) is negative spatial curvature of time and/or positive time curvature of space. Curvature is exponential reality of a constant, while its inversion, which is flattening (non-Euclidean reality) is derivate of exponential (accelerated) reality. Mass is exponent (curvature) of time in space and/or derivate (flattening) of space in time—it is constant of time and variable of space. Energy is exponent (curvature) of space in time and/or derivate of time in space—it is spatial constant and variable of time. Mass is time within (inside) space and/or space without (outside) time, while energy is space within time and/or time without space. One is 3D time as spatial surface, while another is 3D space as surface of time. Mass/ energy are vertical equal opposites of horizontal space and time.

Space/time is inertial context (3rd direction) of dual mass/energy, which is its inversion/double inertion and equally vice versa—mass/energy is inertial framework of dual space/time (inversion/double inertion). Mass/energy and space/time are double duality or 4x1/4 multiplication/division (wave) of inertial here and now. They are 4 linear directions of a surface (+/- space and +/- time). Singular duality of mass/energy is 1D fractal difference from singular (non-differentiated) space/time. Space in time curvature is mass within and energy without, while time in spatial curvature is mass without and energy within. Mass/energy is positive/negative time (variable) of inertial (constant) space and/or negative/positive space (variable) of inertial time (constant). Mass and energy are positive/negative spatial density (size density) and/or negative/positive time density (density of motion), where one is singular (constant), while another one is infinite (variable), because of fractal difference. Infinite number of spatial densities (massive sizes) is only one time density (inverted spatial density that is energy without or motion of massive size), which is light speed. Equally vice versa—only one spatial size (massive universe and or massive atom) is infinity of different motions within.

Mass/energy are space in time context and/or time in spatial context. They are curvatures (exponents) in context of one flat space/time (derivate) and equally vice versa—they are derivates in context of one curvature (exponent). Mass/ energy is 2x1/2 multiplication/division of inertial space/time, just as space/time is 2x1/2 multiplication/division of inertial mass/energy. Singular mass or energy is inertial/inverse space/time duality, just as singular space or time is inertial/inverse mass/energy duality. Each duality is 2x1/2, while all 4 are double duality (4x1/4). Space/time and mass/energy are horizontal and vertical (angular) dualities—one is form of another one's content. Inverted/double space/time is mass/energy and equally vice versa—mass/energy turned inside out is space/time. Space and mass are 1D fractal time difference, while time and energy are 1D fractal spatial difference. Space/energy and/or time/mass are 2D fractal difference. Space/energy are double space or space within space, while time and mass are double time. Mass is time curvature of space (inversion), while energy is its flattening (double inertion). Mass is round and energy is flat (linear) direction of non-differentiated 3D space. Energy is round and mass is linear direction of time. Mass/energy are inert/inverse Euclidean/non-Euclidean curvatures of space/time, where non-Euclidean curvature is two equal and opposite Euclidean curvatures, just as Euclidean curvature is two equal and opposite non-Euclidean curvatures.

Space and time are positive/negative (inertion/inversion) non-Euclidean curvatures of Euclidean space/time and equally vice versa—mass and energy are inertial and inverse non-Euclidean curvatures of Euclidean space/time. Double round curvature is flat curvature in between and double flat curvature is round curvature in between, where round in space is flat in time and equally vice versa. Mass in spatial context is energy in context of time and equally vice versa. Space in massive context is time in context of energy, just as space in context of energy is time in massive context. 3D mass/energy are 2D/4D space/time (inversion/double inertion), just as 3D space/time are 2D/4D mass/energy. Mass/energy of the universe are 4D of 3D inertial space (therefore only 2D time) and equally vice versa—mass/energy of micro-world are 2D space (where any increase of spatial fractal reality is equal and opposite decrease of time and equally vice versa; therefore 2D space is 4D time, just as 4D space is 2D time). Mass/energy are relativity/uncertainty of space/time and equally vice versa. Uncertainty is inverse/double relativity and relativity is inverse/double uncertainty.

Any 2D description of 3D reality is fractal (relative/uncertain), because space/time and mass/energy differentiation is relative. Space in inertial context is mass in inverse and energy in double inertial context and equally vice versa—mass in inertial context is energy/space in inverse/double context. Mass of spatial reality is energy in context of time, just as spatial energy is mass in time context. Spatial motion (constant change in time) is constant (double/inverse) size (no change) in time and equally vice versa—constant spatial size is constant (double/inverse) motion in time. Space/time and mass/energy are 4 surfaces of a cube (and/or 4 linear directions that are boundaries of a square) and any distinction is relative, because all 4 surfaces are same cube. Space in one direction is time when direction of observation is diametrically opposite. In positive/negative vertical direction, same reality is mass/energy. Space/time in context of mass/energy is 2Π curvature divided into two equal fractal opposites or 4 quadrants of surface reality (Π/2 difference, which is singular quantum of 4x1/4 curvature). Space/time is 1D fractal difference or 2x1/2 differentiation (singular quantum of 1D difference—Π), while additional mass/energy is 2D fractal difference or 4x1/4 differentiation from inertial here and now. There is no absolute description of 3D in dual context.

Mass and energy are inverse/double spatial curvatures and/or double/inverse time curvatures, where time curvature is flat space; just as spatial curvature is flat (linear) time. Energy, $\mathbf{E=d^2}$ (within) and/or **d/t** (without), while mass is $\mathbf{m=t^2}$ (without) and/or **t/d** (within)—where within in space is without in time and equally vice versa (d-space and t-time). **D/t=C=E** (motion) and **t/d=1/C=m** (size). Size in motion is mass/energy (particle/wave) duality, which is $\mathbf{(d/t)/(t/d)=(d/t)^2=C^2=E/m}$. Singular mass/energy duality is double space/time duality (equally vice versa in opposite direction). Spatial motion is size in time, just as time motion is spatial size, because double/inverse motion is size and double/inverse size is motion. Mass and energy (size/motion and/or particle/wave) is synchronicity (eigenvalue) of space and time. It is time content of empty space and/or spatial content of empty time, which is simultaneous reality of both and equally vice versa—space and time are content of empty mass/energy (they are content of empty form). Only duality or self-interference (two equal opposites) is real. One without the other is nothingness of empty space or time (empty shell of mass or energy). Only synchronicity of space and time is content of empty form of either. Mass/energy is eigenvalue of space and time.

Space is fractal equal opposite of time and mass is fractal equal opposite of energy; they are 1D/2D fractal difference or horizontal/vertical dualities. Mass in spatial context is energy in context of time and equally vice versa—energy in spatial context is mass in context of time. Space in massive context is time in context of energy and vice versa. Distinction between space/time is 1D fractal difference in one direction of 3D, while distinction between mass/energy is 1D fractal difference in equal and opposite direction of 3D. 1D fractal difference in third direction, which is inertial direction of its two equal and opposite directions, is gravity/antigravity. In context of mass, there is no difference between space and time—they are two equal opposites of its inertial reality (equally in context of energy only in inverse direction). Mass is inclusivity (inertion) of exclusive spatial and time reality (inverse/double inertion) and so is energy only in inverted ratio. Mass and/or energy are inertial 3rd direction of space/time duality; just as space and/or time are inertial 3rd direction of mass/energy duality of two exclusive opposites. Inertion or 3rd direction is always inclusiveness of both exclusive fractal opposites. Inertion is eigenvalue of both—inversion and double inertion (equally vice versa).

Just as space is inertion between its inversion and double inertion that are two equal and opposite times, so is mass inertion between two equal and opposite energies and energy is inertion between two equal and opposite masses (within and without). Mass is kinetic energy without (variable motion) and potential energy within (double or infinite; therefore constant motion, which is not motion but size). Energies within/without mass are 2D fractal difference or 1D fractal difference from inertial mass in between. Energy is fractal (quantum) of mass and mass is fractal (quantum) of energy, just as space is fractal of time and time is fractal of space. Any two equal opposites (duality) are fractal reality of each other. Space/time is fractal mass/energy and mass/energy is fractal space/time—one is flattened Euclidean curvature of another's non-Euclidean curvature (where Euclidean or flat in space is round or non-Euclidean in time and equally vice versa). Any relative (dual) difference is meaningless in context of their inertion (3rd direction), because such exclusive opposites are not opposites, but singularity that is equal opposite of inertion.

Action/reaction in context of inertion is not duality, but singular inversion (equal opposite), while 3rd direction of such dual inertion/inversion is double inertion. Each exponential singularity is duality in derivated context (and double duality in twice derivated context). Derivate in space is exponent in time and equally vice versa. Each of 3 directions—inertion, inversion and double inertion is action/reaction or space/time duality, hence 6 space/time directions. Such 6 directions are yet again 2x1/2 fragmentation, which is 12 direction of fractal ¾ reality (2x3=6 and 3x4=12). Etc. Curvature of double helix is infinite, because it is infinite double repetition of fractal difference ($1.111...^2$=1.23456...). Singular fractal reality is real only in context of its two equal opposites (whole numbers). Only duality has 3rd direction, hence only duality is singular 3D (only inertial difference between concave/convex curvatures is singular reality in between—illustration 2).

Illustration 2 - 1/2 Fractality

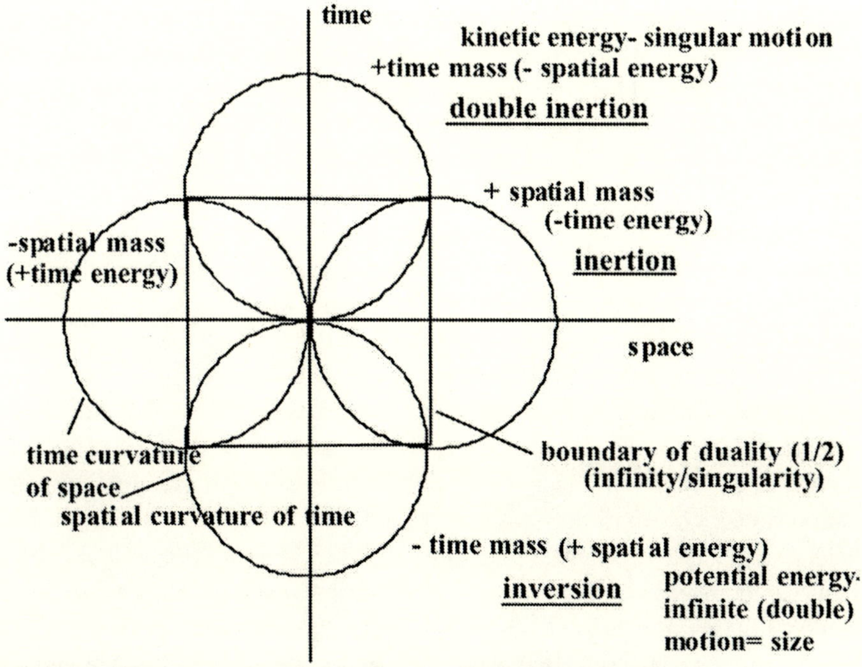

Space/time duality = 2x1/2 - 2 linear directions (Euclid)
Mass/energy double duality =4x1/4 -4 curvatures (non-Euclid)
Mass/energy = accelerated/decelerated space/time

Any odd-numbered reality is fractal of even-numbered reality and equally vice versa—even-numbered reality is fractal of odd-numbered reality (time is fractal space and space is fractal time). Space/time in context of mass/energy are two equal and opposite 2D realities or one 4D reality. 3D in between is 1 ½ reality—it is one and a half duality or fractal 3/2. It is singular mass in context of dual space/time, singular energy in context of space/time, space in context of mass/energy, time in context of mass/energy, mass in context of space/energy etc. Only double duality (4D), which is 3D space in time curvature and/or 3D time in spatial curvature, is whole reality. Third direction of such double duality is gravity, which is 5D, therefore another (double) fractal. Equal opposite of gravity is anti-gravity, which is complete 6D (6 surfaces of a cube). Complete 6D is triple duality or double 3D and third direction of such duality is yet another fractal reality, which is only ½ of its complete reality.

Duality is two equal opposites, of which one is +/- fractal difference from another—it is always fractal symmetry/asymmetry (asymmetry in depth is symmetry on surface, i.e. spatial symmetry is time asymmetry). 3rd direction of fractal ½ is 3D, which is only fractal $^2/_3$ of 4D, which is its 3rd direction (inertion or fractal difference). 4D is only fractal ¾ of 5D, while 5D is only fractal 4/5 of 6D. Odd numbers are fractal time curvatures and even numbers are fractal spatial curvatures (or vice versa). Fractal ¾ of 5D, fractal $^2/_3$ of 4D etc. are 3D inertion/inversion=double inertion or one sequence of Fibonacci curvature that adds past to the present (or small to the large) sequence that is instant future (double) sequence. It is infinite fractal double helix curvature of space/time, where fractal ½ is real only in context of fractal $^2/_3$ and $^2/_3$ only in context of ¾ etc. Increased spatial fractal difference is decreased time fractal difference; therefore it is double anti-parallel curvature, with no beginning or end.

Derivate of exponential singular direction is duality, where all 3 directions are inertial 3D. 2 directions twisting around each other in constant direction of remaining 3rd direction is singular double helix. Because 3D is 3 different combinations in which two directions twist around third directions—it is triple double helix or inertial/inverse and double double helix. Again—double double helix is =2D, while inverse double helix is—1D fractal difference, which is yet another (exponential) double helix. Mass/energy is +/- density of space/time (equally vice versa), while gravity/antigravity is +/- density of mass/energy (double density of space/time). Double helix is double 2x1/2 or 4x1/4. It is space/time in curvature of mass/energy in linear direction of gravity. Double helix, which is double anti-parallel curvature, is only one linear direction of gravity. 3D gravity is only one fractal double helix that is double fractal double helix of mass/energy and 3-fold space/time fractal curvature (equally vice versa, because difference between space/mass/gravity is only relative and/or twice relative or uncertain difference).

3D is double fractal difference (without and within or space and time), which is double derivate. It is double/inverse 2x1/2 multiplication/division or double singularity/infinity, which is 4x1/4 curvature (wave). 3D is exponential curvature with beginning/end in singularity/infinity (singular time = infinite space and vice versa). If space is singular parabolae, then time is double hyperbolae and equally vice versa. Derivate of such variable curvature (parabolae/hyperbolae) is its underlying constancy, which is constant round curvature (Π). Round curvature is 2D space of 3D time and/or 2D time of 3D space. Round curvature is constant variability, while parabolae/hyperbolae is accelerated variability of linear constant. Linear constant is underlying constancy or derivate of round curvature. Double linear constant is round curvature, which is curving of two linear directions around each other. Round curvature is accelerated/decelerated or double/inverse linear curvature, where linear curvature is no curvature at all. Parabolae/hyperbolae are double/inverse or accelerated/decelerated round curvatures, therefore 4-fold linear direction or two-fold acceleration/deceleration, which is 2D fractal difference. Linear direction is curvature of singular non-dimensional point—it is its acceleration/deceleration in space and/or time. Singular point is two equal and opposite linear directions (space/time), while each linear direction is two equal and opposite round curvatures (mass/energy) and each round curvature is two equal and opposite exponential curvatures (gravity/antigravity). It is always duality of singularity, regardless of relative/uncertain form.

All 3 curvatures—exponential, round and linear curvature is fractal double helix. It is two round curvatures, which is two equal and opposite exponential curvature in context of singular linear direction. Fractal double helix is two horizontal (spatial) curvatures that overlap ½ over each other. Such 2x1/2 overlap is their angular curvature (time). Time is real only in context of its double space and equally vice versa—space is real exclusively in context of its double time (illustration 3). Space is fractal ½ of whole (2x1/2) time and equally vice versa. 2x1/2 is real only in context of its inertial ½ (1D fractal difference), which is ½ of larger/smaller (inverse) 2x1/2. Horizontal 2x1/2 duality in context of angular (vertical) 2x1/2 is 4x1/4 sin wave that is 5 inversion points, which is fractal 5/4 reality. Fractal 5/4 is 2D fractal difference of fractal 3/2—it is space of space or double space and/or time of time or double time, which is gravity/antigravity of space/time (spatial attraction is repulsion in time and spatial repulsion is attraction in time). Each 2x1/2 is singular duality (space/time, mass/energy or gravity/antigravity).

Illustration 3 - 3/2 Fractal double helix (one strand - 1.111...)

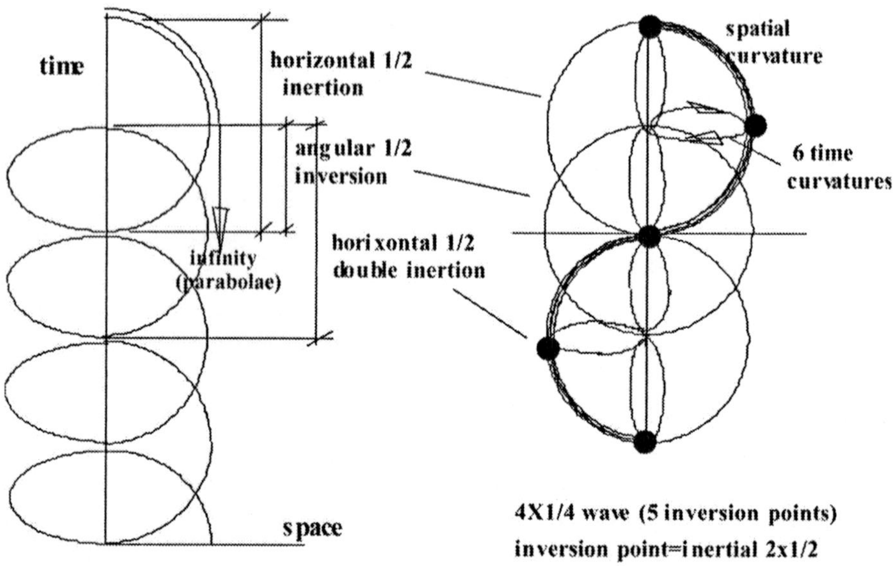

FRACTAL 3/2 FRACTAL 5/4

4x1/4 is double duality, which is space/time in context of mass energy (and/or vice versa), space/time in context of gravity/antigravity and mass/energy in context of gravity/antigravity. Fractal 3/2 is any duality in context of its 3rd direction—i.e. space/time in context of mass or energy, mass/energy in context of space or time, mass/energy in context of gravity or antigravity, space/time in context of gravity or antigravity and equally vice versa. Fractal 5/4 is any 4x1/4 (double duality or 4D) in context of its 5th direction—i.e. space/time and mass/energy in context of gravity and any relatively different combination of double duality in context of its singular inertion. Fractal 3/2 is 3/2 of 2x1/2 wave (curvature) and/or ¾ of 4x1/4 wave. Fractal 5/4 is 5/4 of 4x1/4 wave and/or 5/6 of 6x1/6 wave (curvature). Fractal spatial reality is whole time and fractal time is whole (3D) space, because space is surface of time and time is surface of 3D space. Only infinite number of surfaces is 3D, where such infinity is determined by inertion/inversion/double inertion. Inversion of 2D (surface) is ½; therefore infinite number of fractal inverse linear directions that comprise whole 2D is 4 (4x1/2=2). 4 fractals is boundary of 2D infinity. Double inertion is $2/(1/2)=2^2=4$; therefore its inversion is ¼. 16 inverse surfaces are one whole 4D. 2x1/2 is real only in context of its inertial ½, which is fractal 3/2D. Such 3rd half of duality (2x1/2) is fractal

1/3 of 3x1/3. Inversion of 3D is 1/3 or $9=3^2$ surfaces that are boundary of its infinity. Inversion of 6D is 1/6 or $6^2=36$ fractals that are boundary of 6D infinity. In context of infinite fractal spiraling curvature, only infinite number of fractal surfaces is one whole, where boundary of infinity is inversion that is double inertion. Inertion/inversion = double inertion of fractal 3/2 is $(3/2)/(2/3)=(3/2)^2=6/4$ and $(6/4)/(4/6)=(6/4)^2$.

Inversion is derivation, while double inertion is exponent of inertion. 1D fractal difference is simultaneous exponential space/derivate of time and/or vice versa. Only inertial here and now is non-differentiated space/time, where space=time (no difference). Singular difference is double/inverse difference—it is simultaneous derivation of space in time and/or inverse derivation (exponentiallity) of time in space. Derivate of time is twice larger (square) and twice slower (inverse square) space (slower passage of time). Exponent of time is twice smaller and twice faster space (squared passage of time). Derivate of space is exponent of time and exponent of space is derivate of time. Double derivate/exponent is double fractal difference or space/space difference (gravity). Increase of fractal spatial reality is equal and opposite decrease of its fractal time and equally vice versa. Fractality is infinite curvature of singular here and now, because any differentiation is fractal duality, where two directions of such duality are two infinite inverse/double (anti-parallel) space/time curvatures. Because any singular direction is fractal duality, all 3 directions are double/inverse fractal 3D reality, which is infinite double curvature. Any variable is underlying constant, which is only inverse/double variable. There is always further derivate in one and further exponentiallity in opposite direction. There is always larger/smaller than inertial fragment.

Double/inverse fractal curvature is time curvature of space (i.e. relativity) in additional (double/inverse) curvature of spatial curvature of time (i.e. uncertainty). Spatial and time curvatures are equal and opposite 1D fractal difference each, while their mutual intertwined curvature is singular 2D fractal difference, which is gravity (inertion) of such inverse/double space/time. It is infinite double stranded spiraling curvature in which every new loop is derived from previous inertion/inversion, by adding past sequence to the present in order to generate new sequence. Each strand is ½ of infinite self-interfering or precessional curvature. Exclusive cause/effect of such double curvature is inertial here and now of physical reality. It is infinite repetition through 2x1/2 multiplication/division of inertial here and now. Each further loop of double helix consists of previous two. Each further loop is larger spatial fragmentation in time and equal and opposite larger multiplication of time in space in context of one strand and equal and opposite time fragmentation/spatial multiplication in context of another (equal and opposite) strand. Each strand is perpetual precessional curvature, where one fractal precedes another that is infinitely smaller (fractal) time/larger space and/or vice versa. It is infinite self-interference of inertial here and now. Infinite self-interference is infinity of different possibilities of inertial (singular) possibility here and now.

In context of any singularity only two (equal and opposite) possibilities are simultaneous. Only equality of spatial curvature of time and equal and opposite time curvature of space is singular eigenvalue that is real here and now. Only inertion and equal and opposite inversion are simultaneous reality that is double inertion (here and now). Consequently, any reality is always 3D inertion/inversion=double inertion. Double inertion of inertion/inversion is gravity (force or acceleration) of space/time (size/motion), which is 3rd direction of such duality. All infinity of relatively/uncertainly different possibilities are not real in context of inertial here and now. They are real only in context of all the time and all the space. Here and now, they are empty space and/or empty (constant passage) of time. Infinite space is simultaneous reality of all different time possibilities, while infinite time is simultaneous reality of all different spatial possibilities. Here and now is duality of two fractal and perpendicular 2D slices of such infinite 3D reality. It is one spatial possibility of time that is inverse/double time possibility of space, out of infinite number of different possibilities. Just as all planets in course of its orbital time pass through center of their gravity (i.e. Sun), which is multiple of their fractal realities, so every inertial here and now in context of all its time explores all its spatial possibilities that are different than singular choice, which is real here and now (Kepler). Any different physical reality here and now is only fractal—it is only temporary time and/or space possibility, where boundary of such reality are inversion that is double inertion. Any difference is only fractal (partial, temporary). In context of all the space and time there is no difference between singular inertial here and now and its infinite universe/micro-world.

CHAPTER 2

VARIABLE SIZE/ MOTION OF CONSTANT GRAVITY

Space as time curvature is relative definition. Reaction of time curvature is equal and opposite spatial curvature of time (uncertainty). Inertial time curvature is inverse spatial curvature. 3rd direction of such inert/inverse non-Euclidean curvatures is double inertion that is flat Euclidean gravity in between. Action is real only in context of equal and opposite reaction. Space is real only in context of time and time only in context of space. Non-Euclidean space is Euclidean time and equally vice versa—Euclidean space is non-Euclidean time, which are relative/uncertain perspectives (curvatures) of reality. Curvature is reaction of flat reality, just as flat reality is reaction of curvature. If space is round, than time is flat and equally vice versa, because any dual description of 3D is incomplete. Beginnings/ends of such incomplete duality are singular/infinite distortion of reality here and now. There is no difference between universe/micro-world and perspective here and now. Reality is perspective and perspective is duality of two equal and opposite curvatures. Increased distance of observation is increased size of horizon (boundary) and equal and opposite decrease of any content within its boundaries. Increase of form is equal and opposite decrease of content and equally vice versa. Form is singular size without, while content is infinity of smaller sizes within. It is fractal space/time curvature, where singular flat space is infinity of time curvatures within and/or singular flat time is infinity of spatial curvatures within. It is +/- time curvature or inert/inverse density of singular spatial reality here and now (double inertion).

Increase of horizon is decrease of density of such form (horizon) and simultaneous equal and opposite increase of density of any content within. It is static spatial perspective. Static perspective is flat (exponential) curvature, while dynamic perspective is its round curvature and equally vice versa, because flat space is round time and round time is flat space. Dynamic space is static time and dynamic time is static space. In dynamic spatial perspective (static time) such increase/decrease is not constant but variable (constant change). In perspective of time, it is not constant size but constant motion—it is constant expanding universe. Expansion of universe is reflection of constant passage of time here and now. It is dynamic perspective, which is 1D fractal difference from static spatial perspective. Space is only fractal or variable of constant time (equally vice versa). Its underlying constant (round curvature of infinite parabolae) is time. Time is derivate within and exponent without fractal space in between. Infinite increase of spatial distance is infinity of spatial fragments that are one whole time. Infinity of space is singularity of time—it is boundary that is inversion of inertion that creates double inertion. Spatial universe is inversion of spatial inertion here and now—it is singular time (infinitely large space) of singular space here and now. It is infinity of all different time possibilities of one spatial reality here and now. Universe is infinite space; therefore singular time, just as here and now is singular space and infinite time. Here and now is spatial (static) perspective of time, while universe is time (dynamic) perspective of space. Here and now is increased/decreased size, while universe is increased/decreased motion.

Micro-world is inversion of spatial universe, which is double inertion. It is singular space; therefore infinite time within point of observation (here and now), as opposed to universe, which is its reality without. In context of universe here and now is singular space/infinite time; while in context of micro-world it is singular time/infinite space. Here and now is inversion/inertion of universe/micro-world—it is eigenvalue of both or double inertion (3rd direction of duality). Micro-world is infinity of all different spatial possibilities of one time. Universe is infinity of different motions of one spatial size, while micro-world is only one—infinite or light speed motion of different sizes. Universe is gradually increased linear expansion; proportional with distance, while micro-world is instant (light speed) rotational contraction. Universe is gradual spatial change in context of infinite time here and now, while micro-world is instant spatial reality in context of no time (singular time). Universe is instant time and gradual spatial change, while micro-world is instant space and gradual time change. Universe is constant passage of time, while micro-world is all time at once—there is no passage of time that creates different spatial possibilities. It is all spatial possibilities at once, while universe is all spatial possibilities only in context of infinite passage of time. Universe is all time possibilities at once, while micro-world is all time possibilities only in context of passage of time. Micro-world is spatial reality of time and time reality of space. Space of micro-world is instant time, just as its time is gradual spatial change. Micro-world is inversion of inertial universe in context of double inertion here and now.

Mass in context of universe is energy and energy is mass in context of its inverted micro-world. Space of universe is time and time is space of micro-world. They are fractal inertion/inversion that generates double inertion, which is reality here and now. Size/motion is 1D fractal difference of same inertion. Spatial size is motion in time and spatial motion is constant size in context of time, because space/time are 1D fractal difference. Infinity/singularity are equal and opposite boundaries (inversion/double inertion) of inertial here and now. Infinite motion (light speed C) is singular (0D) size and singular motion (no motion) is infinite size (equally vice versa in time perspective, because spatial light speed is no motion in time context, just as no spatial motion is light speed motion in context of time). Singular size of infinite motion (C) is 1/C (and/or infinite size of singular motion in inverted perspective), which is $C/(1/C)=C^2$, which is gravity or acceleration (double inertion) of such size/motion (inertion/inversion). Fractal difference is singular/infinite difference. It means—all the time is only singular spatial point and/or all the space is only singular instant of time.

Singularity is infinitely small quantity—it is 0. Singularity is instant. All the time of inertial reality is no time at all—it is instant spatial reality of inverted here and now. Likewise—all the space of inertial reality is no space at all, but instant (passage of) time of inverted here and now. Instant time is constant passage of infinite time or infinitely long gradual spatial change. Instant (singular) time is infinite space and instant (singular) space is infinite time. Consequently—spatial motion is not motion in context of time, which is instant space (no allowance for change in constant spatial context)—it is instant fractal size of time and equally vice versa—spatial size is constant change of time size; therefore its constant motion. Fractal difference is difference between constant change and no change (singular constant). Constant spatial change (i.e. motion) is no change of time (i.e. size) and constant time change is no spatial change, but only its constant size (no motion).

Energy of universe is motion of its mass (size) and/or size of its space, while energy of micro-world (atom) is size of its mass and/or motion of its space. Mass of universe is energy and energy of universe is mass, just as space is time and time is space in context of micro-world and equally vice versa. Micro-world is mirror image of universe and equally vice versa—it is flat 2D reflection of 3D spatial reality, which is 3D time of flat 2D time (decreased space = increased time etc). Universe is fractal 3D space/2D time, while micro-world is fractal 3D time/2D space. Universe is inertial time curvature of space (positive non-Euclidean curvature that is relative perspective), while micro-world is inverse spatial curvature of time (negative non-Euclidean curvature that is uncertain perspective). Positive time curvature is negative spatial curvature and equally vice versa. Relative perspective is round space in context of flat (linear) time, while uncertainty is round time in context of flat (linear) space. They are derivate/exponent or variable/constant of each other. One flattens round reality of another and/or curves flat reality of its equal opposite. Round space is flat time and flat space is round time.

Round motion in context of infinitely larger flat reality is no motion at all but instant (round) size and equally vice versa—linear (flat) motion in context of infinitely larger round curvature is instant linear (flat) size in context of infinitely small flat curvature, because of fractality of space/time. Fractality means time of one is no time (instant) of another and space of one is no space but singular point of another. Round motion in flat context and/or flat motion in round context are instant motions in context of no time. It is uncertain reality, where massive object in motion is everywhere at the same time. Instant motion in context of no time is not motion but instant size—it is space and not time. Constant space (size) is uncertain motion, just as constant time (motion) is relative size. Spatial uncertainty is relativity of time and equally vice versa. Space and time are equal and opposite comparisons, where comparison is 3rd direction of such duality that is gravity of its round/flat (relative/uncertain or 1D fractal difference) duality. Round motion (rotation) is reaction of flat size and linear motion is reaction of round size. Round time is reaction of flat space and equally vice versa, where spatial motion is size, just as spatial size is motion in context of its time. Relativity of spatial reality is at the same time uncertainty of its time and uncertainty of spatial reality is relativity of its time. **Relative spatial universe is quantum uncertainty of its time, just as quantum uncertainty of spatial micro-world is relativity of its time.**

Only 3D is real. Only inertial fractal difference between two equal opposites is 3D here and now. Only incremental geometrical difference between two different curvatures is instant 3D of such flat duality. Discrepancy between action and its reaction that is inertial framework of such duality is space and time here and now. Size is fractal difference or discrepancy between two equal and opposite motions, just as motion is geometrical (comparative) difference between two equal and opposite sizes. Singular size is real only in context of dual motion and motion in context of dual size. Size is inertial framework of duality of motion (spatial/time motion) and motion is inertial framework of dual size (spatial/time size). Space is reality between inverse/double time and equally vice versa. Singular space is reaction of double time and singular time is reaction of double space. Space and time are equal and opposite (fractal) curvatures. 3rd direction or inertion of space/time duality is its massive context and equally vice versa—space is inertial framework or discrepancy between equal and opposite curvatures of mass and energy.

There is one mass between equal and opposite infinity of space and time. Reaction of one mass is infinite energy. Equally vice versa—reaction of singular energy is infinite mass. Singular/infinite mass/energy is reaction singular/infinite space/time. One mass between space and time is infinite energy and one energy between space and time is infinity of mass (inversion of inertion). Mass is inertial discrepancy between space and time (spatial and angular curvatures), while reaction of mass is equal and opposite (inverse) energy. Equally vice versa—there is one space between mass/energy and reaction of one space is infinite time. In inverse perspective—one time between mass/energy is infinite space. Mass (or energy) of space/time is 3D, while duality of mass/energy in context of equal and opposite (fractal) space/time duality is 4D—it is 3D space in time curvature (and/or 3D time in spatial curvature). Mass/energy is angular duality of horizontal space/time duality—it is double duality or 2x(2x1/2)=4x1/4 multiplication/division of inertial here and now.

Duality is real only in context of its inertial discrepancy. No inertial discrepancy is one infinitely large duality and/or simultaneous infinitely large number of smaller (singular) dualities in opposite direction (reaction). One duality without is infinite number of dualities within. Larger inertion is reduction of dual possibilities, while smaller inertion is increase of dual possibilities. Increased possibility in horizontal direction is proportionally decreased possibility in angular (vertical) direction. No spatial inertion is infinitely large time inertion and infinitely large spatial inertion is no time inertion. Larger inertion without is smaller inertion within. In inertionless context, duality is infinite. Infinite space/time duality is infinite number of inverse (mass/energy) dualities, all equally real in context of their inertial duality. Infinite mass/energy (singular) duality is infinite number of space/time possibilities, all equally real (Everett's many worlds). Infinite singular duality is reaction of inertionless framework, while infinite number of inverse dualities is reaction of infinitely large inertion, where inertionless space is infinitely large time inertion and equally vice versa.

Singular spatial possibility is infinity of different time possibilities and singular time is infinity of simultaneous spatial possibilities. Universe is one time of infinity of different spatial possibilities, while micro-world is one space of infinity of different time possibilities. Singular time curvature (space) is infinite number of spatial curvatures (time) and equally vice versa. Singular space is infinite number of times and singular time is infinite number of spaces. Singular space/time is infinite number of mass/energy dualities and singular mass/energy is infinite number of space/time dualities. Singular mass is infinite number of different quanta of energy and singular energy is infinite number of different quanta of mass, because infinite horizontal inertion of action is singular angular inertion of its reaction. Singularity is singular quantum of infinity (where spatial singularity is infinity in time and equally vice versa). Nothingness is smallest unit of something, just as something is smallest unit of nothingness. Nothingness is only one out of infinite number of different possibilities of something, just as something is only one out of infinite number of possibilities of nothingness. Something/nothing is meaningless (relative/uncertain) or dual description of 3D. In context of inertial 3rd (inclusive) direction, such dual exclusivity is unreal. Something and nothing are two equal opposites of their inertion—they are inertion/inversion that is double inertion here and now.

Nothing in curvature of something and/or vice versa is inertion/inversion = double inertion. Inertial discrepancy between something and nothing is double something and reaction of double something is double nothing, etc—fractal double helix is infinite curvature of singularity and/or singular curvature of infinity. Reality is real only in its inverse context. Something is real only in context of nothing and nothing in context of something. There is no difference between something and nothing—nothing is only one (smallest) possibility of inertial something and something is smallest possibility of inertial nothing, where small in space is large in time and/or vice versa. Something and nothing are singular/infinite boundaries of inertial here and now. They are only (exclusive) horizons of inertial perspective, without any real (inclusive) 3D meaning. Singular nothing on surface is infinity of different something and singular something is infinity of different nothing in its depth. Infinite number of different something is one nothing and infinite number of different nothing is singular something. Something in 3D context is nothing in 2D and/or 4D context and equally vice versa.

Something in space is nothing in time and something in time is nothing in space. Something/nothing in space/time is double something that is mass of such space/time and reaction of double something (mass) is double nothing (energy) and equally vice versa. Out of infinite number of nothing, only one is real, just as only one out of infinite number of something is real here and now. Consequently, only one space that is infinite time and only one time that is infinite space is real here and now. Only eigenvalue of inertion/inversion = double inertion is simultaneous reality here and now. Rest is empty space and/or time. Rest is infinity of different spatial possibilities that are real only in context of past/future time and infinity of different time possibilities that are real only in context of larger/smaller (distant) space. Past/future time and/or larger/smaller space are not real here and now—they are infinity of different space/time possibilities that are equally real as here and now, but not here and/or not now. Such different space/time (and/or mass/energy) possibilities are real only in context of their inertial background, which is gravity of such space/time (mass/energy). Mass/energy is reaction of space/time, while gravity is inertial discrepancy or fractal difference between such equal and opposite geometrical curvatures. Gravity is 3rd direction of space/time vs. mass/energy duality (fractal 3/2 reality) and/or 5th direction of 4x1/4 reality (fractal 5/4 reality).

There is one gravity between equal and opposite space/time and mass/energy and its reaction is infinity of different antigravitational curvatures. Equally vice versa—reaction of infinity of different gravitational realities is one antigravity. One expansion is infinite number of contractions and one contraction is infinite number of expansions. One gradual spatial expansion without (i.e. universe) is infinite number of instant massive contractions within and equally vice versa—one instant spatial expansion without is infinity of gradual massive contractions within (which are instant energy expansion without and/or gradual time expansion within). Gravity is inertion of dual size/motion. There is one acceleration between equal and opposite size and motion and equally vice versa—there is one size between acceleration/motion and one motion between acceleration/size. It is size/motion= double (accelerated) motion. Reaction of one accelerated (double) motion is infinite number of decelerated (decreased) sizes and equally vice versa—reaction of one double size (density) is infinite number of accelerated/decelerated motions (forces of attraction/repulsion).

Space/time gravity is inversion of mass/energy gravity and spatial gravity is inversion of time gravity, just as massive gravity is inversion of gravity of energy. Spatial gravity is antigravity of time and space/time gravity is antigravity of mass/energy, because space/time difference is no difference in its inertial (mass/energy) context and equally vice versa. Spatial gravity and antigravity is same reality in context of time gravity, which is its fractal inertion and equally vice versa. Spatial attraction/repulsion (two equal opposites) is double time attraction (or repulsion). Spatial attraction/repulsion is only larger/smaller attraction of time, where reaction of singular attraction is infinity of repulsions. Space/time is singular 2x1/2 duality of size/motion—it is rotation of massive particle (2 inversion points). Double (space/time vs. mass/energy) duality is 4x1/4 multiplication/division, which is dual (2D) sin/cos wave of 4 inversion points. Triple duality is double 3D, which is 6x1/6 in 3D context and/or 8x1/8 in 4D context. It is 3D sin/cos or rotating wave that is 6 inversion points. Space/time, mass/energy and gravity/antigravity are 6 surfaces of a cube. They are 6 harmonized eigenvalues of fractal 3/2 reality (3x2=6) or one point of inversion at every 60° of curvature (as opposed to 4 eigenvalues of 4x1/4 wave at 90° intervals).

Illustration 4 Double helix of space/time

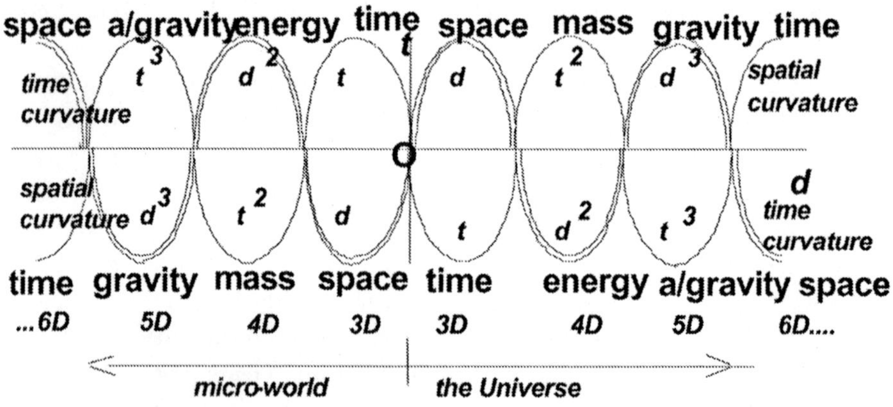

Each one is singular quantum of infinite fractal 3/2, which is infinity/singularity of preceding/succeeding curvature. Infinite/singular difference is light speed difference that is difference between singular and infinite size. It is triple positive and triple negative or 6-fold light speed difference (4-fold in context of 4x1/4 wave). Mass/energy is double infinite/singular space/time, while gravity/antigravity is triple singular/infinite space/time (and/or double singular infinite mass/energy). $E=d^2$ (within) and/or d/t (without), while $m=t^2$ and/or t/d—one is time curvature of space, while another is spatial curvature of time and their inertial geometrical difference is $(d/t)/(t/d)=(d/t)^2=E/m=C^2$ = exclusive gravity (g) or antigravity (ag). $\mathbf{C^3=(d/t)^3=(E/m)^2=g/ag}$. Gravity/antigravity duality is again real only in context of its inertion or incremental discrepancy between the two, which is 3rd direction of such duality. It is double gravity and reaction of singular double gravity is infinite double antigravity and equally vice versa. Double spiraling curvature of singular here and now is infinite double helix, where only beginning/end—only inert/inverse boundary is only here and now (see illustration 4). Any dual differentiation is relative—space in one perspective is time in diagonally opposite perspective and mass and/or energy in perpendicular perspective. Same reality is gravity and/or antigravity in vertical (angular) perspective of such surface 4x1/4 reality. There is no real difference between space, time, mass, gravity etc—any difference is only limitation of inertial perspective here and now, because mass/energy is gravity/antigravity of space/time and equally vice versa. Gravity/antigravity is space/time of mass/energy and mass/energy of space/time, just as space/time is mass/energy of gravity/antigravity and vice versa. 3D is non-differentiated singular inertion/infinite inversion = singular/infinite double inertion, and any difference is infinite relativity/uncertainty (infinite fractal curvature) of only one reality that is no difference.

Singular size is duality of motion, just as singular motion is duality of sizes. There is one size between two equal and opposite motions and there is one motion between two equal and opposite sizes. Size is inertial context of dual motion and motion is inertial context of dual size. Double (accelerated) size is motion and double (accelerated) motion is size. Inertial context is comparison of two equal opposites. Motion is relative comparison (time) between two spatial sizes and/or vice versa—it is space between two motions. Size is comparative spatial reality between two equal and opposite motions (times), just as motion (time) is comparison between two equal and opposite sizes (spaces), where size in time is motion in space and equally vice versa. Size and motion are inert/inverse space/time comparisons. Size in motion (duality) is 3rd direction (comparative inertial background) between space and time. Double space/time comparison is comparison between size and motion that is double/inverse or accelerated/decelerated size/motion (gravity), which is +/- curvature of flat and/or +/- flattening of round size/motion. Accelerated/decelerated size/motion is constant and fractal (equal and opposite) change, which is gravity (density) of such duality (just as size and motion are constant equal and opposite fractal changes of space/time). Change of two equal opposites is always fractal supersymmetry—it is instant change of one and equal and opposite gradual change of its equal opposite. Instant (singular) space is gradual (fractal) change in time and equally vice versa—instant (one) time is gradual (fractal) spatial change. Instant (one) size is gradual (fractal) change of motion and instant (singular) motion is gradual (fractal) change of size.

Instant gravity is gradual (fractal) change of antigravity and equally vice versa. Spatial gravity is time antigravity and equally vice versa. One is instant (whole), while other is fractal (gradual change). Only infinity of fractal (gradual) changes is one whole change. Only infinity of fractal density changes of mass is singular change of spatial density (change of density of empty space) and equally vice versa—only infinity of fractal (gradual) changes of spatial density is singular change of density of its mass. Only infinite number of fractal changes of mass is singular change of energy and infinite number of fractal energy changes is one massive change. Instant gravity of mass (spatial attraction/time repulsion) is infinite number of fractal (gradual) changes of spatial antigravity (spatial repulsion/time attraction). Instant gravity here and now (mass) is gradual repulsion of spatial universe (gradual expansion in course of linear passage of time, where time is repulsion of space, just as space is contraction of time, because one is inverse boundary of inertial reality of another). Instant gravity of mass is gradual antigravity of space and equally vice versa—instant spatial gravity (attraction or contraction) is gradual massive antigravity (repulsion or expansion). Attractive spatial reality is repulsive time and equally vice versa, where one is instant surface of fractal 3D reality of another, because reality is infinite repetition of fractal difference that is inertial here and now. Gradual (fractal) motion without is instant size within and instant size without (i.e. universe) is gradual fractal (motion) within. Uncertainty within reflects as relativity without.

Within (inside) is always fractal spatial reality of without (outside). Fractal (gradual) space is whole (instant or singular) time and equally vice versa. Within in space is without in time and within in time is without in space. Instant size is fractal (double/inverse) motion and instant motion is fractal (double/inverse) size. Instant attraction is double/inverse repulsion and instant repulsion is double/inverse attraction. Instant attraction in between is double/inverse repulsion within and without. Spatial attraction is two equal and opposite spatial repulsions/time attractions and equally vice versa. Any reality is 3D—it is inertion/inversion=double inertion. It is one inversion between two equal and opposite inertions and/or one inertion between two equal and opposite fractal inversions. It is one whole between two equal and opposite fractals and/or one fractal between equal and opposite wholes. It is one round between two flat and/or one flat between two fractal round curvatures. It is one perfection between two imperfections and/or one imperfection between two perfections. Between two symmetries is one asymmetry and between two asymmetries is one symmetry. Between two motions is one size and between two sizes is one fractal motion. Etc.

Between two motions is one accelerated motion and between two accelerated motions is one constant motion. Between two variables is one constant and between two constants is one variable. Between two derivates is one exponent and between two exponents is one derivate. Singularity is real only in context of equal and opposite duality and equally vice versa. Between two changes there is always one changeless reality and between two changeless realities is one change (difference). Between two 3D's there is one surface and between two surfaces, there is one 3D. Between two emptiness there is one density and between two densities there is one emptiness. Something is reality between two nothingness and nothing is in between two something. There is one relative between two uncertain realities and equally vice versa. One beginning is double/inverse end and one end is double/inverse beginning. Between two realities, there is one non-reality, just as there is one reality between two non-realities. Affirmation is double/inverse negation. 3D is duality of inverse/double fractal difference. Between two rotations is one linear motion and between two linear motions is one rotation. Spatial rotation is linear motion in time, just as rotation in time is linear spatial motion, because duality is fractal—infinity of one is singularity of another. Linear motion in flat space is round motion in time, which is its curvature and equally vice versa—rotation in time is linear motion in space, which is its inverse curvature (flattening).

Rotation is double/inverse or spatial (linear) motion accelerated/decelerated in time (equally vice versa—linear time motion accelerated/decelerated in space). Rotation is duality of two linear motions—in space and time. It is curvature (fractal difference) of linear motion and equally vice versa—linear motion is inverse curvature of rotation. Duality is acceleration/deceleration of singularity. Linear motion is time acceleration/deceleration of spatial rotation and/or spatial acceleration/deceleration of rotation in time. Linear motion is duality of spatial and time rotation—it is accelerated rotation. Accelerated (double/inverse) linear motion is equal and opposite constant rotation and accelerated (double/inverse) rotation is equal and opposite constant linear motion (1D fractal difference). Rotation of mass is linear motion of energy (i.e. universe) and equally vice versa—rotation of energy is linear motion of mass (i.e. micro-world), where one is instant; while another one is gradual (fractal) change. Rotation of mass and energy is 1D fractal difference or spin 1 and spin 2 of boson/fermion particles (1D fractal difference between massive and massless particles, which is mass/energy or particle/wave fractal difference). Spin 1 and 2 are fractal difference between size in rotation and rotation that is size. It is difference between spatial rotation of time particle and time rotation of spatial particle, because micro-world is flattened space/rounded time—i.e. it is spatial derivate and time exponent or twice smaller and twice faster space, which is double space in ½ (inverse) time (2x1/2).

Micro-world is two (instant) spatial possibilities of inertial time here and now, while universe is two (gradual) time possibilities of inertial space. Micro-world is eigenvalue of two spatial possibilities—it is two spatial realities at the same time, while universe is two time realities of same space. Two spatial realities, where one is infinity of singularity of another, are instant reality in context of no time (micro-world), while two times that are two possibilities of inertial space (universe) are gradual or fractal change—it is gradual relativity (expansion of universe). Duality is always fractal symmetry/asymmetry of space/time that is gravitational supersymmetry (no difference). Consequence is that micro-world is instant spatial duality, because space and time of micro world are two instant spatial directions in context of its 3rd direction (gravity) that is time of such duality. Instant space/time duality is Pauli's exclusion principle that allows for two spatial realities (electrons) in context of their inertial orbital motion (time). It is duality of electronic pairs, where one is reflection of another—i.e. one electron is spatial particle, while another is time particle in horizontal/vertical (angular) motions. It is one particle in double motion and/or two particles in one motion. Instant space and gradual time of universe are instant double space of micro-world. Its time (inertion of space/time duality) is gravity, which is 1D fractal difference from space/time duality. Universe, on the other hand is gradual (fractal) duality of two time directions—it is gradual exploration of different time possibilities of inertial space here and now.

Space is depth of time and time is depth of space. One is equal and opposite (double/inverse) curvature of another. Micro-world is flattened space; therefore time, which is its depth, is instant reality. It is not fractal passage of time but instant (one or whole) time. Instant time (no passage) is not time but space. Micro-world is 2D space—it has no depth, because its perpendicular spatial depth is inverted (90° difference, which is singular quantum of 2x1/2 reality) into flat horizontal direction. Time of micro-world is instant density of such flattened spatial reality. Density is double instant space or instant size within size. Double size is infinitely smaller/larger size, which is not size at all but singular motion—it is instant light speed motion. Consequently—light speed rotation of such particles is not spatial motion—it is time motion, which is double spatial motion or rotation that is double (accelerated) linear motion. Light speed rotation in time is linear light speed motion in space and equally vice versa—rotational light speed motion in space is linear light speed motion in time. Here and now is duality of such equal and opposite motions—it is $C/(1/C)=C^2=E/m$=gravity. It is simultaneous spatial and time light speed motion, which is not motion at all but constant size. Any distance/difference from here and now is equal and opposite differentiation of such motions, therefore it is not only size (particle) but also increasingly motion (wave).

Spatial difference is distance, while time distance is difference. Spatial distance is horizontal (spatial), while time distance is angular difference. Spatial distance is gradual time difference, while time distance is instant spatial difference—it is infinite variability of spatial reality here and now. Massive surface is infinite time distance, just as is spatial universe (infinite expansion) infinite spatial distance. Reality that is here (i.e. mass) is infinite time (angular) distance (difference) away, just as reality now (i.e. spatial universe) is infinite spatial distance away. It is $C/(1/C)=C^2$ distance/difference that is inertial reality here and now (double inertion of inertion/inversion or gravity of space/time). Universe is real now, only because it is not here, while its micro-world (mass) is real here only because it is not now. Angular time distance is spatial difference, just as horizontal spatial distance is not distance at all but instant differentiation of singular time. Infinite space is no distance at all in context of time, just as micro-world is no distance at all in context of space. Infinite universe is only instant surface of time (no spatial depth), which is infinitely larger/smaller or inverse/double space. Infinitely long period of constant passage of time in context of spatial reality within is not time but instant space of no time in context of its time that is space without.

Space is real only in context of its equal (fractal) opposite—time. Without time it is not space but double/inverse space Likewise—time is real only in spatial context. Without spatial context, time is not constant passage or gradual (fractal) spatial change, but instant or singular spatial change that is not change at all—only an instant spatial reality. Without its time context, time is not time but space, just as without spatial context—space is not space but time. One space is infinite (fractal) time—it is its constant passage. In context of all infinite time or in context of one undivided spatial reality, such time is not time but singularity of space. It is instant reality of no time, where no time is infinitely larger/smaller passage of time. Time of time is accelerated or double time, which is spatial gravity. Gravity is time of time and space of space—it is accelerated size and/or decelerated motion (equally vice versa), where spatial acceleration is time deceleration and equally vice versa. Gravity is 2D fractal difference. In context of gravity, there is no difference between spatial and time reality—they are instant (double/

inverse) space of its time and/or fractal (double/inverse) time. Space and time of gravitational reality are square/inverse square of space/time, which are gravity and antigravity (attraction/repulsion).

Mass/energy are size/motion of space/time and equally vice versa—space/time are fractal size/motion of mass/energy. Gravity/antigravity are space and time of mass/energy; they are its size/motion that is double/inverse space/time size/motion Boundary of space/time motion is light speed that is singular/infinite size of its mass/energy. Boundary of mass/energy motion is double light speed that is singular/infinite size of its gravity/antigravity. Gravity is double light speed motion—it is light speed motion in space and time or linear and angular (rotational) light speed motion, which is instant 2D surface. It is instant reality of double electron that rotates simultaneously in two perpendicular directions and/or two electrons in one 2D motion. It is double uncertainty, where such electron is not instantly everywhere only in its orbital but also in every point on surface of a sphere (where one uncertainty is infinitely larger than another). Mass/energy is linear/rotational (1D) light speed motion in space/time context (2D in mass/energy context), while gravity is 2D light speed motion (which is 1D in context of inertial mass/energy). Gravity is double (squared) spatial light speed motion and inverse time light speed motion.

Space is time curvature and in its context, linear light speed motion of energy is curvature (instant time). Light speed motion of energy is reduced and motion of mass is increased for same amount, proportionally with spatial distance that is angular time curvature. Decreased motion of energy/increased motion of mass is angular perspective of expanding universe. Gravity is double/inverse motion—it is proportional decrease of double light speed motion (mass) in infinite distance to only singular light speed. Mass is double light speed (spatial and time or horizontal and angular light speed) here and now. Such motion is exponentially (tangent of curvature) decreased to only singular light speed motion in infinite spatial distance, which is 90° or ¼ of its time curvature. Beyond such boundary—gravity is not gravity but mass/energy duality or singular light speed difference, which is infinite boundary at 180° or ½ of its time curvature (remaining ½ curvature is space/time context of such mass/gravity duality). Spatial motion of mass increases in distance of infinite spatial horizon from 0 to C (energy decreases from C to 0), while gravitational spatial motion of mass decreases from C^2 to C, which is inversion from infinite attraction (mass) here and now to infinite repulsion (space) that is reality of universe, where spatial attraction is repulsion of time and equally vice versa.

Infinite spatial boundary is eigenvalue of gravitational and spatial motion of mass (C), while here and now is its double/inverse reality—it is double motion, which is not motion at all but its inversion, which is constant size. Gravitational size/motion is inversion of spatial size/motion, which is double size/motion of mass/energy here and now. Mass/energy is eigenvalue or symmetry line between spatial and gravitational reality, where gravity is double space within/without, just as space is double gravity within/without (2D fractal difference). Gravity and space are singular/double space of time in between (equally vice versa—antigravity/time are double time that is singular space in between), where time is only fractal reality of space and space is only fractal time. Only double difference is complete (whole) difference. Space/time is only fractal ½ difference. Only even numbers are spatial reality, while odd numbers in between are its fractal time and equally vice versa. Inversion is equal opposite of inertion—each is only ½ of double inertion, because of fractality between space/time. Double inertion is duality of inertion and its inversion and equally vice versa—double and singular inertion are only fractal ½'s of inversion in between. One vertical (angular) ½ is overlap between two horizontal ½'s. ½ of ½ = ¼ of each horizontal ½ is ½ of their angular ½ (see illustration 3). It is fractal 3/2 in context of fractal 5/4 reality, where 3/2 without is 5/4 within. It is 3D, which is dual action/reaction (2x1/2) that is real only in context of its inertial (angular) overlapping ½ in between. Such 3/2 is 6/4, where inertial 2/4 is singular point between equal and opposite 4x1/4—it is simultaneous beginning/end between two sin waves. 5/4 is 5 points of inversion of 4x1/4 wave, where 5th point is perpetual inertion of such wave or perpetual multiplication/division of inertial here and now (1.111...).

Gravity is angular curvature of horizontal space/time—it is time of both. Space/time is 1D fractal difference, while space/gravity is 2D difference. Reality of both—universe and its micro-world is 6D fractal difference or 6 successive precessional (fractal) curvatures. It is space/time/gravity without and space/time/gravity within, where space without is time within and time without is space within (attraction without is repulsion within and mass without is energy within etc.) It is space/time in context of mass/energy in context of gravity/antigravity or 3 times 2x1/2, one within the other. Such complete singular 3D sequence is $(2/(1/2))^3= 4^3=64$ fractal sequences. Increased space is equally decreased time—it is 6 different time curvatures of singular space and/or 6 different spatial curvatures of inertial time. Each strand of double helix is 6 x 2x1/2 multiplication/division loops, where each loop is singular/infinite (fractal or relative/uncertain light speed) difference. Such complete 3D sequence of double helix is $12/(1/12)=12^2=144$ fractals. Fractal difference between one and two strands of double helix is 144/(2x64)= fractal 9/8 or $3^2/2^3$ (double 3D/triple 2D) difference. It is 1.125, which is abbreviated inverse (1/10) 3D sequence of infinite $1.234567...=1.111...^2$ sequence (perpetual double/inverse repetition of inertial here and now in space and time), where1.111 is singular fractal difference (cube/square) and 1.23456...is double fractal difference.

Only here and now is real and all infinite space/time (physical reality) is its double repetitive curvature. Double repetition is singular/double fractal difference, which is twice infinitely differentiated reality. It is inverse/double matrix of constant motion/constant acceleration of constant size (equally vice versa), in which there is no two same sequences. Fibonacci fractal sequencing is infinite curvature of inclusion of two exclusive opposites, in which past (inverse) sequence is added to present (inert) sequence in order to generate future (double) sequence. Only duality (inertion/inversion=double inertion) is real (3D). The rest of its infinite curvature is empty space and/or empty time, which is constant changeless passage of time. Only eigenvalue of two equal opposites produces 3rd direction. Two fractal opposites—one within and one without are only reality. Projection of that eigenvalue in context of linear reality is φ (golden mean), which is 1D synchronicity ratio of duality and Π, which is 2D synchronicity ratio (flat/curvature or Euclidean/non-Euclidean curvatures). Φ/Π is 1D/2D fractal difference that is 3D here and now. Only reality that is eigenvalue of both is real here and now. It is infinite irrational sequence of singular here and now, where every sequence is relatively different from any other.

Every whole is relatively different combination of infinity of its fractal parts. It is infinite double/inverse repetition in which no singular sequence is repeated twice. It is infinite surface asymmetry (no pattern) that is infinitely larger (inverse/double) symmetry. No pattern is infinitely larger pattern and no symmetry is infinitely larger symmetry. Pattern is dual limitation, while no pattern is infinite 3D freedom, where such infinite freedom is only further limitation of larger/smaller (inverse/double) duality. Every surface duality (symmetry) is limited by its line of symmetry, which is its asymmetry in depth. Line of symmetry is inertial context of symmetrical duality—it is its inversion that creates double inertion. Symmetry is real only in context of its line of symmetry, which is infinitely small discrepancy of such surface symmetry. Line of symmetry is 3D reality of surface symmetry. It is line that breaks surface 2x1/2 symmetry in infinity of different fractals in its depth. Line of symmetry is negation of surface symmetry, because it is its asymmetry in 3rd direction. Any pattern on surface is no pattern in its depth and equally vice versa—no pattern on surface is pattern in its depth, which is infinitely larger/smaller pattern. Symmetry of two equal opposites is real only in context of their inertion, which are time delay and/or spatial discrepancy between such fractal opposites. Symmetry is real only in comparison with its equal and opposite asymmetry, which is its line of symmetry. Discrepancy (no pattern) is inertial context of perfect dual symmetry. It is discrepancy and/or asymmetry only on surface. In depth it is infinitely larger pattern (symmetry).

Motion is asymmetry or discrepancy between two sizes, just as size is asymmetry between two motions. Spatial size is discrepancy (fractal difference) between two motions that are its time boundaries. Spatial motion is discrepancy between two sizes that are its inverse/double spatial boundaries (horizons). Size and/or motion are spatial and/or time discrepancy/delay between two equal opposites. They are line of symmetry that breaks surface symmetry. Curvature is line of symmetry of flat reality and linear direction is line of symmetry of curvature. One is asymmetry of other one's symmetry. Rotation (angular motion) is asymmetry of linear motion and equally vice versa. Acceleration is asymmetry of constant motion that is infinitely larger/smaller (inverse/double) symmetry. Motion is real only in context of its inverse/double motion that are its size and accelerated motion and equally vice versa.

Motion in spatial context here and now (i.e. 3D), is no motion at all in context of its fractal reality of relative universe (4D). Only accelerated motion here and now is constant motion in context of relative space (time) and only twice accelerated motion is constant motion in context of twice relative space (5D or gravity). Accelerated motion in 3D context is constant motion in 4D and constant size in 5D context. Constant motion is 3D size in 4D and only surface size (2D) in 5D context. In 6D context it is only one linear direction of 3D. Constant size of 3D is surface (2D) of 4D, linear direction (1D) of 5D and singular non-dimensional point of 6D. Time of time is not time but instant space, just as space of space is not space but constant passage of time. Spatial motion is not motion of its larger/smaller (relative/uncertain) space, because it is its time—it is its size and equally vice versa. Singularity is boundary of reality. In 7D context, such constant 3D size is beyond boundaries—it is more than absolutely improbable (more than uncertain) reality. 3D fractal difference is boundary of singular 3D (exclusive space or time—1D fractal difference), while 6D is boundary of inclusive space/time (gravity—2D fractal difference). 6D fractal difference is 6 harmonics of inertial here and now or 6 inversion points between space and time, which is 3D wave. In context of 4D reality, which is 3D space + time curvature; reality is 4 synchronized harmonics or 4 inversion points of 4x1/4 sin/cos wave.

Eigenvalue of 4x1/4 wave is 360/4 =90° curvature, while eigenvalue of 6x1/6 wave (3D or duality of sin+cos waves) is 360/6=60° curvature. Eigenvalue of 2x1/

2 wave is 180° curvature. It means Π/4; Π/6 and Π/2 are singular fractals (quantum) of such inertial reality. Only overlap between two fractals, which is their spatial discrepancy and/or time delay or asymmetry of perfect duality, is real here and now. Such overlaps are every 90° of double duality, 180° singular duality and/or 60° of triple duality (double 3D). Infinite space of universe, that is its singular passage of time, is only time delay or overlap between two times of infinitely larger and smaller spatial reality. Time here and now is delay (discrepancy) between spatial action and its reaction, just as space is discrepancy (delay) between action and its reaction in time. Infinitely long passage of time here and now is only infinitely small time delay between 4D spatial action and its reaction. Likewise—infinitely large spatial universe is only small spatial discrepancy between action and reaction of 4D time. It is equally vice versa—an atom that is small spatial difference here and now is infinitely long passage of time in context of its inertial space and/or infinitely large spatial universe in context of its inertial time. Reality is always here and now—infinitely large spatial universe and/or infinitely large period of constant passage of time.

An atom is infinite repetitions of spatial universes in every instant of time here and now and/or infinite repetitions of constant passages of time in every singular point of space. An atom is universe repeated in an instant infinitely many times in space and time. Every spin of an atom is all the time of spatial universe and/or all the space of its time. Every spin of singular electron is fractal ½ reality—it is all the time in 2 out of 3 spatial directions (equally vice versa), while every spin of dual (pair) electrons is complete 3D space in all its time and equally vice versa. Light speed motion is boundary of inertial size, just as infinite/singular size is boundary of inertial motion. One spatial curvature of light speed motion (one rotation of electron) is all the space and all the time of its universe, just as one time curvature is all the space of spatial universe in all its time. Light speed motion is eigenvalue between fractal motion and its size. Every curvature is light speed motion—it is all the space of one time and/or all the time of one space. There is no difference between universe, Earth or an atom—they are singular curvature; therefore light speed motion (either in space or time). They are all space of one time or all time of one space that is their variable density/gravity. Larger size is smaller motion and smaller size is larger motion, where small in space is large in time and equally vice versa. Light speed is curvature of flat reality and/or flattening of round curvature. It is variability of constant and/or derivate of exponent. Light speed is fractal difference or inertion between two equal opposites. It is double inertion that is inversion of inertion. It is reality here and now.

Light is constant, while its size/motion is infinitely relative/uncertain, because it is different in different fractal contexts of space and time. Light (energy) is inversion of mass (size) that creates double inertion (gravity). Light speed motion is infinite motion (energy) that is singular size (mass). Light (energy/motion), mass (size) and gravity (density/acceleration) are inertial 3D context of dual space/time. Light speed is linear motion of energy in context of spatial universe and size of mass (double rotation) in context of its micro-world (atom, which is singular time quantum). Stationary universe in spatial context (infinitely large size) is light speed rotation in context of time (infinitely small size), just as spatial light speed rotation (black hole) is infinite time universe. Spatial universe is black hole of time or black hole from inside, just as black hole is inverted (time) universe or universe from outside. Universe is black hole turned inside out in space, just as black hole is universe turned inside out in time. Space of black hole is time and its time is space. Its spatial attraction is repulsion of time, just as universe is spatial repulsion in context of attractive time. Black hole is time of universe within (space without), just as universe is its time without (space within). Black hole is singular boundary of spatial universe and infinite boundary of time universe, while universe is singular boundary of time and infinite spatial boundary.

Black hole is space/time eigenvalue or inversion of inertion that creates double inertion (space/time reality of universe). It is negation of negation that is double affirmation. It is end of spatial universe that is double/inverse beginning of time universe, which is inverse/double spatial universe. Universe/black hole are infinite/singular space/time, which is exclusive reality of inertial here and now. Physical reality is infinity of fractal reflections of such inertion and has no other meaning. Universe/micro-world is exclusive reality of inertial here and now. In context of relatively different inertial here and now, such universe is totally unreal—it is only its empty space and/or time. Universe/micro-world is inertion/inversion of here and now (double inertion). It is eigenvalue of inertial here and now or one out of infinite number of different possibilities. In context of all its time and all its space, any relatively different space/time is same reality, because space passes through all its time and time through all its space (Kepler). Inertial here and now that is one out of infinite number of different space/time realities (universes) in context of all its time explores all spatial possibilities (all time possibilities of one spatial reality). In context of all its space and time, all infinity of different inertial points of here and now is one reality that is no different from all infinity of different space/time possibilities. It is one space in context of all its time and also one time in context of all its space. It is one gravity in context of all its space and time and one antigravity in context of all its gravity. It is one inertial gravity/antigravity (no difference between attraction/repulsion) in context of all its non-differentiated physical reality. It is one reality in context of all its non-realities and/or one non-reality in context of all its realities.

Non-reality is inverse reality that creates double reality. **Inversion is singular fractal or quantum of its inertion.** Non-reality is singular quantum of reality, just as reality is singular quantum of non-reality. Space is quantum of time and equally vice versa. Mass is quantum of energy, which is infinity of massive quanta and equally vice versa—energy is quantum of mass, which is infinity of energy quanta. Size is quantum of motion and infinity of motion quanta is singular size. Fractal is singular quantum of whole reality. Only infinite quantitative (gradual or fractal) change is singular quantum change and only infinite quantum change is singular fractal change. Infinity of time quanta is singular quantum (fractal) of space and equally vice versa—infinity of spatial quanta is singular time quantum. Infinity of energy quanta is singular quantum of mass and infinity of massive quanta is singular quantum of energy. Only infinite fractal (gradual) quantitative change is instant quantum change and only infinitely many quantum changes is singular quanta of qualitative (fractal or gradual) change. Only all the time is singular spatial change and all the space is singular time change.

CHAPTER 3

FRACTAL 3/2 DENSITY (BODE'S PERIODICITY)

There is no room for accidental or coincidental reality in context of fractal double helix, because any imperfection is double perfection and any asymmetry is double symmetry. Causality of action/reaction is its only principle and any description that deviates from singularity of such principle is not real. Causeless reality is not real, because every effect has equal and opposite (fractal) cause. Quantum uncertainty of micro-world stems from inversion of spatial universe (N. Bohr), because it is its mirror image. Micro-world is universe turned inside out. It is mass of spatial energy and/or energy of spatial mass. It is infinity of inverse quanta (time) in context of inertial singular quantum (space), which is fundamental of Bode's periodicity and Kepler's harmonics (cube of space = square of time or 1.111 fractal difference). Both descriptions are fractal applications of 3/2-space/time reality, in which present adds past to generate future sequence. It is description of double inertion (gravity) that is inversion of inertion (space/time) or infinite 2x1/2 multiplication/division in context of its inertial (angular) ½. Same principle applies universally, because any different size/motion is relative/uncertain differentiation of no consequence in context of inertion of such duality. Physical reality is infinite Fibonacci sequencing of fractal difference that is inertial here and now. An atom is perfect description of fractal Bode's periodicity (and/or Kepler/Newton's conclusions, which are special limited cases of fractal curvature), just as is any size/motion of the universe.

Universe is gradual time curvature, which is 2D time/1D space, while micro-world is instantly flattened time curvature (reaction) that is inverse or spatial curvature of time, which is 2D space/1D time. Universe is gradually flattened time (instantly curved space), while micro-world is instantly flattened space (gradually curved time). Universe is double motion of singular size, while micro-world is double size of singular motion (Pauli's exclusion principle). Universe is two time quanta of singular spatial quantum and micro-world is two spatial quanta of singular time quantum. They are 2x1/2 equal opposites in context of inertial ½ that is their eigenvalue (here and now). Universe/micro-world are inertial/inverse gravitational curvatures—they are equal and opposite (fractal) densities of empty space without and dense mass within. Density is content (time) of space—it is instant empty space without and gradual densification within (infinity of mass/energy quanta in context of singular spatial quantum that is empty space. Micro-world is flat (uncertain) space, which is reaction of curved (relative) space without—it is round time or spatial curvature within of round space or time curvature without. It is spatial reality of time (time is not constant passage or gradual spatial change but instant space that is double/inverse change).

In context of spatial universe, every gravitational nucleus has only one mass in each orbit. In context of micro-world (atom), which is double space (time), each orbit has two rotating masses, as per Pauli's exclusion principle (one orbital = two electrons). Two electrons of each orbital are reflection of double spatial reality in context of its twice-smaller time (twice faster or infinitely faster passage of time). An atom (orbiting electrons) is double (squared) space and only ½ (inverse) time—it is two simultaneous spatial possibilities of one (whole) time (equally vice versa—two time possibilities of one space, because space/time difference is relative/uncertain differentiation, which has meaning only here and now). An atom is two instant fractals of one whole. It is instant double 2D space in context of no time, because time is instant additional spatial direction of an atom. Atom is not gradual, but instant spatial relativity in context of no time, which is its spatial uncertainty. Time of an atom is not time, but gravity, which is 1D fractal difference of time and/or 2D fractal difference of inertial space. Because time here and now is its spatial direction, there is no possibility of linear (spatial), but only rotational (angular) motion.

Singular here and now is boundary of spatial (linear) motion—it is no spatial motion that is infinite (light speed) linear motion in time. Linear motion in time is rotation in space. Any reality smaller than here and now is fractal (partial) curvature (inversion of inertion) of linear motion. Fractal curvature is infinite/singular parabolae. Atom is complete (whole) inversion of inertial here and now (360° or round curvature)—it is perfect duality of no inertion in between (perfect orbital with no eccentricity). Here and now is singular spatial quantum of universe, while an atom is its inversion, which is singular quantum of time. Linear motion in context of universe is rotation in context of an atom ($C/(1/C)=C^2$ here and now). Only rotation, which is double/inverse or accelerated/decelerated; therefore curved spatial (linear) motion, is equal opposite (reaction) of singular gravity (acceleration), which is time context of such spatial reality. Linear motion is possible only without but not within spatial reality of an atom. Slow linear motion in context of universe is fast rotation in context of an atom (spatial singularity = time infinity). Universe is infinity, while atom is singularity of only one—infinite or light speed motion. Universe is simultaneous infinity of different possibilities (velocities) of linear motion, while atom is only one possibility of rotation (1/C). In context of infinite (light) speed rotation, there is no distinction between different positions of orbiting electron—it is everywhere at the same time.

Such instantaneity of all orbital positions is singular uncertainty, which is only surface or 2D uncertainty. 3D uncertainty is double uncertainty or duality of orbital motion (Pauli's exclusion principle). It is horizontal and angular uncertainty, which are 1D fractal difference—i.e. horizontal rotation is infinitely faster than angular rotation and/or vice versa (light speed difference). Horizontal/angular rotations are linear/rotational difference (difference between constant and accelerated motion). If horizontal rotation is curvature of linear motion, angular rotation is double curvature or curvature of rotational motion—it is double rotation, which is not rotation but instant size (density). Angular rotation is precessional or fractal curvature of horizontal rotation, created by inertial difference that is reality of such rotation. Angular rotation is eccentricity (parabolae) of perfect (round) orbital motion. It is inertion of perfect duality or angular ½ of horizontal 2x1/2 duality. I.e. orbiting mass of an electron causes constant eccentricity of such orbital (small discrepancy of perfectly round curvature). In context of time such constant eccentricity gradually shifts its position in orbital plane, which is orbital of eccentricity or orbital of orbital that is double orbital (i.e. perihelion precessional motion). It is angular (2D) rotation or curvature of horizontal (1D) orbital. Horizontal rotation is an instant ring that is an electron in light speed rotation, while angular rotation is vertical rotation of such ring, which is an instant sphere.

Horizontal rotation is motion of electron, while angular (precessional) rotation is not motion of electron, but motion of its orbital. Orbital (rotational motion) of electron is instant (light speed), while angular rotation is gradual reality of quantum steps, whenever precessional (angular) curvature is 90° of rotational motion (orbital), because space/time is 4x1/4 sin wave of 4 inversion points of a curvature (De Broglie eigenvalue). Horizontal curvature is instant, while angular (precessional) curvature is real only in ¼ quantum steps, which are instant changes of orbital motion (quanta of energy of massive particle). Only infinite number of horizontal orbitals is accumulation of gradual shift (eccentricity) that is 90° curvature, which is one fractal step in angular direction (quantum of energy of rotating electron). Horizontal orbital is light speed motion; while angular orbital is inverse/double light speed motion—it is gravitational motion that is inverse reduction of double light speed motion or double reduction of inverse light speed motion (gravity). Double light speed in time context is inverse (less than light speed) spatial motion and equally vice versa.

Angular spatial motion is infinitely slower than horizontal light speed motion, because it is double (squared) time motion. It is not instant but gradual (quantized) motion. In context of angular rotation, horizontal rotation (orbital) is not motion, but instant 2D space (i.e. angular rotation of plane of ecliptic). Spatial rotation (rotation in context of universe) is in context of inverse reality of micro-world not motion but double motion that is instant size. Double (2D) rotation is instant and constant gravity—it is instant size of an atom. In context of inertial atom, linear motion of universe is rotation, while its rotation is instant size. There is no time context to differentiate position of electron in context of horizontal (spatial) motion, which is everywhere at the same time, because such orbital is not time but space of an atom. Its time is angular or time motion, which is instant space of universe. 2D motion (rotation/linear motion) of singular size (universe without) is translated as 1D (angular) motion of double size (duality of one electron). One is inversion of another's inertion, where only point of observation is point in perfect rest (no change of size or motion). It is always reality within/without that changes, where such changes are inertion/inversion, while here and now remains in perfect rest.

Two electrons in one orbital or two orbitals of one electron is only uncertain/relative description, which is reaction (inversion) of inertial (changeless) perspective. Micro-world is two instant spatial eigenvalues of one time (uncertainty or infinite spatial distortion of time), while universe is two gradual time eigenvalues of one space (relativity or infinite time distortion of spatial reality). Here and now is only inertion of such double/inverse fractal curvatures, because it is only point of reality that is singular space/time eigenvalue (no difference between space and time). Micro-world is perspective of inertial space in context of inverse, while universe is space in context of double (squared) time. In 2x1/2 fractal context, one space allows two motions and/or one motion allows two spatial sizes—any difference between the two is only relative/uncertain (fractal) differentiation of a whole (here and now). Any motion is real only comparatively—spatial motion without (action) is time motion within (reaction). Expansion of spatial universe is equal and opposite contraction of its mass within. Spatial motion is at the same time equal and opposite time motion in opposite direction.

Stationary object is stationary only relatively—it is dynamic object in perspective of here and now of another dynamic object, because here and now is always in perfect rest. Rotation in spatial context is linear motion in context of time and spatial size is time motion (equally vice versa). Size/motion are only comparative (dual) or surface reality. In context of their inertion (3rd direction), there is no difference. Size/motion is fractal (relative/uncertain) differentiation of reality that is no difference (here and now). Two electrons in one orbital or one electron in dual orbital—i.e. rotating horizontally around its nucleus and at the same time having additional angular motion (expansion/contraction), are two fractal (partial) perspectives of same reality. They are double/inverse perspectives or double reaction (within and without) of singular action of inertial here and now. Two electrons in 1D rotation are 2D rotation of one electron—it is one electron smeared all over 2D surface of its dual rotation; therefore every electron in 2D rotation and/or duality of one electron in 1D rotation is surface (fractal) reality of such atom.

Every orbital of electron is particular fractal density of an atom or particular gravity of dual size/motion. Every orbital of electron is sphere of particular size and density. Atom is quantized and not gradual uninterrupted change of size/density. It is continuous density here and now (one whole) fractionalized into space/time dualities of particle in motion. Particle (electron) in all its motion (time) and/or all the space of one motion is singular (inertial) density or singular size of infinitely larger/smaller (double/inverse) particle, which is atom without and/or nucleus within. Orbiting electrons are 2x1/2 multiplication/division (fractionalization) of inertial (singular) reality of an atom and inverse multiplication/division of inertial nucleus within. It is continuous (uninterrupted) space or density of an atom (and/or inverse density of nucleus) concentrated into fractal rotating particles, which are its time eigenvalues. Time eigenvalue of an atom is spatial eigenvalue of its nucleus, because orbiting electrons are duality—they are fractal inversion of inertion of an atom and double inertion of its nucleus (vice versa in context of time). Orbiting electrons are space/time inversion of inertial atom that creates double inertion of nucleus within—i.e. there is one fractal between two wholes and/or one whole between two fractals, which is 3D sequence of infinite precessional (fractal) curvature of double helix.

Infinity of orbiting electrons is 1D fractal difference, while singular nucleus within is double fractal difference of inertial atom, which is inversion of two directions of 3D reality. Such orbiting particles in context of all its time are continuous uninterrupted density of inertial time of an atom, because infinite time of electrons is singular or no time of an atom. Constant passage of time of inertial atom is infinitely faster than time of electrons, just as constant passage of time here and now is infinitely faster than time of an atom (smaller size = faster motion = slower passage of time). Electrons between an atom without and nucleus within are fractal 2x1/2 densities of uninterrupted space/time reality. They are uncertain space within an atom and/or relative space without a nucleus. They are positive/negative non-Euclidean curvatures (time and spatial curvatures) of equal and opposite flat realities of an atom without and nucleus within (if atom is horizontal flat reality, nucleus is vertical flat reality and equally vice versa). Round in space is flat in time; therefore they are also positive/negative flattening of equal and opposite round curvatures that are an atom on surface and nucleus in its angular depth. An atom within and/or nucleus without are equal and opposite quantized (fractal) realities, where one is negation of another, which is double negation that is singular affirmation (orbiting electrons or mass/energy of empty space/time).

Density is continuous curvature only here and now, because only here and now, space and time are simultaneous reality (eigenvalue). Linear increase of distance/difference (distance of time is spatial difference and distance of space is time difference) is double/inverse (exponential/derivated) increase of distortion of inertial here and now. Increased distance/difference (which is angular distance) is exponential curvature—it is parabolae or increased eccentricity of round curvature. ½ is boundary of uninterrupted curvature, which is infinitely eccentric (parabolic) curvature. Reality that is larger than ½ inertial distance/difference is distortion larger than 90°, which is reality beyond corner of flat perspective. Only ½ of spatial reality (sphere or a cube) is instant. Another ½ is beyond horizons of 2D perspective—it is time of such space. Another ½ is not real in spatial, but time context. Only in time a sphere rotates, when other ½ is also real. ½ of a sphere is space, while another ½ is its time reality (½ is instant size, while another ½ is motion of such size). ½ is singular gravitational quantum of space/time and/

or singular time quantum of space (singular spatial quantum of time). ½ is inertial context of any 2x1/2 duality.

Infinite number of atoms within every singular quantum of space is fractional (quantized) reality of such inertial space. Infinite numbers of electrons within singular atom (singular time quantum) is fractional (quantized) reality of inertial atom. Infinite number of atoms here and now is infinite fractal spatial curvature or quantized spatial reality, while infinite number of electrons of each atom is double fractal curvature—it is fractionalization (quantization) of density of inertial space. Electrons in context of inertial atom are double fractal spatial curvature (simultaneity of positive and negative non-Euclidean curvatures), which is flattening of round and/or curving of flat curvature. Double spatial curvature is time curvature and double time curvature is spatial curvature. Double time is space and double space is time. If an atom is uncertain reality, its electrons within are twice uncertain or relative reality. **Electrons within an atom are quantized relativity of instant uncertainty on surface, just as mass in motion in context of universe without is quantized uncertainty of instant relativity. Instant relativity (gradual time curvature) is inertial context of quantized uncertainty (gradual spatial curvature) and equally vice versa—instant uncertainty is inertial framework of quantum relativity.**

Universe is quantized time in context of inertial space, while micro-world is quantized space in context of inertial space. Atom is infinity of quantized (fractal) sizes, while universe is infinity of quantized (fractal) periods of time. Universe is infinity of time waves in context of singular spatial wave, which is not wave but particle (mass or space, i.e. size), while micro-world is infinity of spatial waves in context of inertial time wave, which is not a wave, but only constant linear passage of time (energy or time, i.e. motion). Boundary of constant passage of time is its inversion, which is infinitely long period of time, which is size of such motion, just as boundary of size is its motion or constant passage of time. It is double negation that is singular affirmation. An atom is quantized density (space), while universe is quantized inverse density or quantized empty space. Micro-world is quantum density of space (size), while universe is quantum density of time (motion). Empty space without and dense mass within are inversion/double inertion of inertial spatial reality here and now. Empty space without and dense mass within are double/inverse space of inertial space here and now (positive/negative non-Euclidean curvatures of flat space in between). They are time without/within inertial space. Empty space and dense mass are convex/concave time curvatures (relative/uncertain space).

Empty space is inversion of dense mass, just as dense mass is inversion of empty space. They are two equal opposites (action and reaction of inertial here and now). Empty space is motion of sizes, while dense mass is sizes of motions. Universe is quantum de-densification of mass, while micro-world is inverse densification, where dense in space is empty in time and dense in time is empty in space. Universe is quantum spatial relativity or infinity of relative quantum shells, one within the other, while micro-world is quantum uncertainty or infinity of uncertain quantum shells, one within the other. Universe is exponential quantum increase of spatial emptiness, while micro-world is exponential quantum increase of spatial density. Universe is quantum time, while micro-world is quantum space. Space is instant time, just as time is instant space. Micro-world is instant quantum reality (space), while universe is gradual quantum reality (time). Universe is quantized reality only in infinitely long periods of time, because time is quantized reality of inertial space.

Mass of universe is energy of micro-world and mass of micro-world is energy of universe. Mass/energy are inert/inverse density (dense mass/empty space). Energy is infinitely dense empty space or infinitely dense time, just as mass is infinitely dense space. Energy is inversion of inertial mass and equally vice versa. Atom is fractal (quantum) density of mass, while universe is fractal (quantum) density of energy (which is empty space). Micro-world (atom/nucleus duality and fractal orbiting electrons in between) is triple density of mass within, while universe (star/galaxy duality and fractal orbiting planets in between) is triple density of energy or empty space without. Atom/nucleus and fractal electrons in between are exponentially increased and fractionalized density of space (mass). Star/galaxy and fractal orbiting planets in between are exponentially decreased fractionalized spatial density, which is increased density of energy or empty space (time). Universe/micro-world are double/inverse densification/de-densifications of inertial here and now. Micro-world is fractally increased massiveness of empty space, while universe is fractally increased emptiness, which is increased energy (increased massiveness of time). Spatial universe consists of 3 fractal (quantum) steps in time, while micro-world (time) consists of 3 spatial quantum shells, which is 6 quantum shells (steps) of inertial duality, which is gravity ($g=E/m=(d/t)^2=C^2$) of such dual space/time (mass/energy).

Physical reality is 6 gravitational shells of inertial here and now (3 without and 3 within), which is 7th shell in between (inertial here and now). There is no difference between infinite physical reality and an atom—they are singular gravity consisting of 7 space/time shells. Inertial here and now is 4th shell of 3 shells within and/or 3 shells without, which is fractal ¾ reality (and/or 4/3). Fractal ¾ is 3D space in context of linear time and/or 3D time in context of linear space, which is its ¼ inertial (fractal) difference. ¼ fractal difference is time (curvature) of inertial (flat) space and/or space of inertial time. ¼ is inertion here and now, which is fractal difference between space and time. Here and now is space of time and/or time of space (no difference between space and time), which is perfect duality of no inertion. Such perfect duality is inertion of all infinity of its fractal curvatures. It is spatial context of micro-world (time) and/or time context of spatial universe (equally vice versa, because space/time difference is relative/uncertain differentiation). Universe/micro-world is 6-fold repetition in space and time of inertial here and now (6x1.111=6.666). Fractal 3/2 is 6 eigenvalues of such curvature.

Universe and micro-world are inertial/inverse exponentially increased discontinuity between space and time. They are exponential increase/decrease of inertial fractal difference, which is reality here and now (increase in space is decrease in time etc). They are inversely proportional disparity between perfect pace/time duality here and now. Such disparity is quantum interruption of inversion by its inertion and equally vice versa. Space/time is singular/infinite matrix, where infinity is interrupted by singularity and singularity by infinity (space is interrupted by time and time by space). Infinite/singular (space/time) points of interruption are double negation that is affirmation or instant inversion of inertion that creates double inertion. Any change less than infinite (complete) change is only gradual accumulation of surface irregularity. Only infinite spatial change is singular quantum or infinitely small change in time and equally vice versa, because space and time are fractal duality—one is only surface of 3D reality of another. Any change on surface is only gradual increase in tension of 3D and equally vice versa—any change in depth of 3D is only gradual increase in tension on its surface. Gradual increase in tension is only fractal reality of inertial here and now, therefore it is not real. Only when accumulated tension is equal to inertion here and now (90° curvature of 4x1/4 and/or 60° curvature of 6x1/6) is such surface reality also real in 3D context (and/or 3D is real in context of its surface).

Any change less than singular quantum (i.e. ¼) is not real here and now. Such change is real only when accumulation of spatial fractal difference in time (or accumulation of time fractal difference in space) is equal to singular quantum (inertion here and now). Only when time curvature (inversion) of such fractal difference is equal to its spatial (inertial) curvature is such reality instant (double) reality. Fractal reality is only gradual shifting (orbital) of eccentricity (less than ¼) or gradual increase of tension within, which does not affect size/motion without, unless a full quantum. Any change of density of mass within and/or empty space without is not real, unless such change is full 2x1/2 quantum. Consequently, empty space is interrupted by dense mass, just as dense mass is interrupted by empty space. **Continuity of space is interrupted by its time and continuity of time is interrupted by its space. Interruption is double/inverse reality of continuous space and/or time—it is its instant mass/energy (particle/wave or size/motion) duality. Interruptions of continuity are double/inverse continuity.** Mass/energy of universe/micro-world are double/inverse interruptions of space/time density. They are double/inverse density of gradually accumulated differential of density, which is not whole quantum but only infinity of fractal quanta that are not real here and now, but only gradual increase of tension.

Mass/energy is twice denser (squared) space in ½ (inverse) time and/or twice denser (squared) time in ½ (inverse) space. Mass/energy of space/time (universe/micro-world) are its quantized reality, which is contracted (squared) spatial reality in context of expanded time and/or contracted (squared) time in context of expanded space. Contraction within of simultaneous expansion without is double/inverse density. Mass/energy is quantized density of space/time. Only eigenvalue of space and time is real here and now. Fractal space of time and/or fractal time of space is not real—it is only empty space and/or time of such reality. Empty space is empty only on surface, while in its depth it is gradual (fractal) change of its density. Constant density is constant also only on surface, while in depth it is gradual change in space and time (half-life). Constant passage of time is constant also only on surface, while in its depth it is gradual change, which is not real here and now, unless full quantum (i.e. ¼). Constant motion on surface is acceleration in depth and equally vice versa. Constant passage of time is constant only in spatial context (1D fractal difference), which is surface of time.

In time context (2D fractal difference), time is not constant passage but infinitely accelerated/decelerated reality and equally vice versa—space is constant size only in context of time, while in spatial context (2D fractal difference), it is constant motion or constant change. Space is constantly expanding in its spatial context, just as its time (mass) is constantly contracting and equally vice versa—mass is expanding and its space is contracting in context of time. Only duality of action and equal and opposite reaction is real here and now. Fractal spatial action is one reaction only in context of constant passage of time, just as fractal time action is real only in spatial context and equally vice versa. Time is reality of fractal space and space is reality of fractal time. Fractality of time is singular space (size), just as fractality of space is one constant passage of time. Fractal reality of mass/energy is continuous reality of empty space and constant passage of (empty or changeless) time. Quantum interruption is equal opposite of its inverse continuity. Density (gravity) is fractal quantum interruption of space/time, just as space is quantum interruption of time and time is quantum interruption of space.

4x1/4 wave (curvature) has 4 inversion points at ¼ intervals (90°), while 6x1/6 wave has 6 inversion points (60° intervals). 4x1/4 wave is 2x(2x1/2) or perfect double duality (4D), while 6x1/6 is fractal 3/2 reality—it is 2x1/2 in context of inertial (fractal) ½. Fractal 3/2 is 1D fractal difference (precessional curvature) of 4D—it is fractal (odd) 5D between whole (even) 4D and 6D. Singular duality (2x1/2) is only two inversion points (180° apart), while singularity (1) is only one inversion point that is no inversion at all because it is one reality in 360° curvature. Singularity is inertion of all infinity of its fractal inverse curvatures (successive 2x1/2 multiplications/divisions). Double duality is 2x2=4 eigenvalues or harmonics of inertial reality, while fractal 3/2 is 2x3=6 synchronous harmonics. 6 simultaneous harmonics are boundary of Bode's 2x3 periodicity. 1/6 (inversion) is singular quantum (5D) of inertial 6D, which is fractal 3/2 space/time (space as surface of time and/or time as surface of space, which is double motion of singular size and/or double size of singular motion). 6 different spatial possibilities of inertial time and/or 6 different time possibilities of inertial space, which are simultaneous reality (eigenvalue), are infinite boundary (beginning/end) of fractal 3/2 reality. Beginning/end of any fractal curvature is its inertial whole. Begging in time is double/inverse end in space and equally vice versa—spatial beginning is double/inverse end in time. Every end is double beginning and beginning is double end.

Nucleus within is inversion of inertial atom without. If surface of an atom is end of its spatial reality, it is also beginning of its time within. Nucleus is inverse end of time and beginning of space, because nucleus/atom are space/time fractal difference—space of an atom is double/inverse space or double time of nucleus and equally vice versa. If an atom is fractal 3/2, than nucleus is its larger fractal or ½ reality (vice versa in inverse context, because 3/2 in space is ½ in time, just as 3/2 in time is ½ in space—total space/time is 4/2=2 equal opposites). Atom/nucleus duality are equal and opposite beginning/end of space and time. Nucleus is inversion or singular spatial quantum (infinite time) of infinite spatial reality of an atom (equally vice versa). Nucleus is dense mass of empty space (atom), just as atom is dense energy of empty time (nucleus). Atom/nucleus are inert/inverse density. They are positive/negative non-Euclidean curvatures. Time curvature of nucleus without flattens (inverts) equal and opposite time curvature of an atom within and equally vice versa—spatial curvature of an atom flattens spatial curvature of nucleus. One is negation of another, which is double negation that creates affirmation or infinite reality of orbiting electrons in between. Infinity of orbiting electrons in between singular atom/nucleus is fractal space/time curvature in between singular space (i.e. atom) and singular time (i.e. nucleus).

Double singularity is inverse/inertial duality. Orbiting electrons are infinity of motions between two equal and opposite (infinite/singular or inert/inverse) sizes. They are infinite fractal reality in between two wholes (singular space and singular time). They are fractal space/time between singular atom and nucleus (i.e. singular space and time, which are infinite/singular or inert/inverted space that are 1D fractal difference in between). Atom/nucleus are infinite/singular boundaries (infinite time is singular space etc) of orbiting electrons—they are spatial and time context of such fractal duality in between. Fractal 3/2 duality is 6 instant eigenvalues, where precessional (inverse) curvature overtakes inertial curvature. There is 6 points between singular space and singular time (or singular mass and singular energy), where inertion inverts creating double inertion. Fractal space/time curvature is 6 instant points (6 planets of Bode's periodicity), which are size in motion (mass/energy of empty space/time). 4 inversion points of 4x1/4 wave are 2D curvature (circle), while 6 inversion points of 6x1/6 wave or rotating 4x1/4 wave is double (3D) curvature (sphere). 2x1/2 wave is 2 inversion points or one linear direction, which has beginning and end. 4 inversion points is positive/negative 2D coordinate system (xy directions) or double linear direction, while 6 inversion points is positive/negative (space/time) 3D coordinate system (xyz directions), which is double 2D coordinate system or 3 linear directions.

2x1/2 is linear direction, 4x1/4 wave is a square (surface) of 4 sides, while 6x1/6 wave is a cube of 6 sides. One is real only in context of its inverse and double reality without/within. 2D coordinate system is double linear direction and only ½ of 3D coordinate system. It is curvature (i.e. time) of linear direction and flattening (i.e. space) of 3D coordinate system. It is double/inverse reality that is singular fractal reality in between two wholes and/or singular whole in between two fractals, because fractal of time is singular space and fractal of space is singular time. 2D is double/inverse eigenvalue of its 3D/1D and equally vice versa. Eigenvalue is point where inertion = inversion, which is instant double inertion. Only where inertial space is equal to inverse time is there double reality (mass/energy of empty space/time). Only when precessional (angular) curvature is ¼ of its inertial curvature is such reality two equal opposites (size in motion). Only when spatial curvature within is equal to time curvature without (and/or vice versa) is such reality an instant duality—instant mass/energy or particle/wave duality. Only when spatial wave of time is equal to time wave of space, such two waves add up to create double/inverse wave.

Only when space and time are perfect duality or perfect equal opposites of no inertion in between are they singular double reality or instant duality, which is mass/energy of empty space/time. Space resonates in time and time resonates in space only when they are equal. Space and time are equal and not opposite only here and now. Such 6 points, which are eigenvalues of inertial here and now, are its 6-fold multiplication/division. They are 6 simultaneous possibilities of inertial here and now. Only eigenvalue of here and now is real here and now—rest is empty space of time and/or empty time of space (constant passage of time). Such different spatial possibilities (empty space) that are not time eigenvalue of here and now (and/or different time possibilities that are not spatial eigenvalues) are reality of past/future time and/or larger/smaller space, but not right here and now.

Empty space/time is infinity of different realities, which are equally real in context of inertial reality (gravity of space/time), but are not real here and now (not real in space and time), because there is spatial inertion (tension) between its double/inverse time and/or time inertion between its double/inverse space. In gravitational context, there is no difference between empty space and dense mass—dense mass is only one out of infinite number of simultaneous possibilities. Mass/energy of universe/micro-world is reflection of inertial here and now, without any absolute reality. It is one out of infinite number of different fractal space/time possibilities, all equally real. Singular whole of infinity of fractal spatial realities of time and fractal times of spatial reality is singular gravity, which is inertial ½ of 2x1/2 duality. There is infinite number of different relative and/or uncertain possibilities, all equally real. Only one such possibility is real here and now. Only possibility where space= time (no difference or inertion in between) is real here and now. In context of fractal 3/2, such inertial possibility is 6 simultaneous harmonics, while in inertial ½ context; it is only 2 simultaneous harmonics (or 4 harmonics of double duality). It is 6 harmonics of 3D (cube), 4 harmonics of 2D (square), 2 harmonics of 1D (beginning/end of linear direction) and only 1 harmonic of singular point (here and now), where singular in time is infinite in space and equally vice versa.

Fractal double helix is a cube that is Bode's 2x3 periodicity. Cube without is square within (Kepler)—it is curved reality in one and flattened curvature in opposite direction (likewise—square without is linear direction within and linear direction without is singular point within). Fractality of space/time means that one side is flattened (contracted), while opposite side is curved (expanded). It is instant space in context of infinite time in one direction and instant time in context of infinite space in opposite direction. It is inflated/deflated space in context of deflated/inflated time. Inertion is ½ way point between such equal opposites—it is in between deflated (2D) and inflated (3D) space, which is inflated/deflated time. It is fractal 3/2 in one and fractal ½ in equal and opposite direction. Boundary of fractal 3/2 reality is fractal ½, while boundary of ½ is singularity. Consequently—inertial context in which fractal 3/2 is real is ½ reality, which is singular gravitational quantum of infinity of fractal 3/2 (space/time) realities (equally vice versa). 6 harmonics of fractal 3/2 are real only in context of 2x1/2 or two flattened harmonics on each side. It is 6—3D harmonics in between 2 flattened 2D harmonics on each side that are its inertial context (i.e. time of inertial space, space of inertial time or gravity of space/time).

Boundary of flattened 2D harmonics are equal and opposite singularities (singular space and time or singular mass and energy without and within). It is 2+6+2=10 orbiting electrons of 1st/2nd atomic shell and/or 10 planets (including asteroid belt) of solar system, which are real only in context of its larger/smaller (inert/inverse) reality, which is atom without/nucleus within and or Sun within/solar gravitational reality without. It is also reality of 7 atomic shells, because 1st shell is only flattened duality of 3D of its remaining 6 shells, same as Venus/Mercury in context of solar reality, where angular curvature of Mercury's reality is irregularity of surface Bode's periodicity. Mercury/Venus within and Neptune/Pluto without are fractal ½ curvatures of inertial flat periodicity of 6 planets in between. They are positive/negative angular (vertical) curvatures or inert/inverse time of flat spatial periodicity in between (whole multiples of 2x3 periodicity). 1st atomic shell is same horizontal irregularity that is angular space/time curvature, which is flattening of round and/or curving of flat reality of remaining shells. 2nd electron of 1st shell is singular flattening of 3D curvature, which is 2D (2D in space is 4D in time and equally vice versa), while its 1st electron is double flattening or flattening of two directions of 3D reality.

2nd electron is inversion of one and 1st electron is inversion of two directions of 3D—they are 1D and 2D fractal difference from other electrons. Nucleus is inversion of all 3 directions or total inversion of 3D reality—it is space of time and/or time of space (attraction within of repulsion without, where spatial attraction is repulsion in time and equally vice versa). Nucleus within is mass of energy without (orbital motions) and/or energy within of mass without (potential energy within of massive electrons without). Nucleus is space of time of orbiting electrons and/or time of their spatial reality. It is fractal reality of orbiting electrons turned inside out or complete (double/inverse) negation of their reality, where double negation is singular affirmation. Nucleus is one whole of infinity of fractal realities of orbiting electrons—it is their inertial background. Electrons are real only in context of their nucleus, which is singularity of all their space and time. Space and time of orbiting electrons are singular inertial reality of no difference in context of their nucleus (and/or singular inverse reality of no difference in context of singular atom in opposite direction).

Nucleus is gravity of fractal space/time reality of orbiting electrons (and so is an atom without, which is equal and opposite gravity that is repulsion without of attraction within and/or vice versa in opposite direction). Singular nucleus is equal opposite of all infinity of orbiting electrons. Reaction of singular nucleus (1D) is dual electron of first shell (fractal 2x1/2 or 2D), while reaction of 1st shell is fractal 3/2, which is its 3D infinity. 1st shell is singular dual electron that is one electron in two motions and/or one motion of two electrons. It is inertial surface (square) of 3D reality of remaining 6 shells, which are singular cube. Boundary of fractal 3/2 infinity is 6 synchronized (repetitive) harmonics. 1st shell is inertion (1/2) that repeats itself 6 times through repetitive 2x1/2 multiplication/division of fractal 3/2 reality, which are remaining outer 6 shells of fractal 3/2 reality. Each shell is only surface in between its two equal and opposite 3D shells and/or 3D in between equal and opposite surfaces of its preceding/succeeding shells. Each shell is inert/inverse surface boundary (singularity/infinity) of its neighboring shells.

Without time means, it is instant reality. Space is instant, while its change (time) is gradual. In context of gravity, which is double space (2D fractal difference)—time is instant, while its change is gradual. Gradual change of time in context of gravity is less than gradual change in spatial context—it is change in quantum steps, separated by constant passage of time in between. Constant passage of time is empty space of gravitational reality. Quantum spatial change within is quantum change of time without, because in context of gravity, space and time are non-differentiated whole and not two equal opposites (2x1/2). Space and time are instant surface or instant 2 directions of gravitational 3D reality. Space and time are space, while gravity is its time. Space is 2D surface and time is its depth (3rd direction), while both are 3D space in context of gravity, which is its time or 4th direction. Only gradual change of time, which is spatial change in quantum periods of time here and now, is real in context of gravity. Constant passage of time in between two quantum spatial changes is not mass/energy of gravitational (double space) reality, but only gradual accumulation of tension within, which is no change without. It is only its empty space, which is instant mass/energy only in quantum steps, when inertion = inversion that is double inertion.

Quantum spatial change, which is infinitely larger spatial change (equal to infinite gradual spatial change), is instant time, which is instant space of gravitational reality. Quantum spatial change is instant change caused by gradual accumulation of fractal change in course of constant passage of time. Infinity of gradual (fractal) changes within is one instant change without. Constant passage of time, which is gradual (fractal) change, is not real in context of instant time that is space of gravitational reality. Only quanta of inertial reality (gravity) are real in its context and space in period of constant passage of time is less then quantum, therefore not real. Only whole period of constant passage of time, which is singular quantum of time or singular fractal spatial change of gravitational reality are real in context of such inverted (infinitely larger/smaller) reality. **One spatial quantum of double space (gravity) is period of constant passage of time, which is size of its empty space. Relative spatial reality here and now is uncertain or quantum reality in inverse perspective. In context of relative reality (time), reality here and now is its uncertain quantum reality. Space here and now is uncertain space of relative space, which is time.** 3D space here and now is only one quantum of time or an atom of its relative reality without.

There is no difference between an atom within and reality here and now in context of spatial (relative) universe—they are fractal 3/2 reality, which is described by Bode's periodicity (and/or Kepler's laws). An atom within is inversion of time here and now (equal and opposite spatial inversion), just as here and now is inversion of relative universe. Atom turned inside out (exchange of space with time and time with space) is here and now and here and now turned inside out is spatial universe (equally vice versa in opposite direction). Here and now is fractal reality between two wholes (atom/universe), which are one 3D sequence of inertion/inversion= double inertion. Universe is inversion of inertial atom (equally vice versa), which is double inertion here and now and equally in direction of universe and an atom. Universe is double inertion of an atom, while here and now is its fractal inversion, where singular inertion in space is double inertion in time and equally vice versa.

Fractal difference is light speed (C) difference. Difference between inertion/inversion is singular ($C=d/t$), while difference between inertion/double inertion (and/or inversion/double inversion) is double ($C^2=E/m$). Triple fractal difference is difference between gravity and antigravity (attraction and repulsion or contraction and expansion), which is C^3=gravity/antigravity. Triple fractal difference is full curvature—it is complete inversion of inertial 3D sequence (space/mass/gravity or atom-here and now-universe), which creates double inertion. Singular fractal difference (C) is space/time boundary; double fractal difference (C^2) is double space/time boundary (mass/energy boundary), while triple fractal difference (C^3) is triple space/time boundary (attraction/repulsion boundary). Triple space/time boundary is boundary of inertial 3D or spatial boundary. Quadruple space/time boundary is boundary of space and time or boundary of size and motion, which is light speed expansion without and/or Planck constant within. Triple boundary is boundary of spatial reality, while quadruple boundary is also boundary of all infinite time of the universe. It is spatial universe in context of all its time and/or an atom in context of all its quantum changes (infinite half-life of mass).

An atom is instant time, while empty space of an atom, which is orbital of its electrons, is its constant passage of time, which is not time here and now, but only instant space. Time of an atom is gravity (double space); therefore atom is double space or two simultaneous spatial possibilities. Space without time is instant time or uncertain (double) space, which is twice instant space. It is instant 2D (surface), as opposed to instant 1D (linear direction) here and now. Linear direction here and now is surface in context of inertial atom and equally vice versa—linear direction of an atom is surface here and now (1D fractal difference or light speed difference). It is one spatial reality in two directions at the same time. Certain linear direction here and now is uncertain linear direction of an atom that is its **certain surface** and equally vice versa. 1D uncertainty is 2D certainty and equally vice versa. 3D uncertainty is certainty of both—1D and 2D (equally vice versa—3D certainty is 1D/2D uncertainty) and at the same time it is 1D relativity. Here and now is double uncertainty/singular relativity and/or double relativity/singular uncertainty in opposite direction (without/within). It is double space of one time and/or double time of one space. Time here and now is surface (space) of an atom, while space, which is constant passage of time is its depth.

An atom is content or density of empty space. It is its inversion within, which is double inertion without. Shells are content or density of an atom—they are empty space of its orbiting electrons, which are double density. Shells are 1D, while electrons are 2D fractal difference (singular/double light speed difference). Each shell is inversion of inertial reality of its succeeding/preceding shell, which is double inertion (inversion in space is double inertion in time and equally vice versa). Each shell is singular quantum (inversion) of its succeeding/preceding shell (where spatial succession is time precession and time precession is spatial succession). Singular spatial reality of one is infinite space of another shell. Shells are real only in context of its two equal and opposite neighboring shells, which are its double/inverse context. Succeeding/preceding shells are infinite/singular boundaries of inertial (fractal) shell in between—they are infinite space and infinite time of fractal space/time shell in between. Neighboring shells are equal and opposite 2D surfaces of inertial 3D in between and/or two equal and opposite 3D of a surface in between. Certainty of one shell is uncertainty of another, because one is negation of another, while both are double negation that creates affirmation (succeeding shell of previous two, which are fractal 3/2 or space/time reality). 3rd shell is inertion of its preceding dual shell—it is future sequence created by addition of past to present sequence or inertial context of dual action/reaction.

Boundary of such Fibonacci sequencing of 3/2 reality is 3x2=6 synchronized (simultaneous) harmonics of 3D sin wave (sphere). An atom is 6—3D shells in context of one inertial 2D shell (2D wave). In context of curvature within (atom), 7th shell is its 1st inertial (flattened) shell, while in inverse context of nucleus within, 1st shell is such inertial shell. In curvature within (atom), successive 6 shells are successive 2x1/2 multiplications/divisions of inertial (flat) shell, while in curvature without (nucleus), 6 shells without 1st shell are its 6-fold repetition. 1st/7th shells are singular/infinite boundaries or space/time of inertial atom/nucleus duality. First shell is singular surface (2D wave), while remaining 6 shells are its angular curvature, which are 6 surfaces of a cube (3D wave). 6 shells are rotation of 2D sin wave (1st shell) in all 3 directions. 1st shell is space of time, which is remaining 6 shells and/or time of their spatial reality. 1st shell is equal opposite (½) of all other shells (½)—it is their inertion. 6 shells are equal and opposite reaction of 1st shell, which is singular pair of electrons, just as 1st shell is reaction of nucleus within. Reaction of nucleus is 1st electron, which is only linear direction (1D), while reaction of 1st is 2nd electron, which is surface duality of two electrons. Nucleus and two electrons of 1st shell are singular 3D sequence (inertion/inversion=double inertion).

Nucleus is singularity of spatial reality of orbiting electrons—it is massive center of their orbiting motions (energy) and/or energy of their massive reality. Nucleus is singularity of dual size/motion (mass/energy)—it is inertial context of such space/time duality (just as atom without is its inverse inertial context). Nucleus is inertion of 1st electron (its size and orbital), which is its inversion (equal and opposite ½). Nucleus is 0D of 1D of first electron, where reduced spatial reality is increased time and equally vice versa. Double inertion of dual inertion/inversion (nucleus/1st electron) is second electron, which is inversion of nucleus/1st electron duality (2D). Inversion of first pair of electrons (1st shell) is 2nd shell and their double inertion is 3rd shell, which is inversion of 2nd shell. Duality of 2nd/3rd shell produces 4th shell, which is inversion of 3rd shell etc. It is

6-fold spiraling curvature, where present sequence (inertion) adds past sequence (inversion) in order to create future sequence (double inertion). Line of symmetry of such spiraling curvature is 4th shell, which is ½ way between equal opposites. 4th shell divides an atom into two equal opposites (2x1/2). Reality inside 4th shell is attractive reality of nucleus within, while reality outside 4th shell is its repulsive reality and/or attractive reality within an atom. 1st shell is surface, while remaining 3 shells are fractal spatial reality without nucleus, while outer 3 shells are its relative space (time or expanding universe).

7th shell is inner surface of an atom, while following 3 shells are fractal density of its mass within. Beginning of 4th shell is end of massive reality of an atom within—it is center of its gravity. 4th shell is line of symmetry between equal and opposite densities of nucleus within and atom without. It is line that is infinite empty space of nucleus without (its infinite universe) and/or infinitely dense center of massive atom within. Reality beyond 4th shell is more than totally uncertain reality of an atom, just as it is relative reality without nucleus (twice uncertain within and/or once relative without, because of fractality or 1D difference between space and time, which is inertion of reality here and now). 4th shell is dividing surface between infinite density within and infinite emptiness without—it is inversion between empty space and dense mass in opposite direction or inversion between attraction/repulsion. Nucleus and surface of atom are its beginnings and equal and opposite ends in space and/or time, while 4th shell is its singular beginning and end in gravitational (inertial) context of such duality. Likewise—if nucleus and surface of atom are spatial beginning/end (singularity/infinity), then 4th shell is duality of such curvature—it is its beginning and end in time, because double space is singular time and double time is singular space. Two equal and opposite spatial beginnings are one end in time and equally vice versa.

Consequence is that additional electrons cluster in zigzag pattern around line of symmetry, because it divides atom into two equal ½'s. Electrons within 4th shell are time curvature of space, while electrons without are spatial curvature of time—they are convex/concave (+/- non-Euclidean) curvatures. Electrons within are fractal reality without nucleus, while electrons without are inverse fractal reality within an atom. Electrons within are uncertain reality of electrons without, just as electrons without are relative reality of electrons within. They are negation of each other, which is double negation that is affirmation of reality of atom. Inversion of atom is its nucleus within. Atom/nucleus divided by two triple shells without and within along their line of symmetry (4th shell) is double duality or 4x1/4 multiplication/division of inertial reality here and now (sin wave), which is space/time in context of its gravity/antigravity (contraction/expansion). 4th shell is 5th quarter or inertion of such 4x1/4 wave. It is 1D fractal difference of double duality. 3 shells within/without their line of symmetry are triple fractal spatial curvature of time and/or triple fractal time curvature of space. They are double 3D and/or triple 2D, whose curvature is $3^2/2^3=9/8$ or $1/8^{th}$ difference. 1/8=0.125 or 3D numerical sequence of infinite double helix (0.123456...)

Shells are equal and opposite density of atom/nucleus, while electrons are density of such empty shells. They are inverse density within that is double density without. Shells are fractal reality in between equal and opposite surfaces of atom and nucleus, while electrons are fractal space/time reality between two equal and opposite surfaces of a shell. Shells are fractal content of an atom, while electrons are fractal content of shells. Shell is singular fractal difference—it is space of time and/or time of space (2x1/2), while its electrons are double fractal difference—they are gravity of such dual space/time and/or space/time of inverse attractive/ repulsive duality of nucleus/atom (4x1/4). Shells are 2D surfaces of a cube (3D atom), while electrons are linear directions of which such surfaces consist. Number of linear directions, which comprise one surface, is inversion of such inertial surface, just as number of surfaces of a cube is inversion of inertial reality of a cube. I.e. inversion of 6D, which is infinite/singular or finite (fractal—inverse/ double) 3D, is 6x1/6; inversion of infinite 3D is 3x1/3, therefore double 3D is again 6x1/6; inversion of 2D is 2x½ or 2 equal opposites etc.

Fractality or quantization is only relative/uncertain (limited or finite) space/ time reality. Reality is unlimited (infinite)—only inertion here and now is its inverse/double (space/time) boundary. 2 equal opposites are only 2/3 of 3D, while 3 directions of 3D are only ¾ of 4D and 4 space/time directions are only 4/ 5 of 5D etc. Only two equal opposites are eigenvalue of 3D and only 3 directions are eigenvalue of singular 4D. Fractal difference is always inertion or equal opposite of such fractal reality (½). ¼ which is inertial difference of ¾ reality is equal and opposite (inverse) ¾. ¾ within and ¾ without are boundaries (space and time) of inertial here and now in between (3/4x4/3=1). Inertion is double/inverse limitation of physical reality (dual space/time). Larger fractality is larger number of directions or larger quantization of inertial here and now. Singular quantum of fractality is 2x1/2 or two equal opposites (two linear directions of a surface and/or 3 surfaces of a cube). It is largest quantization or largest limitation of unlimited (infinite) reality. It is reduction of infinite freedom of choice to only one dual choice. In context of any relative/uncertain duality, only ½ of such reality is simultaneous (instant), while another ½ is time of such choice, which is infinity of all relatively different choices. In context of larger fractality (i.e. ¾, 4/5 etc), it is instant multiple choice (3 out of 4, 4 out of 5 etc).

Multiple choice is only larger/smaller duality, because fractal difference is always equal opposite (½) of any choice. Equal and opposite ½ is infinite fractal curvature, because it is increasingly larger/smaller fractal reality (larger quantization). Equal opposite is always inversion of inertion and every exclusive duality (inertion/inversion) has 3rd inclusive (singular) direction. Every singular fractal is whole of larger/smaller fractal, infinitely many times. Equal opposite is infinite fractal (½) curvature of singular choice, regardless of its relative/uncertain fractionalization (quantization). Time is infinite fractal curvature of singular space, just as space is infinite curvature of singular time. Any choice is dual (relative or fractal) choice and any reality is fractal 3/2 reality—2x1/2 in context of angular ½, which is inertion of such dual action/reaction. Any choice is fractal, because it is choice of reality, which is no choice. Reality is always fractal curvature of double helix, with boundaries that are equal and opposite inversions of inertial here and now.

2 electrons of 1st, 8 electrons of 2nd, 18 electrons of 3rd shell, etc are only fractal (½) interruptions of one (1) uninterrupted reality. Reality is always fractal interruption of uninterrupted reality, which is 1+½ =3/2. It is only progressive fractionalization in distance of space/time of one uninterrupted density, which is progressive fractal curvature. It is infinity of inversion of inertion that creates double inertion or infinity of interruptions of inertial reality (i.e. time) with its equal opposite (i.e. space), which creates double duality. Every fractal duality is real in context of inertial discrepancy between action/reaction, which is its 3rd precessional curvature or their line of symmetry—infinitely many times in directions of space and time. Empty spatial reality without nucleus is dense spatial reality within an atom, where any determination is relative comparison with inertial density here and now. Infinite universe or empty space without massive nucleus is dense mass within an atom. They are action/reaction of inertial here and now. Atom is equal and opposite fractal 3/2 curvature from empty space of the universe the universe. Atom is dense space or dense mass, while universe is dense time (empty space) or dense energy (empty mass). Both are infinite double spiraling curvature, where one strand is fractal reality of its equal and opposite strand.

Fibonacci sequencing that is reality of double helix is double repetition of singularity in its first two sequences, because of its fractal reality. Fibonacci sequencing starts with 1,1,2,3,5,8…etc, where double 1 are horizontal and angular 1. Horizontal singularity (1) is angular (vertical) infinity. It is one principal of infinite number of subordinates (Balmer). Double one is spatial and time singularity, while all the rest of infinite sequencing is infinite number of their fractal curvatures. Double one is inversion, which is double inertion of inertial here and now, which is double singularity or 0 of such infinite curvature. Equal opposite of 0 are two equal and opposite (inverse/double) 1's—they are space and time of inertial here and now. 0 is 0 only in context of two equal and opposite 1's, just as 1 is real only in context of two equal and opposite 0's. 0 is 3rd direction (inertion) of double 1 and 1 is inertion of double 0—one is real only in context of another. 0 in space is 1 in time and equally vice versa. 0 and 1 in space or time are meaningless (fractal) or infinitely relative/uncertain description in context of their gravity. 0 and 1 are double 1 in gravitational context and/or double 0 in context of gravity.

0 is infinitely small fractal (quantum) of 1 and equally vice versa. Difference between 0 and 1 is relative, because they are real only in context of one another. They are two equal and opposite singularities of infinite number of fractals in between. There is no 0, because 0 is an illusion of 1 and there is no 1, because 1 is an illusion of 0. There is only duality of both, because only double shadow is 3D reality and only double shadow is shadow with content and not an empty shell. Only something in context of nothing and nothing in context of something is fractal 3D. One without the other is non-reality. 1st electron is inversion or equal opposite (1) of its inertial nucleus within (0), where 0 in space is 1 in time and equally vice versa. Inertion of such duality (inertion/inversion= double inertion) is 2nd electron. 2nd electron is symmetry line or discrepancy between two equal and opposite curvatures (nucleus/electron). It is inertial background or eccentricity between two equal and opposite perfect curvatures. It is future sequence generated by present/past sequences. 1st electron is precessional (fractal) curvature of dual nucleus, while 2nd electron is precessional (fractal) curvature of nucleus/electron duality. Each one is temporary balance between two equal opposites. Small inertial difference between two equal opposites (action/reaction), which is less than singular quantum, is real only in time. Only accumulation of such difference in time is instant additional singular quantum or additional electron. Because time here and now is instant space of such inverted reality (atom), all infinity of different electrons are instant spatial reality in inertial context here and now. Same reality here and now (infinite number of electrons) is real only in context of infinitely long period of time.

Duality (symmetry) is real only in context of their inertial system, which is its line of symmetry. Line of symmetry is infinitely small discrepancy between symmetry of action/reaction—it is infinitely small reality of infinitely larger reality here and now. It is time of space and/or space of time. Line of symmetry is discrepancy between two equal opposites—it is their asymmetry (eccentricity of round curvature). Line of symmetry between two motions (rotation/orbital) is their synodic/sidereal difference, which is fractal (inversion) of inertial precession of such duality. Precessional motion of dual motion and/or precessional motion of rotating size is its 3rd direction, which is double inertion created by such dual inertion/inversion. Precessional motion is line of symmetry between two equal opposites, which is their fractal asymmetry that is double/inverse symmetry. Every duality is gradual (fractal) curvature in 3rd direction, which is less than singular quantum. Only ¼ of precessional motion, which is singular quantum of double duality (fractal surface reality or 4x1/4 wave), is singular quantum here and now. Consequently, only gradual accumulation of synodic/sidereal difference (inertial difference between two equal opposites) is real here and now. Only ¼ of precessional time between two equal opposites (i.e. nucleus/electron), creates additional (2nd) electron, and only ¼ of precessional time between 1st/2nd electron, which is infinitely larger period of time, creates 3rd electron etc. Only nucleus/1st electron are instant reality (space), while rest of electrons are exponentially larger periods of time in which they are real. They are only progressively larger quantized reality of inertial duality between nucleus and its 1st electron.

Just as all infinity of numbers larger than 1 are only infinity of fractal curvatures of inertial 1,1 Fibonacci sequence, so is all infinity of electrons only fractal repetition of such inertial duality. It is not instant reality, but reality that is real only in exponentially larger periods of time. Only nucleus/electron duality is instant space, while rest of electrons are its time. Rests of electrons are only fractal (relative/uncertain) spatial reality and not instant space. Because atom is 1D fractal difference from here and now, its time is space here and now; therefore all its fractals are simultaneous (instant) reality. Their fractal space/time difference is fractal mass/energy difference here and now. Larger fractals (larger number of electrons) are larger mass/smaller energy (smaller half-life) or smaller energy needed for their complete change from mass to energy (larger instability). Only small (fractal) change of space/time here and now (only small period of time) is cause of deterioration of infinitely large number of electrons (short half-life), while only all infinite time of universe is deterioration of 1st electron (infinitely long half-life).

2nd electron is precessional curvature (time) of spatial duality of 1st electron/nucleus duality; 3rd electron is precessional (angular) curvature of 1st/2nd electron; 4th electron is inertial ½ of 2nd/3rd electrons (2x1/2); 5th electron is their inversion/inertion=double inertion; 6th electron is yet another 3D sequence of previous duality etc. Nucleus is singularity (0), while 1st/2nd electrons are its two equal opposites (1,1). Nucleus is angular (time) ½ of 2x1/2 (space) of dual electrons of first shell. First shell of two electrons is equal opposite of nucleus—it is its double negation, which is singular affirmation. 1st/2nd electrons are horizontal/vertical or space/time duality of inertial gravity (nucleus). They are double singularity, which is singular space/infinite time and/or singular time/infinite space. 1st/2nd electrons are only 1D/2D reality, while only all infinity of rest of electrons are fractal 3D realities. They are infinite number of different fractal (space/time) possibilities between duality of first two electrons. They are different fragmentations and their recombinations of inertial 2x1/2 duality of first shell. 2nd electron is all the time of 1st electron (and/or all space of its time, because it is inversion of its inertion). 2nd electron is time of 1st electron (space). Exponentially larger fractal repetitions (larger number of electrons) are exponentially larger fractal realities of its time. While 1st/2nd electrons are simultaneous (instant or no time difference), rests of electrons are real only as geometrical progression in time.

Only one pair, regardless which one is instant (simultaneous) reality, all the rest are positive/negative exponential fractality of its space within (micro-world) and/or fractality of its time (universe) without. Electrons within inertial pair of electrons are its uncertain (i.e. negative fractal) reality, while electrons without are its relative (i.e. positive fractal) reality. Electrons within inertial pair (space/time or rotating size) are spatial reality of such pair (its mass or uncertain space within), while electrons without are its time (its empty or relative space without, which is dense mass only in exponentially larger periods of time, where frequency of such periods is ¼ of its subsequent precessional periods). 6 shells of exponentially larger (double) number of electrons, without 1st inertial shell, which is singular spatial duality, are 6x1/6 fractal 3/2 (space/time) curvatures of inertial 4x1/4 curvature of first shell (double size of singular motion and/or double motion of singular size). 6 shells are repetitive or subsequent precessional curvatures of inertial duality of 1st shell (exponentially larger periods of time). They are 6 surface possibilities of a cube (see illustration 5). Each surface is one relative sequence of inertial singularity—each is exponentially larger quantum period of time that is exponentially larger spatial change.

6 shells of inertial 1st shell are 7 simultaneous spatial possibilities, which are eigenvalues only in time and not instant here and now. Likewise—only one pair of electrons is instant here and now, while all infinity of electrons within/without are its equal and opposite quantum (fractal) realities (relativity of empty space and/or uncertainty of dense mass). Space/time reality within/without is infinite fractal curvature of double helix, which is infinite double precessional curvature of inertion/inversion (gradual/instant or quantum change), which generates dou-

ble inertion. It is infinite 2x1/2 multiplication/division in context of inertial (angular) ½ or infinite 3/2 fractality of space/time in context of its singular gravity, where time is precessional curvature of space and gravity precedes time (double spatial precession).

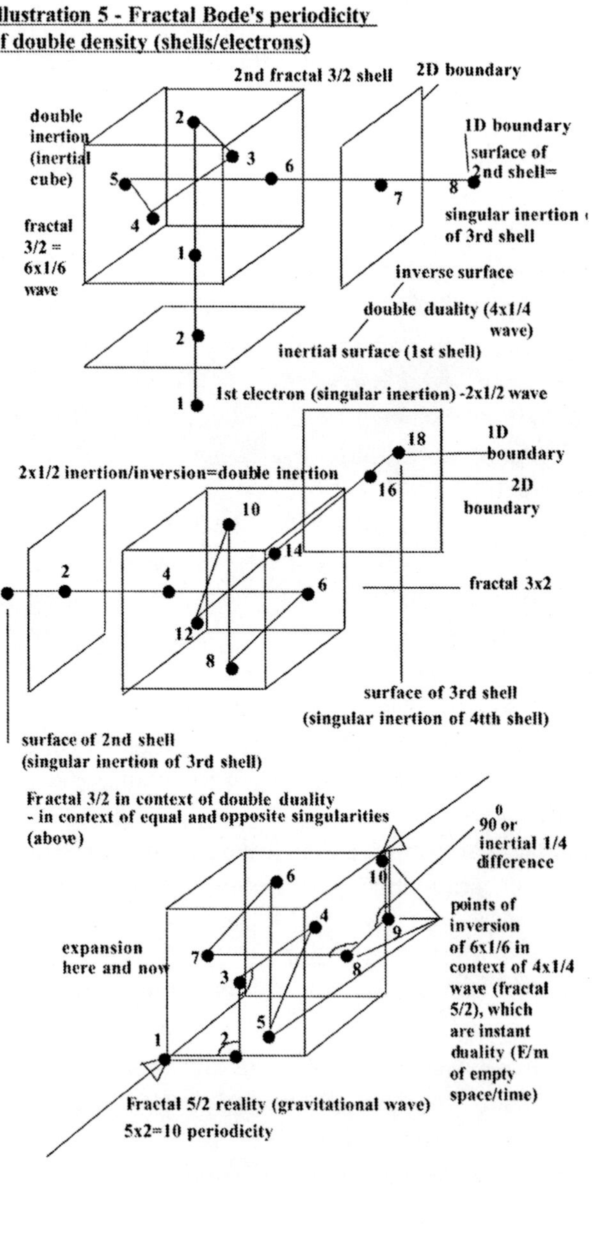

Atom is contracted (singular), while universe is expanded (infinite) space. Atom is expanded (infinite), while universe is contracted (singular) time. Universe is all the space in one instant of time, while atom is all the time in one spatial instant, which is no time in any instant of time here and now (double fractal difference). Atom is negation of time of the universe, just as universe is negation of spatial reality of an atom. Contracted space is double/inverse space (time), where linear spatial direction is singular point and round curvature is linear direction (equally vice versa). Linear spatial direction in context of universe is singular point in context of an atom. Bode's periodicity describes linear spatial distance between planets in context of spatial universe (relativity) and/or singular distance or number of electrons in context of micro-world, which is inverse space or time of spatial universe (uncertainty). $1^{st}/2^{nd}$ electrons are Mercury/Venus of Bode's periodicity, where position of Mercury is only fractal ½ reality, while reality of both is inertion of periodicity of remaining 6 planets (except Neptune/Pluto, which are inverse fractal ½ curvature in opposite direction). Mercury (0.4) and Venus (0.7) are 1.1 or double/inverse 1 (1+1/10) of inertial dual Fibonacci sequence, which is inertial reality here and now (Earth, which is 1 of such reality, where 1 is real only in context of its equal opposite—1/10, which is gravity of such dual space/time).

Mercury/Venus are fractal spatial reality, while outer planets are fractal time of inertial Earth (micro-world/universe duality). 10 planets are 2x3=6 in context of equal and opposite 2z1/2 or 2+6+2=10 planets and/or 10 electrons of $1^{st}/2^{nd}$ shell. It is fractal 6x1/6 (3/2) wave in context of 4x1/4 (dual) wave (two equal and opposite 2x1/2). Outer 6 planets are inversion of inner 2 planets that create double inertion (reaction) of additional 2 planets without inner 8 planets (action). Inversion (inside out) of linear direction is surface, which creates cube (double inertion) and equally vice versa. Fractal 3D is simultaneous reality of singularity (linear direction), its precessional duality (surface) and double precessional 3D (cube). It is singular massive nucleus, its dual electrons (size/motion) of first shell, which is its inversion, and following 6 electrons, which are its double inertion. 6 electrons are fractal reality of equal and opposite two electrons (double duality) on each side. 2 electrons of 1st shell and/or 2 last electrons of 2nd shell are equal and opposite (spatial and time) reaction of inertial (fractal) 6 electrons (planets) in between.

Such 2 electrons (planets) on each side are boundary of inertial cube in between—they are its two equal and opposite surfaces (inner surface turned inside out). 2 planets (electrons) are two equal and opposite dualities that are boundaries of fractal 3/2 reality, while boundaries of duality is singularity—it is massive nucleus (Sun) within and/or surface of shell (atom or solar reality) without. Surface of shell is inversion of such 2+6+2 reality within, which is double inertion without. Surface is symmetry line between two equal and opposite shells, which is instant 2x1/2 multiplication/division of previous shell (instant inversion of inertion that creates double inertion). Only one dual shell and/or one pair of electrons are instant reality here and now—rest are its micro-world or uncertain space (relative time) within and/or relative space (uncertain or quantum time) without. Numbers of electrons in each shell are 2x1/2 multiplication/division of space in time and/or time in space—they are further (double) duality of previous inertial and inverse duality. Inertial shell and/or inertial electron (planet) is product of its previous and successive two electrons. If previous two electrons (planets) are its spatial inversion/double inertion, than successive two electrons are its time inversion/double inertion 4 successive/preceding electrons/planets/shells etc are 4 neighboring (angular) surfaces of any inertial surface here and now. Equal opposite of such 5 surfaces is equal and opposite (6th) side of a cube, which is yet another 5 surfaces or total of 2x5=10 surfaces of fractal 5/2 reality, which is double fractal 3/2. **3/2 is fractal space/time, while 5/2 is fractal gravity** of such inertial space/time—it is inclusive reality of exclusive space and time, which is mass/energy of space/time and equally vice versa.

Unequal things are equal only if treated unequally. Micro-world is unequal equality of universe and equally vice versa. Space is unequal equality of time and expansion is unequal equality of contraction. Unequal equality or asymmetrical symmetry is fractal supersymmetry of no difference between symmetry and asymmetry. In its context there is no difference between fractal equal opposites, because inversion of inertion is double inertion or infinite spiraling curvature of double helix, which is infinite dual (fractal) repetition of singular here and now. Just as atom reflects universe (N. Bohr), so also universe reflects an atom, because they are equal and opposite (unequal) fractal reflections of inertial here and now. Only law of physical reality is perspective, which is double inertion of inertial/invert reality of space/time (C=d/t=const.). Space and time are action/reaction of inertial perspective. There is no different realities, only different (fractal) perspectives of one reality, which is no difference.

One certainty here and now is 6 different certainties of inversion that is double inertion (universe/micro-world of inertial here and now). More than one certainty is uncertainty. Fractality of 6x1/6 wave (3/2 reality) is fragmentation of one constant reality in its 6 underlying derivates (6 sides of a cube, which is 4x1/4 wave or round curvature rotated for +/-90°). Fractal curvature is differentiation between space and time—there is no time of spatial reality, or space of time, without fractality (duality of equal opposites). Fractal spatial difference is its time, just as fractal time difference is its space. Fractal spatial curvature in context of its time, what is passage of all its time, passes through all infinity of inverse possibilities of any inertial possibility here and now (Kepler). In total curvature, any inertial (singular) reality is reality of no difference, because time is infinity of excluded spatial possibilities, just as space is infinity of excluded time possibilities of singular inertion here and now. Infinite space and infinite time are infinity of different space/time possibilities of inertial here and now (Everett's many worlds). One in curvature of another is reality of no difference, because all infinity of different possibilities are only fractal reflections of singular reality, which is no difference. Differentiation is only fractal or surface reality. It is space, taken out of its context—time and/or time taken out of its spatial context, which is meaningless infinite/singular (relative/uncertain) distortion of reality here and now.

Reality is real only in context of its negation (equal opposite), which is its non-reality. Reality is duality—it is fractal reflection of non-differentiated singularity. Consequently—space in context of its time is equal to time in context of its space. Any inertial space in course of all its time passes through all its excluded possibilities, because, without such excluded possibilities, there is no such inertial reality either. Any choice is double (positive/negative) choice—it is simultaneous choice of singular (inertial) reality here and now, as well as its boundaries (infinite inversion), in which such inertial reality is real. Boundary or context of inertial spatial reality is its time (equally vice versa), while boundary of both—space and time is their gravity. Boundaries of inertial reality are infinity of all choices (possibilities), excluded by inertial choice (possibility) here and now. In context of its time, singular spatial possibility here and now passes through all infinity of such excluded possibilities and equally vice versa—time passes through all infinity of its excluded possibilities, because reality is double fractal curvature. Time equalizes all spatial possibilities and space equalizes all time possibilities, while gravity equalizes all possibilities of space and time.

Consequence is that any increased observation reveals in equal and opposite proportions underlying constancy, as well as its increased uncertainty/relativity, because any symmetry is broken by its line of symmetry, which is 3rd direction of dual symmetry. Increased uncertainty here and now is increased certainty of such different/distant reality and equally vice versa. Difference between empty space and dense mass is only relative/uncertain. Any fractal shift in inertial perspective is equal and opposite shift of fractal space/time reality, which is instant change of empty space into dense mass and dense mass into empty space (or instant change of mass into energy and energy into mass; size into motion and motion into size; contraction into expansion and expansion into contraction etc). As long as perspective remains unchanged, fractality of space/time remains—only difference is what is space and what is time; what are shells/electrons/planets etc; what is mass and what is its inertial framework—empty space/time within/without. Change of inertial perspective is instant change of fractality of space/time reality, not only change of position of mass/energy in context of empty space/time. Change of dual perspective is instant different quantization of space/time, which is not only different quantitative, but also different qualitative change.

4th side of 4th shell is 4th quarter of 4x1/4 reality—it is fractal difference of fractal ¾ reality (and/or 5/4, which is its equal opposite), which is time of inertial space in one direction and space of inertial time in opposite direction. ¼ is inversion of inertial 4D—it is inertial context of such reality, which is perpetual repetition of 4x1/4 wave. ¼ is difference between space and time and mass and energy or space and time and attraction and repulsion. It is difference that is inversion of inertion, which creates double inertion (and double inversion). ¼ is difference or singular quantum of double duality (4D), while 1/6 is difference of fractal 3/2 reality, which is no difference between all 6 inertial directions (space/time, mass/energy and attraction/repulsion). Inversion/double inertion is equal and opposite (space/time) quantization of inertion, because singular quantum of time is infinite space and singular spatial quantum is infinite time. Quantification is inversion of horizontal infinity into angular singularity and/or vice versa. Singular variability (i.e. constancy) is singular quantum of infinite inverse/double variability (Balmer—each principle is infinite number of subordinates). Singular quantum without is infinity of quanta within and equally vice versa, where such infinite quantification is only gradual qualitative change in one direction and instant (double/inverse) change in opposite direction (singular quantum).

Singularity/infinity are two equal and opposite ½ of inertial here and now. Singularity is only ½ of duality and duality is ½ of singularity—one is cause/effect of another. There is no nothing or something, because something is ½ of nothing, just as nothing is ½ of something. 0 is an illusion of 1 and 1 is an illusion of 0. Double illusion is fractal reality. It is infinite fractal Fibonacci sequencing where something in curvature of nothing and/or vice versa, generates double something and where double something is double nothing in inverse curvature. Inversion of nothing is something and inversion of something is nothing. 3rd direction of such duality is square/inverse square or denser something/nothing, where such density is real only in context of its equal opposite. There is no difference between something and nothing or space and time, because it is always singular consciousness of point of observation, which is singularity/infinity of any duality, because it has no equal (comparable) opposite.

Chapter 4

GRADUAL/INSTANT (QUANTUM) CHANGE

Difference between space/time, mass/energy and gravity/antigravity here and now is angular. They are 6 corners of a hexagon, where two equal opposites resonate one within the other (eigenvalue). Angular spatial difference is spatial distance in context of time (inversion). They are instant reality here and now, while in context of relative universe (time), they are infinite spatial distance apart. Only infinite space is singular time and equally vice versa. Infinitely small angular difference is infinitely large spatial distance. Infinity/singularity is light speed or 1D fractal difference—it is instant inversion that creates double inertion. Every duality (space/time, mass/energy and gravity/antigravity) is two equal and opposite corners of a hexagon (180° difference), while its double/inverse dualities are 2 corners in between (60° difference). Space/mass/gravity are round (angular) curvatures here and now, while in context of spatial universe, they are round spheres, one within the other (linear spatial distance apart). Infinite size of time is its space and infinite size of space is its time. Space and time are boundaries within/without such 3D sphere. Such inertial sphere consists of 3 fractal spheres, which are space in context of mass, mass in context of gravity and gravity in context of inverse/double space (time) and equally vice versa in opposite direction; therefore 6 fractal spheres or 6 synchronized harmonics of inertial here and now. They are 3 fractal spheres of the universe (Earth in context of solar reality, solar reality in context of galaxy and galaxy in context of expanding universe) and 3 fractal sphere of its micro-world (atom/nucleus/Planck constant). Each sphere is space/ time duality or size in motion.

Illustration 6 – Fractal permutations of 3D

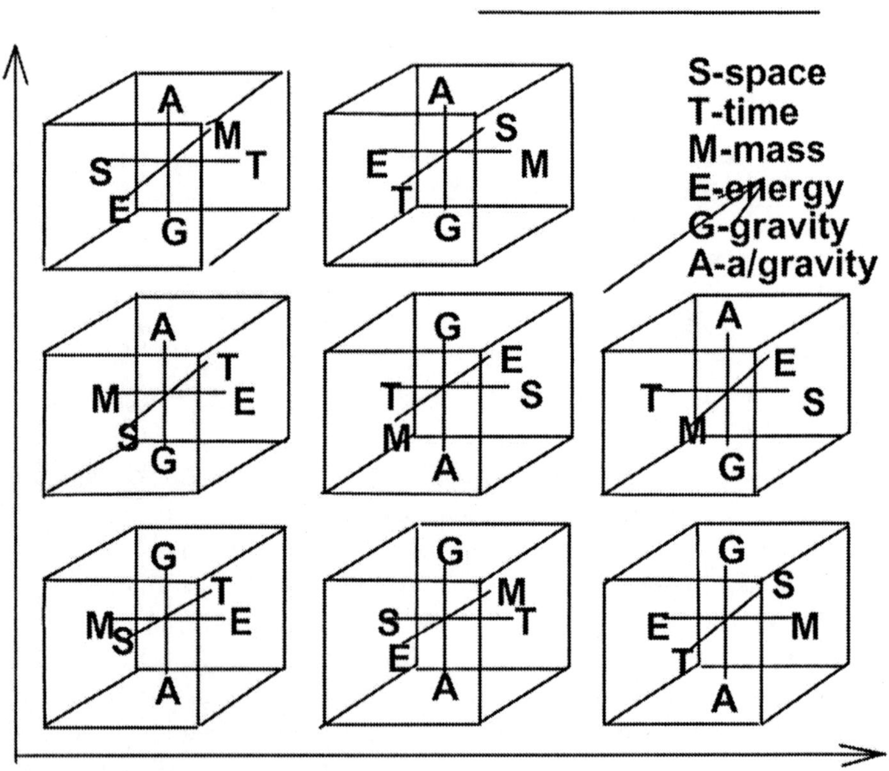

Double 3D is 3^2=9D, while its fractal opposite (inversion) is 3D. Fractal 3/2 is **3^2-3=6D** here and now. Fractality is fractal ½ reality, which is odd numbered between two even numbered realities and/or even-numbered between two odd-numbered realities. Here and now is fractal 3/2—it is 3D space (odd), which is at the same time only surface of time (2D time—even). Every reality is space/ time duality; therefore it is simultaneously odd and even numbered reality (super-symmetry). Odd in space is even in time and equally vice versa. Such ½ fractality is space in context of constant passage of time, which is its inversion. It is motion of size without, which is equal and opposite size of motion within (i.e.

whole orbital of Earth without, which is fractal size of solar reality within). One is inertial context (whole) of its equal opposite (fractal of infinitely larger space/smaller time) and equally vice versa—whole in inverse context is only infinitely small fractal of such larger reality. Singular inertion is infinite inversion and equally vice versa. Only infinite number of fractals (quanta) of one is singular quantum (whole) of its equal opposite.

Double 3D is triple 2D in context of inertial here and now. There are 2^3=8 different permutations of fractal 3/2 reality (see illustration 6). Double 3D without (3^2=9) is triple duality within (2^3=8). There are 8 different permutations of inertial here and now (total of 9) in context of fractal double helix. Inertial difference of $3^2/2^3$=9/8 duality (3rd direction) is inertial here and now, which is inertial 1/9 and/or inverse 1/8 difference. 1/9 without is 1/8 within or 9x8=72° of full (360°) curvature, which is 1/5th difference (pentagon) of 4/5 fractal reality. 72 is ¼ of 288, where 288=360-72. **72=9^2-9**, which is fractal inertion between double/inverse 3D realities. 6 is number of fractal surfaces of 3D (fractal 3/2), while 72 is number of fractal surfaces of double (squared) 3D (fractal 9/8 reality). 1/9 inertial difference is 0.111…or singular strand of double helix (infinite repetition of singular here and now), while 1/8=0.125, which is singular (abbreviated) sequence of fractal duality that is 1/8.1 or 10/81=10/9^2 =0.1234567…which is double infinite repetitive curvature or two strings of double helix (two strings are 1D fractal difference from one string, which is decimal difference). Double 3D is 9, and its fractal reality is 6 different sequences; double 9D is 81 and its fractal reality is 72 different sequences, while double 81 is 6561 and its fractal reality is 6480 different sequences (**81^2-81**), which is number of years of ¼ precessional period. ¼ precessional period is triple inversion/inertion=double inertion (triple 2x1/2 multiplication/division) of inertial 3D here and now, which is time of 3D space and/or 3D space of time, where time is fractal difference of spatial reality (3/4), therefore inversion of inertial 3D (1D or linear time is infinitely larger 3D, which is relative space).

Reality is fractal duality of two equal opposites, of which Newton's inverse square is fractal gravity/space or fractal ½ sequence; Kepler's harmonics is fractal 2/3 (time/space), while Einstein's relativity is fractal ¾ (space/time) sequence. One is inverse non-Euclidean curvature of another. Fractal 4/5 (time/space) is double or quantum relativity (precessional curvature of space/time duality) in context of fractal 5/6 (double precessional curvature). Fractal double helix is always $\mathbf{x^{(x+1)}/(x+1)^x=const}$, which is infinite 2x1/2 curvature (2-1/2=3/2 within is 2+1/2=5/2 without, which is infinite repetition of singular 1 here and now). 3^2/2^3=9/8=1.125/0.888=0.79/1.26=1.6/0.62=0.39/2.56 (successive inertion/inversion). Inertion/inversion of $2^3/3^2$ is $3^4/4^3$, while inertion/inversion of $3^4/4^3$ is $4^5/5^4$ and inversion of $4^5/5^4$ is $5^6/6^5$ (double exponential increase from 3^2 to 3^4 in context of equal and opposite singular increase of duality from 2^3 to 4^3; from 4^3 to 4^5 in context of equal and opposite increase from 3^4 to 5^4; from 5^4 to 5^6 in context of increase from 4^5 to 6^5). Each additional duality is additional double fractal difference or additional 2x1/2 multiplication/division.

Only inversion/inertion of $2^3/3^2$ is perfect $3^4/4^3$, while further inertion/inversion is exponential increase of discrepancy, which is relativity/uncertainty (fractal difference) between such equal opposites. Only 1D fractal difference is absolute time of inertial space and/or absolute space of inertial time, because only 1D fractal difference is space/time duality. Increased fractal difference is proportionally increased relativity/uncertainty of such space/time, which are increased discrepancy of perfect duality and/or increased eccentricity of perfect curvature. 4/5 and 5/6 are singular/double relativity/uncertainty of inertial space and time here and now. $5^6/6^5$=15625/7776 = 2/(1/2) or perfect 2x1/2 duality, which is boundary of inertial here and now—2/(1/2)=4D.

1D fractal difference is space and time (inertion/inversion) of inertial reality here and now. Here and now is where two curvatures meet. It is singularity of two equal opposites (2x1/2), which are double/inverse 1D fractal differences that are its space and time. Here and now is singular point of two equal and opposite infinites and equally vice versa. It is asymmetrical symmetry or supersymmetry (perfect imperfection) of both. Space and time are fractal horizontal and angular boundaries of inertial here and now. 1D fractal difference in space is double fractal difference in time and equally vice versa. 1D fractal difference is space of time and/or time of space, which are infinite/singular spatial size in context of equal and opposite infinite/singular motion (light speed). 1D fractal difference is light speed, which is constant duality of space/time—it is its mass/energy (double/inverse light speed of energy is singular light speed of mass and double/inverse light speed of mass is singular light speed of energy). Only motion larger than light speed is inversion of inertion, which is double inertion. Only larger than light speed change is change of inertial context and not only its fractal action/reaction. Larger than light speed is infinite change, which is singular quantum (fractal) change of its equal opposite. Complete (whole) change of inertion is only singular quantum (fractal) change of its inversion, which is double inertion.

Spatial light speed is no motion at all in context of time and equally vice versa; therefore any change is light speed change. Infinitely small spatial change without is infinitely large change of time within. It is only fractal spatial change that is complete (whole) change of time; therefore light speed change in time. Likewise—complete spatial quantum change, which is infinite period of constant passage of time, is only fractal (gradual) change of time itself. Only infinite change of content is fractal (singular quantum) change of its form and equally vice versa. Only infinite change of volume is changing of its surface and equally vice versa. Only infinite (total) 3D change is singular change of 2D and equally vice versa. Only infinite (light speed) motion is change of size and equally vice versa—only infinite change of size is singular change of motion. Any change is change of light speed or change of inertial context of duality. Infinitely large spatial change (within) is spatial light speed change, which is not change of motion in time (without) at all, while infinitely small spatial change, which is not change of spatial motion at all is light speed change in context of its time within. Light speed change of uncertain reality is no change at all here and now, just as light speed change here and now is not change at all in context of relative universe.

Only total shift, which is total inversion of boundaries (size/motion), is inversion/double inertion of space/time reality. Only full curvature, which is no change at all, is total change. An atom/universe is singular inertial/inverse quantum of space/time and only infinite (light speed) change is its change. Any change less than infinite is less than singular quantum (fractal) of its equal opposite, because any change less then complete is only fractal change, therefore only qualitative and not quantitative change. It is only change of frequency and/or change of spectral lines, not specter itself. It is only gradual (fractal) change and not change of context of dual reality itself. Only infinite spatial change is singular quantum of time change and equally vice versa—only infinite time change within is singular quantum of spatial change without (different spectral line). Less than infinite (total) is gradual spatial change, which is constant passage of time, just as less than infinite time change is only constant size. Infinite gradual spatial change is boundary of constant passage of time, which is its instant inversion that creates double inertion. Such instant inversion is only fractal (singular quantum) of gradual change of time. Likewise—only infinite time change is singular quantum of spatial gradual change, which is its instant change. Only infinite action in time is singular spatial reaction and only infinite spatial action is singular time reaction, which is 2x1/2 larger/smaller reaction—it is twice larger spatial change in ½ time and/or twice larger (square) time change in context of only ½ (inverse) space. Only infinite angular difference is singular quantum of spatial distance and infinite spatial distance is singular quantum of angular (time) difference in equal and opposite direction (micro-world/universe).

Gradual spatial change in between two quantum periods of time, which are its double/inverse (spatial) boundaries, is constant passage of time. Space is angular

½ of time and time is equal and opposite angular ½ of space. Reality of both is fractal space/time, which is 3/2 reality in between. Space is surface (fractal) of time and time is surface of 3D space. Each is singular 3D, while both are double/inverse 3D. Fractal space/time (3/2) is again only angular ½ of its equal opposite (gravity). Space is ½ of time (equally vice versa), while both are fractal 3/2 or ½ of double 3D. Double 3D is gravity of singular space and antigravity of its time. Double 3D is only ½ of quadruple 3D, which is singularity between dual gravity/antigravity. Double (squared) reality without is only ½ (inverse) within (2/(1/2)=4) and equally vice versa. Quadruple reality without is only ¼ within (4/(1/4)=16) etc, which is infinite double fractal curvature of inertion/inversion = double inertion or $C/(1/C)=C^2$. Double without is ½ within of inertial 1 here and now. 3D space is only fractal 3/2 of its inverse/double time and equally vice versa. 3D space and/or time is only 3/2 of 3D space/time, which is 6D double/inverse space/time (gravity/antigravity).

Space within is instant change, while without; it is gradual change in time. Mass is gradual change within and/or instant change without, because within/without are inertial/inverse frameworks of reality here and now. Time of one is space of another; infinity of one is singularity of another. Spatial change within is instant, while same change without is infinitely long passage of time between two quantum (fractal) changes. Micro-world is infinity of spatial changes in context of no time, while universe is only one change in context of infinite period of constant passage of time. Infinitely long period of time without (universe) is singular time within (micro-world), just as infinite space without is singular space within. Micro-world is infinity of light speed changes in no time at all, while universe is only one change in context of infinite period of time, because period of time is its constant size. Every change is full curvature, where time curvature is singularity of space within and/or infinity of space without. Only infinite number of light speed changes in time is singular light speed change of spatial reality, just as infinity of light speed changes of spatial reality is only one light speed change of time (relative/uncertain space). Every spin of an atom is infinity of light speed changes, just as universe in all its space and all its time is only singular spin of an atom in all 3 directions of inertial 3D, therefore only singular light speed change. Only infinite action is singular reaction and equally vice versa, where infinity/singularity are inverse/double boundaries of inertial here and now. Singular space is double/inverse time—it is 3rd half of two equal and opposite ½'s. There is only one reality—here and now, where its inversion is double inertion, which are space and time or its infinite/singular framework or boundaries. Such equal and opposite boundaries are equal and opposite fractal realities of a whole (here and now), and their singular duality is fractal 3/2.

Only total inversion or full curvature (infinity of spectral lines) is change of inertial context and not only gradual change of its action and equal and opposite reaction. Mass/energy is infinite content of singular space/time, and any change of mass/energy is only qualitative and not quantitative change of space/time and equally vice versa. Qualitative change of space/time is quantitative change of mass/energy and qualitative change of space/time is quantitative change of mass/energy. Qualitative spatial change is quantitative change of time and equally vice versa. Infinity of mass/energy quanta is singular spatial and/or time (inverse) quantum and equally vice versa—infinity of space/time quanta is singular massive and/or energetic (inverse) quantum. Only total qualitative change of infinity of massive/energetic quanta is one quantitative spatial or time change—where one in direction without is infinity in direction within and equally vice versa—one whole number is infinite number of fractals. Quantum change of reality is change of context or change of inertial system, not only its relative/uncertain content. It is not only qualitative (gradual), but instant (quantum) change. Change of content is only qualitative change (larger/smaller size/motion) of such inertial system. Only change of its form, which is infinitely larger/smaller (2x1/2) space/time change, is change of space/time reality itself, which is its instant inversion that creates double inertion or instant inversion of size into motion and motion into size, which is instant gravity of such duality.

Formalization of space/time is its fractal quantization, where any subsequent 2x½ form is infinite double/inverse quantization without and within, and only total inversion, as per $E/m=C^2$ in one direction, $d/t=C$ in another, and gravity/antigravity=C^3 in 3rd direction, is quantitative change or change of reality (inertion) itself, and not only its fractal (space or time) content. Qualitative (i.e. gradual or fractal) change in direction without (universe) is quantitative (whole) change in direction within (atom) and equally vice versa. Content (inertion) of the universe is form (inversion) of an atom, and content of an atom is form of the universe. Surface (2D) of 3D universe (i.e. time curvature of space) is 3D reality of an atom, and 3D of universe is surface of an atom (i.e. spatial curvature of time). It is reduction/increase of one of its directions in positive/negative direction (contraction/expansion of space and time in anti-parallel directions) It is simultaneous spatial expansion/time contraction in one and spatial contraction/time expansion in equal and opposite direction. Inflated (round) time of an atom is deflated to only surface (instant or flat) reality here and now and again inversely inflated in opposite direction of the universe. Likewise—inflated space of the universe is deflated to flat or instant reality here and now, and again inflated in inverse direction of an atom, where spatial inflation/deflation is in constant time and time inflation/deflation is in constant space. Only totally deflated reality can be inflated again in opposite direction, therefore only total change is change, because it is no change at all. One inflation without is infinite number of fractal deflations within and equally vice versa.

Fractal difference is boundary between infinity and singularity; space and time—it is light speed difference. It is infinite gradual change (infinite constant passage of time) on one side and instant change (no time), which is 2x1/2 gradual change, on another side. Instant change is in context of no time, because time is spatial (instant) reality of inverted spatial reality. Time in context of time is double/inverse time, which is space of time, just as space in spatial context is double/inverse space, which is not space but 2D time (two equal and opposite surface boundaries) of such 3D space. Only 1D fractal difference or space/time difference is constant passage of time, which is gradual change. 2D fractal difference is space/space (space/gravity) and/or time/time difference, which is not fractal (gradual), but complete (whole) change. Space/time is only fractal 3/2, while space/space and/or time/time, which is gravity/antigravity of space/time is instant change. Space in context of time and time in context of space is fractal (1/2) reality, while space in spatial context and/or time in context of time is whole (quantum) difference. 1D fractal difference is gradual, while 2D is instant change, where instant in space is gradual in time and instant in time is gradual in space, because space/time are asymmetrical symmetry or fractal reality, where singular fractal difference in one direction, reflects as double fractal difference in opposite direction (any pattern on surface is randomness in its depth and equally vice versa). Even-numbered reality of one is odd-numbered reality of another and instant surface of one is infinite depth of such surface in opposite direction. An odd number is inclusion or wholeness, while even number is exclusion or fragmentation, where odd in space is even in time and equally vice versa.

If inversion is 1D fractal difference than double reality, which is equal and opposite inversion, is 2D fractal difference, because it includes inertion here and now in its context (equally vice versa). Only excluding inertion here and now from such dual equation, is reality perfect duality of two equal (symmetrical) opposites, while here and now is their inertion/inversion, which is double inertion. Here and now is singular inertial difference, while its space and time (two equal and opposite light speed difference) are inverse/double inertial difference. Here and now is double—spatial and time light speed. Inertion/inversions of here and now (double inertion) are exponential/derivated curvatures of such inertial here and now. One is rounding of flat (inertial) reality and another is equal and opposite flattening of its inversion. Fractal difference is decimal difference—space is $10^1=10$, while its time is 1/10 of inertial here and now (1), while space/time is double difference of $10/(1/10)=10^2=100$. 10 is singular fractal difference between two equal opposites (light speed difference)—it is one quantum of space/time reality. There is no difference between here and now ($10^o=1$), its infinite space ($10^1=10$), time (1/10) and their gravity/antigravity (10^2 and $1/10^2$)—it is infinite self-interfering curvature of inertial here and now—its infinite 2x1/2 multiplication/fragmentation. Complete 3D inversion is inversion/inertion=double inertion in all 3 directions, which is triple fractal difference (10^3 and/or $1/10^3$).

Complete spatial or time change is 3 fractal (quantum) changes each, while complete space/time (gravitational) change is 6 fractal changes. Space/time is exclusive (fractal) attraction (contraction) or expansion (repulsion), which are two equal opposites of double 3D. Double 3D is singularity of gravity/antigravity, which is $3^2=9$ fractal quanta. Double 3D is 9 quanta of space/time in context of inertial quantum here and now, which is its fractal difference (1). Total reality is 9+1=10 fractals (quanta), where inversion of inertial 9 sequences is 1/9=0.111 or infinite repetitive fractal curvature of inertial here and now (1). 9 space/time quanta are equal opposite (infinite inversion of singular inertion) of inertial here and now (1). 10 fractal curvatures are inertial 4x1/4 in context of equal and opposite 2x1/2 or 2+6+2=10 planets of solar system and/or 10 electrons of 1st/2nd shell. 10 fractal curvatures within is singular fractal difference without. 3 fractal spheres is boundary of spatial reality (1 direction of 3D), 6 spheres are space and time boundaries (2 directions), and 10 spheres is complete 3D boundary (gravity/antigravity) of inertial here and now.

Any inertion is one clog of clockwork observed as separate and partial (fractal) reality. Inertion is 3rd inclusive direction of dual exclusion (inversion/double inertion), which connects such dual clogs with yet another smaller/larger duality, of which singular inertion is only ½. Every clog has two equal and opposite clogs in all 3 directions (and/or 3 clogs in 2 directions or 6 clogs in only one direction)—therefore total of 6 clogs around every singular clog (6+1=7), where every clog is ½ of 3 equal and opposite dualities. 6+1=7 is inertial space/time sequence (singular fractal difference), while 9+1=10 is complete gravitational sequence (double fractal difference, which is boundary of 3D). 3x2+1=7 is triple duality (6 surfaces of a cube) in context of inertial here and now, while 3x3+1=10 is triple 3D (triple space/mass/gravity sequence, of which 2 are reality of universe and 1 is reality of its micro-world and/or vice versa) in context of inertial here and now. Only 2/3 of 9 gravitational (mass/energy) sequences are simultaneous space/time reality (eigenvalue). There is discrepancy of any surface symmetry and/or eccentricity of any perfect curvature, which is 1D fractal difference that is inertial here and now, which is singular/infinite point of any physical reality. 1D fractal difference here and now is beginning/end of all infinity of different fractal curvatures.

Every singular inertion is cause/effect of 6 space/time and/or 9 gravitational harmonics or quantum vibrations. Once such reality is observed as complete—i.e. it is observed as singular (1) and separate clockwork, it means such clockwork misses its equal opposite or its reaction (fractal ½), and therefore also its 3rd direction which connects it to neighboring 6 clogs, because space/time is limitless reality—any boundary is only perspective. 2x6=12 clogs, which is 9 clogs of double 3D + inertial here and now (its 10th fractal), where inertial here and now is equal and opposite 3D; therefore 9+3=12 clogs. Only 5/6 x12 =10 is gravitational (fractal) eigenvalue of double gravitational reality that is 12 simultaneous sequences. It is infinite curvature, because it is reaction of singular point. Any point here and now is equal to any other, because it is relatively different/distant size/motion of same reality! Any difference of size/motion (and its resultant force) is only relative/uncertain differentiation, without any real meaning, apart as particular relative/uncertain reality of equally relative/uncertain reality (size/motion) here and now. Any difference is only dual illusion.

Double helix is infinite fractal curvature of singularity (0) here and now. It is infinite repetition of 1D fractal difference in context of different sizes/motions, where present sequence adds past fractal sequence in order to generate future (larger/smaller) sequence, where each sequence is always 1D fractal difference. It is: $(x-1)(x+1)/x^2$, which is infinite pattern: $(1/2)/(2/3)=3/4=(1x3)/2^2$, $(2/3)(3/4)=8/9=(2x4)/3^2$, $(3/4)/(4/5)=15/16=(3x5)/4^2$, $(4/5)/(5/6)=24/25=(4x6)/5^2$, $(5/6)/(6/7)=35/36=(5x7)/6^2$, $(6/7)/(7/8)=48/49=(6x8)/7^2$, $(7/8)/(8/9)=63/64=(7x9)/8^2$, $(8/9)/(9/10)=80/81=(8x10)/9^2$ etc. 1D fractal difference is infinite singularity of infinity of relatively different fractals. Space/time (fractal) reality is infinity of different space/time waves, where only waves that resonate are simultaneous—rest is empty space—i.e. only 4/16=1/4, 8/16=2/4 and 12/16=3/4 of 15/16-fractal reality are also real in context of fractal 3/4. Smaller fractals are only gradual accumulation of tension and only when such surface tension is equal to ¼, it causes an instant inversion that creates double inertion in 3D context. Only multiples (wholes) of fractal inertion are real here and now. Only such multiples are quanta of inertial here and now, while any different fractal change is only gradual change in depth of surface reality (only gradual angular differentiation of spatial reality).

Triple fragmentation of 3D is at the same time 9 fold fragmentation of 3 times larger/smaller 9D, but only 3 of its fractals are at the same time quanta of 3D (3,6,9). There is 100 fractals of number 100, where all are eigenvalue of number 1, which is singularity of such number—its space and/or time, and equally vice versa. Only even fractals (quanta) are real in context of duality (2D), because 2D is only ½ of such inertion. Only multiples of 3 are 3D quanta, multiples of 4 are 4D quanta, etc. There is only 4 multiples of number 25, which is 25 times larger/smaller than 1 and/or 100, and there is only 1 multiple of number 50. Any reality larger than 50 (½) is larger than context of such reality, therefore beyond its inverse/double boundaries. Any variation larger than ½ is no change at all in framework of inertial reality (100), because ½ is fractal boundary between two equal opposites, and any reality larger/smaller than ½ is not inertial, but inverse reality, where space is time and time is space.

Physical reality is action/reaction or duality of **2** equal opposites. Double/inverse boundaries of inertial duality (2) are ½ within and/or 20 without (1D fractal difference). ½ change within is 20-fold change without, just as 20-fold change within is only ½ change without. ½ and 20 frequencies are inversion and double inertion of duality in between. Absolute duality is only relative/uncertain fractal of 3D. Double/inverse boundaries of 3D are 1/3 and/or 30. 4D (relative/uncertain reality) is double duality or 4x1/4 wave. Relative reality is double/inverse boundary of duality here and now. Double/inverse boundaries of 4D are ¼ and/or 40. Relative (4D) is only fractal of 6D reality and equally vice versa—6D is only fractal of twice relative reality (2x4=8D or 8x1/8 wave). Boundaries of inertial 6D are 1/6 and/or 60, while boundaries of 8D are 1/8 and/or 80. Fractal 3/2 within is 6D without—it is 3x20=60 without and/or 1/3x2=1/6 within. Fractal ¾ is 3x40=120 and/or 1/120, while double ¾ is 240 etc. Inertial/inverse quantum of duality is singular quantum of relative reality, while inertial/inverse quantum of relative reality is singular quantum of twice relative reality (20/(1/2)=40 and/or 40/(1/2)= 80). 3D and 6 D are only fractals of infinite multiplication/division of duality, while 5D in between 4D and 6D is double fractal (1/4 reality). Singular quantum of duality is only ½ of relative and/or ¼ of twice relative reality, just as singular quantum of twice relative reality is two quanta of relative and/or 4 quanta of dual reality. Only multiples of inertion here and now are real, where multiple in space is fractal in time and multiple in time is fractal in space.

Reality which is only ½ fractal repeats in units of 20, reality which is ¼ fractal repeats in units of 40, etc, because only whole quantum is real. Smaller fractal within is larger repetition without, before such fractal is instant reality here and now. Any fractal repetition less than whole is only constant passage of time and/or empty space. Only full curvature is instant reality, which is real here and now. Space is singular (infinitely small) fractal of time and equally vice versa. Space is singularity/infinity of double/inverse time and equally vice versa. Only infinite repetition of spatial sequences is singular time quantum and only infinite repetition of time sequences is singular spatial quantum. Only infinite gradual spatial change and/or infinite constant passage of time are singular quantum of spatial reality of time (4D or relative space), just as all infinite time change is singular spatial quantum that is real here and now (atom). **Constant passage of time is discrepancy or fractal difference between two fractal spatial realities (duality and/or relative reality), just as constant size of space is discrepancy or fractal difference between double and inverse (anti-parallel) passages of time.**

Time is fractal between double/inverse spaces, just as space is fractal between double/inverse times. Infinite passage of time is angular difference between two spatial realities without/within and singular size is angular difference between two anti-parallel passages of time. Space is eccentricity of time curvature and time is eccentricity of spatial curvature. Mass/energy of empty space/time is overlap between such two +/- non-Euclidean curvatures. Mass/energy is inertion or discrepancy between space and time, because it is reality of both. Space and time are double/inverse boundary of mass and inverse/double boundary of energy, where energy is double/inverse boundary of mass and equally vice versa. Only overlapping fractals of space/time curvatures are reality of inertial mass/energy in between—rest is its empty space that is inversion of constant passage of time. Such overlapping fractal space/time curvature is inertion of singular mass/energy and equally vice versa—mass/energy overlap (duality) is inertion of its space/time without and within.

Passage of time is negation of spatial reality and equally vice versa. Infinite negation is singular affirmation, because infinity is instant inverse singularity. Infinite passage of time is infinite spatial negation, which is instant inversion that creates double inertion or double spatial reality. Infinite passage of time is instant change from absolute to relative spatial reality or instant change from time reality of space to spatial reality of time. Infinite passage of time is double/inverse spatial reality here and now. Inversion that is double inertion of spatial reality here and now is infinitely long period of time. Such period of time is singular size of relative spatial reality (size of relative atom). Period of time is instant spatial size of time, just as period of space is instant time size of spatial reality. Space is instant period of time, which is double/inverse (accelerated/decelerated) time of spatial reality here and now. Space is instant (double/inverse) time, just as time is instant (double/inverse) space. Infinite period of constant passage of time is instant inversion of space into double (relative/uncertain) space, which is singular quantum of spatial reality of time—singular quantum of relative reality, which is infinite passage of time in context of uncertain reality within (here and now). Here and now in context of relative universe is only uncertain quantum reality within and its infinite passage of time is instant singular space or instant atom of relative spatial reality. Constant passage of time here and now is instant space or density within relative spatial reality, which is infinitely accelerated constant passage of time. Space is boundary of time and its infinite gradual change (passage of time) is instant (quantum) change, which is double/inverse or twice as large spatial change in only ½ time.

Only singularity has no double/inverse boundaries, because it itself is boundary of all infinity of its fractal physical realities (dualities of different sizes/motions). Singularity is double/inverse boundary of space and time, which does not change itself in context of space and time. It is changeless cause of infinity of double/inverse changes. Singularity is not fractal physical reality—it is its equal opposite or action that causes infinity of different physical realities (reaction). Singularity is not real in context of space and time, yet it limits any physical (fractal) reality and equally vice versa—physical reality is boundary of singularity. It is boundary (constant passage time) of any relative spatial reality, as well as boundary (spatial size) of any relative constant passage of time. It is size of motion, as well as motion of size, and their singular force of attraction/repulsion. Singularity is infinity of different space/time realities, as well as their double/inverse boundary.

Singularity is double/inverse negation, which is affirmation of physical reality. Singularity is unreal in fractal context of physical reality, because it is whole (quantum) of two equal opposites (2x1/2), just as physical reality is unreal in context of singularity. Singularity is twice larger physical reality, just as physical reality is double singularity—one is equal and opposite ½ of each other, while 3rd direction of such duality is fractal 3/2 that is reality here and now. Here and now is both—singularity and its fractal physical reality, which is $2/(1/2)=2^2=4$ or 3D space in context of its passage of time (fractal ¾). 4D is 4x1/4 sin wave, which is instant change of inertial here and now every ¼ period of its constant passage of time. ¼ period is instant or quantum spatial change of reality here and now, which is singular quantum of relative space or singular quantum of spatial reality of time (singular quantum of inverse/double reality). ¼ of constant passage of time is singular quantum of change of inertion and not only is dual action/reaction—it is not only passage of time/gradual spatial change, but instant change of both, which is instant (quantum) change of reality here and now. It is not change of space in context of time and/or change of time in context of space (1D fractal difference that is gradual change), but change of inertial context of such dual reality, which is instant change in context of no time (accelerated time, which is gravity of space and time).

Instant (quantum) change is instant 2x1/2 multiplication/division of double helix, when present resonates with past sequence in order to create double (future) sequence. It is instant duality or space/time eigenvalue, which is double/inverse (gravitational) duality of inertial here and now (where double in space is inverse in time and equally vice versa). It is instant overlap between two equal and opposite curvatures, which is mass/energy of empty space/constant passage of time. It is instant density of previously empty space and instant space of previous empty passage of time. It is instant (whole) gravity of gradual (fractal) space/time, which is instant mass/energy of space/time. It is quantum leap in time that creates double space or mass of empty space. Reality is self-interference—it is only eigenvalue between two equal opposites (inertion/inversion)—rest is infinity of different possibilities of space/time (Everett's many worlds), which are equally real—only not here and now. Duality that is not eigenvalue (singularity) is not 3D; there fore it is only its emptiness or nothingness. Such nothingness is reality of past/future time and/or larger/smaller space, but not right here and now. Every quantum change is 2D fractal difference—it is infinitely accelerated gradual change (1D fractal difference), which is instant change.

1D is fractal reality of 2D—it is only ½ or only one of two equal opposites, which is either space of time or time of space. Only 2D fractal difference is 3D space/time, because it is synchronized duality of two equal opposites. 1D fractal difference is inertion (here and now), while 2D fractal difference is duality of inversion/double inertion. 1D is continuous (gradual) curvature, while 2D fractal difference is its expansion and equal and opposite contraction, therefore a matrix of periodic interruptions of one in context of another. 2D fractal difference is quantum reality—it is quantized spatial reality of time (relative universe) and/or quantized time of uncertain space (micro-world). 1D fractal difference is space/time difference, which is gradual change, while 2D is space/space or time/time difference, which is instant change—it is instant spatial change in context of no time in between and/or instant time of no space in between. 1D fractal difference is singular space/time, while 2D fractal difference is perfect duality—it is two equal and opposite spatial realities, without context of time and/or two times without spatial context. Fractal space/time is one time between two spaces and/or one space between two times—it is 3D or dual action/reaction in context of their inertion.

Space is distortion (expansion/contraction) of time and time is equal and opposite distortion (expansion/contraction) of space. Space/time is two equal and opposite expansions, and their overlap (eigenvalue) is inertial space/time here and now. Space is eccentricity or discrepancy between action and reaction in time, while time is discrepancy or eccentricity between spatial action and its reaction, where one is always gradual and its equal opposite is instant. 2D fractal difference is only two spatial realities without time in between and/or two times, without space in between. 2D fractal difference is perfect duality without distortion (eccentricity) of equal opposite. Such perfect duality is singular duality, without its 3rd direction, which is inertion between action and reaction. 2D fractal difference is inertionless or instant reality. There is no delay (inertion) between action and its reaction. There is no passage of time between two equal and opposite spatial realities, just as there is no size of spatial reality between two anti-parallel passages of time. 2D fractal difference is inversion/double inertion of inertial 1D fractal difference, which is gradual spatial change in context of constant passage of time. 2D fractal difference is negation of space/time reality, which is its gravity. 2D fractal difference is double relativity/uncertainty (double space/time), while singular relativity/uncertainty is its singular quantum, just as space is singular quantum of time and equally vice versa. 1D fractal difference is inversion and 2D fractal difference is double inertion. Only infinity of one is singular quantum of another.

1D fractal difference is singular infinity of space in context of one time and/or singular infinity of time in context of one spatial reality. 2D fractal difference is double/inverse infinity. 1D fractal difference in one direction is infinite fractal differentiation in equal and opposite direction. 2D fractal difference is infinite fractal differentiation in 2 out of directions of 3D, which is double relativity as opposed to singular relativity. 1D fractal difference is freedom of interpretation or infinity of different choices in only one, while 2D fractal difference is double freedom of different choices in two directions of 3D. 3D is 1D fractal difference in one, 2D fractal difference in another and infinite fractal difference in 3rd direction. Singular fractal difference (singular duality) without is infinity of different fractal differences within. In each instant of time, only one spatial fractal is real here and now and in each instant of space, only one time fractal is real. In context of all the time, all spatial fractals are real and in context of all space, all time fractals are real. Larger difference/distance between two fractal space/time realities (two different space/time possibilities) is larger period of time and/or larger distance of space in between. Only spatial multiples (wholes) of inertial time and/or time multiples of inertial space here and now are eigenvalue with such inertion (here and now). Rest is empty space/time, which is its reality only in quantum steps, when gradual accumulation of fractal difference is whole quantum.

Larger spatial and/or angular (time) distance/difference between here and now and such infinitely different fractal space/time possibility means there is proportionally decreased number of quanta which are real here and now—it is increasingly empty space and/or slower passage of time, because proportionally smaller number of quanta of distant/different reality are eigenvalue here and now. Larger fragmentation means larger period of time and/or larger spatial size in which gradual accumulation of fractal difference is singular quantum (whole), which is real here and now. Fractals are inverse quanta of multiples, where multiple of space is fractal of time and multiple in time is spatial fractal. Larger fractal difference between them is proportionally decreased synchronicity, which is increased period of time and/or increased size (equally vice versa in opposite direction—smaller fractal difference is smaller period of time and/or smaller size). Spatial universe is largest fractal differentiation of time, which is smallest or singular time (½), just as micro-world is largest spatial fractal that is real here and now. Spatial universe is infinite time fractionalization, which is double/inverse boundary of space/time here and now, and only one of its fractals is real here and now. Infinite number of different fractal (time) possibilities of spatial universe is real only in quantum leaps, when gradual accumulation of spatial change is instant inversion that creates double inertion. Singular space here and now is likewise only one (smallest) spatial fractal of time of the universe (which is its 4D or relative space), which is simultaneous reality of infinity of such different possibilities. Here and now is uncertain (quantum) micro-world in context of spatial universe, and different space/time possibilities of its relative reality are real only in quantum steps.

Only 2x1/2 multiplication/division is real in context of duality. Only its fractal ½, ¼, 1/8 etc in one direction and 20, 40, 80 multiples in equal and opposite direction are quanta of such inertial reality—any different fragmentation/multiplication is only its infinite empty space and/or constant passage of time of no density. Any different multiplication/fragmentation is only gradual change in context of its inertial quantum, while only eigenvalue (self-interference), when accumulation of fractal (gradual) change is equal to one quantum is instant (quantum) change. One quantum change without is infinite gradual change within and one gradual change without is infinity of quantum changes within. Quantum change is double/inverse boundary of gradual change and equally vice versa—gradual change is double/inverse boundary of quantum change. Gradual spatial change (passage of time) is real only in context of its double/inverse boundary, which is its quantum change and equally vice versa. Quantum change is instant change in context of no time (accelerated or double/inverse time), which is equal and opposite change or reversal of gradual change in context of constant passage of time. Quantum change is infinity of qualitative change in context of no time, just as any fractal step of qualitative (gradual) change is infinity of inverse quantum changes, which are only gradual change in context of inverse (anti-parallel) time.

Quantum change is instant change. It is beginning/end of infinity of fractal (gradual) changes. Gradual change is reversible, while quantum change is irreversible or permanent change, because it is change of reality and not only fractal change of its space or time. It is change of inertion here and now. Only infinite gradual change is one quantum change. Only infinite change of size is singular quantum of motion and only infinite change of motion is one spatial change of size. Only infinite gradual change of tension (inertion) within (depth) is singular quantum of change without (surface) and equally vice versa, because infinity of inertion is singular quantum of inversion. Only quantum (whole) is eigenvalue of two equal opposites (2D fractal difference). Only quantum is self-interference between inertion and inversion that creates double inertion. A drop of water does not fall, unless tension within is equal to tension without (gravity). Only infinite gradual increase in size of drop is one instant motion (fall). An object only gradually bends (reversible change of size), unless applied force without is equal to tension within, when it instantly breaks into two fractal parts (irreversible change of size). Only gradual degradation leads to an instant collapse and equally vice versa—only gradual effort results in an instant (complete) structure. Quantum changes are space/time boundaries of inertial here and now, which are its instant and irreversible changes. Quantum change is light speed change or change from within to without.

Space and time are double/inverse boundaries of mass and energy and equally vice versa. Gravity/antigravity are double/inverse boundaries of mass/energy and equally vice versa. Quantum space/time change is 2x1/2 quantum change of mass/energy and/or 4x1/4 change of gravity/antigravity and equally vice versa. Quantum change of space/time is only ½ (fractal) change of mass/energy within and/or 20 times larger change without (equally vice versa). Quantum change of mass/energy is only ½ fractal change and/or 20 times change of its attractive/repulsive reality, which is ¼ and/or 400 times space/time change. Only 2x10=20 space/time quantum changes (C=d/t) are one quantum change of its mass/energy (C^2=E/m) without (size/motion) and only $(2x10)^2$=400 space/time changes are one quantum change of its gravitational reality (C^3=gravity/antigravity or expansion/contraction). 400 times without is ¼ change within and 20 times without is ½ change within. Larger change without is smaller change within. Infinitely large change without is no change within at all—only constant space/time reality of constant size/motion or constant passage of time of spatial reality here and now. Infinitely large change without is no change within and equally vice versa, because infinite space is singular time and equally vice versa. Infinite gradual change is singular quantum (instant) change and infinity of quantum changes is only gradual (fractal) change of double/inverse realities that are boundaries (fractals) of inertial here and now.

Duality of fractal/quantum changes is only once and twice infinitely smaller/larger fractal (gradual) change in context of twice-larger quantum reality, where infinity is only a number that is double/inverse number of inertial here and now. Boundary of spatial Earth is its time without/within, where time is inversion/double inertion (relative/uncertain space) of its size. Boundary of its time is double/inverse or relative/uncertain time, which are space of its time. Space of time is period of time that is inversion/double inertion of inertial time here and now, which is time of one rotational and/or orbital period. Rotational/orbital time are 2D time that is 3D space—it is surface reality of relative (4D) space, which is 3D time (larger/smaller space), while constant passage of time is discrepancy or eccentricity between such two equal opposites of rotational and orbital times, which are double/inverse spatial reality of time (boundaries). Discrepancy between rotation/orbital motions is their precessional motion and ¼ of such time is singular quantum of change of inertial here and now (spatial Earth). ¼ precessional change is De Broglie's eigenvalue or instant quantum change, where change of E/m in context of an atom is change of space/time in context of universe. ¼ period of precessional motion, which is singularity of two equal opposites (inertion of action/reaction), is singular period that is not gradual reversible, but instant irreversible (quantum) change. It is beginning/end of constant passage of time, which is singular quantum of spatial reality of time or singular quantum of relative spatial reality.

CHAPTER 5

DOUBLE/INVERSE INERTION

An atom within is universe without—they are double/inverse time boundaries of inertial space and/or double/inverse spatial boundaries of inertial time here and now. Universe without and atom within are action/reaction in context of inertial here and now (surface of Earth). Universe is inverse atom and atom is inverse universe—they are inversion/double inertion of inertial here and now (constant passage of time on surface of Earth). Solar reality without is an atom within, while galaxy is its 2D fractal difference or double inversion that is singular inversion of solar reality. There is no difference between universe and an atom—they are fractal double helix of inertion/inversion= double inertion or fractal 3/2 Bode's periodicity, which is infinite Fibonacci sequencing—$C/(1/C)=C^2$. Each relative/uncertain shell (size/motion) is infinity of double/inverse (2x1/2) spectral lines in context of inertial frequency here and now. Reality within is fragmentation of space in time (spatial quantization), while reality without (universe) is its multiplication, which is fragmentation of time in space (time quantization). Atom is fragmentation of space/multiplication of time, while universe is fragmentation of time/multiplication of space (Bode's 2/3 periodicity). Atom is infinity of spatial waves in context of inertial time wave here and now, while universe is infinity of time waves in context of inertial spatial wave here and now.

Solar reality and an atom are positive/negative (convex/concave) non-Euclidean time curvatures of inertial space, where positive time curvature is negative spatial curvature—i.e. positive relativity is negative uncertainty. An atom is 1D fractal difference within (1/C), while solar reality is equal and opposite 1D fractal difference without (C). Reality of both is double fractal difference, which is $E/m=C^2$ of such empty space/time. Atom is inversion of reality here and now in direction within, while solar reality is its inversion without. Solar reality/atom are double/inverse energy of inertial mass (surface of Earth) and/or equally vice versa—they are equal and opposite mass of inertial energy (motion) here and now. Surface of Sun without and surface of atom within are equal and opposite reactions (energy) of inertial mass here and now. Solar reality is infinitely slower motion and/or infinitely larger space (double or accelerated passage of time), while an atom is infinitely smaller size and infinitely faster motion (inverse or decelerated passage of time). Difference is only comparative—atom and/or solar reality are same reality as here and now in double/inverse perspective. Space here and now is +/- time of atom/solar reality, and its time is their -/+ space. Space here and now, solar and atomic time are triangulated 3D or one fractal sequence.

Here and now (Earth) is line of symmetry between such equal opposites, which is inertion of such duality. Line of symmetry breaks surface symmetry, because it is their discrepancy or eccentricity between two curvatures, which is overlap (eigenvalue) of both. It is surface between two inverse 3D and/or 3D between such two surfaces. One is boundary or horizon of another. Space of solar reality and an atom are equal and opposite (convex/concave) reactions of inertial time of Earth, just as space of Earth is +/- (dual) reaction of time of solar reality/atom. Atom/Earth/solar reality is inertial 3D—it is $3/4 \times 2/3 = 1/2$. Earth's empty space/time is dense E/m of solar reality/atom, while empty space/time of solar reality/atom are double/inverse (dense) E/m of Earth's reality. Constant passage of time is empty space of solar reality, which is eigenvalue of its dense reality only in ¼ precessional periods. Spatial reality of time, which is dense reality of its empty space, is real only in quantum periods.

Earth is ¾ fractal space/time—it is ¾ fractal of solar reality. Light speed (C) is 300.000 km/sec and circumference of Earth (d) is 40.000 km. Light speed is space/time constant, which is 1D fractal difference—any change of space is equal and opposite change of time and equally vice versa. $C/d=300.000/40.000=(3/4)x10=1/t$ (inertial time t=0.1333. sec), where 1/t is time curvature of such inertial space and/or energy of its inertial mass. Time curvature of Earth is ¾x10 (1D difference). Spatial reality (3D) of Earth is ¾ time wave (relative 4D) and equally vice versa. Earth's orbital in context of light speed is $(3x10^8 x\Pi)/C = \Pi x10^3 = t$

=3140 sec, which is curvature (Π) of triple fractal difference ($3D=10^3$). They are two equal and opposite times or synodic/sidereal difference between orbital/rotational motion (3140/0.133=23 550, which is $235.5x10^2$, where 235.5 sec. is synodic/sidereal difference). Time curvature of spatial Earth (variability of constant space) is spatial curvature of time (solar reality), which is variability of constant time, where constant passage of time here and now is instant space of such inverse (solar) reality. Duality of constant Earth's space and equal and opposite constant solar space (which is time here and now) is t (solar reality) x d (Earth) = $4\Pi x10^7$, which is double curvature of 7D fractal difference (6 fractal shells of inertial 3/2 here and now).

Space/time here and now (Earth) is 4x1/4 multiplication/division of its inertial (gravitational) reality. It is inversion of solar spatial reality (constant passage of time here and now), which is double inertion of an atom (spatial or instant reality of such constant passage of time). Infinity of different curvatures without/within (empty space/dense mass) is infinite repetition of inertial fractal difference here and now in infinitely different number of relative size/motions, which are relatively different densities and/or relatively different passage of time (smaller size = faster motion = slower passage of time). It is double inverse/fractal infinity of singular whole here and now, where fractal of time is spatial multiple and spatial fractal is time multiple. Duality of multiples (inverse/double fractals) is duality of quantum reality—it is instant quantum reality of mass within and/or gradual quantum reality of empty space without. It is instant quantum uncertainty and gradual quantum relativity, where gradual in time is instant in space. Gradual spatial relativity or quantum periods of constant passage of time between two spatial quanta are instant quantum uncertainty in context of solar space. Passage of time here and now is not real in context of solar space—it is only its instant density that does not vary in context of its time (double or accelerated time, which is gravity here and now). Constant passage of time here and now is not reality of the universe—it is exclusive Earth's reality. Rotation/orbital motion and their discrepancy (precessional motion), which is constant passage of time here and now, are not motions—they are infinite or light speed (singular and double) rotation, which is not motion, but size. Orbital is not time in context of solar reality, but instant orbital (spatial) size.

Atom within and solar reality without are equal and opposite reactions of inertial Earth (here and now). Double fractal difference/distance between atom/solar realities is inertial 1D fractal difference here and now (C) and equally vice versa. Inertial difference between spatial reality of an atom and solar space is time (motion) here and now, while difference between their inverse/double times is spatial reality (size) here and now. Singular size is double motion (rotation/orbital) and equally vice versa—singular motion is double size (i.e. orbital is discrepancy between solar and Earth's size). Spatial size is double/inverse light speed motion and/or vice versa—spatial motion is double/inverse size, where size in space is motion in time and equally vice versa. Size of Earth is rotational light speed of an atom within and/or linear light speed in context of solar reality without. Size/motion of an atom/Earth/solar reality is fractal 3/2 (space/time) reality. Two are action/reaction of inertial 3rd. Any one size or motion is inertion of infinity of relatively different space/time quanta, because infinity is only repetitive fractal curvature of double helix. One relative size or motion is total certainty of infinity of different sizes/motions (where spatial certainty is uncertainty in time and equally vice versa). One inertion (one size or motion) determines infinite physical reality—all its relative/uncertain sizes and motions, because it is infinite causal curvature of double helix, where every action is double/inverse reaction.

An atom within and a star without is double/inverse or positive/negative 1D fractal difference from inertial here and now (rotational and linear or round and flat light speed difference). Each is equal and opposite 1D variability of inertial 3D here and now. Variability in 2nd direction, which is infinitely larger/smaller (double/inverse) space/time, is variability of the universe beyond boundaries of solar reality without and an atom within (2D fractal difference). Nucleus within an atom and/or galaxy without solar reality is 1D fractal difference (double/inverse) of solar reality/an atom, which is double fractal difference (2D) from inertial here and now. Singular fractal difference is $C=d/t$, while double fractal difference is $C^2=E/m$ (and/or vice versa in opposite direction). 1D fractal difference is singular relativity/uncertainty, while 2D fractal difference is double relativity/uncertainty, where relative space without is uncertain (quantum) time within, just as uncertain (quantum) space within is relative time without. Reality is fractal supersymmetry, which does not differentiate between symmetry and asymmetry, because they are two equal opposites, without real distinction. 1D fractal difference (relativity/uncertainty) is singular, while 2D fractal difference is double density of inertial space/time—it is singular/double density of mass within and singular/double empty space without. Double density is square or accelerated density and so is double empty space twice empty or twice de-densified mass.

1D fractal difference is instant reality of one direction of time, which is additional density of inertial space within (mass). Its reaction is equal and opposite de-densification in opposite direction, where instant time (density) is not instant, but constant passage of time in context of its empty space. 2D fractal difference is double density or instant spatial reality of two directions of time, which is double density of mass. Its reaction is twice emptier space without, which is twice slower motion or twice-faster passage of time (accelerated passage of time in context of galaxy versus constant passage of time in context of solar reality). **Increased density is decreased size (mass), while decreased density is larger size (empty space); decreased size is faster motion, while increased size is slower motion; faster motion is slower and slower motion is faster passage of time.** They are only relative/uncertain fractal difference of inertial here and now, which is the only whole quantum of both—space and time fragmentation. Atom/star is singular (space/time) fractal difference, while galaxy and space within an atom (electrons orbiting their nucleus) is double (gravitational) fractal difference.

Universe without galaxy and/or space within nucleus are triple fractal difference (triple relativity/uncertainty or triple density of mass within and/or triple empty space without), which is gravity/antigravity (contraction/expansion) difference (C^3). The universe is singular relative variability of inertial galaxy within (C), just as galaxy is inverse (uncertain) variability of inertial universe. Universe is double relative variability of inertial star within a galaxy (C^2) and/or triple variability of Earth's mass, just as star is its double uncertain (quantum) variability within, while here and now is its triple quantum (uncertain) reality. Universe is fractal triple relative space (3D time) and equally vice versa—Earth is triple uncertain (quantum) reality within. Triple fractal variability is total variability of all 3 spatial directions or complete 3D inversion that creates double inertion. Universe/micro-world are inertial/inverse uncertainty/relativity or inertial/inverse triple fractal difference, which is total of 6 uncertain/relative (fractal) shells of inertial here and now—3 without and 3 within (3 time curvatures of inertial space and 3 spatial curvatures of inertial time (which is inverse/double space). Physical reality is 6 quantum shells of inertial here and now, which is its 7th shell in between. 6 shells in 2 directions is 9 shells in 3 directions of inertial 10th shell here and now (3 shell without and/or $2^3=8^{th}$ shell within). Reality here and now is fractal ¾, where 3D space is only ¾ of its time, which is its fractal difference or 4th direction. 4D is 3x4=12 shells of inertial 13th shell in between (4D or fractal ¾ is double fractal 3/2). Here and now (Earth) is always very center or line of symmetry of all equal opposites of its fractal space/time spheres. It is inversion point of all infinite fractal space/time directions without/within. Here and now is only singularity/infinity of physical reality.

Atom/nucleus within and/or star/galaxy without are double differentiation or double inertial and inverse density (C^2). They are double/inverse or horizontal (spatial) and angular density (time), which is fractal ¾ in context of fractal $^2/_3$

reality in double/inverse context of here and now, which is fractal ½ reality. ¾, $^2/_3$ and ½ are triple fractal boundary of inertial here and now (Earth), which is triple time curvature of inertial space in context of equal and opposite triple spatial curvature of time. It is triple duality of space/time, mass/energy and gravity/antigravity, where mass/energy is 2x1/2 of space/time and gravity/antigravity is 2x1/2 of mass/energy. Mass/energy is twice contracted (inverse square) reality within in context of twice expanded (square) boundaries without, while gravity/antigravity is 4-fold contraction within and 4-fold expansion without. Mass/energy is simultaneous infinite expansion of space and equal contraction of its time (equally vice versa in opposite direction). Gravity/antigravity is double infinite expansion of space and equal contraction of its time, where gravity within (i.e. micro-world) is antigravity without (i.e. universe).

Space/time are 1D fractal difference in between, mass/energy are 2D fractal difference in between and gravity/antigravity are 3D fractal difference in between, which is spatial reality here and now. They are 3 spheres, one within the other, where infinitely small is synonymous with infinitely large. Mass/energy is reality within and without space/time, where mass without is energy within and equally vice versa. Gravity/antigravity is within and without mass/energy (therefore twice within/without space/time), where gravity within is antigravity without, because spatial gravity is time antigravity and equally vice versa. ½ (i.e. mass/energy) within is 20 times without, while ¼ within (i.e. gravity/antigravity) is 400 times without. Mass/energy is $20/(1/2)=40=2^2x10$, while gravity/antigravity is $400/(1/4)=1600=4^2x10^2$ repetitive curvature. Mass/energy is 2x1/2 quantum reality of space/time and equally vice versa—space and time are double/inverse quantum context of mass/energy. Gravity/antigravity is 4x1/4 double/inverse boundary of both. Only space in context of time and/or time in context of space are gradual spatial change, which is constant passage of time. Mass/energy of inertial space/time (gradual change) are its double/inverse (equal and opposite) quantum changes—one is instant change within (i.e. instant mass), while another is instant change without (i.e. instant energy). Mass is instant (double/inverse) time, while energy is instant (double/inverse) space and equally vice versa. Mass is instant time, while space is its gradual change (passage of time, which is instant space); energy is instant space, while mass is its gradual change (instant time), etc. Gravity/antigravity is double quantum framework of space/time—it is double instant reality, because it is instant reality of space/time and inverse instant reality of mass/energy.

Only space in context of time and time in context of space, which is less than 1D fractal difference, is gradual (fractal) change. Change of context is instant (whole) reality. Singular/double quantum change of context is boundary of inertial here and now—it is mass/energy of empty space/time (and/or vice versa) and its gravity, which is double boundary. Instant (double/inverse time) is space and instant (double/inverse) space is time. Double fractal difference is infinite/singular boundary of inertial 3D here and now, which is Planck length within and/or black hole without in context of infinite spatial universe. Such black hole is never total black hole, because there is always 3rd direction of any duality in context of infinite 3/2 fractal curvature, therefore, there is always 1D fractal difference of further reality within or without. Black hole is only an empty space/time without any further content—i.e. without mass/energy (size/motion) and equally vice versa—it is mass or time without space within, which is infinite spatial universe without. Reaction of singularity is infinity in equal and opposite direction. No content (space) within is infinite content without. Infinite universe without galaxy is reaction of its singularity (black hole) within—universe is black hole turned inside out, just as black hole is inverted universe.

Black hole is reduction of mass but not gravity of inertial space to 0. It is reduction of infinite (universe) to singular size, which is equal increase of singular to infinite (light) speed motion (rotation). It is reduction of 1D fractal difference to 0 in direction of black hole and equal infinite expansion of universe in opposite direction. Instant spatial black hole within is gradual spatial change (constant passage of time) without. Black hole is infinitely dense mass, which is its inversion, therefore no mass at all that is double mass or gravity of inertial space. Singular space (black hole and/or Planck length) is reduction of 3D to perfect space/time duality (2D) without any discrepancy or inertial content within (which is infinite content of its inversion). It is perfect curvature of infinite eccentricity without, where perfect spatial curvature, is infinite time distortion and equally vice versa. It is infinitely inflated space without in context of singular time (universe) and/or infinitely inflated time in context of singular space (black hole). Line of symmetry or 3rd direction of such infinite/singular duality (space/time) is gravity—it is infinite instant gravity within and gradual antigravity (expansion) without, where gradual in space is instant in time and equally vice versa. Black hole is inverse universe or universe of inverse spatial reality (time). There is no difference between infinite universe and singular black hole—they are same reality in diametrically opposite angular perspective.

Infinite distance between here and now and boundary of spatial universe is singular angular difference (time), just as singular distance to black hole is infinite angular difference (infinite time distance). Here and now is inertial ½ of such duality, which is ½ way between infinite linearly expanding universe and core of spiraling galaxy. Spatial distance between core and Earth is equal to time difference between Earth and boundary (light speed) of expanding universe and equally vice versa—spatial distance of infinite universe is equal to time difference between here and now and core of galaxy. Earth is inertial (angular) ½ of such 2x1/2 (double/inverse) reality. Galaxy is spatial curvature of time, while universe is equal and opposite non-Euclidean time curvature of space. Here and now is flat Euclidean reality between such opposite curvatures—it is their overlap (eigenvalue) or discrepancy. It is equal and opposite eccentricity of such double/inverse perfect curvatures. Galaxy is densified or contracted empty space of universe, while universe is expanded or de-densified dense galaxy, which are two anti-parallel strands of fractal double helix (action/reaction). Arrow of time of galaxy is exponentially inverted in opposite direction in context of universe (equally vice versa). Time of universe is positive accelerated time of galaxy, while time of galaxy is negative accelerated (decelerated) time of universe. Likewise—time of galaxy is positive acceleration of solar time and solar time is negative acceleration (deceleration) of galactic time. Solar time is accelerated time here and now, while here and now is decelerated solar time (1D or light speed fractal difference). Micro-world is yet again triple time deceleration, where any acceleration is exponential/derivated or infinite difference.

Black hole, Planck length and light speed (singularity/infinity) are exclusive reality of inertial here and now, without any reality in context of relatively different space or time coordinates (larger/smaller size or motion). They are relative/uncertain boundaries of equally relative/uncertain here and now. Every relative size/motion of space/time is black hole of its double/inverse reality—it is infinite/singular boundary of infinitely larger/smaller size and/or slower/faster motion. Black hole is quantum boundary (and so is light speed). Any here and now is singular black hole, which is infinity of black holes; one within the other, where black hole of space is infinite time and black hole of time is infinite space. Black hole is totally relative/uncertain boundary of inertial density here and now (1D fractal difference E/m). Any change of density is equal and opposite (double/inverse) changes of its boundaries—black hole within and size of infinite universe without. Black hole in infinitely smaller or faster perspective (atom) is infinite spatial universe and equally vice versa—in context of infinitely larger and slower perspective (universe), size/density of universe is its black hole. Different size/motion is only fractal (gradual) change, while only infinite change is quantum change of size into motion and motion into size. Infinity of fractal changes is real only in inertial space/time context. In context of their gravity, infinity of such changes is no changing at all. Any change of size less than singular infinite (black hole) is only fractal (relative) reality of inertial gravity. Only complete change is instant (singular) inversion that is double inertion or change of context (gravity) of relative space/time coordinates here and now and equally vice versa—only complete change of gravity is one fractal space/time (E/m) frequency.

Size/motion and their density are 3 coordinates of inertial 3D. Size/motion is empty duality without content. Content of any size/motion is its density, where spatial density is double/inverse time density (E/m of empty space/time, where spatial mass is energy of time and mass of time is spatial energy). Depth of surface space (universe/micro-world) is +/- variable density in context of inertial density here and now. Space and time are equal and opposite variables of constant (inertial) density. Density is time or spatial depth (4th direction), while inverse/double density is space, which is depth of time. There is no difference between stars without and atoms within (and/or nucleus/galaxy) in context of inertial density here and now—they are two equal opposites and any difference is only exclusivity of perspective. Density is angular direction of surface (2D) space/time reality, where density of mass is inverse/double density of energy and equally vice versa. Reality without and within is only comparatively different, therefore such difference is only its fractal surface reality. Space and time are double/inverse variable or double/inverse density of each other. Inertial 3D is only surface of larger 4D, just as 4D is only surface of 5D. It means, entire infinite 3D spatial depth of the universe is only flat (instant) surface of such larger (relative) reality without, while its instant depth (3rd direction) is constant passage of time here and now. In context of relative universe (4D), 3D universe is only thin surface or boundary of such different space/time quantum.

Consequently, infinite 3D empty space of universe is fractal surface of dense mass of 4D (relative) universe. Likewise—infinite empty space of 4D universe is fractal surface of dense mass of 5D universe, just as dense mass of 3D here and now is only infinity of fractal surfaces that are infinity of different empty spaces of its micro-world. Universe here and now is only one out of infinite number of relatively different possibilities, all equally real (but not here and now). Only infinity of all possibilities (all space in context of all its passage of time) is instant 3D spatial reality of 4D (relative) universe. Only one space/time eigenvalue, which is real in context of inertial here and now, is only one fractal surface of instant 4D. 4D is instant reality of infinity of all different 3D fractals (surfaces). 4D is spatial reality of time or infinity of all different spatial possibilities (Everett's many worlds). Universe and micro-world are triple relativity/uncertainty or 3 fractal quanta of relative reality, which are 4D/5D/6D space. Solar reality is singular relativity (4D), galaxy is double relativity (5D) and universe is triple relative reality

(6D). Star/galaxy/universe are 3 time shells of inertial space and/or 3 spatial shells of inertial time (fractal ¾, 4/5 and 5/6 realities).

Each fractal shell is double/inverse or two equal and opposite fractal shells (2x1/2), which are space and time of such inertial fractal shell, which is singular fractal gravity. Universe is 3 sets of double shells with anti-parallel motions and overlapping reality between them is mass/energy of such empty space in context of constant passage of time. Mass/energy is double/inverse fractal difference (double/inverse light speed) between space/time shells (spheres). Double/inverse fractal difference is infinite/singular or whole (quantum) difference, which are full curvature of flat reality and/or complete flattening (infinite eccentricity) of round curvature. There is double/inverse light speed difference (C) between any whole action and its reaction. It is same inertial fractal difference (which is reality here and now) between Earth and solar reality, solar reality and galaxy, atom and its nucleus, mass and energy, space and time, universe and its Planck length, time of one orbital of electron and time of the universe, infinite something and singular nothing or between any two heartbeats, because 3rd inclusive direction of fractal surface duality is always same fractal difference of no size/motion, which is indistinct (non-physical) here and now.

Fractal reality is double +/- precessional curvature of double/inverse densities of successive space/time shells, where infinite 3D space is instant surface of time and infinite 3D time is instant spatial surface. Surface of 3D is quantum reality of its 3D depth and equally vice versa. Singular (quantum) reality one is infinity of fractal quanta of another. Singular possibility of one is infinity of possibilities of its double/inverse opposite. Singular gravitational sequence of a star is infinity (10) of orbiting planets, while infinity of different gravitational (E/m) sequences is double/inverse infinity or 2x1/2 multiplication/division of 10-planet solar systems (atomic shells). Double/inverse infinity is constant passage of time, which is gradual quantum change of such inertial sequence or passage through all different possibilities that are not real right now (Kepler), but are instant space of solar reality, because time is only quantum interruption or empty space (quantum discontinuity) of its instant density. Double infinity is constant density (mass) of solar reality, which is empty space here and now.

Singular gravitational sequence of galaxy is infinity of its orbiting stars, which are its constant density in context of infinite passage of time, just as singular universe is infinity of galaxies embedded in its expanding space. Expanding universe in context of all its time is inverse density of empty space, which is reaction of double fractal (precessional or exponentially denser) realities within. Universe is one empty shell without of progressively denser shells within, which are its gravitational (E/m) sequences. Different shells of universe are only infinitely small fractal angular difference (multiple light speed difference), which are multiple infinite spatial distances. Different shells of micro-world are infinitely large angular differences that are no spatial distance, because micro-world is double/inverse universe and equally vice versa. Eigenvalue of both is here and now (surface of spatial Earth in context of its quantum period of constant passage of time). It is infinity of equal realities in unequal perspective or infinite asymmetry of perfect symmetry. Unequal perspective or asymmetry of perfect symmetry is inertion of reality here and now, which is cause of double/inverse (space/time) infinite distortion of reality, which is no difference (C=const). It is distortion of no difference with equal and opposite no difference, which is infinite double/inverse differentiation ($1.111..^2=1.23456..$). Double singularity is inverse infinity.

Empty space of reality within is dense mass of reality without and equally vice versa—empty space of dense mass within is dense mass of empty space without, which are two anti-parallel (accelerated/decelerated) directions of constant passage of time. Future time of empty space is past time of dense mass, just as future time of dense mass is past time of empty space. Singular space/time sequence here and now is only fractal surface duality of double/inverse dense 3D. It is only inertial (7th) out of 6 different fractal surfaces of a cube, all equally real. Any horizontal surface of a cube excludes 4 angular surfaces, which are its double/inverse 4x1/4 time difference (1D=C) and one equal and opposite horizontal surface, which is its double angular (time) or singular spatial difference ($2D=C^2$). Here and now is inertial (11th) surface between two equal and opposite (spatial and time) cubes, which are 10 different surfaces and/or linear direction, which is line of symmetry between two equal and opposite squares (7th side in between two equal and opposite 3 sides). Likewise—inertial here and now is inertial cube between 6 equal and opposite cubes, which are 6x6=36 different surfaces or 30 different surfaces without inertial 6 surfaces here and now (6^2-6=30, which are 5 cubes, because one inverse cube is equal opposite of inertial cube).

Inversion is always inertion of double inertion and equally vice versa. 30 surfaces are 60 and 36 surfaces are 72 linear directions, which are 2D fractal difference or quantum gravitational sequences of inertial (fractal) space/time. 60/72 linear directions is 40/48 corners, which are fractal 2/3 antigravity of gravitational sequences and/or ½ of gravitational sequences of time sequences (60/72) of inertial space here and now (fractal 3/2=6 surfaces of a cube). 3/2 gravitational sequences are 2x40/48=80/96 inverse sequences of inertial 60/72 sequences of double inertion here and now (6 sequences). Between singular space and time, there is double/inverse mass/energy and vice versa. Between 2D space and time there is double/inverse 3D/1D mass/energy (4x1/4) and equally vice versa, which is double/inverse density of inertial density here and now.

Geometrical or numerical abstraction is reduction of infinite relativity/uncertainty to singular certainty. Abstraction is double/inverse or infinite relativity that is singular certainty and/or infinite uncertainty that is no relativity. It is singular inertion of infinity of different fractal dualities or singularity of infinite uncertainty/relativity. It is singular point of infinity of relative/uncertain theories, because it is non-physical reality, which is double/inverse physical reality, which is no surface difference. No difference is infinitely large inverse differentiation, which is larger difference than any of its fractal (relative/uncertain) inertial framework, therefore such large difference is not difference but inertial framework of infinity of fractal differences. It is quantum of fractal infinities or reality of infinity of relatively different space/time sequences (possibilities). Abstraction transcends its relative/uncertain limitation, because it is inertionless physical reality, which is infinity of relatively different inertions in context of dual space/time. Numerical abstraction is singular ½ of infinity of fractal 2x1/2. It is singular possibility of infinite number of dual (physical) possibilities. Numerical abstraction and geometry are non-physical only relatively—they are double/inverse or quintessence of all physical (space/time) realities.

Any relative/uncertain physical reality is only fractal surface of abstract reality. Abstraction is equal opposite of physical reality—it is its double/inverse negation, which is its singular affirmation. Abstraction is fractal supersymmetry, which does not differentiate between surface symmetry and asymmetry or superficial inequality between equals. Fractal (dual) differentiation into space/time, mass/energy, size/motion etc. is surface exclusivity of singular inclusive abstraction. Fractal duality (differentiation) is infinite complexity of singular abstraction. Complexity is not shortest direction of least resistance—it is a circular route. Complexity of fractal differentiation is equal opposite of singular eigenvalue—it is infinity of all different possibilities, which are not real here and now, by excluding only possibility that is real. Fractal complexity is inversion of simplicity of numerical abstraction, which is singular eigenvalue (quantum) of all such different complexities. Abstraction is 3D of infinite number of fractal surfaces, which are duality of action/reaction taken out of its inertial (singular) context. Inertial context is cause of both—action and reaction. Singular relative sequences of action and reaction are only one of infinite number of reactions of singular inertion. Shape of a cube is singular derivate of infinity of relative space/time sequences. It is inversion/double inertion of infinity of relative/uncertain theories, which are only fractal sequences of its abstract (double) physical reality.

Boundaries of infinite universe without and/or infinite micro-world within (Planck length) are diametrically opposite surfaces of a cube, which are double/ inverse or space/time fractal difference. Two opposite sides of a cube are light speed or size/motion (quantum) difference of any relative/uncertain (fractal) reality. Reality is infinite fractal curvature; therefore infinite 2x1/2 multiplication/ division of different directions, because here and now is singular fractal difference that repeats itself in infinite numbers of different fractal realities, where one relative sequence precedes another, 2x1/2 smaller/larger sequence. Inertial here and now is inertial fractal difference of all relatively different sequences, because it is odd (asymmetrical) sequence of surface symmetry and/or symmetrical sequence of asymmetrical surface. It is asymmetrical 3rd half of 2x1/2, which is only 3rd quarter of symmetrical 4x1/4. It is asymmetrical 5th quarter, which is 5th 1/6 of symmetrical 6x1/6. It is linear (7th) direction between 2x3 directions of equal and opposite squares. 7 is fractal ½ asymmetry of 8 directions, which is its symmetrical reaction and 9 is asymmetrical reaction of fractal symmetry of 8 directions.

Reaction of asymmetrical triangle is symmetrical square and its reaction is asymmetrical pentagon, while reaction of pentagon is symmetrical hexagon etc. Inertial here and now is one out of 6 surfaces of symmetrical cube and/or 7th surface of asymmetrical cube, because one of the surfaces is double/inverse surface of singular reality of other surfaces. It is fractal 5/6 within and fractal 7/6 without. It is fractal curvature in anti-parallel perpetuity of space and time. Here and now is singular relative/uncertain difference that perpetuates itself in infinity of different fractal curvatures, because only self-interference is real (eigenvalue of inertion and inversion, which are two equal opposites or two fractal reflections of singular whole).

Any change of one is equal and opposite change of its double/inverse directions. Any singular change of size is cause of double/inverse change of motion and density (4x1/4 wave) and equally vice versa. Density within is acceleration without. Density is double/inverse or accelerated/decelerated size/motion. Double size is singular motion and equally vice versa. Double motion is density of size and double size is density of motion, which is double/inverse size. Any singularity (inertion) is 4x1/4 wave or double/inverse duality of singular here and now, which is 5th quarter or overlapping end/beginning of such sin wave. 4x1/4 wave is double/inverse density of singular fractal difference (perpetual inertion), which is double/singular fractal difference of inertial 3D. It is singular positive/negative (linear) change in context of double/inverse (positive/negative) accelerated change or singular/double infinity of singular here and now. Size/motion/density are 3 coordinates of inertial here an now, which is its double/inverse content.

Inertial here and now defines coordinate system, just as coordinate system inversely defines here and now. Any relative point of fractal space/time is in perfect symmetry with infinity of all other relative points. Space is only variable (relative) surface or instant 3D time and time is variable (uncertain) surface of instant 3D space—it is everywhere at the same time. Gravity without and/or density within is surface or double/inverse instant reality of both—space and time, and equally vice versa (where gravity without space is density within time and gravity without time is density within space).

Universe/micro-world are only double/inverse surfaces without its angular (vertical) depth (density). Depth or density is same scalars, but opposite vectors in context of double/inverse universe/micro-world. 3D universe is fractal density of larger 4D universe, and 2D universe (micro-world) is fractal density of 3D universe. Fractal density is only angular rotation of same vector, which is also double vector or tensor between two equal opposites (their fractal difference or overlapping eigenvalue), as well as spinor in double/inverse context. Any relatively different coordinates (different size/motion/density) is relatively different double/inverse angular perspective of the universe—it is universe of relatively different size/motion/density. Singular universe on the surface is infinity of relatively different universes in angular depth, which are reality of different fractal (dual) 3D points. Perspective of 3D coordinate system differs from different angular points, because such point is singularity (0D) or only point where all 3 directions meet. Relatively different point in one of its linear directions is same space, but different time/gravity and any relatively different point in another directions is same time, but different space/gravity and/or same gravity—different space/time. Point that is same surface is same space/time and different density or same density and different space/time. It is same spatial density of different time and/or same time density of different space. Point, which is same 3D cube, is same inertial 3D—it is universe/micro-world of such point. Point that is same linear direction is 1D eigenvalue, point that is same surface is 2D eigenvalue and point that is same cube is 3D eigenvalue.

It is infinite double helix, because any inertion has two different depths—without and within, which are infinite anti-parallel (fractal) Fibonacci sequencing. Any inertial reality is only surface, which is line of symmetry between two equal 3D opposites (equally vice versa). Numerical abstraction and its geometry is singular description of infinite fractality (infinity of preceding/succeeding relative/uncertain curvatures). Any fractal (dual) description is ultimately an infinite complexity. Inertial here and now is the only equal and opposite surface of its universe/micro-world and equally vice versa—relative universe and uncertain micro-world are equal and opposite exclusive surfaces of reality here and now. Infinity of different space/time surfaces, which are not eigenvalue of inertial here and now are its empty space in context of its constant passage of time, which are different fractal densities that are equally real, just not here and now. Such different spatial possibilities of time and/or inverse time possibilities of space are real only in quantum steps, which gradually reveal different densities of inertial universe/micro-world. In context of all its time, any space is same density as here and now, only in context of relatively different fractal space/time sequence.

Empty space and dense mass are double/inverse density of inertial here and now. Time curvature of spatial universe is reaction of equal and opposite spatial curvature of Earth. Universe without is positive and Earth within is negative non-Euclidean time curvature, where negative time curvature is positive spatial curvature. Eigenvalue of such fractal opposites is flat surface reality here and now. Spatial universe is massive Earth turned inside out, where center of gravity within is infinite boundary (antigravity) of spatial universe without. Universe is gradual expansion of equal and opposite instant contraction within dense mass, where spatial expansion is time contraction and equally vice versa. Singular quantum of spatial reality is time and singular quantum of time is space. Singular quantum of spatial density is mass of an atom. Infinite density of mass of any spatial object (concave curvature) is no distance between atoms and at the same time, their exponentially increased angular differentiation (heavier atoms). Singular density of empty space is infinite spatial distance between singular atoms that are no angular difference (fractal ¾ reality of singular and/or one pair of electrons, which are only one orbital).

1D fractal difference between dense mass here and now and empty space, which is dense mass of solar reality is once-infinite distance between singular atoms, while 2D fractal difference between solar dense mass (empty space here and now) and its empty space, which is dense mass of galaxy, is double infinite distance between atoms. Interstellar space is twice empty space of dense mass here and now and/or singular empty space outside solar reality (solar system). Universe outside galaxy is triple fractal difference or triple empty space—it is empty in all 3 directions of inertial 3D space or triple infinite distance between singular atoms. Reality beyond boundary of 3D space is 4th quarter of fractal ¾ reality, which is its time. Singular quantum of massive reality beyond spatial boundaries is not an atom but its fractal particles or nucleus/electron pair (positive/negative duality or double/inverse spatial curvatures of time). Reality beyond spatial boundaries is infinitely empty space, which is instant inversion that is double inertion (successive Fibonacci sequence of physical reality). Infinitely empty space is singular quantum of time density, which is singular quantum of energy. No spatial density is infinitely dense time and no mass is dense energy. Dense energy is double/inverse density of mass, which is twice empty space without, as well as twice dense mass within (no difference).

Energy without is context of mass within and equally vice versa. All different density of mass from infinitely empty space to infinitely dense mass is only content in context of its energetic reality without, which is its inertial framework (1D fractal or quantum difference). Energy is double and inverse reality of singular mass here and now—it is its reality without and within. Energy is double/inverse negation of mass, which is its singular affirmation in between, because infinitely small size in context of time is infinitely large size in spatial context and equally vice versa. Micro-world (mass) and universe (space) are inversion of each other's inertion—time of one is space of another and equally vice versa. Infinitely dense spatial mass is infinitely empty space in double/inverse (time) context and equally vice versa. Dense energy is core of dense mass, as well as empty space, because it is inversion of empty space that is double inertion of mass and inversion of mass that is double inertion of empty space (eigenvalue of no difference). Energy is singular eigenvalue of space/mass duality (3rd inclusive direction of exclusive fractal opposites). Dense core of solar reality (Sun) is twice infinitely empty space, which is singular density of time (dense energy), while dense core of galaxy (black hole) is three times empty space, which is double density of time (twice denser energy).

Sun is inversion of empty space, which is dense energy and black hole is yet another inversion (successive sequence) of such energy, which is double energy. Solar dense energy is fractal reality in between dense mass here and now, which is fractal reduction of size of energy and twice dense energy (core of galaxy), which is complete inversion from size to motion. Fractal solar density is duality of mass and energy (size and motion)—it is mass on surface and energy in depth, while galactic density is singular energy of no mass (motion of no size). Sun is no mass but twice empty space, which accounts for its resonant capacity, while core of galaxy is twice empty space, which is twice resonant capability. Sun is mass (singular atoms of hydrogen fusing into helium, both of which are singular orbital of nucleus) only on surface—in its depth it is not mass but energy (neutron mass), which are singular particles (nuclei). It is reduction of no distance between atoms (dense mass) in less than no distance (more than uncertain reality), which is exponential increase of density of nucleus and equal and opposite decrease of density of empty space without nucleus (orbiting electrons). Singular massive density of an atom is fractal ½ density of only one orbiting electron (hydrogen) and/or singular density of one pair of electrons in singular orbital (Pauli's exclusion).

Increased number of electrons is increased density of such empty space without nucleus, which is increased massive density. 1 electron is boundary between mass and energy. Nucleus (neutron, which is duality of positive/negative particles) with no orbiting electrons, is not mass, but energy (inversion), because nucleus is double/inverse (dense) reality of an atom (empty space) without. Only fusion of nuclei is an atom or duality of nucleus and orbiting electron (mass), because in context of energy, there is no difference between nucleus and its electrons—they are fused into singular neutron. Depth of solar reality is inverse fusion or fission, which densifies core of atom (nucleus) and de-densifies its empty space. Galactic reality is double inverse fusion or double fission, which densifies size of nucleus into no size at all, because in context of double energy, nucleus is not dense, but empty space and only its core, which is spatial singularity (Planck length) is dense mass of such empty space without. Surface of energy (neutron mass of Sun) is mass (atoms), while surface of black hole are singular particles (energy). Massive surface of solar reality is boundary between two empty 3D shells, which are double/inverse mass without and within. Double mass without is empty space and inverse mass within is dense energy. Likewise—surface of black hole (energy) is boundary between two empty 3D shells, which are its double/inverse energy. Double/inverse energy is its infinite gravity without and infinite antigravity within, where spatial gravity is antigravity of time and equally vice versa.

Constant (instant) surface is gradual change of its equal opposites within/without. Massive surface is linear energy without and rotating energy within, while surface of energy is linear gravity without (attraction of mass, repulsion of energy) and rotational gravity within (repulsion of mass, attraction of energy). Discrepancy between attraction and repulsion (overlap or eigenvalue) of mass and energy is space/time reality here and now, which is complete attraction of mass and only fractal, infinitely small attraction of energy in solar context and vice versa in galactic context. Space/time is particle/wave duality—it is infinitely large particle and infinitely small wave and/or vice versa. Particle/wave overlap is eigenvalue of space and time, which is singular space/time reality here and now. In context of overlap (discrepancy or asymmetry of surface symmetry) both equal opposites are the same—i.e. both, mass and energy are attracted by solar/galactic reality, where massive attraction is infinitely large and energy attraction is singularly small and/or vice versa in galactic context. Overlap or eigenvalue is distortion of perfect curvature, which is space/time reality here and now. Eigenvalue is point of self-interference.

Sun is massive surface between two energies (rotational and linear light speed within and without) and black hole of galaxy is energetic surface between two gravities (double light speed, which is infinite space without and infinite time within. Sun is relativity of inertial space here and now, while galaxy is relativity of solar reality, which is quantum uncertainty of its solar reality within. In inertial spatial context Sun continuously radiates energy and black hole radiates gravity. Gradual reality of inertial context (space) is quantum reality in double/inverse context (time). Gradual spatial change is not reality of time, because time is double/inverse space—it is relative space in context of no time, because time here and now is spatial direction of relative reality. Gradual reality is real only in context of constant passage of time—without time, it is instant density. In context of time (relative space), Sun does not radiate energy, which is its instant density. In relative (time) context, Sun radiates mass and black hole radiates energy. Radiation of energy (Sun) and/or gravity (core of galaxy) is uninterrupted spatial (gradual) reality here and now. Radiation of mass of solar reality and/or energy of galactic core is not gradual, but instant reality in infinite periods of constant passage of time. Radiation of mass of solar reality and/or energy of black hole is quantum reality only here and now (constant passage of time).

In context of relative reality, quantum changes (infinite passage of time in between) is its gradual spatial change, because gradual change of double/inverse reality is quantum change here and now (equally vice versa), which interrupts its own gradual change in context of constant passage of time. Quantum change is double/inverse gradual change, which is not gradual spatial, but gradual time change in context of gravity, which is time of time (double/inverse time). Only infinite period of radiation of energy (Sun) is accumulation of energy, which is singular massive quantum and only infinite period of rotation of gravity is accumulation of gravity that is singular quantum of energy. Radiation of energy (Sun) and radiation of gravity (black hole) is only fractal (less than quantum) relative reality (time). Only infinity of fractals is one whole quantum. Fragmentation of relative reality (time) is multiplication of spatial reality and equally vice versa. Infinite period (time) of radiation of energy from solar reality are instant formation of planets (mass) of solar reality, while infinite period of radiation of gravity (black hole) are instant formations of stars (energy) of galaxy. Formations of planets of solar reality and/or stars of galaxy are quantum events, which follow infinitely long passage of time, which is only gradual accumulation of tension preceding such formation.

Uninterrupted radiation of energy (Sun) and gravity (core of galaxy) are not real in relative context—they are only gradual increase of tension in its density.

Infinite increase of tension is instant inversion that causes double inertion—formation of massive reality of orbiting planets and/or orbiting stars. Such singular and double quantum events are instant interruptions of singular and double periods of constant passage of time in between (here and now). They are interruptions (beginnings and ends) of constant passage of time (quantum solar relativity), as well as interruptions of constant passage of time of relative (solar) reality, which is its gravity (quantum galactic reality). Solar reality is singular fractal relativity and galaxy is double/inverse relativity. Constant passage of time of solar reality is galactic space (which is gravity here and now), while constant passage of time here and now is solar space. Gradual spatial change of galactic reality is double, while gradual spatial change of solar space is singular quantum reality here and now. Galactic spatial reality is gravity and solar spatial reality is time of inertial space here and now. Gradual spatial changes of solar/galactic reality are singular and double quantum events here and now, which are infinite and twice-infinite periods of time interrupted by such quantum (instant) changes of time and/or gravity, because passage of time of solar and galactic reality is once and twice-infinitely faster (light speed difference).

Instant (quantum) time/gravity changes are 2x1/2 and/or 4x1/4 infinitely larger spatial changes here and now in context of accelerated and twice accelerated passage of time. Uninterrupted radiations of energy (Sun) and/or gravity are exclusive reality here and now (local passage of time) and not solar/galactic reality. They are instant density of such double/inverse spatial reality. In context of solar/galactic inertial time, gradual spatial change, which is quantum event here and now (De Broglie) are radiations of mass (planets of solar reality) and/or energy (stars of galaxy). Beginnings/ends of constant passage of time are quantum changes of double/inverse reality or changes of boundary of inertial time. Quantum change is cause/effect of constant passage of time and/or vice versa—passage of time is cause/effect of quantum change.

Nucleus is core—it is Sun of solar reality and/or core of galaxy. Nucleus is energy of massive atom (and/or mass of its energy). Solar reality is infinity of singular nucleus, while core of galaxy is infinity of singular stars, where infinity is double/inverse reality. Nucleus is singular quantum of energy, while solar reality is infinity of quanta and galaxy is double infinity. Nucleus is solar reality turned inside out and equally vice versa. Galactic core is nucleus of solar reality (Sun) turned inside out and equally vice versa. Nucleus/Sun are 1D fractal difference, while nucleus/galaxy is 2D fractal difference of inertial 3D energy. Fractal difference between nucleus and Sun and/or Sun and core of galaxy are space and time of such inertial reality, while 2D fractal difference between nucleus and core of galaxy is their gravity. 3D fractal difference between nucleus and infinite expanding universe is only fractal ½ difference, without spatial reality if its double/ inverse reality, which is three times dense energy beyond event horizons of black hole.

3D is only surface (2D density) of its 3D without/within. It is inversion of empty space within into dense mass without and vice versa. Singular density is double duality of space/time and its mass/energy without and within (inertion/ inversion). It is singular 4x1/4 wave. Density is 5th or inertial fractal of 4x1/4 reality. Infinite space in context of infinite constant passage of time, which are its mass/energy without and within is singular density of inertial reality here and now. Density is double eigenvalue or eigenvalue of double/inverse duality. Change of density is equal and opposite change of its space and time and inverse equal and opposite change of its mass and energy within and without, where gradual change of space/time is instant change of mass/energy and equally vice versa. Only double—gradual and quantum change, which is infinitely slower gradual change (change of space/time and mass/energy) is singular quantum of density change and equally vice versa—only infinite gradual change of density is singular change of space/time and twice infinite change is singular quantum change of mass/energy (equally vice versa).

Change of size/motion is equal change of its density, where any qualitative or gradual (fractal) change is no quantitative change at all in context of double/ inverse reality. Change of size/motion, less than total (C) change is no change at all in double/inverse context. Any change less than complete change is only surface change that does not affect depth (density) of double/inverse 3D. Any change in depth, which is less than complete (C) is not eigenvalue of its surface and equally vice versa. Gradual qualitative change of surface is no change of 3D—only total quantum change of surface is also fractal 3D change. Only infinite change or complete inversion of size into motion and/or motion into size is one quantum or irreversible change of its double/inverse reality. Such qualitative or fractal gradual change is only increased tension in context of its double/inverse reality, which does not affect depth or density of such fractal surface reality. I.e. fractional turn of small wheel only increases tension within larger wheel, which is not enough to cause its motion, because such fractal tension is less than wheel's inertion.

Only complete turn of infinitely smaller wheel (period of time) is enough accumulated tension that causes singular (instant), infinitely small fractional motion of larger wheel. Quantum period of constant passage of time of smaller wheel (turning) is instant spatial change (no time) of larger wheel. Motion (time) is inertial context of both wheels-it is gradual reality of one and instant reality of infinitely larger wheel. Equally vice versa—infinitely small shift of larger wheel, which is only instant of its time causes all infinite time of smaller wheel. 2x1/2 is double/inverse reality of both—2 and ½, which is 2/(1/2)=4 or ¼ quantum difference. One turn of twice-smaller wheel is only ¼ turn (singular quantum) of twice-larger wheel. Only 4 turns of smaller wheel is complete turn of larger wheel (1D fractal difference). Complete turn of larger wheel is only ¼ turn of yet larger wheel (2D fractal difference), which is complete turn only in 16 turns of smaller wheel (4/(1/4)=16). Slower motion of larger wheels is their comparatively faster (accelerated) passage of time. 2x1/2 increase in size (space) is double or 4x1/4 increase in time, which is double/inverse duality. 4x1/4 sin wave of double/inverse duality is Newton's inverse square law.

Spatial motion less than light speed is no motion at all in double/inverse context of time, but only gradual change of its size (equally vice versa—change of spatial size is change of motion in time). Only infinite (light speed) motion is singular fractal change of its inverse/double size and equally vice versa. Only accelerated/decelerated (double/inverse) spatial motion (change of density) is motion in time. Such accelerated motion is likewise no motion at all in context of gravity, which is 2D fractal difference, but only its gradual change of size. Only 3D acceleration/deceleration is 4D motion and only double acceleration is 5D motion etc. Only acceleration is constant motion in context of relative spatial universe and only twice accelerated motion is motion in context of its expansion. Spatial motion is only surface (2D) change of 3D time (4D space), which does not affect its depth and only linear change of 3D gravity (5D space) and equally vice versa. Infinity of inertial quanta is singular quantum of double/inverse reality, and equally vice versa. Spatial motion is only fractal change of surface tension of time and equally vice versa. At the same time it is only fractal change of one direction of surface tension of gravity and vice versa. Only infinite repetition of spatial motion is singular quantum of motion in time and only twice infinite number of repetitive spatial motions is singular quantum of motion in context of gravity. Only infinite repetition of motion is singular quantum of instant density of its double/inverse reality. Singular quantum of double/inverse change is instant change, because it is change in context of no constant passage of time here and now.

It is change in context of double/inverse time, which is gravity/antigravity (contraction/expansion) of spatial reality here and now. Infinite number of repetitive spatial motions is instant density or instant space/time duality in inverse/double context. Constant passage of time that is infinity of such repetitive spatial motions (i.e. rotation/orbital) is not real in double/inverse (time) context—it is only its empty space within of no density. Massive particle in orbital motion is not real in context of its relative universe, because it is equally everywhere in same instant of time (uncertainty), which is 1D fractal time difference. Time of orbiting particle is infinitely faster motion and/or infinitely slower passage of time (light speed difference). Time of orbiting particle is infinitely expanded passage of time, just as passage of time of its relative universe, which is its inertial context without, is infinitely contracted. All the time of uncertain orbital motion within—i.e. all the time of infinite number of orbitals is not time but instant surface reality. Orbital is not motion in context of relative universe—it is instant surface (size) of dense mass within.

Change is inertial perspective—it is not size/motion here and now that changes, but its reality without/within (expansion/contraction). Inertial here and now is in perfect rest (no change of size and/or motion), which is only relative perspective, because change without is equal and opposite change within, which is double/inverse or no change at all. Fractal expansion/contraction, which is result of size in motion, is duality of instant change in context of gradual change and equally vice versa. It is gradual expansion of space without (universe), which is reaction of instant contraction within (dense mass) and/or gradual passage of time without, which is instant space within. Space/time is uncertain/relative duality. It is gradual spatial change in context of passage of time, followed by instant change in context of no time, followed by yet another period of relatively different sequential passage of time, which is effect of relatively different (changed) density (because quantum change is irreversible). Infinity of quantum spatial changes is only gradual change of such double/inverse reality (i.e. time), just as gradual spatial change is infinity of quantum time changes within, which are gradual change only on surface, which is infinitely accelerated (contracted) passage of time of its micro-world within (atom).

Only double infinite change (C^2) of universe is singular inverse change of its micro-world and equally vice versa, because universe/micro-world (empty space/dense mass) are equal opposites of inertial here and now. Double infinite change of size/motion without is inverse (½) change within and equally vice versa (inverse square). Without/within are fractal asymmetry (supersymmetry) and only infinity of inverse fractals is singular inertial fractal. Only double infinity without is singular infinity within. Only infinite change of space and time is complete massive change and equally vice versa—only complete change of mass and energy is exclusive complete spatial or time change. One inclusivity is reality between two exclusive realities and equally vice versa. There is an odd number is between two even and an even number is between two odd numbers, where odd in space is even in time and equally vice versa. Complete $E/m=C^2$ is only ¾ change—it is exclusive spatial change, which is either infinite expansion or contraction, but not both. Only double E/m change is complete change of spatial and time reality, which is complete expansion and equal and opposite contraction. Double change without is ½ change within (2/(1/2)=4) and quadruple change without is ¼ change within (4/(1/4)=16).

Reality is singular/double acceleration of inertial here and now, which is 5th quarter of such sin wave (inverse square), where singular and double (horizontal and angular) accelerations are size and motion of inertial here and now and/or its space and time, which are equal and opposite densities of inertial density in between. Infinite change without (¾) is only ½ change within of inertial $^2/_3$ in between. Infinite change of size/motion of universe without (empty space) is only ½ change of size/motion of its mass within and equally vice versa—one is infinite fractal curvature of singular (whole) another. Both are double/inverse fractal curvature of double helix, where fragmentation in space is multiplication in time and equally vice versa. It is infinite double/inverse multiplication/division (self-interference) of inertial here and now. Total relativity is only ½ uncertainty and/or equally vice versa—one relative curvature without is infinity of fractal uncertain curvatures within and equally vice versa, where relative space is uncertain time and relative time is uncertain space. They are asymmetrical fractal reality only in context of inertial here and now. Without such inertion in between, which is double/inverse angular distortion (time), they are perfect symmetry (duality). Instant relativity without is gradual uncertainty within and instant uncertainty is gradual (fractal) relativity.

Relative spatial reality is gradual (uninterrupted) curvature only in context of quantum (interrupted) spatial reality within, while in context of uninterrupted space within, relativity is interrupted fractal curvature. Relativity is gradual only in context of constant passage of time. In context of no time—it is interrupted fractal (quantum) reality. Relativity is gradual only without—within it is reality of quantum steps (instant interruptions), because infinite (complete) change in one direction is only ½ change in opposite direction. Inverse square law is interruption of gradual reality in quantum steps, which are points of instant self-interference or instant symmetry of two equal opposites, because it is eigenvalue or instant when gradually accumulated tension within is equal to inertion of surface (and/or vice versa). Quantum change of relative reality is reality in inertionless context, when inertion here and now is 0, because equal and opposite double/inverse inertions are perfect symmetry—one perfectly resonates in context of another. Double/inverse inertion are space and time or boundaries of inertial here and now.

Quantum change is eigenvalue of such equal opposites, which are eigenvalue of inertial here and now only when they are perfect symmetry (no difference). Eigenvalue of space and time is also eigenvalue of such space/time with inertial here and now. Quantum change is perfect symmetry of inertial here and now with its double/inverse boundaries, when space here and now is all infinite space of relative universe in context of all its infinite time. It is instant relative (4D) reality here and now, which is instant irreversible change of its density. Such irreversible change is singular Fibonacci sequence when present adds past in order to generate future 2x1/2 sequence. It is instant multiplication/division that adds yet another space/time sequence in evolutionary spiraling of double helix. It is instant creation of different physical reality.

Universe is singular spatial quantum or instant space in gradual change (passage of time), while micro-world (its mass) is instant time (singular time quantum) in gradual change (half-life—i.e. density change in 2x1/2 quanta). Universe is one spatial quantum (one time possibility), while its micro-world within is infinity of different spatial quanta (different time possibilities)—2x1/2 one within the other. Each spatial quantum is universe. It is one universe without, which is infinite number of universes within or infinity of spatial universes, one within the other, where relative universe of inertial here and now is boundary of inertial perspective without any real difference from any different universe within or without. Relative universe is only relatively different, because it is infinite (90°) angular distortion of inertial perspective. Infinitely smaller spatial reality (micro-world) on surface is infinitely larger double/inverse spatial reality in its depth and equally vice versa. Likewise—infinitely slow passage of time on surface is infinitely faster double/inverse passages of time and equally vice versa. Each singular spatial universe is double/inverse infinity of universes within and without. Each spatial universe is double/inverse infinity of time universes and each time universe is double/inverse infinity of gravitational universes, where such triple differentiation is only one relative 3D sequence in infinite curvature of double helix, which curves in infinity of different directions.

Infinity of singular here and now is infinite, infinitely many times. Only boundary of such infinity is inertion here and now, which is its fractal differentiation into quantum periods and sizes, which are singular density in context of its infinite empty space (relative/uncertain space). Every uncertain perspective is infinity of relative curvatures and equally vice versa. Every symmetry is infinity of asymmetrical curvatures and equally vice versa, where any difference is fractal difference, which is singular inertion here and now. Boundary of space is double/inverse time and boundary of time is double/inverse space. Boundary of space/time is gravity. Boundary is time of inertial space and/or space of inertial time—one is content of form of another. Boundary or horizon (infinite/singular end) of inertial here and now is only beginning of double/inverse (relative/uncertain) space/time, which is twice larger/smaller and/or twice faster/slower space/time sequence, which is inertial context without/within inertial space/time sequence here and now. It is double/inverse infinity of relatively different fractal sequences of physical reality, of which only one is eigenvalue here and now. Quantum change of space/time sequence is also quantum change of inertion here and now, because they are action/reaction of each other and change of one is equal and opposite change of another.

Each symmetrical 2x1/2 sequence on surface is progressive fractal (asymmetrical) sequence in depth. Each additional fractal (light speed) difference is additional asymmetrical (fractal) sequence, which is fractally increased angular curvature of surface symmetry. Such exponential curvature is exponential increase of fractality from surface ½ to $^2/_3$, ¾, 4/5, 5/6 etc in its angular depth, where fractal difference is always singular here and now (1) in relatively different size/motion, which is relatively different space/time sequence, which is increased randomness of surface symmetry. Surface symmetry is total randomness in its depth and equally vice versa—randomness on surface is perfect symmetry in depth, which is inertial context of surface reality (its double/inverse reality). Fractal supersymmetry is both at the same time—perfect symmetry and chaos are only its two equal opposites, without real difference between them, except angle of perspective in which such reality realizes itself. Difference between symmetry and randomness; certainty and uncertainty are only relative or surface fractal differences of inertial 3D here and now, because uncertain space is certain time and spatial randomness is symmetry in time (equally vice versa).

Time distance is spatial difference and spatial distance is time difference, where difference is angular curvature of flat (linear) spatial reality. Infinitely distant space is only singular (no) time difference and infinite time difference is only singular (no) spatial distance. Mass, which is no spatial distance at all, is infinite distance in time (infinite past/future), therefore an infinite number of spatial universes within. Likewise—infinite spatial distance is no time distance at all—it is same mass (inertial atom) everywhere within such infinite spatial universe. Mass (atom) is infinity of spatial universes in depth of inertial surface, which is singular time, while universe is infinity of time universes in depth of its surface, which is singular space. Universe is one time atom consisting of infinity of fractal (relatively different) spatial atoms, while atom is one spatial universe, consisting of infinity of time universes. Atom/universe are action/reaction or two equal opposites in context of inertial space/time sequence here and now. Here and now is inertial density or inertial space/time sequence of any physical reality, which is meaningless (unreal) without such framework.

Any sequence of physical reality, which is double/inverse space/time sequence is triangulation between inertion, which is 0 or relative nothingness and its equal and opposite space and time (inertion/inversion), which are two equal and opposite relative something of absolute nothing and/or equal and opposite relative nothing of absolute something, because there is no difference between something and nothing—they are perfect mirror image of each other and one is real only in context of another. Reality is supersymmetry (singularity) of any surface duality. It is reality of no difference. Reality without equal opposite is not comparative surface reality; therefore it is not physical reality here and now, where non-physical in space is physical in time and equally vice versa. Non-physical of singular fractal difference is physical reality of double fractal difference, which is infinite double/inverse (anti-parallel) double helix of infinite Fibonacci sequencing of singular here and now.

Every singularity is double/inverse infinity within infinity—infinitely many times, because reality is constant (C) and its perspective is equal and opposite constant (1/C)—one is 3D, the other one is 2D. Perspective of reality is fractal $^2/_3$, which is its flat mirror image (surface), without depth (content). Perspective curves flat (Euclidean) reality and/or flattens round (non-Euclidean) reality. Perspective of reality is its inversion, which is double inertion. **Perspective is equal opposite—it is reaction of reality and not reality itself, which is always 1D fractal difference from its perspective.** Perspective is duality (physical reality) of inertial 3D, which is non-differentiated reality. Physical reality here and is only one out of infinite number of equal possibilities, which is eigenvalue of inertial perspective. It is one angular curvature out of infinite number. It is one relative surface out of infinite number, which is particular choice of inertial here and now. Physical reality is real only in 2D context, as surface of infinite depth, which is its inertial context (3D). Physical reality is singular surface of infinite depth within and without, which is its double/inverse physical reality (equally vice versa). Double/inverse physical reality (1D fractal difference each) is relative/ uncertain space, which is 2D time of 3D space here and now. 3D is horizon (boundary) of 2D perspective, meaning any change is triple echo or triple vibration.

It is triple 2x1/2 curvature (3x(2x1/2)=6x1/6), repeated infinitely many times in space and inversely in time. Every triple double/inverse infinite repetition is one space/time sequence, which is singular point of yet another double/inverse (6x1/6) sequence. Each preceding/succeeding sequence is relative/uncertain physical reality of inertial here and now in between, which is fractal of such equal and opposite multiples and/or singular multiple (quantum) of equal and opposite fractals. **Physical reality is fractal—it is only relative/uncertain surface of reality.** Physical reality is only ½ of a whole, where ½ + ½ is not 1, but 3/2—it is 2x1/2 in inertial curvature of angular ½, which is inertion (here and now) of such duality ("+" in between two ½'s is their 3rd ½, which is singularity of their duality). It is infinite relative fractal curvature in one (infinity of exponential relative sequences), and infinite inverse relative curvature (which is uncertain reality) in opposite direction, which is infinite double fractal curvature of spiraling double helix, which is infinite 2x1/2 multiplication/division in anti-parallel directions. Each infinite curvature is only fractal of a whole. Physical reality is infinite double/inverse repetition of 6-fold sequence (quantization), which is 6, 6^2=36, 6^3=216, etc, where odd exponent is fractal of even-numbered exponent and equally vice versa. Perspective is infinite, because reality is only one.

CHAPTER 6

FRACTAL (QUANTUM) SPACE/TIME SEQUENCING

Physical reality is perspective or surface of 3D reality. Relative/uncertain realities are consequence of diametrically opposite perspectives. Universe is reality in perspective of round Earth. Round Earth is inertial context of relative universe, which is its double/inverse reality. Time curvature of empty space (relativity) without is reaction of inverse (spatial) curvature of dense mass, which is its uncertain reality within. Space of universe is gradual relative reality (gradual time curvature) of equal and opposite instant uncertain (quantum) reality of massive Earth within. Time of relative space is uncertain quantum reality in context of massive Earth within. It changes in quantum steps, as opposed to gradual spatial change here and now, interrupting it with 2x1/2 (infinitely) larger (quantum) spatial changes. Singular quantum change of time (infinitely large spatial quantum change) here and now is smallest fractal of gradual change of spatial reality of time (gradual change of relative space). Relative space without (universe) is simultaneously uncertain or quantum time within (Earth), just as uncertain or quantum space of an atom within is instant relative time without (everywhere at the same time). Universe is non-Euclidean curvature (relativity) of equal and opposite non-Euclidean reality here and now (uncertain Earth), which is its inertion. Relative universe is real exclusively as consequence of round Earth.

An atom is inverse perspective of universe, because inertial framework of its reality is not round, but flat Earth here and now. Universe (empty space) without is **double** curvature—it is time curvature in context of inverse (spatial) curvature within (dense mass), therefore double fractal difference from flat Earth here and now. An atom is singular **inverse** curvature in opposite direction. Atom/universe are 1D/2D (inverse/double) fractal difference in context of **inertion** here and now, which is remaining 1D fractal difference or eigenvalue in between such duality (balance, resonance, synchronicity etc). Total space/time reality is 1+1+2=4D space/time of inertial (singular) point here and now, which is ¾ fractal reality of 3D space (instant reality) in context of its gradual change (passage of time), which is its ¼ relative/uncertain content. Inertion of such reality (here and now), which is discrepancy or distortion of inertionless reality, is singular ¼ quantum. It is double duality or 2x (2x1/2)=4x1/4 sin wave, which is gradual change of instant (spatial) reality in context of equal and opposite instant (spatial) change of gradual reality (time), which are action/reaction of inertionless (non-differentiated) reality. Relativity/uncertainty are equal and opposite (double/inverse) fractal realities of inertial here and now in between, which is their equal and opposite fractal distortion (double/inverse inertion of such waves). They are set of two opposite surfaces of a cube, which are one out of 3 such sets.

Relativity/uncertainty are triple fractal curvature each, where difference between fractals is infinite/singular or linear light speed difference. Relativity and/or uncertainty are fractal curvatures in each of inertial 3 directions. It is double/inverse relativity of mass and gravity in context of inertial spatial relativity and/or vice versa. Relativity of mass is ½ of spatial relativity, while relativity of gravity is ½ of massive relativity and/or ¼ of spatial relativity. Uncertainty is mirror image of fractal relativity or relativity turned inside out. Physical reality is triple fractal relativity in direction without (increase from singular to infinite size), which is simultaneous inversely increased uncertainty within and equally vice versa in opposite direction of micro-world, which is increased uncertainty and inversely increased relativity. Relativity and uncertainty are inversely proportional—larger spatial relativity is larger uncertainty of its time (which is decreased relativity) and equally vice versa. It is triple relativity/uncertainty in each linear direction, which is $3^2=9$ different eigenvalues in all 3 spatial directions and/or 3x4=12 different eigenvalues in all 4 directions of relative space (space and time). Because time is fractal equal opposite of space, it is also 3D, of which only one direction is time here and now (space as surface of time and vice versa, which is

90° angular difference). Time is inertion (overlap or delay) between space here and now and relative space—it is double/inverse surface of such equal and opposite spatial realities that are 1D fractal difference in between (equally vice versa).

Consequently, 12-9=3 is only fractal reality in between two equal and opposite 3^2=9 inertial realities. Time is double/inverse surface boundary between two spatial realities that are light speed difference apart, which is difference between singular and infinite size (equally vice versa). Duality of space and time (relative space) is 2x9=18 relatively different fractal (relative/uncertain) curvatures, which is only 2D surface of 3D gravity, which is eigenvalue of both. Inertial gravity of dual space/time is 27 different fractal curvatures or 27-3=24, where 3 is inertial here and now, which is overlap or distortion of inertionless gravity (likewise: 9=8+1, where 1 is inertial overlap or distortion of triple duality-2^3=8; therefore 3x8=24—see illustration 8). Gravity is only double/inverse surface between its equal and opposite fractal antigravities, which are 2x18=36 fractal antigravitational curvatures and/or total of 72 gravitational/antigravitational curvatures (triple light speed of no difference C^3=const). 9 fractal curvatures are 2/3-fractal reality of 12 curvatures (eigenvalues) and/or ½-fractal of 18 curvatures (and/or 3/2 of 6 curvatures). 18 curvatures is 2/3-fractal of 24 and/or ½-fractal of 36, while 12 is 2/3-fractal of 18 and/or ½-fractal of 24. It is infinite double fractal curvature in which one fractal precedes another—i.e. space is fractal of time and time is fractal of double/inverse space in infinitè intertwined double precessional curvature of double helix.

Relativity/uncertainty are double/inverse fractal reality, which are boundaries or inertial context in which here and now is real. They are equal and opposite non-Euclidean curvatures of inertial Euclidean (flat) reality here and now and equally vice versa, because round in time is flat in space. Relativity is gradual spatial change in context of inertial time, while uncertainty is perspective of gradual time change in context of inertial (changeless) space here and now. Gradual time change within (mass/atom) is instant time change without (space/universe), just as gradual spatial change without is instant spatial change within. Gradual change is 1D fractal difference, while instant change is 2D fractal difference, which is infinite double/inverse fractal (precessional) curvature. Singular spatial relativity (universe) is infinitely fragmented uncertainty within, which is infinitely fragmented relative time, just as singular uncertainty (Earth) is infinitely fragmented relativity within (atom), which is infinitely fragmented uncertain time. One relative fractal curvature on surface is infinity of fractal uncertain curvatures in depth and equally vice versa. One instant (spatial) reality is infinity of fractal changes and equally vice versa—one fractal change is infinity of instant (spatial) realities.

One space is infinitely fragmented time and one time is infinitely fragmented space. One changeless reality here and now is double/inverse infinity of quantum changes of space and time (within and without) where quantum changes of space within (atom) is instant space without (here and now) and quantum changes of space here and now is instant spatial reality of relative universe, which is spatial (instant) reality of time. Gradual passage of time here and now is only one direction of its spatial reality and gravity here and now, which is double/inverse time, is two simultaneous directions or fractal 2/3 relative reality. Only gravity (mass) and its equal opposite—antigravity (space) are singular quantum of relative spatial reality (universe). Only all infinite time in context of all infinite space is total spatial expansion without and/or total massive contraction within (equally vice versa), which is instant spatial density (instant massive reality) of relative universe.

Because they are consequence of diametrically opposite perspectives, relativity and uncertainty are negation of each other. They are equal and opposite (double/inverse) inertial distortions of inertionless reality. Distortion is 3rd direction of such duality—it is inertial here and now in between (surface of Earth). Here and now is their eigenvalue or context of their reality (equally vice versa). Out of infinite number of different relative/uncertain distortions (Everett's many worlds), only distortion, which is eigenvalue with here and now is real—rest is empty space within/without its massive shells and/or constant passage of time. Any perspective (physical reality) is only ½ of total curvature, which is its equal and opposite negation (inversion). Consequently, two equal opposites are simultaneously real only when they are perfect inertionless duality (no inertion in between). Only when each is perfect (symmetrical) ½ of its equal and opposite ½, are they simultaneously real. Only self-interference, which is infinite fractal 2x1/2 multiplication/division of inertial here and now, is real. Relative/uncertain realities are synchronized (they resonate in inertial framework here and now) only when they are perfect 2x1/2. Two waves resonate only when spatial wave in time is equal to its double/inverse time wave in space. Only when two waves are synchronous (crest-to-crest matching) they combine to produce double/inverse singular wave, which is 2x1/2 larger/smaller then its incremental double/inverse waves.

Only resonance is real. Resonance is perfect duality or eigenvalue, when two equal opposites are not opposites but singular 2x1/2 reality. Resonance is singularity of duality. It is inertion/inversion=double inertion. Eigenvalue is singularity or line of symmetry of two equal opposites—it is singular inertial context of infinity of its relative (dual) actions/reactions. Eigenvalue between relativity/

uncertainty is here and now. Eigenvalue is instant inertionless reality—it is perfect curvature of no eccentricity (time delay or spatial distortion) of two eccentric curvatures of relative and uncertain perspectives. If relativity is positive inertial distortion, than uncertainty is negative inertial distortion. When such two distorted curvatures are perfect mirror image of each other, their eigenvalue is 3rd (precessional) curvature, which is perfect curvature of such equal and opposite distortions, because positive distortion balances equal and opposite negative distortion. They are mirror image of each other only when inertion of one is equal to inversion of another. Only infinite inertion is equal and opposite infinite inverse inertion, which produces instant non-inertion, which is double inertion (double negation is affirmation). Only perfect duality of inertion/inversion produces eigenvalue that is instant double inertion, where double inertion is negation of singular inertion (equal opposite). Double inertion is inertionless reality. Only eigenvalue is reality of double helix, where singular spatial eigenvalue is infinity of different time eigenvalues and singular time eigenvalue is infinity of different spatial eigenvalues.

Relativity/uncertainty (positive/negative fractal realities) are eigenvalue only as perfect inertionless duality, which is instant self-interference (realization of preceding/succeeding inertionless fractal space/time sequence, which is 1D fractal difference from inertial here and now). Relativity and/or uncertainty are equal and opposite ½ distortions of inertial here and now, which are positive/negative 3rd (inertial) ½ of inertionless duality (and/or non-inertial ½ of inertial duality). Relativity is positive and uncertainty is negative 2x1/2 non-inertial fractal reality of inertial (neutral) ½ here and now and/or vice versa. Relativity is positive and uncertainty is negative reaction of inertial here and now. They are positive/negative 2x1/2, while eigenvalue of both is 4x1/4 reality (inertial quantum is ¼). Inertial relativity and/or uncertainty are ½ inertionless reality and ½ inertial here and now, which is double/inverse surface between such opposites. Inertial here and now is in-built ½ of any perspective, because it is its equal and opposite cause/effect. Perspective (i.e. relativity/uncertainty) is only ½ of reality—it is only its inversion, while inertion is its here and now (reality of point of observation, which is singularity of infinite and/or infinity of singular physical reality; therefore it is space of time and/or time of space or ½ of any inertial perspective). Point of observation is double/inverse boundary (singularity/infinity) of its physical reality (perspective), just as physical reality is its inverse/double boundary. Point of observation and its space/time are 2x1/2 reality. There is no physical reality without its boundaries (here and now), which is its inertial framework.

Any reality includes inertion of here and now—it is reality of both—point of observation and its space/time.

Perspective of space/time, which is physical reality, is consequently not inertionless or objective reality. It is inertial or subjective reality of two equal opposites—physical reality (space/time) and its inertion (here and now of such perspective). Physical reality within/without point of observation is only ½—point of observation is another ½. Perfect duality between space and time or perfect resonance between relative and uncertain realities, which are 2x1/2 each other is at the same time perfect 4x1/4 inertial reality (where inertion in space is non inertion in time etc). It is 4x1/4 multiplication/division of inertial here and now. Inertial here and now is singularity (eigenvalue) or singular quantum of any infinite duality that is real in its context. It is non-comparative, therefore beyond boundary of its relative/uncertain physical realities (dualities). Non-comparative means it cannot be formulated in dual context, except as eigenvalue of two equal opposites. Non-comparative reality is supersymmetry (eigenvalue) of its equal opposites. It is reality of no difference, because it is simultaneous inertion and inversion (space and time). Its infinite spatial inertion is singular or non-inertion in context of time and equally vice versa. Infinite inertion is singular non-inertial reality. There is no difference between inertial and non-inertial (symmetry and asymmetry) here and now. Infinite space/time inertion, which is cause of infinite fractal space/time (relative/uncertain) curvature (double helix), is perfect non-inertion in double/inverse context and equally vice versa. Inertion of physical reality is perfect non-inertion of its non-physical reality (1D fractal difference) and equally vice versa, because perspective (2D) and its reality (3D) are double/inverse (fractal) reality of each other.

Space and time are equal and opposite distortions of singular reality, just as singular reality is duality of their equal and opposite distortions. There is no difference between space and time or space/time and inertial here and now. They are inertion/inversion = double inertion of same (non-comparable) reality. Space is inversion of time and time is equal and opposite inversion of space. Here and now is inversion of both, which is inertial gravity/antigravity (contraction/expansion or acceleration/deceleration) and equally vice versa. Each one separately is only fractal ½ of reality. Only perfect duality of two equal opposites is real. Space and time of inertial here and now are 4x1/4 or fractal reality that is ¼ inertial difference, which is inertion of space/time duality. Space and time (relative/uncertain reality) are consequently 2x1/2 each of inertial/non-inertial here and now, while singular space/time, which is gradual spatial change in context of constant passage of time is fractal 4x1/4 reality. Space is ½ of time and time is ½ of space, while their eigenvalue (here and now, which is inertial gravity of space/time) is fractal ¼. Double/inverse ½ fractals of inertial here and now (1) are fractal ½ and/or 3/2, which are +/- ½ difference (relativity/uncertainty of inertial here and now). Double/inverse fractals of dual here and now are 2 +/- ½, which are fractal 3/2 and/or 5/2. Double/inverse fractals of 3D are 3 +/- ½, which are fractal 5/2 and 7/2, while fractals of 4D (space/time) are 7/2 and/or 9/2.

Inertial quantum of fractal duality is ½, while inertial quantum of double duality is ¼. Double/inverse reality of singular here and now (1) is fractal ¾ and/or fractal 5/4, while double/inverse reality of dual here and now is 7/4 and/or 9/4 etc. Fractal ½ is two equal opposites of inertial ½ in between, while fractal ¼ is two equal opposites of inertial ¼ in between. Fractality of such double/inverse opposites are infinite relative/uncertain curvatures (double helix). 3/2 of singular wave (inertion) is at the same time only ¾ of double wave (space/time duality) and/or ½ of triple wave (3D). Between two (double/inverse) fractal ½ realities, there is one fractal ¼ and equally vice versa—between two fractal ¼, there is one fractal ½. Between two fractal ¼'s is one fractal 1/8 and equally vice versa, while between two fractal 1/8's, there is one fractal 1/16 etc, which is infinite fractal 2x1/2 curvature. Fractal reality of singular (inertial) wave is ½; fractal reality of double wave (duality), which is 2x1/2 singular wave, is ¼; fractal reality of quadruple wave (4D), which is 2x1/2 duality, is 1/8 etc. Fractal 2x1/2 frequency of precessional waves, which are double inertion or eigenvalue between inertion/inversion, is $3/2^n$, where fractal in time is multiple in space and equally vice versa, which is Bode's periodicity of $3x2^n$. I.e. 3/2, ¾, 3/6=1/2, 3/8, etc in one direction is mirror image of 3/1=3, 3/(1/2)=3x2=6, 3/(1/4)=3x4=12 etc in opposite (inverse) direction.

Precessional wave is eigenvalue of its inertial/inverse waves, which is instant duality in context of no time; therefore singular reality of space/time duality. **Precessional waves are instant (spatial) reality of both—space and time (relative space), which is instant double space of no time in between. Double space is not only empty, but dense space (mass), just as double time of no space in between is not only empty passage of time, but energy.** Precessional wave is instant duality of no difference, where space and time are not opposites, but same reality. Precessional wave is eigenvalue or resonance between two waves, which is perfect wave of two imperfect (distorted) waves and equally vice versa. Precessional wave is surface line of symmetry of duality, which breaks their symmetry in depth. It is perfect inertionless duality of no inertion in between, which is real here and now in quantum ¼ periods. Precessional wave are 4 sides of surface reality here and now (quantum boundary). It reconciles two equal and opposite fractal ½ differences (and/or 4x1/4 differences) into a singular whole. Precessional wave is singular reality of relative/uncertain space/time curvature. In context of precession, relativity and uncertainty are not fractal opposites, but same reality of no difference. Precessional wave reconciles fractal difference between equal opposites of space and time. Precession is therefore double reality. It is double space of no time (mass) and/or double time of no spatial reality (energy), where double time in spatial context is double space in context of time (within/without difference). Mass is instant overlap between two spaces and energy is overlap between two times (and/or vice versa). Mass/energy are instant precessional curvatures of space/time. They are spatial and time reality of precessional space/time curvatures or instant duality.

Double precessional curvature is eigenvalue of two (double/inverse) precessional waves, which reconciles double fractal difference. It is not density of space and/or time but density of mass and/or energy, which is their gravity/antigravity. 2D precessional wave is 4x1/4 reality, because it does not reconcile only space and time (relative space, which is 1D fractal difference), but also double/inverse space/time, which are mass and energy of empty space/time. It reconciles 1D fractal difference between mass and energy, when they are not opposites, but same inertial reality (as here and now). Gravity and antigravity are instant double/inverse densities of mass/energy with no space and/or time in between. Gravitational (2D) precessional wave is 4 corners of surface reality, which are their singular (changeless) boundaries, just as space/time precessional waves are its 4 linear sides that are constant passage of time in between two corners. 1D (space/time) precessional wave inverts time difference between two spatial realities in direction without, which is constant passage of time of such mass, while 2D (gravitational) precessional wave inverts spatial reality between two times in direction within, which is its constant gravity. Any difference less than ¼ curvature (Π/2) is not spatial reality of time, but its constant linear passage. Only corners, which are 90° points of inversion (instant reality or overlap of both—double and inverse linear directions of time), are instant spatial reality of time (relative space), which are instant mass/energy of space/time and/or instant gravity of mass/energy.

Precessional wave is gradual (fractal) curvature of instant space here and now, which is only its constant passage of time and/or gradual spatial change. **Precessional wave is time wave** of inertial space (universe), just as inverse precessional wave is spatial wave of inertial time (micro-world) here and now. Precessional wave is only empty space and/or constant passage of time in quanta less than ¼ curvature. Only ¼ curvature is perfect duality of no inertion in between or instant spatial reality of time (reality of relative space) and/or instant time of no space in between in context of micro-world, which is singular quantum of energy (motion). Quantum points (4 corners) are instant density of empty (relative) space in context of no time, which is singular quantum of mass (size). **Double precessional wave is gravitational wave** of inertial mass (and/or energy in context of micro-world). Gravitational wave is gradual curvature of time, which is gradual change of relative spatial reality (4D) and/or its constant passage of time. 90° of gravitational wave is instant quantum change of time, which is double quantum change of space. It is singular quantum of time change and/or infinity of quanta of spatial change in no time. Time wave is instant change of size/motion, while gravitational wave is instant change of density of such size/motion. Time wave and gravitational wave (1D and 2D precessional curvatures) are 1D fractal difference; therefore they are fractal 3/2 and/or fractal ¾ reality of each other. Here and now, such waves are 1D/2D fractal difference, which is fractal 5/4 and/or fractal 6/5 reality.

3rd half of duality (and/or 5th quarter of double duality) is its singularity (1D fractal difference), which is flattening of round and/or rounding of flat curvature. It is contracted or expanded space or time of its angular duality. 3rd half and 2x1/2 (duality) are 90° angular difference. Angular ½ is overlap (eigenvalue) between beginning and end of two successive waves, which is depth or density of surface duality. It is end of one and beginning of another sequence, which is double reality of its preceding/succeeding inertion/inversion. Such singular overlap is perpetuity that generates infinite 2x1/2 repetition. It is inertial background of infinite wave (curvature). Inertial overlap is +½ of preceding and -½ of succeeding wave (curvature), therefore ½ in context of one (i.e. action) and 3/2 in context of equal and opposite wave (i.e. reaction) and/or vice versa, because inertion (action) is in-built in its inversion (reaction)—i.e. singular space is singular quantum of time and equally vice versa. Wave is real only including its inertion. Each preceding wave is inertion of its successive wave or curvature, which is infinite fractal ½ curvature—½ within/without 3/2 within/without 5/2 etc. It is infinite multiplication/division of fractal ½ here and now, where multiplication in space is division in time and equally vice versa. 2x½ is cause/effect of 4x¼ (double duality), which is fractal surface of 3D (6x1/6). +/- ½ of fractal 2x1/2 and/or +/- ¼ of fractal 4x1/4 are relativity/uncertainty between equal and opposite waves (curvatures), where relativity of one is uncertainty of another.

¼ overlap is relativity of 4x1/4 wave, which is fractal ¾ within (passage of time as fractal difference) and 5/4 without (instant time or its spatial reality as fractal difference). 4th quarter is time of ¾, while 5th quarter is space of following (inverse) ¾. ¾ within is 5/4 without and 4/3 within is 4/5 without. 5th quarter (fractal difference or inertion of duality) is end of one and beginning of another wave or curvature. It is space of one and time of another or attraction of one and repulsion of another, because positive difference in context of one is negative difference in context of its equal opposite. Fractal ½ and/or 3/2, as well as ¾ and/or 5/4, are double/inverse time of inertial space or double/inverse space of inertial time in between. They are relative/uncertain space of inertial space here and now, which are inversely relative/uncertain times. ½ is exclusive spatial and/or time relativity/uncertainty, while ¼ is singular relativity/uncertainty (no difference between inertion/inversion), which is space/time or gravitational relativity/uncertainty. Gravitational relativity is double/inverse relativity, which is simultaneous relativity and uncertainty. It is simultaneous relativity of space and time.

Inversion of singular ½ is 2 and inversion of singular ¼ is 4. ½ is singular quantum of surface duality, while ¼ is singular quantum of 4D or singular quantum of relative/uncertain reality. ¼ is singular quantum of space in context of infinite time and/or singular quantum of time in context of infinite space. ¼ is eigenvalue of 4D space/time (relative/uncertain) reality, while ½ is singular quantum of their space and/or time, which are gravity/antigravity of space/time (mass/energy, which are double/inverse space/time). ½ is singular quantum of singular curvature (Π) or 1D fractal difference, while ¼ is singular quantum of double curvature (2Π) or 2D fractal difference, where singular fractal difference in space is double difference in time and equally vice versa. ½ x Π=90° is singular quantum of fractal reality. **Fractal (precessional) curvature, less than Π/2 is not instant eigenvalue here and now, but only gradual accumulation of tension (passage of time) in depth of surface reality (De Broglie).**

¼ (90° of curvature) is singular quantum of surface reality in its depth and equally vice versa. It is singular quantum of fractal reality (1D or light speed difference), which is real here and now. It is singular quantum of time of inertial space, as well as singular quantum of space of inertial time or singular quantum of gravity of inertial space/time. Inertial reality here and now is 2x1/2, which is particle/wave duality (where particle is overlap between double/inverse waves and wave is overlap between two particles) or size in rotation (where size is overlap between two motions and motion is overlap between two sizes). It is duality between inertial here and now and its physical reality. 2x1/2 is size/motion duality, while 4x1/4 is its relative/uncertain reality, which is double/inverse size and motion. 4x1/4 is relative/uncertain reality, which is real here and now (2x1/2) only in quantum ¼ periods. Inertial 2x1/2 wave (i.e. space) is in context of its time gradually curved in angular direction. Such gradual curvature is not real, unless accumulated gradual curvature is 90°, which is instant 2x1/2 change of gradual 2x1/2 change. 2x1/2 is exclusive gradual or quantum change, while 4x1/4 is duality of gradual/instant change. 2x1/2 is exclusive relativity or uncertainty, while 4x1/4 is both. It is instant inversion of gradual inertion, which is double gradual inertion (equally vice versa). Likewise 4x1/4 (double) duality is gradually curved in angular direction, which is instant inversion of gradual inertion. 2x1/2 is 1D, 4x1/4 is 2D and 6x1/6 is 3D wave (duality). 3D wave is rotation (curvature) of 2D wave, just as 2D wave is rotation of 1D wave and 1D wave is rotation of linear direction, which is no wave. Only points of instant inversion (90°) are real in context of preceding wave and equally vice versa.

Gradual accumulation (change) in context of inverse reality, less than singular quantum (¼), is not instantly real in context of its inertion and equally vice versa. Delay between spatial action and reaction (accumulation of fractal difference) is time between such double/inverse spatial realities, while delay between action and reaction in time is instant spatial size in between. ¼ is singular quantum of relative and/or uncertain (fractal) reality. It is singular spatial quantum of micro-world (dense mass) and/or singular quantum of time of the universe (empty space). It is singular quantum of instant reality, which is twice larger/smaller in context of ½ (inverse) gradual reality, which is constant passage of time. ¼ is singular quantum of spatial reality of time (relativity) and/or time reality of space (uncertainty). ¼ is quantum of energy in context of inertial mass as well as quantum of mass in context of inertial energy. 4 inversion points of 4x1/4 wave and/or 4x90° of curvature are points of instant change of direction. In perspective of succeeding/preceding inverse points, point in between is infinite double/inverse distortion of their inertial reality and equally vice versa. Point in between two points is infinite/singular point of instant inversion of inertion, which is double inertion. Only such quantum points of dense mass are real in context of empty space and equally vice versa. Density that is different than perfect double duality (not 90° curvature) is empty space of the universe/atom.

Time, which is less than ¼ curvature is not spatial (relative) reality but only empty passage of time. Likewise—gravity, which is less than ¼ curvature is changeless constant. Only ¼ of gravitational curvature (wave) is double/inverse reality, when gravity is not singular, but linear constant—i.e. not changeless constant but singular quantum of its change, which is its time reality (gravity as accelerated time or gravity as constant passage, which is instant double/inverse spatial change here and now). **¼ is singular quantum of fractal reality.** Any reality in between is only gradual accumulation of fractal difference, which is less than whole (singular quantum), because it is less than inertion here and now;

therefore not its eigenvalue. Gradual change in context of double/inverse time (relative/uncertain reality) is real here and now only in ¼ quanta. Only ¼ of spatial curvature is real in inertial time and only ¼ of time curvature is reality of inertial space. Gradual spatial relativity (universe), which is constant passage of time, is interrupted by quantum changes of time in ¼ periods, just as gradual spatial uncertainty (atom) is interrupted by quantum spatial changes (spectral lines or frequencies, which are singular quanta of its E/m) in ¼ periods. Eigenvalue of space/time (fractal) reality is always ¼ of their precessional curvature, which is its inertion (here and now). Space/time is fractal ¾, which is reaction of ¼ here and now. It is 3 fractal parts, which are 3D space in context of ¼ fractal part, which is its time and equally vice versa. Singular quantum of inertion is its inversion and equally vice versa. Singular quantum of duality (2D) is ½ and singular quantum of double duality, which is singularity between inertion and inversion (i.e. gravity of space/time) is ¼.

Only here and now is gradual spatial change, which is its constant passage of time. Such time is not time, but space of its double/inverse (fractal) reality. Such time is not constant passage, but instant space (density) of its fractal reality. Reality, which is double fractal difference, is constant (changeless) gravity here and now, which is constant passage of time in ¼ quantum intervals and/or instant spatial reality in 1/16 intervals. Time is spatial (relative) reality in ¼ intervals and gravity is spatial (double relative reality or 5D space) in twice ¼ intervals. Time is singular relative/uncertain space and gravity is double relative/uncertain space (infinite variation of 1 and/or 2 directions of inertial 3D). ¼ intervals are begging/end of constant passage of time. They are 4 inversion points of 4x1/4 reality, when relative/uncertain reality is not only different/distant universe/micro-world, but instant reality here and now, which is instant inversion of inertial here and now that creates double inertion (reversal or instant reaction). Only quantum points of inversion of fractal (different/distant) reality are real here and now. Such instant inversion is instant change of space/time sequence, which is real here and now. It is instant change of sequence of physical reality, which is also inverse instant change of inertial here and now. It is instant transition between two preceding/succeeding (inertial/inverse) space/time sequences (possibilities).

Quantum change is instant (inertionless) duality—it is eigenvalue between inertion and its space/time boundaries, when there is no difference between here and now and all its infinite space and time. It is here and now in context of all its space and all its time. It is 3D/4D eigenvalue or eigenvalue between here and now and its relative/uncertain reality. Double quantum change is instant eigenvalue not only between space and time, but also their gravity (which is fractal boundary—infinity/singularity of space/time), which is 4D/5D eigenvalue. Time is constant linear passage, which is gradual spatial change only in less than ¼ precessional periods, while ¼ points of inversion are accelerated time, which are instant double/inverse spatial change. Such double change is twice larger change in twice faster time, than its infinite gradual change in context of all its constant passage of time. It is instant reversal of time or instant spatial reaction of gradual action in time). Quantum change is singular quantum of relative (and/or uncertain) spatial reality. It is singular fractal of its constant passage of time, which is gravity (attraction/repulsion) here and now. Infinite period of passage of time here and now is only singularity of constant passage of time of relative reality. Infinite period of passage of time is singular spatial quantum, while its instant inversion is singular quantum of spatial reality of time and/or time reality of gravity (relative reality). It is singular quantum of gradual gravitational change, which is inertion (eigenvalue) of space and time. It is change of space/time (double/inverse) sequences of inertial gravity. Double helix is infinite fractal curvature, where each successive/precessional curvature is real only in ¼ periods. Space and time are anti-parallel (horizontal and angular) fractal 2D curvatures, curving in double/inverse linear direction of their gravity.

Each successive precessional sequence is eigenvalue of its previous/future two sequences. Each space/time sequence of inertial gravity (and/or vice versa) is instant incorporation of past into present sequence, which generates future 2x1/2 space/time sequence. It is infinite 2x1/2 positive/negative multiplication/division of inertial gravity (1/2), where each space/time sequence is ½ and each spatial and/or time sequences are equal and opposite ¼ of inertial (gravitational) sequence. Space here and now and relative space are two successive (double/inverse) space/time sequences of inertial gravitational sequence. Space and time (relative space) are fractal 2x1/2 of singular gravitational sequence. 4 space/time sequences are 2 gravitational sequences that are singular 4D sequence of inertial here and now. Each sequence is gradual angular curvature of present sequence (constant passage of time), which is real here and now as its instant boundary in ¼ intervals that interrupt gradual change with instant reversal of all accumulated spatial change (passage of time).

Eigenvalue of duality is instant precessional curvature. Spatial size is singular eigenvalue or discrepancy between massive attraction within and spatial repulsion without. Massive attraction without is linear reaction of singular repulsion within and equally vice versa. Size is difference between repulsion of empty space without and attraction of dense mass within, which are singularity (center of attraction) and equal and opposite linear direction. Linear attraction of mass is reaction of singular repulsion of empty space and/or vice versa—one is equal and opposite fractal curvature of another. Size without is singular eigenvalue of dual attraction/repulsion within. Attraction precedes repulsion and/or vice versa. Size precedes density and motion precedes size. Space precedes gravity and time precedes space (equally vice versa in opposite direction). Spatial size of Earth is precessional (instant) curvature of its gravity (density), while precessional curvature of spatial size is its rotation, which is discrepancy between equal and opposite spatial and time curvatures or instant eigenvalue between size and its density (gravity). Precessional motion is inertion of double/inverse reality—it is gradual curvature of such duality in angular direction, which is instant quantum change in ¼ periods (De Broglie). Precessional motion of rotation is eigenvalue between spatial size and its time (rotation), which is its orbital motion (rotation is real in context of orbital motion only in ¼ quanta).

Likewise—discrepancy or eccentricity between rotational and orbital motion is its precessional motion, which is instant synodic/sidereal time difference. Synodic/sidereal difference is eigenvalue or inertial context of such dual motion—it is time of their spatial realities and/or space of such double/inverse (rotational/orbital) times. Synodic/sidereal difference is fractal difference between space and time or fractal difference between inertial reality here and now and its relative (solar) spatial reality. It is difference between dense mass (Earth) and empty (solar) space. It is spatial reaction of constant passage of time, which is gradual precessional motion of Earth's reality. Gradual precessional motion is accumulation of fractal (synodic/sidereal time) difference, which is constant passage of time of Earth's reality. It is gradual angular curvature of spatial reality here and now, which is its gradual spatial change. ¼ of precessional motion is instant spatial reality of time or instant relative (solar) spatial reality, when time is not constant passage but instant double/inverse space. ¼ precessional motion of Earth is boundary of constant passage of time of inertial spatial sequence. ¼ precessional time is singular sequence of physical reality on surface of Earth.

Gravity is 1D, size 2D and rotating size is 3D. Fractal 3/2 is size/motion or particle/wave duality. Precessional motion of rotation is orbital motion, which is discrepancy or overlap between size and its rotation. Rotation is time of spatial size of Earth and/or instant density of space of orbital motion and equally vice versa. Rotation is time of size and orbital is time of rotation. Time is overlap between double/inverse spatial realities and space is overlap between double/inverse times. Rotation is overlap between two spatial realities—dense mass of Earth's reality and empty space of solar reality, which is orbital motion here and now. Motion of Earth's reality is size of solar (double/inverse) reality. Orbital motion is time only in context of inertial Earth. In context of inertial solar reality—it is its instant spatial density (which is real only in ¼ precessional periods). Space is surface of time and equally vice versa. 3D space is simultaneously only 2D of time, which is relative space. 3rd direction of relative space (instant time) is constant passage of time here and now, which is instant depth or density of relative space. Instant 2D time of 3D spatial reality of Earth are its rotation/orbital motion, while constant passage of time is its precessional motion, which is eigenvalue or 3rd direction of orbital/rotational times. Precessional motion is 3rd direction of spatial reality of time (relative space), which is 4th spatial direction or passage of time here and now.

Precessional motion is constant passage of time, which is gradual increase of tension in depth of relative spatial reality, which is instant reality here and now only in ¼ periods. ¼ period of precessional motion is singular quantum of solar (relative) spatial reality. It is singular quantum of spatial reality of time, which is instant inversion of inertial reality that creates double inertion. ¼ precessional motion is boundary of constant passage of time, because it is its 90° curvature, which is instant change of direction or instant singularity of space/time duality, when space and time are not equal opposites (fractals), but same inertial reality as here and now. ¼ precessional motion is space/time resonance, when spatial and time waves add up creating double/inverse wave. ¼ precessional period is instant spatial reality of motion and/or instant density of empty (solar) space. It is instant spatial reality of time in context of accelerated time (gravity) or instant reality, which is 1D or light speed fractal difference from inertial here and now. It is instant quantum change of uncertain Earth's reality in context of its relative universe.

Orbital motion is motion in context of Earth's reality and/or density of space in context of solar (relative) reality. Orbital motion is fractal surface of size within such motion. Orbital time of Earth's reality is overlap between solar and Earth's (empty and dense) spatial realities. It is time of Earth's reality, which is instant space of solar reality. Rotation/orbital (solar space) are fractal ¾ reality or eigenvalue between inertial and fractal reality, which is relative reality of Earth's inertion and uncertain reality of solar (inverse) inertion. Orbital precedes rotation and precessional motion precedes orbital. Precession is discrepancy (eigenvalue) between rotation and orbital or eigenvalue between Earth's space and time, which is eigenvalue between Earth's and solar space (Earth's orbital motion). Orbital motion is instant solar space, while precessional motion is its rotation (time). Precessional motion (solar rotation) is gradual (fractal) spatial change only in quantum periods (less than ¼). Only ¼ of such motion is instant spatial reality here and now. Fractal rotation of empty solar space, which is less than ¼ curvature, is only gradual increase in tension, which is passage of time here and now. Rotation is 3D, orbital (solar space) 4D and precession (solar rotation) is 5D. Solar space/time duality, which is rotation of solar (empty) space (orbital/precessional motion here and now) is fractal 4/5 reality. Earth's rotation is motion within motion, which is its precessional motion. It is Earth's rotation of in context of solar rotation. Their discrepancy (synodic/sidereal difference) is constant passage of time here and now in ¼ periods of solar rotation.

Illustration 7 - 3D space/time sequences (harmonics)

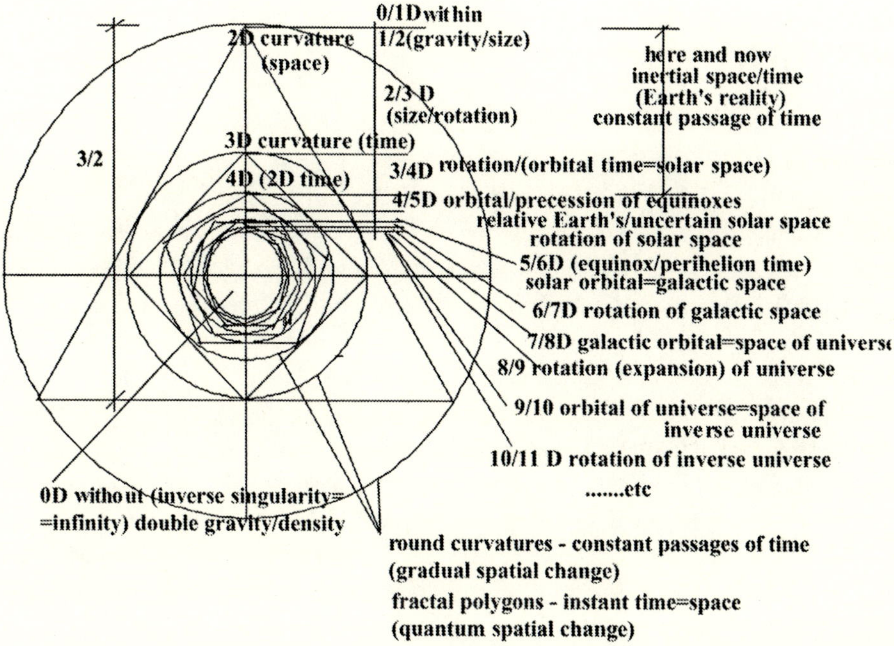

Precessional curvature between orbital (solar space) and precessional motion (solar rotation) is solar orbital (6D) in context of galaxy, which is precessional motion of precessional motion or double precessional motion here and now. Solar orbital is eigenvalue between solar and galactic times, which is instant density of galactic space. Solar space is time (1D fractal difference) and galactic space is gravity (2D fractal difference) here and now, where gravity is accelerated or time of time. Gravity (galactic space) is time of solar space. It is repulsion of empty space, which is equal opposite of attraction of dense mass here and now. Repulsion of empty space is contraction in context of its massive reality and equally vice versa. Gradual repulsion (expansion) of galactic space is instant attraction in context of its massive reality (2D fractal difference). Galactic space (size), which is solar orbital (motion), is precessional curvature of solar rotation, which is fractal 5/6 reality or double (2D) precession here and now.

2D precession (fractal5/6) is reaction of 1D precession (4/5) in context of inertial reality here and now (orbital time, which is fractal 3/4). Fractal 5/6 (double precession or solar orbital) is real here and now (inertial Earth) in 20x¼ quantum periods of Earth's reality, which are 5 precessional quanta (eigenvalue of fractal 5/4 is 5x4=20 fractal curvatures of one whole). Double/inverse quanta of inertial duality (2) are ½ within and 20 without. Precessional motion of solar reality, which is eigenvalue between its rotation/orbital (solar time/galactic space), is 3D precessional motion here and now, which is constant passage of time of solar reality in its ¼ quantum periods. Precessional motion of solar reality is galactic rotation (fractal 6/7) and precession of galactic rotation is galactic orbital (fractal 7/8), which is instant density of spatial universe. Precessional (fractal 8/9) curvature of inertial size of universe (which is galactic orbital) is its linear expansion, which is double/inverse rotation (inversion between rotation and linear motion, which is 3D or attraction/repulsion inversion). Expansion of spatial universe, which is its double/inverse rotation or rotation in time that is linear spatial motion here and now is boundary of inertial 3D here and now.

Fractal 8/9 is boundary of inertial 3D, which is 1/9=0.111...fractal difference. Infinite surface (2D boundary) of spatial universe is fractal 10/9=1.111.. (infinite repetitive strand of double helix). It is limitation of unlimited 0.111.. curvature within to only limited surface without (fractal difference 1/10=0.1 or singular sequence of infinite repetition). Fractal 10/9 is boundary between two equal and opposite universes (one is relative or expanding reality of uncertain or contracting reality of another). Expansion of universe here and now is contraction in perspective of such double/inverse universe and equally vice versa—contraction of such double/inverse universe here and now is its expansion, because space of inertial universe is time of inverse universe and space of inverse universe is time of inertial universe. Duality of double/inverse (space/time) universes is 9+9=18D reality, while double duality (gravitational/antigravitational universes) is 18+18=36D reality. Each preceding universe within is uncertain micro-world of its succeeding relative universe without, while surface boundary in between divides its dense mass within from empty space without, which is rotating massive particle (particle/wave duality). ¼ precessional motion of its rotation is its quantum change, because ¼ is infinite time boundary of singular spatial sequence and equally vice versa.

Time of previous curvature (motion) is instant space of successive (precessional) curvature (motion), just as space (size) of one sequence is time (motion) of previous sequence. Earth's rotation is instant space of its orbital motion and orbital time (motion) is instant space (size) of solar rotating reality. Precessional motion is time (rotation) of solar space, therefore time or additional motion of orbital motion. Precessional motion (solar rotation) is real here and now only in ¼ quanta. Precessional motion (solar rotation) is space of its orbital motion, which is instant galactic space, while precessional motion of solar reality is galactic time. Reality is always duality of size/motion (particle/wave) in context of their fractal difference in between, which is its precessional motion that is spatial reality of passage of time (relative spatial reality). Time is real only in its spatial context and equally vice versa. Inertial difference in between is perfect duality (singularity), where space and time are same reality—i.e. time is space (and/or space is time). Such fractal difference is their eigenvalue or inertial context of their double/inverse (dual) reality. Here and now is real only in its relative background, which is infinity of different possibilities of inertial here and now (time of inertial space). Such relative space/time sequences are real only in ¼ quanta of their precessional time. In context of all passage of time, here and now passes through all such relative space/time sequences (Kepler), which is instant spatial reality of universe in context of no time (instant density). Fractal difference is dual (inertial/inverse) surface between two 3D's and/or double 3D between two surfaces. It is equal and opposite angular (90°) curvature of space and time. Space, time and inertial here and now is 3 surfaces or ½ of a cube, which are 3 coordinates of any reality that is its content (E/m). Content is inversion of such formal inertion, which is duality of double/inverse 3D sequence (2x3=6) or rotating particle in inertial context of its precessional (inertial) curvature.

Orbital is time of rotation, while rotation is space of orbital. Orbital is time only in context of inertial Earth. It is instant spatial density of solar reality, which is 1D fractal difference, while precessional motion is its time. Precessional motion is double orbital motion, which is double (accelerated) or time of time. Precessional motion is time in which time here and now is instant spatial (relative) reality. Precessional motion, which is double time, is again only spatial reality of its (2D) precessional motion that is solar orbital in context of galaxy. Precessional curvature is infinite fractal curvature, because every duality is real only in context of its inertion, which is 3rd direction of such duality. 3rd direction is equal opposite of 2nd, which is equal opposite of 1st direction. Equal opposite of 3rd direction is 4th direction etc, where every previous direction is duality of present/past sequence. 0 and 1st are cause of 2nd; 2nd and 1st are cause of 3rd; 3rd and 2nd are cause of 4th direction etc. Precessional motion is infinite fractal curvature, which is one strand of double helix. Double helix of space/time is double/inverse (anti-parallel) precessional curvature, which is infinite Fibonacci sequencing of space/time. Points of inversion, which create double inertion, are points of intersection of such curvatures on flat 2D surface (in 3D context it is changeless reality, which is always same distance/difference apart).

Each strand of double helix is infinite repetition of singular fractal difference ($1.111...=10^{0}+10^{-1}+10^{-2}+...$where inverse or negative fractal difference in space is inertial or positive fractal difference in time and equally vice versa) in equal and opposite directions, which is infinitely differentiated matrix of $1.111/(1/1.111)=1.111^2=1.234567...$, which is physical reality here and now. 1.111...is numerical formulation of infinite fractal reality (1.111 is difference between diagonal of 3D cube and diagonal of 2D square). 1/1.111..=0.9, which is 1/10 fractal difference. Physical reality is infinite repetition of 1/10 fractal difference in ¼ intervals (quanta). Double duality of 4x1/4 reality is 10/4 or fractal 5/2 reality, which is duality of both—½ and 3/2 fractal reality. It is 3x2=6 in context of 2x2=4 (10 planets/electrons), which is inertial 3D here and now. Fractal 5/2 is eigenvalue of space and time in their inertial (gravitational) context, while fractal 3/2 is eigenvalue of space in context of time and/or equally vice versa. Fractal 5/2 is supersymmetry or duality of Euclidean and non-Euclidean curvatures. It is instant reality (eigenvalue) of two equal and opposite fractal 3/2 realities in context of inertial ½ here and now (2x3/2=6/2–½, which is double/inverse surface or instant duality of two equal and opposite 3D—i.e. it is here and now, which is inertial difference of any duality). Reality is always fractal, because it is infinite vibration of fractal ½, which is difference of any fractal perspective. There is ½ remainder of any fractal space/time sequence, which is time (equal opposite) of such spatial sequence and/or space of time sequence. ½ is relativity/uncertainty of any space/time (dual) sequence, which is instant inversion of gradual inertion and/or gradual inversion of instant inertion in ¼ intervals. ¼ intervals are double inertion or successive Fibonacci sequence of physical reality here and now (where successive in time is precessional in space and vice versa).

Every odd curvature is fractal of even curvature and equally vice versa. Full curvature of space is fractal 1+0.5=1.5 curvature of time and equally vice versa. Full curvature of space and time is double duality, which is 2+0.5=2.5. 4x1/4 wave of double duality (gravity) is 4x2.5=10, while 4x1/4 wave of space/time duality is 4x1.5=6. Fractal reality in between is fractal 10/6=5/3 reality, which is time of such double/inverse spaces and/or space of double/inverse times. 10/2.5=4 instant points of inversion, while 10/1.5=6.666...or 6x1.111, which is infinite 6-fold repetition of singular here and now. 4x1/4 (5/2) is finite curvature of infinite 2x1/2 (3/2). It is its boundary (equally vice versa). Fractal ¼ is infinite/singular boundary of space/time (½)—it is beginning/end or singular curvature (limitation) of any infinite spatial size and/or infinite passage of time. Limitation of infinite reality is its quantization, which is equal opposite or instant (singular) inversion of infinite gradual change. Quantum reality is instant self-interference of equal and opposite gradual self-interference in periods of constant passage of time. Instant in space is gradual in time, while instant in time is gradual in space, because space and time are double/inverse or fractal reality—one is surface of 3D of another (one is angular curvature of another).

Reaction is always larger than its action, because it is inclusive of inertion, where larger in space is smaller in time and equally vice versa. Reaction=action+inertion, where inertion is discrepancy between perfect duality of action and reaction—it is eccentricity of spatial curvature and/or delay of time curvature. Infinitely large discrepancy between action and reaction is inertial reality of singular here and now, which is instant space in context of passage of time. Boundary of instant space here and now is infinite time of the universe, which is its complete change, while boundary of passage of time is infinite spatial size of the universe, which is its complete change (inversion). Here and now is discrepancy (inertion or overlap) between space and time, which are two different spatial or time realities (light speed difference in between), just as space and time are double/inverse discrepancies between right here and right now. Infinite space of universe is now, because it is not here, just as its infinite time is here, because it is not now. Here and now is relative/uncertain difference between double/inverse spaces—it is double or linear/rotational light speed difference, which is space/time difference. Overlap (eigenvalue) between two equal opposites is inertion of such duality, which is their fractal (partial) difference. It is simultaneous fractal reality of two equal and opposite (whole) curvatures. Inertion is instant or quantum reality of its double/inverse or space/time reality, because it is gravity of both, which transcends horizontal differentiation between space and time.

Inertion is +/-1D fractal difference of inertial space and/or time; therefore it is quantum reality of equal and opposite gradual changes of both. Inertion is eigenvalue between inertial and relative space (time), which are gradual spatial change in context of constant passage of time and/or gradual time change in context of gravity, which is its constant passage of time. Inertial space here and now is quantum of relative space, just as relative space (time) is quantum of inertial space. Now is singular quantum of infinite here and here is singular quantum of infinite now. Relative/inertial spaces are double/inverse quantum realities of inertial here and now. Relative space is quantum spatial reality of time (uncertain time), while inertial space is quantum time reality of space (uncertain space). Double/inverse quantum reality, which is quantum relativity (i.e. space) without and/or quantum uncertainty within (i.e. mass) are equal and opposite 4x1/4 curvatures. They are two equal and opposite (sin/cos) waves, which interrupt each other's reality in ¼ periods. Gradual spatial change (gradual curvature in context of constant passage of time) is only singular point of quantum inversion in context of its time (relative space), just as gradual change of time in context of its gravity is only singular point in context of inertial space. Linear direction is only singular point of its double/inverse reality and equally vice versa.

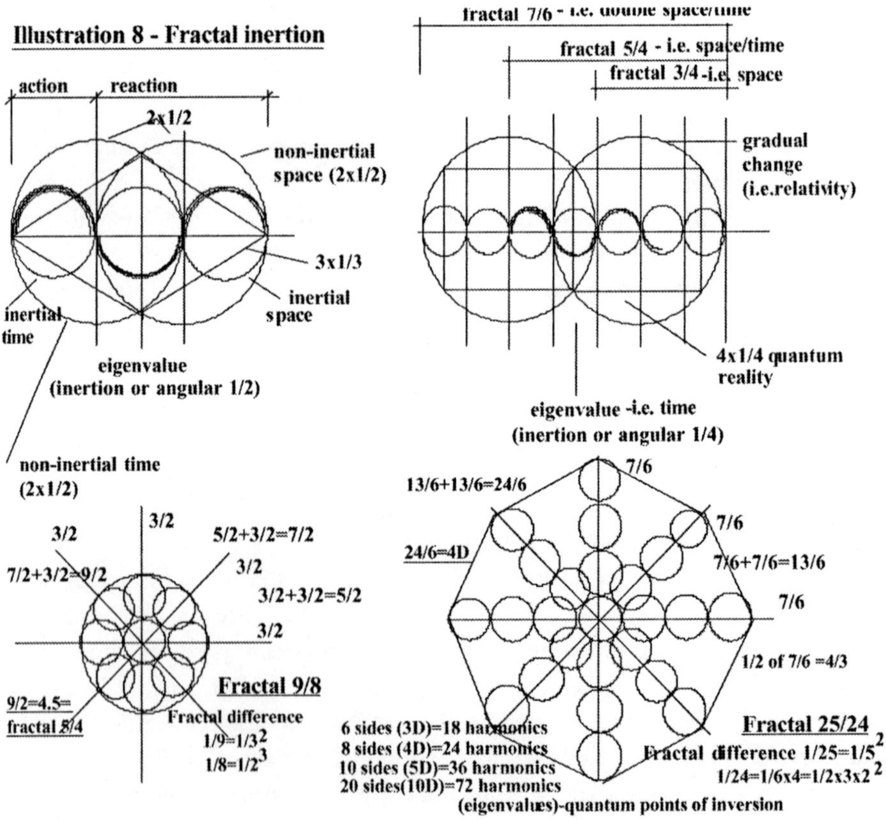

Illustration 8 - Fractal inertion

Linear direction in inertial context is only singular point in double/inverse (relative/uncertain) context. Linear passage of time in context of inertial space here and now is singular point of relative spatial reality and equally vice versa. 4

instant inversion points (quantum change) of 4x1/4 inertial wave are not points, but linear directions (passage of time) of double/inverse wave and equally vice versa—linear direction (passage of time) is only singular point of its relative reality. Gradual spatial change here and now is only 4 instant (quantum) points in context of its fractal time, because only gradual accumulation, which is ¼ or 90° are instantly real in its context. Rest is only gradual (fractal) accumulation of tension in depth of such surface reality, which is its empty space.

4 instant inversion points of 4x1/4 spatial wave are not instant reality in context of time, but gradual 90° curvature (see illustration 8), because instant in space is gradual (passage of time) in context of its fractal time (relative space), just as instant in time is gradual (passage of time) in context of fractal space. Quantum change is instant only in context of inertial time here and now. In context of spatial reality of time, such instant change is not instant but infinitely long period of constant passage of time, which is inertial (instant) space here and now. Such double quantum reality is double 4x1/4 or fractal $8/9=2^3/3^2$ reality (inertial difference 1/8 and/or 1/9). Double 4x1/4 is sin wave, which is 4 gradual periods in between 5 quantum points, which are its instant inversion that is double inertion, where difference between 4x1/4 gradual periods and its instant 4 points of inversion is only relative/uncertain (fractal) difference. 4 instant spatial points are wave in time, just as 4 instant points in time are spatial wave. 5th quantum point is 5th quarter, which is inertion of such reality or perpetuity of such wave (point in time is linear direction in space etc). Quantum point (no time) is gradual reality or passage of time in context of its fractal reality. Quantum (instant) spatial change is infinitely long period of time in context of relative/uncertain space (double/inverse time) and equally vice versa—quantum change of time is infinitely long period of time in between in context of spatial reality.

Time is only fractal reality of space and vice versa—space is fractal of time. All infinite passage of time in context of one is no time but instant spatial density in context of another. Linear direction (time) in context of one is singular point (gravity) in context of another. Such double 4x1/4 or double fractal 5/4 reality is 4 gradual periods interrupted by 5 quantum changes and/or 5 gradual periods (passage of time) interrupted by 4 quantum changes. It is 2x5/2x4=10/8-1/8=9/8, where 1/8 is its inertion or eigenvalue (overlap or discrepancy which is simultaneous end of one and beginning of another period). Such two waves are 90° angular difference in between—one is sinus and another one is cosines wave. Double duality of sin/cos waves intertwined around linear direction of their gravity is spiraling fractal curvature of double helix, which adds new sequence in 90° intervals. Cos wave is gradual angular rotation (curvature) of sin wave and equally vice versa. Cos wave is inertion of sin wave and equally vice versa—one is time of spatial reality of another. They are eigenvalue only in ¼ periods (90° difference, which is 9x10 or 3^2x10 fractal difference. Duality of sin/cos wave is 3D wave, which is 2D sin wave, gradually curved in angular direction, which is its cos wave.

3D wave is fractal 6x1/6 or 6 quantum points of inversion, which is 60° or 6x10= 2x3x10 fractal difference, where 10 in space is 1/10 in time and equally vice versa (spatial inertion is time inversion and vice versa). Cos wave is precessional curvature of sin wave and equally vice versa. Each additional precessional wave is 90°-fractal difference from previous wave, which is additional fractal duality. 2x1/2 is preceded by 4x1/4, which is preceded by 6x1/6, which is preceded by 8x1/8 or 4D wave. Each wave in between its successive/preceding waves is their fractal (double/inverse) reality and each duality is fractal duality of double/inverse duality etc, which is infinite 2x1/2 spiraling. Precessional wave in space is successive wave in time and equally vice versa. Each wave in between is fractal time of double/inverse space and/or fractal space of double/inverse time. Each wave (curvature) in between is quantum reality of its double/inverse or two equal and opposite precessional waves and equally vice versa. **Space is quantum reality of its double/inverse time and time is quantum reality of its double/ inverse space.**

Largest distance is smallest difference and smallest distance is largest (angular) difference. Largest distance is no difference and largest difference is no distance. Largest spatial change is no time change and largest time change is no spatial change. Infinite spatial change is change in context of singular or no time, which is its quantum change (i.e. universe). Infinite time change is change in context of singular (no) space, which is its quantum change (i.e. atom). Quantum change is complete (whole) and not only fractal (gradual) change. It is complete inversion or reversal of preceding period of constant passage of time. 4 quantum changes are full curvature. Because each quantum change is complete reversal in context of inertionless duality (eigenvalue) it is largest change, which is inversion of inertion. Preceding and/or following periods of constant passage of time are its positive/negative fractals, which are infinitely smaller changes. Infinity of numbers larger than 1 are only inverse fractals in between 0 and 1 (inertion and its inversion), because fractal time is multiple space and equally vice versa. Quantum (instant) change in context of no time and gradual change in context of passage of time are 2x1/2 or two equal opposites—one is inertial context of another. Quantum change is reaction of gradual change and gradual change is reaction of quantum change. Quantum change is largest gradual change (boundary) of inertial reality, just as gradual change is largest quantum change of its double/inverse reality.

Infinite number of quantum changes of an atom (mass) is singular fractal of gradual change of universe (space), just as infinite number of quantum changes of universe is singular fractal of gradual change of an atom, which are double/ inverse densities of inertial density here and now. Quantum change is surface between two periods of gradual change, just as gradual change is surface between two instant (quantum) changes, because instant in space is passage of time in time (relative space), just as instant in time is passage of time in space. Surface between two 3D's and/or 3D between two surfaces is light speed or fractal difference, which is inversion of inertion (instant turning inside out). Following period of constant passage of time is only gradual (fractal) reversal of such instant change, which leads to yet another (equal and opposite) quantum change in ¼ interval of its precessional motion. Gradual change or constant passage of time and its double/inverse quantum changes are diametrically opposite vectors of time. If quantum change is in future direction, gradual change is in past direction and equally vice versa. Likewise—time directions of two gradual periods of time are positive/negative or past/future time directions, which are infinite vibrations of exponentially smaller fractals, where smaller spatial fractal is larger time fractal and equally vice versa.

Quantum change is instant inversion between round and flat (non-Euclidean and Euclidean) space/time curvatures. It is instant inversion from gradual round curvature into 4x1/4 square (see illustration 7). Each further fractal difference is gradual deterioration of a square into progressively larger polygons, with larger number of sizes, because fractal difference or inertion (relativity) of 4x1/4 wave (fractal 5/4) is 5th quarter, which is inertial 5x1/5 (pentagon); inertion (relativity) of 5x1/5 (fractal 6/5) is 1/6, which is inertial hexagon etc (see illustration 7). Change from round to flat is instant, while inverse change from flat to round is gradual (where round in space is flat in time and equally vice versa). Difference between round (gradual) curvature and 4x1/4 (instant) curvature of a square is 1/ 10 fractal difference or 1.111..., which is instant self-interference of double helix that adds additional space/time sequence (i.e. difference between circumference and square is $\frac{1}{2}x\Pi/\sqrt{2}$=1.111, while double difference is $\Pi^2/2^3$=1.23456...Larger set of circle/square without is $\sqrt{8}x\Pi/4$=2.222...and smaller set within is $\sqrt{2}x\Pi$=4.444, which is infinite multiplication/division of inertial 1/10 fractal difference without and within). Such gradual difference in time is instant spatial difference and equally vice versa.

Infinite precessional curvature is pentagon without a square, hexagon without a pentagon etc, where without in space is within in time. It is fractal (relative/uncertain) 3/2 reality in context of inertial 4/3, which is relative reality in context of inertial 5/4, which is relative reality in context of inertial 6/5 etc, where relative in space is inertial (non-relative or non-fractal) in time and equally vice versa. Fractal double helix is precession of Euclidean curvature of non-Euclidean reality within, which is Euclidean curvature of larger/smaller non-Euclidean reality within, infinitely many times, where any distinction between Euclidean/non-Euclidean (inertial/fractal or flat/round; relative/uncertain; space/time, dense/empty; size/motion, perfect/imperfect, symmetry/asymmetry etc.) is only relative or comparative duality of two equal opposites. Comparison is possible only in context of its inertion, which is eigenvalue of double/inverse comparison (singular here and now of such duality). Every fractal is inertion (whole) of further fractal, where fractal in space is whole in time and equally vice versa. Double helix of space/time is infinite inertion within/without double/inverse inertion or infinite precessional curvature of gradual (fractal)/whole (quantum) self-interfering inertial reality.

CHAPTER 7

LINE OF SYMMETRY (EIGENVALUE)

Action/reaction is real only in inertial context, which is line of symmetry in between two equal opposites. Line of symmetry is eigenvalue, which is singular reality (whole) of both fractals of action and reaction ($2x\frac{1}{2}=1$). Action and reaction are ½ of duality each, while their inertion is simultaneous reality of both, which is singular reality of such duality. Inertion is double/inverse (square/inverse square) reality of both—action and reaction (space and time). Inertion is 1D fractal difference in between—it is relative/uncertain reality of such duality (double negation that is singular affirmation). Inertion is relative reality of dual action/reaction, which is its uncertain reality within and/or vice versa—it is uncertain reality within of relative action/reaction without. Likewise—in context of inertion—action and reaction are relative/uncertain reality of each other. Reality is C=d/t=const (d-space, t—time), where one is constant (inertion) of its double/inverse variables. Light speed is constant of space/time, which are its relative/uncertain reality; time is constant of space/light speed and space is constant of variable time/light speed. Light speed is eigenvalue of relative/uncertain (fractal) reality—it is energy of space and or mass of time. Light speed is singular reality of dual space and time—it is its mass and energy or angular ½ of horizontal 2x1/2 symmetry.

Inertion is relative/uncertain reality of any surface reality, which is always duality between two equal opposites. Inertion is relative reality without of uncertain reality within—it is space of time and/or time of space, which is light speed difference. Light speed is eigenvalue between space and time or mass and energy (gravity and antigravity). 3D is equivalent triangle between action/reaction and its inertion (light speed) and any change of one is equal and opposite change of its two equal opposites. Singular reality of mass/energy is 3rd ½ of space/time duality, just as singular space/time is 3rd ½ of mass/energy duality. Mass and/or energy are double/inverse eigenvalues of space/time and equally vice versa. Gravity and/or antigravity are double/inverse eigenvalues between singular reality of mass/energy and space/time. Each duality is 1D fractal difference between equal opposites and each successive eigenvalue is further 1D fractal (light speed) difference, which is inertial 3D here and now. 1D fractal difference is gradual change of instant reality and/or instant change of gradual change. Solid surface of mass is instant fractal difference between relatively dense mass without and relatively empty space without, which are double/inverse realities of inertial here and now, which is reality of both (overlap). Solid surface of mass is eigenvalue (line of symmetry) between dense mass and empty space, which are double/inverse spatial realities. Instant solid surface between such inverse spatial densities is their double/inverse time.

Inertion is 3rd (angular) half of horizontal 2x1/2 duality, which is singular reality of such fractals. Singularity is only ½ of its equal opposite. Inertion is singular reality of action/reaction, which is only ½ (action) of its equal and opposite reaction (inverse inertion). Line of symmetry between such larger action/reaction of inertion/inversion is double inertion, which is again only ½ of double inversion etc. Singular quantum of such infinite curvature is 3D sequence of action/reaction=inertion and/or inertion/inversion= double inertion. Action/reaction and/or inertion/inversion (duality) is only relative/uncertain surface of inertial 3D. Only triangulation between 3 equals is 3D, which is again only fractal of larger/smaller 4D, which is surface fractal of larger/smaller 5D, etc. 3D sequence of two equal opposites in context of their inertion, which is 3rd opposite is singular quantum of 3D space/time, which is simultaneous reality of mass and energy in inertial spatial context and/or space and time in inertial context of mass (or mass/energy in context of time and space/time in context of energy). Likewise—it is simultaneous reality of space/time duality in context of mass/energy in context of gravity/antigravity, which is triple duality and/or double 3D or fractal 9/8 reality, which is 1/8 and/or 1/9 inertial difference.

Inertial difference is here and now, which is surface fractal of such reality, just as reality is inverse fractal of here and now. Fractal 9/8 is instant 2x3=6D and equally vice versa—such 6D is only fractal in between 9D/8D. Fractal triple duality is triple density in inertial context of here and now, which is its fractal difference or inversion of such density. Overlap between triple inertial and singular inverse density is fractal ¾ reality, which is instant 3D space in context of time (its gradual change) and/or vice versa. Singular 3D quantum of fractal ¾ reality is 3 relatively different densities of mass (solid, liquid, gas) in context of inertial here and now, which is its equal opposite (energy or empty space of such mass). Fractal difference is always light speed difference. **3D is always double fractal relativity/uncertainty (double light speed difference) in context of its inertion, which is always triple fractal relativity/uncertainty.** 3D space is double fractal time difference, just as double fractal spatial difference is 3D time. Inertial 3D is double/inverse relativity (relativity/uncertainty without/within). 1D fractal difference is time of space and/or space/of time (+/- ¼ of fractal ¾ and/or 5/4 reality, +/- ½ of fractal 3/2 and ½ etc). Relative/uncertain (1D fractal difference) duality of action/reaction and its inertion (fractal 3/2) is cause/effect of larger/smaller fractal ¾, which is cause/effect of larger/smaller fractal 5/4 etc, which is infinite fractal curvature of spiraling double helix, because any horizontal duality (2x1/2) is angular ½ and any horizontal singularity is only ½ of angular 2x1/2 duality. One cause is infinity of fractal effects.

Inversion of inertion (equal and opposite reaction) is succeeding line of symmetry (eigenvalue) of preceding duality of action/reaction, which is addition of past to present sequence, in order to generate future sequence of inertial Fibonacci curvature. Inertial context of duality is its negation—its relative/uncertain reality, which is time of such spatial reality and/or space of such time. Inertial context is angular curvature—it is equal opposite or inversion of inertion, which is infinitely small discrepancy of infinitely large reality. It is mall spatial discrepancy and/or time delay between action and its reaction, which is eccentricity of perfect curvature and where such eccentricity is preceding/succeeding perfect curvature. Eccentricity (discrepancy) is fractal in between two perfect curvatures, just as perfect curvature is fractal between two equal and opposite eccentricities. Perfect spatial curvature is infinite eccentricity or double/inverse linear motion in time and equally vice versa—infinite distortion of linear direction is curvature. Eccentricity and/or discrepancy is asymmetry of surface symmetry, which is infinitely larger symmetry in angular depth of such surface.

Eccentricity is relativity and/or uncertainty (1D fractal difference) of perfect curvature here and now, just as perfect curvature is double/inverse infi-

nitely larger, equal and opposite eccentricity. Infinitely large spatial reality (universe) is discrepancy or fractal overlap between double/inverse (convex/concave) times and equally vice versa—time is delay between spatial action and reaction, where delay in time is spatial discrepancy (size) in its double/inverse (time) context. Smaller period of time is larger spatial discrepancy and larger period of time is smaller spatial discrepancy. Infinitely small or singular passage of time is infinitely large spatial discrepancy, which is spatial universe here and now, just as infinitely long passage of time is spatial singularity (0D point). In context of all its time, space is singular non-dimensional point, just as all infinite space is singular point of time (relative space, which is infinitely larger space or space that is instant time). In context of its time, any massive particle (inertial gravity) passes through its double/inverse center of gravity (Kepler). In context of all the space and all the time, only reality there is, is here and now.

Spatial eccentricity is synodic/sidereal time difference, which is eccentricity in time. Synodic/sidereal difference or distortion of perfect curvature is inertial context between two equal opposites. Synodic/sidereal difference is singular quantum of precessional curvature in between two equal and opposite curvatures. Larger size in motion is relatively larger, and smaller size is relatively smaller synodic/sidereal discrepancy. Any duality is sidereal/synodic difference in between, which is its inertial context. Synodic/sidereal difference is fractal difference or relativity/uncertainty between two perfect curvatures. Any duality is inertion in between and any two surface curvatures are angular curvature in between, which is its infinitely larger/smaller reality. Angular curvature of surface curvature is only infinitely small synodic/sidereal difference. Synodic/sidereal time difference is all infinite space here and now, while its spatial reflection is all infinite time (where space of time and/or time of space is 1D fractal difference—i.e. successive curvature is time of preceding curvature and equally vice versa). Smaller spatial synodic/sidereal difference is larger period of constant passage of time, while larger size of such difference is smaller uninterrupted passage of time. Smaller time difference reflects larger spatial reality and equally vice versa. Synodic/sidereal difference is line of symmetry between two equal opposites. It is time between double/inverse spaces and/or space between double inverse times (and/or gravity between space and time or mass/energy between space/time, which are double/inverse or inertial and relative space of mass/energy).

Such mass/energy that is synodic/sidereal difference between space and time is not real here and now, unless it is singular ¼ quantum (De Broglie eigenvalue). Less than singular quantum is only constant passage of time in context of empty space. ¼ is point of inversion of inertion, which creates double inertion. It is infinity of singularity or infinite accumulation of such difference, which is instant singular duality—mass and/or energy of empty space and/or time. It is instant energy jump of massive particle or emission of a photon, which is double/inverse massive particle, therefore infinitely small (massless) particle or wave of such particle (boson/fermion or singular/double spin duality). Synodic/sidereal difference is thickness of relative/uncertain (angular) reality, which is real on surface (here and now). Mass/energy is singular (instant) duality between space and time (relative space) and vice versa in inverse context of micro-world, where space/time is instant duality between mass and energy. One is inertion of its equal opposite, which is its inversion. Only infinitely large thickness of relative reality, when relative and inertial space (or time) are perfect 2x1/2 duality is such duality perfect (inertionless). Only infinitely long period in which synodic/sidereal difference repeats itself is singular quantum of spatial reality of time, which is mass/energy of empty space here and now.

Surface of Earth is line of symmetry (eigenvalue) between dense mass and empty space. It is instant synodic/sidereal difference or discrepancy between relative universe without and uncertain mass within. Surface of Earth is uncertain reality of inertial universe without and/or relative (expanding) reality of inertial mass within. Earth is discrepancy or eccentricity (inertion) between two equal and opposite perfect curvatures without and within (time curvature of spatial universe and/or inverse time curvature or spatial curvature of time, which is uncertain mass within). Earth is double/inverse distortion of universe without and its mass within. It is fractal reality between such double/inverse space/time (whole) realities and/or singular whole of two equal and opposite fractal realities. Earth is double/inverse time of spatial universe, just as universe is double/inverse time of massive Earth. Earth is angular (fractal) ½ of 2x1/2 space/mass symmetry of universe/micro-world and equally vice versa. Universe without and atom within are double/inverse realities of inertial surface of Earth here and now and equally vice versa, because there is one fractal between two wholes and one whole between two fractals. One is real only in its double/inverse backgrounds, which are its double negation that is singular affirmation. Inertion is real only in context of its double/inverse boundaries, which are its space and time (relative/uncertain) reality.

Instant synodic/sidereal difference here and now is precessional curvature in context of passage of time, which is infinite repetition of such difference. Difference is constancy of its angular curvature or underlying constant time of variable space and/or vice versa. Size of Earth is synodic/sidereal difference in all its passage of time; therefore instant precessional curvature of its double/inverse reality without/within. It is instant precessional curvature of its instant spatial reality without and/or massive (inverse space) reality within, because it is instant time of both (and/or instant space of their double/inverse times). It is instant surface space of double space within and/or inverse space without, which are inverse/double times. Surface of Earth is instant space/time of equal and opposite gravity of dense mass within and antigravity (repulsion) of empty space without. Space and time on surface of Earth are singular discrepancy of equal and opposite massive/spatial or attractive and repulsive realities without and within. Universe without and micro-world within are double/inverse mirror images of surface of Earth here and now—one is real only in context of another. Surface of Earth is instant time of double spatial reality (mass) within and/or instant space of double time (empty space) without. It is their eigenvalue or line of symmetry, therefore uncertain reality within relative universe and/or relative reality without uncertain mass. Singular space here and now is surface of double space within (mass) and./or surface of inverse space (time) without and equally vice versa. Surface of Earth is singular reality of both—universe and its uncertain mass. It is only eigenvalue between infinite space and time here and now.

Earth is double inertion of its inertial/inverse (relative/uncertain) realities. It is only point, where space and time here and now are simultaneous reality—it is singular point of double/inverse coordinate systems, which are its double/inverse infinite boundaries (space and time of massive Earth here and now). Space/time here and now is meaningless (totally relative/uncertain) reality in perspective of any relatively different density of space/time (different coordinates). Earth is instant precessional curvature of its double/inverse density. It is inertial precessional curvature of gravity of its mass and/or inverse precessional curvature of antigravity (expansion) of its space, where one is always instant reality of gradual reality of another. Gradual expansion of spatial universe without is instant contraction of its mass within. Instant reality does not change gradually, but in quantum steps, because instant is inversion of relative reality—i.e. time of instant reality is space. Gradual expansion of relative spatial universe is gradual half-life of its uncertain mass within, which are equal and opposite instant quantum contractions of spatial universe and/or instant expansion of its mass within. Quantum leaps are infinite period of constant passage of time in between, which are its inverse reality—gradual change in equal and opposite direction.

Quantum leaps are double/inverse boundary of gradual change in context of constant passage of time in between, just as constant passages of time are double/inverse boundaries of singular quantum leap in between. Double/inverse gradual changes (passage of time) are relative/uncertain (fractal) realities of singular quantum change in between, just as double/inverse changes on beginning and end of each period of constant passage of time are its double/inverse fractal reality. Larger uninterrupted periods of passage of time are larger quantum changes on its beginning and end. Instant quantum change of constant passage of time (gradual change) is spatial reality of time—it is singular quantum of relative spatial reality or eigenvalue (space) between double/inverse times. Instant quantum change, which is point of inversion that creates double inertion, is singular quantum of gradual change of relative spatial reality (time). Earth is instant spatial reality of its double/inverse (infinitely faster/slower) passages of time without/within. Spatial size of Earth is instant relative curvature of its uncertain mass within and/or instant uncertain curvature of its relative space without. Size of Earth is instant precessional curvature of its double/inverse (relative/uncertain) space or instant spatial reality of its double/inverse time. It is instant precessional curvature of its gravity/density within and/or its antigravity and emptiness of universe without. It is sidereal/synodic difference between space and time of its mass within and/or empty space without in context of all its time, because their time is instant surface of spatial Earth.

Reaction of linear gravity is surface size, which is its instant precessional curvature, just as linear gravity is instant precessional curvature of its singularity (gravitational center, which is singular nothingness within that causes infinity of something without). Size of Earth is discrepancy between spatial repulsion and massive attraction, just as gravity of massive Earth is discrepancy (instant overlap) between spatial and time sizes. Gravity/size of Earth is fractal ½ reality, where one precedes another (precessional time curvature is successive spatial curvature and equally vice versa). Precessional curvature is relative/uncertain reaction of its previous curvatures. Reaction of spatial size of massive Earth is its rotation, which is synodic/sidereal discrepancy between such equal and opposite spatial sizes of universe and massive size of Earth. Rotation (time) is precessional curvature of size (space), which is inertial discrepancy in between that is gravity/antigravity of its mass/space etc. Rotation of massive Earth is singular fractal 2/3 reality and inertion or precessional curvature of such double/inverse reality is its orbital motion, which is fractal ¾ reality. Fractal difference is always time of inertial space (equally vice versa), which is relative reality without of uncertain reality within. Spatial size of Earth is relative reality of its gravity/density within, which is its uncertain reality, just as rotation is relative reality (time) of spatial size. Orbital motion is relative reality (time) of uncertain rotation within, which is its instant space (while size of Earth is its instant density/gravity).

Space is eigenvalue of double/inverse time. Singular 3D space is simultaneously 2 instant (double/inverse) directions of time (rotational and orbital motion). 2 directions of time are instant spatial reality here and now, which is only surface of infinitely larger (relative) spatial reality—4D space and/or 3D spatial reality of time. Double time or double motion without (empty space) is reaction of instant double time within, which is gravity of massive Earth. Mass is instant duality of space and time—it is double/inverse space in context of singular time and/or double/inverse time in singular spatial context. Two separate directions of time (orbital/rotational motions) in context of empty space is singular direction of accelerated time in opposite direction in inverse context of dense mass within. Double rotation without is singular linear acceleration within. Double time without is reaction of singular accelerated time within. Gravity of dense mass within is instant duality of two time directions of empty space without, just as antigravity of empty space (expansion) is instant duality of two spatial directions within (density of empty space, which is its mass). Orbital and rotational motions are relative/uncertain reality of each other and at the same time, they are singular duality or time of inertial space (fractal ¼ of ¾ reality). Orbital motion is boundary of constant passage of time of inertial space here and now. Precessional motion of orbital motion, which is discrepancy or overlap between rotation and orbital time is fractal reality between constant passage of time here and now and its spatial (instant) reality.

Earth orbits Sun in spatial context. Reaction of such orbital is equal and opposite solar orbital around Earth in context of time. Solar reality is infinitely larger space, which is time or relative space here and now. Rotation/orbital motions are 2, while constant passage of time is 3rd instant (spatial) direction of solar reality. Time here and now is instant spatial density of relative solar reality. Orbital motion is eigenvalue (synodic/sidereal difference) between double/inverse spatial densities of massive Earth and empty space of solar reality within/without. It is constant delay (time) between spatial action and its double/inverse (relative/uncertain) reaction. In context of solar reality—infinite number of orbitals are instant density of its surface size. Orbital is time here and now and also instant spatial size of solar reality. It is 3rd half of inertial 2x1/2, which is time wave created by interference of such two particles. It is also 5th quarter between two equal and opposite dualities—rotation of massive Earth within and rotation of relatively empty solar space without, which is precession of Earth's rotation. Orbital motion is discrepancy (delay) between Earth's and solar (double/inverse) rotations. It is eigenvalue or line of symmetry between constant passage of time here and now (3D space) and relative spatial reality of time (4D), which is fractal ¾ duality of rotation/orbital motion within and/or fractal 4/5 duality of orbital/precessional motion without. It is surface between inertial space/time sequence here and now (constant passage of time) and its double/inverse (relative) sequence, which is inertial solar reality in context of its own time, which is double or accelerated time (gravity) here and now.

Reaction (precession) of orbital motion is equal and opposite orbital of solar reality around Earth in context of time (which is its instant spatial reality). Two massive objects orbit each other. Infinitely smaller object (i.e. Earth) is infinitely large and fast orbital around infinitely larger object (i.e. Sun), while infinitely larger object is infinitely small and slow orbital around smaller object. Small and slow orbital of larger object is eccentricity or synodic/sidereal difference of perfect orbital of smaller object (and equally vice versa). Smaller orbital of larger object is time of larger orbital of smaller object, which is its instant space. Equal and opposite solar orbital around Earth, which is infinitely small and slow, is time of equal and opposite Earth's orbital—it is its boundary. Solar inverse (small) orbital in time is limitation of Earth's inertial (large) spatial orbital—it is time of its spatial reality and/or space of its time (its relative boundary). **Solar inverse orbital around Earth is precessional motion of Earth's rotation**, which is gradual accumulation of synodic/sidereal difference. Sidereal/synodic difference is discrepancy between rotation and orbital motion, which is singular fractal (quantum) of precessional motion.

Time orbital of solar reality is precession of Earth's rotation, which is fractal 4/5 reality. Precessional motion is 3rd direction of dual (rotational/orbital) time. Surface (2D) time is 3D space here and now. Its 3rd direction is depth of such surface time, which is constant passage of time here and now. Constant passage of time is discrepancy between inertial (3D) and relative (4D) space. It is gradual spatial change in its depth, which is change of its density/gravity. Such gradual change in depth is real on surface only in ¼ quanta, when tension in depth of massive Earth is equal to its surface inertion. Time is constant passage here and now and instant (spatial) 3rd direction of solar (relative) reality. Precessional motion is 4x1/4 time wave (passage of time) interrupted by 5 points of inversion (2 are beginning/end and 3 are in between). ¼ precessional period is singular quantum of double/inverse or accelerated passage of time, which is time of relative spatial reality (5th spatial direction). Precessional motion is 4x1/4 time wave, while its inversion points are uninterrupted curvature of double/inverse 4x1/4 spatial wave, which is precessional or time of time. Time of time is space. Such 4x1/4 double/inverse spatial wave is all constant passage of time between two quantum points compressed into such points, while singular points are expanded into a full wave (1D fractal difference). Boundary of constant passage of time here and now is singular ¼ quantum of precessional motion (De Broglie eigenvalue), which is 90° angular curvature; therefore instant inversion of constant passage of time in its relative spatial reality. Constant passage of time is 4x1/4 periods of precessional motion. It is 4 periods of constant passage of time, 6480 years each, which are interrupted by 5 quantum changes in between, which are instant spatial reality of time (instant relative space).

Quantum change of time, which is uncertain (quantum) reality of relative (solar) space, is beginning/end of constant passage of time in between. Quantum change is instant inversion of inertion, which creates double inertion. It is instant and complete reversal of direction and speed of rotation/orbital motion and equal and opposite change of size and gravity here and now. It is inversion of one inertial space/time sequence, which creates double sequence. It is time boundary of inertial spatial sequence, as well as spatial boundary of equal and opposite inertial time sequence. ¼ period of precessional motion is surface boundary between two equal and opposite inertial sequences. It is boundary between two relative spatial sequences, which are 1D (1/10) fractal difference. Inertial time sequence (6480) years is singular quantum of relative spatial reality. It is singular inertial quantum of space/time or singular quantum of constant (changeless) gravity. Infinity of such quantum changes is gradual change of spatial reality of time (gradual change of relative spatial reality) in context of its passage of time, which is inertial gravity here and now.

Every spatial motion of Earth is equal and opposite (time) motion here and now (surface of Earth). Precessional motion of orbital motion is its inversion within. It is its reflection in backward (time) direction, which is precession of Earth's rotation, just as reflection of orbital motion is equal and opposite vertical motion (change of seasons) here and now. 3D space is 2D time. Only two spatial motions reflect instantly on surface of spatial Earth—rotation and orbital motion (horizontal and angular rotations). Precessional motion therefore reflects as double/inverse or precession of rotation (also—precession of orbital, which is perihelion orbital, reflects as double/inverse angular motion, which is precession or gradual shifting of line of ecliptic). Precessional motion is discrepancy or synodic/sidereal spatial difference, which reflects equal and opposite synodic/sidereal time difference between orbital and rotation. Precessional motion is time of both—spatial rotation and orbital motion. It is their line of symmetry, which breaks surface symmetry in its depth. Precessional motion is inertial coordinate system of variable rotation and orbital motion. Spatial reactions of rotation and orbital motion are gradual day/night and seasonal changes. There is no instant spatial reaction of precessional motion, because such motion is less than one quantum (fractal). Instant reaction of precessional motion is passage of time in periods less than singular ¼ quantum. Only ¼ precession is instant spatial reaction here and now, which is not passage of time but infinitely larger spatial change, which is instant reversal of all accumulated gradual spatial change in previous period of passage of time. Gradual spatial change in context of passage of time is only built-up of tension, which is instant quantum change on surface only in ¼ intervals of precessional motion. ¼ interval is eigenvalue between space and time or relative and inertial spatial sequence. It is eigenvalue of two equal and opposite waves that are fractal difference in between.

Rotation and orbital are singular time only in context of inertial space (here and now), which is its eigenvalue (equal opposite). In context of each another, they are equal opposites—one is time of spatial reality of another, because they are relative/uncertain (fractal) reality of each other. If Earth's rotation is spatial, then its orbital is in time and equally vice versa, because context of rotation is inertial spatial sequence (here and now), while inertial context of its orbital is solar spatial reality, which is relative space (time). **Each successive motion of inertial motion in between is precessional motion of its previous motion.** Reaction of orbital motion (inertion) is precessional rotation here and now, which is inverse (time) orbital of solar (empty) space (fractal 4/5). Precessional motion of rotation is rotation of rotation or double (squared) rotation. Rotation of Earth is real only in context of such larger and slower (time) rotation. Earth's rotation is not motion (time) but instant space of precessional rotation. Reaction of precessional rotation is equal and opposite precession of orbital motion, which is fractal 5/6 reality. Precessional motion of orbital motion is orbital of orbital, where orbital is not motion (time) but constant size in context of such larger orbital, which is spatial rotation of solar space. Orbital of orbital or double (squared) orbital is orbital motion of perihelion. Gradual shift of perihelion in context of inertial orbital is gradual rotation of orbital motion, which is double precessional motion of Earth's time or double quantum change here and now. Perihelion motion is time of orbital and precessional motion, which is its instant space; just as precessional motion is time of orbital and rotation (or orbital as time of rotation and spatial reality here and now etc). Orbital and precessional motions are fractal (double/inverse) space in between equal and opposite Earth's rotation within and rotation of solar space (perihelion motion) without.

Distinction between size and motion is only relative/uncertain. Each preceding curvature is instant size of succeeding curvature, which is its gradual motion (change). One size within is infinity of fractal motions without and equally vice versa—one motion is double/inverse infinity of different sizes. In context of inertial size here and now (Earth)—its relative reality without is infinity of motions, which are infinite reaction of singular size. It is fractal 2/3 (rotation) in context of fractal ¾ (orbital motion) in context of precessional rotation (fractal 4/5) in context of precessional orbital motion (fractal 5/6). It is infinite repetition of fractal difference here and now (1.1111), which is eigenvalue of any two preceding/succeeding curvatures. It is infinite repetition, where one is relative reality without of uncertain (quantum) reality of its equal opposite within. Size is relativity of gravity, while gravity (density) is quantum uncertainty of size. Rotation is relativity of size and size is quantum uncertainty of relativity. Orbital is relative reality of rotation, which is its quantum uncertainty within. Relative reality of orbital is precessional rotation and relativity of precessional rotation is precessional orbital (perihelion) motion. One is time of fractal spatial reality of another and equally vice versa. Consequence is that any inertial reality here and now is double quantum uncertainty in context of double spatial relativity.

Inertial here and now is eigenvalue between double/inverse dualities or 5th fractal part between two equal and opposite 2x1/2 multiplication/division. Inertial here and now is 5th ½ of double duality (4/2=2), which is fractal 5/2-reality and/or 5th part of singular 4x1/4 wave (fractal-5/4). Inertial 3D is eigenvalue between singular and double fractal difference. It is instant reality between singular and double relativity without, which is singular/double quantum uncertainty within. Rotation and orbital motion are double quantum uncertainty in context of inertial precessional rotation and equally vice versa—precessional rotation and precessional orbital motion (fractal 4/5 in context of fractal 5/6) are singular/double relativity of inertial orbital motion (time). Time is twice relative space and space is twice uncertain time. Constant passage of time in context of inertial ¼ precessional period are relatively different times in context of rotation, than in context of its orbital motion and equally vice versa—precessional rotation and precessional orbital (perihelion) are relatively different periods of time in context of inertial orbital rotation (inertial time here and now).

Each fractal reality before is relatively different quantum reality of its successive sequence. Quantum periods of rotation and/or orbital motion are relatively larger/smaller periods of constant passage of time in between, which is inertial ¼ period of precessional motion, just as relative periods of rotational and orbital precessions (fractal 4/5 and 5/6) are relatively different number of orbital periods in context of inertial rotation and equally vice versa. Relatively different passage of orbital and/or rotational time in context of inertial precessional period is relative/uncertain difference between rotation and orbital. Smaller inertion is larger inversion. Smaller spatial reality within is larger time (larger fractal difference) without. 3D is double fractal difference of inertial here and now—it is double relativity without, which is double quantum uncertainty within. Larger spatial discrepancy (difference of sizes) is larger period of time and equally vice versa. Relativity of orbital motion is larger than relativity of rotation in context of inertial space here and now, which is larger quantum uncertainty of rotation and smaller uncertainty of orbital motion (larger passage of time in context of orbital motion than passage of time in context of rotation). Eigenvalue or line of symmetry between different relative/uncertain realities of orbital/rotational motions is their precessional motion, which is spatial reality of both. Such double/inverse quantum relativity is inertial reality of no difference in context of their precessional motion. Precessional motion is eigenvalue between orbital and rotational motion, while precessional orbital motion (perihelion orbital) is its equal (fractal) opposite. Rotation/orbital/precessional rotation/precessional orbital are fractal 3/4, where one is always time of inertial 3D spatial reality of remaining 3, which is inertial 4D space/time.

Time of each rotation (1 day) is larger than the previous one for infinitely small amount of time, because of synodic/sidereal asymmetry between Earth's rotation and its orbital (as well as between rotation and size and/or size and its density within). Such infinitely small repetitive difference is gradual change or gradual densification of time, which is gradual angular curvature of constant (linear) passage of time here and now. Such fractal angular curvature of time is instant inversion in ¼ precessional periods. Period of one year (one orbital) is double/inverse repetition of inertial 2x3=6, which is inertial here and now of such reality (rotating Earth). It is 6^2x10^1+6^1x10^0 in context of its ¾ inertion, which is a period of 365.25 days. 6^2x10=360 is perfect spatial orbital of Earth, while 6 is its discrepancy, which is equal and opposite time orbital of solar reality around Earth. 6 days out of 366 is eccentricity or discrepancy of perfect curvature, which is inertionless orbital, without equal and opposite reaction of solar orbital around Earth. 6 additional days is relativity/uncertainty of inertial time here and now, which is double fractal relativity/uncertainty of additional 5 days, which is precessional rotation and/or ¼ day, which is precessional orbital (perihelion) motion. 5+1/4 days are double/inverse eccentricity of perfect 360-day orbital. 5 days is of additional time per year is consequence of synodic/sidereal difference between rotation and orbital motion, which is gradual precessional rotation, while ¼ day is discrepancy between orbital and precessional rotation, which is gradual precessional motion of orbital.

360 is perfect curvature (no eccentricity). It is 400-40, where $4x10^2$=400 is inertion in context of its inversion (1D or 1/10 fractal difference), which is 4x10=40. $4x10^2$ and 4x10 are double/inverse reality of inertial 4D here and now (relative space) in context of infinite 1.111.. repetition of such sequence. Equally vice versa—360+36=396, which is 4 less than perfect 400. 4 (here and now) is relativity/uncertainty between two space/time sequences. 360 is perfect surface duality of no 3rd direction, which is its inertion (gradual angular curvature in depth). 5 ¼ days of additional time is inertial context of such perfect curvature, reflected as double/inverse eccentricity of its orbital (Milankovich cycles). 5-¼ day is fractal 5/4 reality, which repeats in sequences of 20 (common denominator of ¼ and/or 1/5 fractal difference, which is 5x4=20). 5-¼ day is precessional curvature of perfect orbital of 360 days, which is fractal 5/4 reality. 5/4 is fractal difference between inertial and relative space/time sequences. Fractal difference between orbital (solar size) and rotational diameter (size of Earth) is $3x10^8x\Pi/3x4x10^3x\Pi$ = **$¼x10^5$**, which is inverse 4 and exponential 5 (double/inverse) or fractal 5/4 of inertial orbital in between. Instant duality between solar and Earth's realities is fractal 5/4 reality. It is 5+¼ additional time quanta or total of 21/4, which is 20/4=5 in context of its inertion—¼ (De Broglie eigenvalue). $¼x10^5$=1/12 x C (light speed), where 12=3x4 or fractal ¾ reality. 5/4 without is ¾ within.

5 days is synodic/sidereal difference between 360-days inertial orbital of Earth and orbital of orbital, which is rotation of solar (empty) space, while ¼ day is additional orbital of such 5 days difference. 360/5=72, which means that every 72 days of Earth's orbital, there is one additional day created by rotation of solar space. 360x4=1440, which is additional day every 4 years (1440/72=20). There is one additional day for every 72 days of Earth's orbital and one day for every additional 20 solar rotations. Orbital motion is eigenvalue between solar and Earth's double/inverse spatial realities. 1 day per 72 days in context of 360-day orbital is 1° every 72 years in context of spatial 360° curvature of Earth. ¼ of additional precessional rotation is 90x72=6480 years, which is instant quantum change of constant passage of time or instant inversion of present space/time sequence. It is instant reversal of east/west rotation as well as full ½ curvature of gradual angular tilt (inversion of changes of seasons). It is instant inversion between north/south and east/west directions (horizontal rotation and angular tilt). Equal opposite of north/south reversal is fractal shift of orbital motion. 1 day per every 1440 days (4 full curvatures) is 1° every 1440 years here and now, which is precession (time or boundary) of precessional motion. 90x1440=129 600 years, which is singular period of infinitely larger (2D) quantum change. 6480 years is ¼ of 25 920, which is singular time wave (fractal 5/4), while 129 600 is ¼ of 518 400, which is

double time wave or singular gravitational wave (fractal 6/5). 6480 is period of constant passage of time between two equal and opposite quantum changes, while 129 600 is period of constant gravity in between two equal and opposite quantum changes of gravitational sequences.

Size and its density is fractal ½, which is synchronicity of only one out of two directions. Rotation of Earth is fractal 2/3 reality, which is eigenvalue (synchronicity) of 2 out of 3 directions, while its orbital motion is fractal ¾, which is eigenvalue of 3 out of 4 directions. Fractal 4/5 (precession) is eigenvalue of 4 out of 5 directions, which is not only spatial (3D) synchronicity but also synchronicity of its time. It is instant (spatial) reality of time in context of gravity as double (accelerated) passage of time. Fractal 5/6 (double precessional curvature) is eigenvalue of 5 out of 6 directions, which is synchronicity of 4 space/time directions and their gravitational (5th) direction. It is instant (spatial) reality of gravity in context of twice accelerated time, which is repulsion of attraction and/or attraction of repulsion. Smaller fractal discrepancy here and now is larger period of time between quantum changes of such inertial reality. Infinitely small eccentricity is infinitely long passage of time and infinitely large eccentricity is infinitely small passage of time.

Fractal 4/5 is instant duality between double/inverse Earth and solar realities, which is orbiting Earth (time) in context of rotating solar space. Fractal 5/6 is instant duality between double/inverse galactic and solar realities, which is orbiting solar reality in context of rotating galaxy. Fractal 5/6 is singular inertion of 4/5 duality and equally vice versa. Earth is inertial eigenvalue between galactic and solar space/time sequences, which are its double/inverse relative realities without and equally vice versa—here and now is double/inverse quantum uncertainty within solar and galactic reality. Solar reality is eigenvalue between its relative (galactic) sequence without and uncertain (Earth) sequence within and galaxy is eigenvalue between its positive/negative fractal sequences of solar reality within and/or universe without. Time here and now is solar space (instant size) and/or gravity (instant density) of galaxy, while space here and now is solar gravity (density) and/or galactic antigravity (inverse density). Earth/Sun is space/time duality, which is double/inverse sequence of inertial galactic sequence—one is mass of energy or space of time of another. Solar size is twice larger than Earth's size and its motion is twice slower (2x1/2), which is infinitely faster passage of time. Solar passage of time is acceleration of passage of time here and now. 1 precessional period is 1 day of solar reality, while 1 day here and now is no time at all, but instant uncertain space of its mass within. Orbiting planets are its instant spatial surface of no time difference. There is no certainty of their positions in any moment of solar time—they are smeared all over their orbits. Their size/motions are surface shells of instant quantum density in between. Solar reality is an atom, planets are its orbiting electrons and Sun is its nucleus.

Galaxy is inertial 3rd direction of inertial/relative sequences (Earth/solar duality), which is inertial gravity of such double/inverse space/time sequences. Solar reality is relative sequence of inertial here and now and galaxy is relative sequence of inertial solar reality, which is singular/double quantum uncertainty here and now. Galaxy is inertial context of both equal opposites—solar reality and Earth. Galaxy is inertial (singular) ½ of horizontal 2x1/2 duality (Earth/solar reality). It is only ½ of infinitely larger/smaller 2x1/2 duality of galaxy/Moon. Moon is reaction within of galactic action without. Galaxy is gravitational discrepancy between solar reality and Earth in direction without, while Moon is equal and opposite (inverse) discrepancy within. Moon is ¼ of fractal ¾ reality—it is time of spatial reality. Moon is massive attraction within of equal and opposite repulsion of empty space without (attraction of galactic energy). Every infinitely large action without is infinitely small reaction within, because large in inertial context (i.e. space) is small in inverse context (i.e. time). Moon/galaxy is infinitely larger spatial discrepancy from Earth/solar discrepancy, because it is infinitely smaller time discrepancy. Moon and galaxy are double/inverse time within and without inertial space of Earth/solar duality.

If Sun is within than Moon is without Earth—i.e. Earth orbits Sun, while Moon orbits Earth. Moon's orbital is 1D fractal difference from Earth's orbital. Moon and Sun are double/inverse or space/time realities of inertial Earth in between. If direction within (Sun) is time than direction without (Moon) is space and equally vice versa. Gravity of Earth is double/inverse context of such opposite realities—it is fractal attraction of one and equal and opposite fractal repulsion of another. Earth/Sun is equal opposite of Earth/Moon duality in context of inertial gravity, which is their line of symmetry (eigenvalue). **Gravity of Earth x its orbital time is time or round curvature (1/C) in context of inertial Earth and space or flat curvature (C) in inertial solar context**—they are fractal equal opposites. No change (inertion here and now) = Double change/inverse change. It means C=g/(1/t), therefore gravity (double time) is inertial space of solar reality and inverse time is its time. If Earth's orbit is time curvature in space than Moon's orbit is spatial curvature in time (and/or vice versa), because Earth is subordinate of Sun and/or principal of Moon's reality (Balmer). I.e. g x t(r)=852 km/sec, what is total motion generated by Earth's gravity, where t(r) is rotational period of 1 day. Its orbital motion is t(o)=30 km/sec; therefore 852/30=28.4—or Moon's orbital period in days. Moon's orbital period (space) is equal opposite of

Earth's orbital velocity (time). Small discrepancy ($1.6=4^2/10$) in between is inertial here and now of such duality. Perfect (inertionless) duality is 864/30=28.8, which is 2x14.4 and/or 4x7.2 (double and singular inversion of fractal 5/6 and 4/5, which are double/singular precessional or quantum reality here and now).

Orbital motion of Moon is refraction point (instant eigenvalue) between equal and opposite gravity of Sun and Earth. It is point of inversion between Earth and solar space/time curvatures, which balances such equal opposites. Moon is instant mass formed in corner (¼) between double/inverse (space/time) surfaces, just as galaxy is its equal and opposite energy without (¾). Moon and galaxy are double/inverse sides of such surface realities—they are mass within and energy without (which is double/inverse mass) of double/inverse space/time curvature between Earth and Sun. Likewise—Moon and galaxy are space and time of inertial massive reality here and now (Earth/solar duality). In gravitational context, Moon is ½ way between Sun and Earth, because small spatial distance is large angular (time) distance and equally vice versa. Consequently, there is no difference in precessional motion being eigenvalue between solar/galactic duality and inverse eigenvalue between Moon/Sun duality. Moon and galaxy are double/inverse action, which is singular reaction of precessional motion here and now (which is solar rotation in context of galactic space, which is inertial gravity here and now). Precessional motion is double eigenvalue—it is line of symmetry between solar and Earth realities, as well as Moon and galaxy. It is 5th quarter of 4x1/4 wave (inertion).

Moon (and its orbital motion) is instant boundary where gravity of Earth and Sun are equal. Equal gravity is double/inverse space/time, which is reason why is Moon same size as Sun in spatial perspective (they overlap during eclipse)—i.e. spatial distance between Earth and Moon is equal time distance between Moon and Sun and equally vice versa (large spatial distance is small time curvature and small distance is large curvature). Any change of one is equal and opposite change of its double/inverse opposites, which ensures that Moon is always same relative size as Sun, regardless of relative distance, because change of gravity is equal and opposite change of density—it is expansion of one that is equal and opposite contraction of another. Same size in spatial perspective is gravitational eigenvalue between space and time (repulsion/attraction). Moon is always same relative distance, which is inversely proportional to their sizes and motions, not by accident, but by necessity. There is no coincidence in context of double helix, which is infinite double/inverse causal curvature. Causeless reality is not real. If Earth/Moon distance is spatial curvature than Earth/Sun distance is time curvature—one is uncertain quantum reality of another's relative reality (where relative space is uncertain time and equally vice versa).

Small spatial distance is large time distance and equally vice versa, where space/time differentiation is meaningless in inertial (gravitational) 3D context. Such double/inverse space/time infinity/singularity is ½-way (eigenvalue) between gravitational equal opposites, in the same way that Earth, which is singular proximity to Sun and infinitely larger distance from its outer boundary is ½-way between gravitational equal opposites of solar system; or same as solar position in galaxy—right at its edge, which is singular proximity to its dense core and infinite distance from its outer boundaries (expanding empty space of universe). Singular proximity of inertion is infinitely large distance of inversion, which is ½-way in context of their double inertion. Infinitely far from one and infinitely close to equal and opposite polarity is unequal equality, because infinitely small spatial distance is infinitely large time distance and equally vice versa, where space/time difference is difference between within and without (inside/outside inertial line of symmetry). 3D is not exclusivity of spatial reality—it is inclusivity of two equal opposites in 3rd direction, which is space/time duality in gravitational context. Space and time are only fractal surfaces of each other in context of their 3D gravity. Surface differentiation of sizes or motions is meaningless in context of gravity—only instant duality of both is inertial (gravitational) 3D.

3D is infinite fractal double curvature, where any surface (comparative) reality is totally relative/uncertain in its depth. Physical reality is instant duality (surface) in double/inverse curvature of inertial here and now. Any dual differentiation (i.e. size/motion, mass/energy etc) is only its surface reality, without meaning in 3D context. It is same fractal curvature, regardless of relative sizes, motions or densities. Just as planets or electrons' space/time distances and their sizes, motions and densities are determined by fractal curvature of double helix—so is Moon's, or any other reality there is, because there is no causeless or accidental reality. Only cause is always inertial here and now. Universe/micro-world is no coincidence—it is infinite double/inverse reaction of inertial action here and now and absolutely any distance/size/motion is the way it is, because it is only possibility (eigenvalue) that is real here and now. Numbers, distances, sizes and motions of all mass and energy of universe and its micro-world are determined absolutely by inertion here and now. All infinity of different fractal parts are always in perfect balance with each other, which is not dual symmetry between equal opposites, but infinite symmetry that is complete dual asymmetry. Singular eigenvalue of such infinite reality (0-point of coordinate system) is only here and now. Point of observation is only point of infinite universe, which is in perfect rest (no change of size/motion). Here and now is perfect singular center of infinite space/time sphere, because it is its only source. Earth's inertial reality is prism that creates entire cosmic reality through its distortion of non-inertial (non-differentiated) reality here and now. Moon, planets, stars, galaxies and their micro-world are only infinite refraction of singular white light through prism of inertial here and now into infinity of relatively different colors.

Moon/Earth distance is $3x10^5/C=1$ second, which is spatial projection of time. Earth/Sun distance is $3/2 \times 10^8/C= ½ \times 10^3$ or $5x10^2$ sec, which is fractal 3/2 and/or inverse fractal 5/2 reality of inertial here and now (10), where derivates of 10^3 =3x2x1=6 or 6/2=3, therefore it is 3D curvature and/or all 6 surfaces of a cube, while derivates of 10^2 are 1x2=2 (two equal opposites of duality). Moon (1 sec angular distance) is ½ or inversion within of solar reality without ($500=10^3/2$ sec angular distance) and so is galaxy in equal and opposite direction. Only Earth has orbiting Moon, because only Earth is inertial 3D, while all the other planets are reduced/increased space or time, which are its fractals (Bode's periodicity). Without Moon there is no 3D Earth, but only its surface (2D) projection (only spatial shell without time context, which is mass/energy of such space). Moon is additional gravitational stability, which is an uneven reality—it is variable positive/negative stability, which has singular inverse eigenvalue when Moon is between Sun and Earth (i.e. eclipse). Orbital in time is 3rd direction of such fractal Earth/Moon duality, which varies gravity of Earth. Only fractal 2/3 is space/time duality of size and motion. ½ is singular gravitational quantum, which is exclusive size of no motion (and/or vice versa)—consequently Moon has no rotation and no time of its own (similarly as Mercury, which is gravitational balance between Sun and all outer planets). Moon is outer shell of gravitational sequence here and now—it is boundary between space within and time (relative space) without.

Space/time between Earth and Sun is saturated with singular twist (inversion) of one direction of Earth's gravity, because it is singular fractional difference. Such twist is additional massive quantum of Moon. **Only Earth is gt=C** (g-gravity, t—orbital time, C-light speed). All the other inner planets are $gt<C$, which is less than singular gravitational quantum (½ curvature); they are only fractals of singular gravity here and now, which is only whole number (1). Outer planets (Jupiter through Uranus) are inverse (negative) space/time curvatures (eigenvalue or symmetry line is asteroid belt), where $gt>C$, therefore they are multiples (inverse fractals) of Earth's inertion. They are instant inversion of inertial reality of inner planets. Inner/outer planets are 1D fractal difference—one is attraction of repulsion of another. Outer planets are additional gravitational freedom or additional gravitational direction, which is inversion of solar reality (inner planets are fractals in between such double/inverse gravity). Outer planets are additional repulsion of equal and opposite additional attraction of solar reality. In order to be in balance with solar gravity, space/time or linear gravity without such planets twists from infinitely many times (Jupiter=10) towards decreased number of such twists, proportional with their spatial distance. Each space/time twist creates instant duality, which is massive moon orbiting its planetary reality. Because they are additional gravitational direction, moons of outer planets are same reality as inner planets (they rotate).

Space of Earth is solar time and its time is solar space. Earth/Sun distance d $=3/2 \times 10^8$ (km), and time of its orbit t = $\Pi x10^7$ (sec), therefore **d/t = 3/2 x 10/ Π. Earth/Sun is fractal 3/2 duality—Earth is flat spatial reality (10) and Sun is its time curvature (Π).** Difference between 10 and Π is 3rd direction of such duality (minor discrepancy, which is inertial here and now). I.e. in context of perfect duality of no inertion, $10=\Pi^2$—flat reality (10) is double—relative and uncertain curvature ($\Pi/(1/\Pi)=\Pi^2$). $10-\Pi^2=9.87$, which is inversion of 1/10 fractal difference, is inertion or 3rd direction of such duality, which is discrepancy or sidereal/synodic fractal difference between two equal opposites. It is eccentricity of perfect curvature. Π is perfect curvature only here and now. Increased distance is inversely increased angular curvature, which is proportionally distorted perfect curvature (Π). Infinitely large distance is inversion between perfect curvature and its infinite eccentricity—i.e. perfect curvature of infinitely distant/different reality is infinite eccentricity (double linear direction) here and now and equally vice versa—infinite eccentric distortion is perfect curvature here and now. Eccentricity and perfect curvatures are only relative determination—perfect curvature (Π) in space is eccentric curvature in time and equally vice versa. D/t=3/2 x10/Π is derivate of Kepler's harmonic law (cube of spatial or linear distance = square of round time), which is its inversion—2d/3t=10/Π. In context of perfect inertionless duality, where $10=\Pi^2$, 2d/3t=Π (perfect curvature).

Space and time (size/motion of Earth) is $12x10^3x864x10^2=10\ 368x10^5$ (kmsec), where 10368 is ½ precessional period in context of its ¼ inertion (12 960x4/5). $12x10^3$ km is perfect diameter of Earth, which is perfect ¾ fractal reality. Inertion of multiplication is addition and inertion of addition is multiplication. Inertion of 3x4=12 is 3+4=7, which is $3x4x10^3+(3+4)x10^2=12.7x10^3$ km diameter and $12.7x864x10^5/C/10$ =365.76 (sec^2), which is number of rotation per orbital (year) in context of inertial difference of 0.51 day (inertial ½ quantum in context of inertial difference $1/10^2$). 365.76 is duality between orbital motion of 360 days and its perihelion motion, which 57.6 year per additional rotation, which is 1° per 57.6 years. 57.6x90°=5184 years, which is inertionless period of constant passage of time 5184x5/4=6480 years of ¼ precession. 2x5184=10368 years, where 12960-10368=2592 and 6480-5184=1296 is relative/uncertain time between fractal 4/5 realities). 86400 sec/12 000 km =7.2, which is fractal 7/2 reality (3.5) or inertial duality here and now (spatial size of Earth) in context of galaxy (7D). 7/2 is 6-fold repetition of singular ½ quantum, which is inertial here and now. It is 3D in context of inertial ½ and/or duality of equal and opposite fractal 3/2 in context of its inertion (½).

7.2=1/10 fractal difference of 360/5=72 synodic/sidereal difference between inertial (Earth) and solar time (orbital motion of Earth and rotation of such orbital). Rotation, orbital and precessional motions are reaction of inertial size and equally vice versa. Singular inertion is infinity of fractal inversions and equally vice versa. It is 4x1/4 or 4D reality of double inertion/inversion, which is a square of 4 equal and opposite sides and change of any one is equal and opposite change of remaining 3 sides. C/86 400 sec=3.47222...km/sec^2 gravitational (galactic) sequence (and/or inverse 86400/C=0.288, which is double 144 and/or 4x72 sequence of double precessional motion or double light speed difference) 3.47222 is fractal 7/2=3.5 in context of its fractal difference 3.5-3.47222= 0.02777..., which is inversion of inertial 7/2. C/12 000 km =25=5^2 and/or 12 000/C=4/10^2, which is inertial 4D here and now. 4D here and now is 6x4/ 5^2=24/25 fractal gravitational reality without (see text on page 69 and illustration 8). Space/time (physical) reality is infinite double/inverse fractal curvature of inertial (singular) here and now, which is only whole number of such double/ inverse fractal infinity. Fractalization means that one is always double/inverse precessional curvature of equal opposites within/without, which is infinite fractal self-interference or infinite repetition of singular eigenvalue here and now in infinity of different relative/uncertain curvatures.

There is no difference between micro-world (atom) and its universe—they are inversely proportional. They are both quantized relative reality and/or relative quantum reality. It is infinity of motions of singular size and/or vice versa, where each successive (precessional) motion is 2x1/2 its previous motion. Each successive motion is ½ (inverse) size in context of twice larger (squared) period of time, which is infinite 2x1/2 echoing of inertial here and now. It is infinity of inverse waves in context of inertial wave, which is infinity of spatial waves in context of singular time wave (i.e. atom) and/or infinity of time waves in context of singular spatial wave (i.e. universe). Static or no wave (Wheeler—DeWitt formulation) is infinity of inverse waves, because **any singular particle of time is infinity of**

spatial waves and any spatial particle is infinity of time waves—i.e. any multiple is infinity of fragments and any fragment is infinity of multiples (multiple in time is fragment in space). It is infinite double helix, where inertion inverts into double inertion through an infinite precessional curvature of double helix, and what is instant in space is gradual in time, just as instant in time is gradual in space. Inertion of inversion is double inertion, which is singular quantum of such reality. Singular quantum of duality is ½, while singular quantum of 4D is ¼. Spatial reality, which is less than its singular quantum is its constant passage of time, just as constant passage of time, which is less than singular quantum is space. Double helix is infinite potential, where any direction of its curvature is determined exclusively by inertion of singular choice here and now. Here and now is only vector of such scalar and it is the only form of its content.

Time is gradual double/inverse change of inertial choice here and now. Inertial choice is total (3D) freedom, because double helix is infinite potential—it is singular reality of all relative choices, regardless of their relative reality. Such gradual change in depth is not real on surface, unless it is singular quantum of ¼ curvature, which is instant inversion (change) of such inertial choice. Passage of time in between quantum changes of inertial here and now is gradual accumulation of inversion of inertial choice (equal opposite, which is its negation), which overtakes (precedes) inertial choice in ¼ intervals. ¼ interval is instant incorporation of both—inertion and its inversion, which creates instant double inertion in context of equal and opposite double inversion, which is its singular fractal quantum. It is double/inverse negation, which creates singular affirmation or instant self-interference of two successive space/time sequences. It is instant eigenvalue between space and time (double/inverse boundaries) in context of inertial here and now, when space and time are not equal opposites (angular/horizontal curvatures), but same reality as its here and now. There is no other reality in instant of quantum change—no space or time, except inertial reality of point of observation. **It is instant of disillusionment, when all surface (relative/uncertain) illusions dissolve instantly in context of its only (3D) reality. It is instant inversion of uncertainty into certainty and infinity of different relative theories into singular reality. It is instant supersymmetry or eigenvalue of fractal duality, in which there is no past or future, distance or difference—only here and now.**

It is moment of revelation of indescribable singularity, when there is no other reality but singular here and now. It is closing of a curvature or full swing of pendulum, which is its return to starting point in order to incorporate previously excluded choice (inversion) into inertial choice, which creates infinitely larger (inclusive) choice, which is following space/time sequence of inertial here and now. It is double (accelerated) time, which is instant end of one and beginning of another time wave. It is instant reaction of infinite accumulated gradual action. Such moment is instant revision of infinity of accumulated changes of previous cycle of time, which determines future reincarnation of physical reality (fractal space/time). It is judgment day, which is day of changing (inverting) choice of what is real and what is not. It is instant quantum leap in time, which is twice as large evolutionary progression than all infinity of its gradual changes in previous period. In such instant linear passage of time stops, inverts in itself and reverses direction (change of vectors or arrow of time). Such instant inversion is moment when future becomes past and past future, when time without becomes instant space within and space without passage of time within. It is eigenvalue of inertial (singular) spatial sequence here and now and its relative (4D) reality, which is fractal infinity of singularity within.

It is synchronization of solar and Earth's reality in their galactic context, which is common denominator of both and equally vice versa. Such quantum reaction is instant gravitational change or change of density here and now. It is simultaneous change or relatively different sequence of size/motion (rotation/orbital). It is instant eigenvalue of all different (excluded) possibilities (Everett's many worlds) of singular choice here and now. Such change is possible only when Earth's reality is 3D eigenvalue of solar reality (and not only surface synchronicity of 2 out of 3 directions), which is at the same time 2D eigenvalue of galactic reality (and not only linear eigenvalue, which is constant passage of time), because galaxy is gravity of Earth/Sun (space/time) duality. It is 1D fractal shift of space/time inertion, which is real here and now and instant when present incorporates past to generate future sequence, which is singular sequence of physical reality relatively different from any other sequence (irrationality of Π/φ). Irrationality of numbers φ and/or Π is fractal eigenvalue of two double/inverse wholes (singularity/infinity of such infinite repetitive curvature in between), because there is infinite number of fractals in between any two multiples, which are its double/inverse context. Φ and Π are 1D and/or 2D fractal inertion (eigenvalues) between double/inverse realities.

Φ is singular eigenvalue between double/inverse (within/without) linear directions, just as Π is only possibility between flat and round (Euclidean/non-Euclidean) curvatures that is real here and now. Φ/Π are 2D/1D projections (derivates) of 3D double helix. Π is 2x1/2 (duality) of singular φ (equally vice versa) in context of small inertial discrepancy, which is inertion here and now between such singular/double fractal difference Their irrationality is reaction of infinite uncertainty/relativity (fractality) of any surface duality, which in its turn ensures relative uniqueness of different sequences. Fibonacci curvature is fractal asymmetry, which is double/inverse symmetry. It is double/inverse infinite repetition of singular here and now, which is infinite double/inverse variability of one in context of another ($1.111/(1/1.111)=1.111^2=1.23456...$). Asymmetry is double symmetry and symmetry is double asymmetry, because double negation is affirmation. Double infinite fractal repetition is 2D matrix of infinity of relatively different sequences (no two same sequences). Symmetry is real only in context of asymmetry and equally vice versa. One is real only in context of its double/inverse negation (boundaries). Only repetition of repetition—i.e. only twice-infinite non-variability is equal and opposite infinite variability, because only duality of two equal opposites is real. Symmetrical surface is asymmetrical depth and asymmetrical surface is symmetrical depth.

Gradual change of surface is instant change in its depth and gradual change (passage of time) in depth is instant change of its surface. Gradual spatial change is instant time change and gradual time change is instant spatial change—infinitely many times. Gradual spatial change is instant passage of time and gradual change of passage of time is instant space, because space/time is angular 90° difference. Equal and opposite gradual change of space and time is instant mass/energy of empty space/time and equally vice versa (double/inverse space/time), which is infinite 2x1/2 spiraling curvature of double helix in context of its inertial 3rd direction, which is inertial here and now. Space/time is infinite double/inverse repetitive duality of inertial here and now, which is only constant underlying infinity of all variable curvatures. It is infinite self-comparison of such inertion, where any comparison is only temporary. Time is temporary space, just as space is temporary time—space/time (both) is only temporary 2x1/2 (double/inverse) space/time in context of singular constant here and now. When temporary reality is equal to its constancy—it is instant of inversion of inertion, which creates 2x1/2 inertion. Every action has its equal and opposite reaction only in context of its inertion. Such two equal and opposite temporary realities are not real in context of its inertion—temporary space in context of equally temporary time is only singular inertion, which is reality of no space and no time, therefore reality of no change.

Size is temporary motion and motion is temporary size—one is time of instant reality of another. All different sizes in context of their inversion (motion) are only one inertial size and equally vice versa—all motions in context of their equal and opposite sizes are only one motion ($C=d/t$ =const). Such singular size/motion is inertial here and now. Mass/energy of empty space/time is double (4x1/4) duality. Energy is spatial (light speed) motion and/or time size, while mass is time (light speed) motion and spatial size. Different sizes are relatively different curvatures of inertial light speed and reactions of different sizes are their inversely different motions. Light speed is 4x1/4 eigenvalue—it is eigenvalue of all different spatial and time sizes and motions, which is inertial here and now. Light speed motion is infinite/singular motion, which is not motion but size (and equally vice versa—it is not size but motion). It is $3x10^5$or fractal 5/3 reality, which is infinite triple repetition (1/3=0.333 or 3x0.111) of singular here and now (10^5-5D fractal difference of inertial here and now, which is 5+1=2x3=6D space/time reality). Light speed is neither size nor motion, by being both. It is eigenvalue of all different space/time sequences, which is rotational curvature within and/or linear direction without (equally vice versa in inverse context). Linear/rotational light speed (E/m) is double/inverse reaction of inertial here and now in between, which is equal and opposite double/inverse light speed—it is linear light speed of rotational light speed of its micro-world within and/or rotational light speed of its universe without. Here and now is inertial eigenvalue between Euclidean and non-Euclidean curvatures of light speed. It is instant duality of such double/ inverse curvatures, which is double light speed (double/inverse mass/energy in context of double/inverse space/time). Double light speed is $E/m=C^2$ or mass/ energy duality, which is instant gravity or 5th quarter (inertion) of 4x1/4 wave. Double light speed in context of inertion is only ½ light speed in context of inversion, which is 4-fold light speed in context of double inertion, etc.

All objects in course of their time pass through every singular point of its space and equally vice versa. In context of constant passage of time of the universe, which is its singular gravitational quantum, there is no difference between its different objects, because totality of such objects and their motions are only surface of larger (relative) spatial reality. Sizes and their motions (here and now) are only its instant surface size and its angular density. Linear light speed, which is infinite spatial size of universe, is light speed rotation, which is finite size (mass of micro-world) of instant density within. Every spin of an atom is one complete expansion of the universe; just as infinite time of expansion is only one spin of an atom, because it is all time there is, only in relatively different (double/inverse) curvatures. Discrepancy (overlap) between such equal and opposite curvatures is their inertial eigenvalue, which is size/motion of Earth here and now (inertial space/time sequence, which is 6480 years of constant passage of time). Earth in its motion is singular difference between universe and its micro-world. Universe and its micro-world are double/inverse reflections of such inertion here and now, which is fractal 3/2 and change of any one ½ is equal and opposite change of its

2x1/2 duality. Universe is spatial repulsion of equal and opposite massive attraction within in context of perfect rest of inertial here and now in between (line of symmetry).

Mass/energy is double/inverse limitation of infinite space/time (round curvature in context of parabolic or exponential infinite/singular curvature). Mass/energy is time of spatial reality (i.e. universe) and/or space of time (i.e. micro-world). It is singular curvature of infinite reality. Universe/micro-world is singular quantum (C) of infinity of inverse (E/m) quanta without/within. Mass/energy are double/inverse infinite multiplication/division of inertial space/time (universe/micro-world) and equally vice versa. Mass/energy of space/time are double duality or double inversion of inertial massive Earth here and now. Double difference means such different/distant mass/energy is not inertial gravitational sequence as Earth and its singular universe/micro-world but its gradual double/inverse variability, which is relatively different attraction/repulsion (different sizes/motions). Different sizes/motions and their densities are real only in context of inertial Earth here and now. Spatial distance is equal and opposite angular time difference, which is reality of different sizes/motions. In context of inertial here and now (no spatial distance or angular time difference), there is only one reality of no difference (same inertial density). Different density, which is relatively different size/motion, is real only because it is not here and now. Universe/micro-world are fractal difference in between—one is flat reality of another one's curvature. Universe is gradual change of density, while micro-world is instant (infinite) double/inverse change of density (light speed difference).

Micro-world is no spatial distance that is infinite angular (time) difference, therefore infinitely different density of spatial reality in context of infinitely slower passage of time (infinitely fast motion/infinitely small size). Universe is equal and opposite gradual (fractal) change of density, proportional with spatial distance. Larger distance is larger angular curvature, therefore relatively larger discrepancy between passage of time here and now and such distant/different reality. Relatively larger space is slower motion and faster passage of time (slower motion). There is double/inverse discrepancy between inertial passage of time here and now and different mass/energy of its universe. Earth is only reality that is inertial passage of time. Any different size/motion is accelerated/decelerated passage of time (different density/gravity). Massive objects of universe and their motions (energy) are exclusive reality of inertial Earth—it is not their inertial reality here and now. Inertial reality of different sizes/motions is shifted in time proportionally with their spatial distance. Fractal spatial distance is different motion of constant passage of times. Reality of different massive (and/or energetic) objects of universe here and now is only their surface spatial reality taken out its time context.

Such reality is not inertial reality (constant passage of time) of different mass/energy—it is exclusive reality here and now, which is increasingly only empty shell of its space/time reality, without its inertial context, which is reason for increasingly different reality of distant objects. Difference is only surface of 3D reality, because difference is fractal (comparative) duality. Only Earth is orbital time x gravity =light speed (inertion). Any different/distant spatial reality is accelerated/decelerated light speed (constant passage of time). Earth is singular eigenvalue of its infinite spatial reality. Universe and micro-world are real only in its context, which is prism of distortion (discrepancy) of their inertial reality, which is same here and now as here and now. Earth is 3D reality, which is triple fractal curvature of double/inverse distortions. Curvature of one direction is space/time, another is mass/energy and 3rd is gravity/antigravity of such physical reality. Solar empty space is inversion of empty mass here and now in direction without, while mass here and now is double inversion within of galactic space, which is its gravity here and now.

Solar and galactic space is double inversion (double light speed difference) of inertial here and now (double/inverse quantum of inertial density here and now). Singular fractal difference is synchronicity of two out of 3 directions, while double difference is synchronicity of only one direction. Mass/energy is surface synchronicity of inertial space/time, while its gravity/antigravity is only linear synchronicity. In context of singular fractal difference (i.e. solar reality), 3D difference is 2D distance, just as 3D distance is 2D difference. In context of double difference (i.e. galaxy), 3D difference is only linear distance, just as 3D distance is only linear difference. Mass/energy here and now and mass/energy of galaxy are double/inverse (within/without) mass of inertial solar space in between (eigenvalue). Galaxy is relative space (empty universe), while here and now is micro-world (atom) of inertial solar reality. Mass is instant size only on surface and infinite gradual change within (inverse spatial expansion, which is its contraction or half-life). Singular quantum on surface is infinity of inverse quanta in its depth. Instant reality or singular quantum is always only surface of larger/smaller 3D, which is its angular depth. Mass is surface eigenvalue of double/inverse 3D densities without and within, while its gravity is linear eigenvalue of twice double/inverse density (twice without/within).

Different sequences of mass/energy of empty space (proportional with distance/difference) are reaction of singular (inertial) choice here and now. Infinite differentiation is real only in background of its reaction, which is singular reality of no difference here and now. Different mass/energy sequences are time of inertial choice. **Choice is real only in context of its double/inverse (fractal) boundaries, which are its space (size) and time (duration).** Consequently, universe and micro-world are positive/negative (future/past) time of inertial Earth. They are its beginning and end. Universe is background or time of singular spatial choice, which is Earth. Earth is real only in background of its universe, which is infinity of different possibilities (Everett's many worlds) that are excluded through singularity of Earth's reality. It means that in context of any relatively different inertial here and now (different density of passage of time), there is only one Earth in entire universe, which is always right here and now. Universe is its reaction (equal and opposite ½). Relatively different inertial here and now is relatively different density, which is relatively different universe/micro-world. Empty space of one inertial here and now (inertial Earth) is dense mass of another and equally vice versa—dense mass of one is empty space of another, which is relatively different passage of time. There is no other point like Earth in universe of inertial here and now, because universe is equal opposite of Earth—it is its reaction or negation. Larger distance/difference is larger negation of inertial Earth.

Different mass of universe/micro-world are past/present space/time sequences of inertial Earth. Relative universe is instant infinity of all relatively different inertial points, because any fractal difference of constant passage of time here and now is not real in its context. All relatively different passages of time, which are relatively different densities, are singular instant density of its fractal (relative) reality. Different inertial here and now, which is infinity of different universes (different densities) are only singular atoms within of even density on surface of such relative reality. Inertial 3D universe here and now is exclusive reality of **one** Earth, which is singular space/time sequence of infinite fractal Fibonacci curvature. Boundary of such inertial spatial sequence is its time, which is ¼ precessional period of its rotation.

C=d/t=const means infinite spatial change in context of its time. It is infinite gradual change or infinite comparison of singular difference here and now against all infinity of excluded choices (Everett's many worlds). **Time is infinity excluded in any singular spatial choice** (equally vice versa—space is infinity of any singular choice of time). In the course of its time, such singular choice passes through all infinity of different (excluded) choices, just as any planet passes through gravitational center of its larger (fractal) reality (star); a star through black hole of a galaxy, etc (Kepler), because here and now is only one possibility out of infinite number that are singular (instant) possibility of such larger choice. In the course of all the time: space=time. Such point of equality is point of inversion (eigenvalue), because quantum of reality is always C=d=t (no difference). It is point of inversion, when time of such choice expired—when its instant spatial reality is equal to all its time (content is equal to its boundaries). Relative difference here and now is inertion (overlap) between space and time and/or their gravity Time is equal and opposite expansion of such inertial difference from within to without and/or equal and opposite contraction of its boundaries. Instant symmetry of such fractal surfaces is their instant inversion or incorporation into double (angular) sequence. Surface boundary (space/time) of inertial quantum is infinite depth of its inverse reality. Instant surface of solid mass and/or empty space is infinite depth in angular direction (where one is always instant transition, while another is gradual change, because symmetry or space of one is asymmetry or time of another). Time is eigenvalue between inertial choice and its negation.

Inertion (i.e. space) is constant, while inversion (i.e. time) is its equal and opposite constant change (derivate). They are equal/unequal (fractal) reality—they are equal on surface, but not in depth and/or equally vice versa. Inertion is instant—it does not change, while change is its equal opposite (inversion). If they are same in space, then they are opposite (singular/infinite) in time and equally vice versa. No spatial change is infinite change (passage) of time and no change (passage) of time is infinite (instant) spatial change. Singular spatial action is infinite reaction in time and singular action in time is inverse infinite spatial reaction. Any surface reality (constant size or constant passage of time) is only fractal ½ of reality, which is always dual (2x1/2) comparison in context of inertial here and now, which is angular (inertial) ½ of such duality. ½ of reality is always beyond infinite/singular boundaries of its equal opposite (opposite side of rotating sphere). It is always duration (time) of inertial spatial sequence and/or size of inertial time sequence, which is equal and opposite fractal ½ of physical reality here and now. Space/time or physical reality, which is real here and now, is only fractal surface or shadow of 3D, which is singular quantum (whole) of reality. Reality is always in opposite direction of its shadow. Reality of surface is in its depth and reality of depth is on its surface. Reality of both is ½ way in between, which is right here and now.

CHAPTER 8

QUANTUM TIME BOUNDARY

Boundary of space is time and boundary of time is space, which is double/inverse (accelerated/decelerated) passage of time (light speed fractal difference). Infinitely accelerated passage of time is infinitely slower motion of infinitely larger (relative) spatial size. Infinitely accelerated passage of time is motion of motion or double/inverse motion, which is precessional motion of inertial motion. It is rotation of rotation, which is not motion but instant space of such precessional rotation. Boundary of constant passage of time is singular ¼ quantum of precessional rotation, which is instant inversion of inertial space/time sequence in context of its relative (fractal) reality. It is instant reversal of direction of constant passage of time in context of its relative (spatial) reality. **Constant passage of time is duality of orbital/rotational motions, which are 2x1/2 multiplication/division (singularity/infinity) of each other. Precessional rotation is inertial framework of such duality (3rd ½).** It is its boundary, which is double/inverse (accelerated) passage of time here and now, which is constant (linear) passage of time of such fractal (relative) reality. One precessional period is singular quantum of constant passage of time of relative (solar) reality, which is singular rotation of solar (empty) space around Earth. ¼ of solar rotation is boundary of constant passage of time here and now, just as ¼ of Earth's rotation is singular quantum (inertion) of orbital motion (¼ of a day, which is additional time of 360+5-days year that is

4x360=1440 additional days of orbital time which create additional quantum of time here and now). Quantum of solar time is boundary of Earth's constant passage of time.

Physical reality is 2x1/2 multiplication/division, which is singular here and now in context of 2x1/2 larger/smaller (relative/uncertain) 2x1/2 multiplication/division. 2x1/2 in context of its larger 2x1/2 is singular 4x1/4 (double) duality. Smaller 2x1/2 is double/inverse inertion (singular quantum) of such relative (larger) 2x1/2, which is 4x1/4 here and now, and 4x1/4 is double/inverse inertion of yet larger 2x1/2, which is twice relative/uncertain 8x1/8 reality. It is infinite double/inverse (2x1/2) spiraling curvature of double helix, where larger spatial division (smaller fractal) is larger passage of time and vice versa, because multiplication is inertial framework (boundary) of division and equally vice versa. Smaller inertial fragment is larger and larger fragment is smaller period of constant (uninterrupted) passage of time. Smallest (singular) division/multiplication is inertial reality here and now, which is 2x1/2 that is 4D fractal difference (2/(1/2)=4, i.e. inertion is 4 times its inversion). Smallest fractal 2x1/2 division is 2 equal opposites of one rotational period (day and night, which are double/inverse sides of a sphere). Each ½ period (12 hrs) is separated by 2 equal and opposite singular points of sunset/sunrise in between, which is double 2x1/2 or 4x1/4 reality. Furthermore, each 12 hrs period (day) is two equal opposites in which Sun rises with exponential velocity (parabolae) towards its apex, which is ½ of such ½ period (6 hrs), followed by equal and opposite setting in opposite direction. Consequently one day is 4x1/4, where each ¼ is 6 hours, which is singular quantum of constant passage of time here and now.

Only orbital/rotation are gradual change, because they are double/inverse reality of each other (in context of inertial size of Earth). Orbital is infinite boundary (time) without of its singular rotation (space) within, which is 4x360=1440 singular ¼ quanta. Any change of orbital is equal and opposite (inverse) change of rotation, which is inversely proportional change of both. Any further motion of Earth is not gradual but quantum (instant) change of time. Only inertion between rotation/orbital, which is time of both, is gradual change or constant passage of time. Inertion between orbital/rotation is relative reality of inertial here and now, which is 4D space. 4D space is not space, but time only in ¼ periods, which are fractal change of its spatial reality; therefore unreal in its spatial context, because time is fractal overlap (delay) between double/inverse spaces. Just as Earth's rotation is constant passage of time here and now, so is inverse or solar rotation constant passage of time of solar reality, which is inverse density of empty space in context of inertial Earth. Inverse rotation is precessional rotation and ¼ of precessional rotation is singular quantum of passage of time of relative solar reality, just as ¼ Earth's rotation is singular quantum of time here and now.

Singular quantum of passage of time of relative (solar) reality is infinity of quanta here and now. 6480 years of ¼ precessional period x 1440 of singular quanta of Earth's time = 9 331 200 Earth's quanta per singular solar quantum, which is $1/10^2$ or double fractal difference of Earth's orbital of 933 120 000 km (which is $3\Pi x10^8$ or fractal 3/8 reality, which is 9.4245-0.0933=9.3312, where $0.0933=9.3312x1/10^2$ is inertial difference of such curvature). Spatial orbital is reaction of its time and equally vice versa, because space is boundary of time and time is boundary of space. Equally vice versa—additional 5 ¼ days, which are inertion (quanta) of precessional rotation, are 5x86 400 (1 day) x 30 (orbital velocity) = 129 $600x10^2$, 1x86 400x30=25 $920x10^2$ and ¼x86 400x30=6 $480x10^2$ km, which are singular ¼ quantum of precession (6 480), whole precessional period (25 920) and double precessional period (5x25 920 or 20x6 480=129 600 years). Fragments of spatial orbital are reflection of quantum periods of its time (quanta of relative spatial reality), while whole orbital is infinity of quantum ¼ periods of inertial time here and now, because orbital (space) is inertion between rotation/precession, which are its double/inverse times without and within.

One orbital is 1440 fragments, each $6480x10^2$ km, which is its singular quantum. Km/year (space/time) is 2D fractal difference (10^2=100, which is 100 km=1 year), therefore one orbital is 144 000 singular quantum periods of constant passage time (6 480 years). 933 120 000 km orbital/25 920 years of one precessional period $=360x10^2$, which is period of one year. One year (time) is inertion (eigenvalue) between spatial orbital and time of precessional motion. Orbital/precessional times are double/inverse times of inertial spatial orbital in between. Spatial orbital is singular eigenvalue between double/inverse passages of time, which is inertial passage here and now (duality of rotation/orbital motion) and its boundary, which is passage of time of solar (relative) reality. Passage of time of solar reality is precessional motion of both—orbital and rotation, which is also derivated (decelerated) passage here and now only in fractal periods less than singular ¼ quantum.

Time of Earth is its rotation and time of rotation is its orbital motion, while time of both is precessional motion, where each successive quantum of time (motion) is accelerated passage of time without and decelerated passage within. Orbital is eigenvalue in between (angular ½). Its double/inverse reactions are vertical tilt (change of seasons) and/or precessional motion in context of spatial earth (size). Vertical tilt is inverse reaction of orbital motion (1D fractal difference), while precessional rotation is reaction of vertical tilt and/or double reaction (2D difference) of orbital motion. Infinitely faster orbital than rotational velocity is reflected as infinitely slower passage of time of vertical motion (change of seasons). Vertical tilt is 2x1/2 or ¼ difference from rotational motion. Rotation is singular 360°, while vertical tilt is 2x45=90°. 4 changes of season (one year) are 360 days. Vertical tilt (inversion) is 360x4=1440 slower motion than rotation (eigenvalue), which is reflection of equal and opposite 30/0.463=64.8 faster orbital velocity (inertion), where $64.8=6480/10^2$ (1/4 precession) and 0.463-rotational velocity (again: 6480x1440=9 331 200 or 1/100 of orbital size, where 6480/1440=4.5, which is fractal 9/2 and/or 4/5 reality). Vertical tilt is space of orbital motion (time) and/or vice versa, which is its inversion in context of inertial rotation in between. One is double/inverse passage of time (boundary) of inertion (space) of another.

Inversion of inertion is time of space and/or space of time. If orbital motion is spatial motion, then vertical tilt is time motion or motion in context of relative (4D) spatial reality, which is constant passage of time here and now. Rotational velocity of Earth is 0.463 km/sec, which is inversion of 2.16 sec/km=86 400 sec (day)/40 000 km (size). 0.463 is its spatial velocity and $2.16x10^2=216$ sec/km is its time velocity or velocity in context of relative spatial reality. Relative spatial velocity is velocity (density) of constant passage of time here and now. $216=6^3$ is inertion (quantum) of constant passage of time here and now. Vertical tilt is 463/1440=0.33 m/sec, which is 10-multiple (fractal difference between inversion and inertion) of inverse orbital velocity—1/30=0.0333...sec/km. Inverse orbital velocity is its time velocity or velocity in context of relative universe (30/0.333=900 or 3^2x10^2, which is fractal 3/2). Difference between inertial and inverse or spatial and time velocity is fractal difference of passage of time between such relative space/time sequences. Duality between orbital/rotational spatial and time velocities is their eigenvalue, which is precessional motion (30/463=64.8, as well as 2.16/0.0333=64.8 and/or 30 000/463=64 800, which is 10 and/or $1/10^2$ or double/inverse fractal difference of 6480). Simultaneous reality of both is 30x0.463=13.89 and 1/13.89 (inversion) is 0.072, which is $72/10^3$ or inertion (quantum) of rotational/orbital duality, which is rotation of solar space (additional 5 days of perfect 360-days year, which is 360/5=72, see page 124/5).

Velocity of precessional motion is 10 000/6480 =1.54321 km/year, which is also 0.333x10/2.16 and/or inversion of $64.8/10^2=0.648$. Orbital velocity is $9.33x10^8$km/year, which is $9.33/1.54=6x10^8$ faster motion (slower passage of time) or fractal 8/6=4/3 reality. $6x10^8/C=2x10^3$, which is fractal 3/2 reality. Precessional motion is 10 000 km in 6 480 years, while vertical tilt is 10 000 km in 1 year. Precessional motion is 6 480 times slower passage of time than change of seasons, while vertical tilt is 1440 slower passage than rotation. Precessional motion is 6480x1440=9 331 200 faster passage of time than here and now. 9 331 200/C=31.104 sec/km, which is 31 + inertial 1/10 difference 1.04=25/24 fractal reality, which is its inertion. 31.104 sec/km is time velocity of precessional motion, which is 31.104/0.00216=14 400 faster than Earth's velocity. Precessional motion in time (30 sec/km without inertion) is same velocity as orbital motion in space (30 km/sec). Inertial reality is only one—here and now, while its physical (space/time) sequences are its infinite repetition (1.111), which is infinite fractal 2x1/2 (double/inverse) curvature, which is infinite succession of ½ (gravity/size) within 2/3 (rotating size) within ¾ (rotation/orbital), within 4/5 (orbital/precessional rotation) etc within and without. Each successive inertial sequence is 2x1/2 multiplication/division of its past (inverse) and/or 4x1/4 of its twice-past (double) sequence.

Each successive sequence is time (boundary) of its past (spatial) sequence. Space expands in time and time expands in space, because in context of all its time, there is no difference between space and time. Time is only delay (inertion) between action and its reaction. Passage of time is gradual reaction of instant action, which is also gradual action of double/inverse instant reaction. Consequently, every successive fractal sequence is 2x1/2 expansion of time in its inertial spatial context and simultaneous, equal and opposite (inverse) expansion of space in context of inertial passage of time. Rotation is inertial 2x1/2 of two equal opposites (day/night), which are two quantum points of inversion in between (sunset/sunrise). Day and night are passages of time (linear 1D direction or time), while its inversion points in between are singularities (1D fractal difference, which is its past or spatial sequence). 2x1/2 (rotating Earth) is fractal 3/2 or 2x1/2 time in context of its inertial size, while following sequence of vertical tilt (orbital motion) is 4x1/4 time or fractal 4/3. Vertical tilt expands singular points (space) of inversion of previous sequence into equal and opposite passages of time. Rotation consists of 2 gradual opposite polarities in context of 2 instant opposite polarities in between (day/night in context of its sunset/sunrise), while vertical tilt, which is infinitely slower passage of time, expands such two points into spring/autumn between summer/winter.

It is division of singular direction of rotation, which is 360° curvature into double (positive/negative) linear direction, which is only 45° curvature each or total of 90° (2x1/2 in context of 4x1/4, which is 2x1/2 difference). Rotation is singular east/west motion, while vertical tilt is double/inverse north/south motion (i.e. north winter, which is south summer follows north autumn, which is south spring, which in its turn follows north summer, which is south winter etc). Singular quantum of inertion is its inversion. Singular quantum of rotation is singular point of its inversion between day and night, which is its synodic/sidereal difference. One day is two equal opposites (day/night) in context of 3 singular points (fractal 3/2), which are its sunrise, sunset and following sunrise, which is overlap between synodic and sidereal times that is full cycle of 24 hours. Synodic/sidereal difference is 4 min per day, which is 90x360=1440 minutes or one day variability per year divided into 4 seasons. Instant synodic/sidereal difference (singular point of rotation) in context of its time is gradual angular curvature (expansion/contraction) of linear rotation in two equal and opposite directions. Instant (inertial) point in time is double time of future sequence, while passage of time here and now is instant space of future sequence (inertion/inversion=double inertion). Rotation and vertical tilt are 1D time each, while both are surface (2D) time. Equal and opposite time motions of such spatial motions are instant surface space. Singular point within is linear direction without. Synodic/sidereal difference in context of passage of time is not singular point, but double/inverse passage of time, which is gradual change of seasons.

Rotation (day/night duality) is inertion of its larger duality (vertical tilt), which is 4x1/4 change of seasons. Inversion of rotation (singular quantum point) is inertion (passage of time) of vertical tilt, where inertion in time is inversion in space and equally vice versa. 4x1/4 change of season is 4 points of instant (quantum) inversion in between (solstice/equinox). Solstice and/or equinox are singular quantum of vertical tilt (inversion of inertion). Just as one day is fractal 3/2 reality of two equal and opposite passages of time in between 3 instant points, where 3rd point is its synodic/sidereal difference, so is one year fractal 5/4 reality of 4 seasons in between 5 quantum points, where 5th point is its synodic/sidereal difference, which is inertion of such 4x1/4 wave. Synodic/sidereal difference is overlap (eigenvalue) between orbital and rotation. 4 min (240 sec) is difference in context of perfect curvature (360 days x 240 sec= 86 400 sec of one day), which is its eccentricity (vertical tilt of rotation). A year of 365.25 days is 235.6 sec difference, which is its 23 hrs 56 min in context of 4 min difference. 240 is 2D sequence (perfect duality of no inertion or discrepancy), while 235.6 is fractal 4D sequence. Inertial 3D sequence is 236 sec, which is 4 sec less then perfect 4 min (4th side of 4th shell of an atom, which is its gravitational line of symmetry). 240 sec=6x4x10 or 60x4 sec. Perfect difference is 240 and imperfect difference is 236, which is 1/60 inertial difference (fractal 59/60). There is one eccentric between two perfect curvatures and equally vice versa (one space between two times and one time between two spaces). 1/60 is inertion (quantum) of synodic/sidereal difference (1 min=60 sec), just as synodic/sidereal difference is inertion (quantum) of one day (24 hours = 1 day).

Synodic/sidereal difference is 4x1min, where 1 min is singular ¼ quantum of such reality, while its double/inverse difference is 4x1=4sec (240-236), where 1 sec is its ¼ singular quantum. Synodic/sidereal difference is singular quantum of rotation (time of 1 day) and/or infinity of quanta of vertical motion (tilt or one year). It is singular point of rotation (inversion) and/or infinite passage of time of its orbital motion (inertion). Synodic/sidereal difference is rotational/orbital eigenvalue—it is singularity of one and infinity of other or surface double/inverse quantum boundary between two equal opposites (space and time). Difference within difference is double/inverse difference, which is 4 sec of 4 min or singular ¼ quantum of 1 sec, which is 1/60 difference. Such double/inverse difference is eigenvalue between orbital and precessional time or vertical tilt (change of seasons) and precessional rotation, which is 2x1/2 larger/smaller (relative) change. Just as 4 min synodic/sidereal difference gradually curves rotation in angular direction, so 4 sec double/inverse difference gradually curves vertical motion in angular direction, which is precessional motion of rotation. Double difference is difference in both directions—it is difference in forward and backward direction, which is precessional motion of inertial motion. Vertical tilt is eigenvalue between rotation and precessional motion or mirror that reflects back. 86 400/240=360 days, which is gradual curvature of rotation in context of one year, caused by inertial synodic/sidereal difference. 86 400/4= 21 600 years, which is gradual curvature of vertical (orbital) motion, caused by double/inverse 4-sec difference. 21 600 years is period of orbital motion of perihelion (motion of orbital motion, which is solar rotation).

21 600 years is precessional period in context of orbital time, as opposed to 25 920 yeas in context of precessional/rotational time. 25 920/21 600=fractal 5/6 reality, which is 1/5 discrepancy between precessional and orbital time. 1/5 discrepancy in time is 1/6 and/or ¼ discrepancy in double/inverse times, which are relative/uncertain spatial realities without and within such inertial time in between and equally vice versa, because 1/6 attraction within is 1/5 repulsion without and 1/5 attraction within is ¼ repulsion without, which is fractal difference between positive/negative non-Euclidean curvatures. 1/5 time discrepancy between precessional and orbital times is reaction of instant (spatial) 5 days of additional time per year. Inertial 5/6 is 6/5 without and 5/4 within. 1/5 is eigenvalue between fractal 1/6 and ¼. ¼ and 1/6 are double/inverse realities of inertial 1/5 in between—they are two wholes (even numbered fractals) of fractal in between (odd numbered fractal) and/or two fractals of one whole in between,

where fractal in time is whole (multiple) in space and equally vice versa. 1/5 inertial difference in context of rotation is 1/6 in orbital context and ¼ difference in context of spatial Earth (size) and equally vice versa. 25 920-21 600=4 320 years is relative/uncertain difference between such fractal opposites (2x2 160 or two zodiacs, where one zodiac is $2.16x10^3$ and/or 216x10).

21 600/4=5 400 is singular ¼ quantum of precessional motion in orbital context. 5 400/5=1 080 is inertial 1/5 difference between precessional motion in rotational (6 480 years) and orbital context (5 400 years). Precessional motion (eigenvalue) of both—rotation and orbital motions is fractal 5/4 reality, which is 1/5 difference within and ¼ difference without (and/or vice versa). 1/6 is eigenvalue without of both fractal differences within—1/5 and ¼. 1/6 in context of ¼ is 1/(6x4)=1/24 (synodic/sidereal difference), which is $5^2/(6x4)=25/24$ fractal sequence, which is 1/24 difference without and 1/25 difference within (see text on page 69). Inertial fractal 5/6 is fractal 24/25 in inverse context. Fractal 5/6 in space is fractal 24/25 in time and equally vice versa. Singular fractal difference of inertial 2/3 (inertial space/time here and now) is fractal 4/5=0.8 (and/or 0.2 or 2/10=1/5), while double difference is fractal 24/25=0.96 difference (and/or 0.04 or $(2/10)^2$= 4/100=1/25, where 4/100 is 4D space/time here and now in context of double fractal difference). Likewise—inertial $5^2/(6x4)$ is followed by $6^2/7x5$ sequence. Inertial 6/7 is 7/6 without and 5/6 within, which is fractal 35/36 or 1/6^2 difference in context of its twice-past sequence (3/4) and $30^2/(36x24)=900/864$ in context of its three times past sequence.

Orbit of perihelion is 25 minutes (5^2) discrepancy of inertial orbital motion, which is its angular eccentricity or precessional motion (rotation of solar empty space, which is inertial context of orbital motion—i.e. orbital motion within is motion within motion). 25x60=1500 sec, which is fractal $3/2x10^3$ and/or 1/2.4 of an hour. 25 min, which is 2.4 of an hour, is fractal 25/24=0.96 fractal reality. 1440 min (one day)/25 minutes of relative discrepancy = 57.6 years in which orbital of perihelion creates additional day here and now. 57.6=60x0.96 or 60x24/25, which is fractal 5/6 in context of its inversion (24/25) here and now. Fractal 5/6 is 6D or 360°/6=60° less 2.4°, which is its inertion here and now. 60° is perfect curvature, while 2.4° is its synodic/sidereal difference, which is size/motion of Earth in such curvature. Size/motion of Earth is inertion or discrepancy of perfect 60°curvature. 60=25x2.4, while 57.6=24x2.4, therefore 60/57.6 is fractal 25/24. **24^2=576** or 57.6/10 (1D fractal difference, i.e. uncertainty), just as **60^2=3600** or 360x10 (inverse 1D fractal difference, i.e. relativity). Difference between 60 and 57.6, which is 2.4, is difference between orbital period of one year (360 days) and additional rotational period in context of its inertial orbital, which is additional day of orbital time every 57.6 years. Just as 72 days creates 1 additional day through additional 5 days of a perfect 360-days year, which is 1° curvature of precessional motion here and now, every 72 years or singular quantum of time that is 90x72=**6480**-years, so 57.6x90=**5184**-year period and/or 60x90=**5400**-year period, where **5184=72^2**

6480, 5400 and 5184 are relative/uncertain or double/inverse times of inertial time here and now (6480=6/5x5400=5/4x5184 and/or 5184=24/25x5400=4/5x6480). 6480 years of inertial precessional time is 5400 years of orbital time and/or 5184 years here and now (rotational time), because one is relative/uncertain (fractal) reality of another. 5184=24x216, while 5400=25x216 or 216 years discrepancy between orbital and rotational time, which is singular ¼ quantum of space/time duality here and now (inversion of rotational velocity, 86 400 sec/40 000 km=216/10^2). 216 is triple 6 or 6^3, which is 3D time (curvature) of 6D space and/or vice versa. 6D is fractal reality between double/inverse 3D, i.e. 3^2-3=6, while 72 is fractal reality between 2x1/2 multiplied/fragmented double/inverse 3D, i.e. 9^2-9=72. Multiplied/fragmented yet again, it is 81^2-81=6480 years of passage of time, which is its absolute boundary. 6480 years is triple 2x1/2 fractalization of inertial 3D here and now, which is time boundary of inertial space and/or spatial boundary of inertial time sequence. Such instant (uncertain) triple curvature here and now is simultaneous reality of mass and energy in context of empty space and time and its forces of attraction/repulsion without and within, because 6480 years in context of passage of time (universe) is instant reality within in context of no time (uncertainty within of relativity without).

6480 years is triple time curvature of inertial space and/or triple spatial curvature of time and its singular quantum is 216. 216x4=864 or 86 400 sec of one day in context of double fractal difference (1/10^2). **216 years is discrepancy between space and time in context of ¼ precessional period, while 864 years, which are constant passage of time (86 400 sec) here and now (rotation) is constant (inertial) discrepancy between relative (solar) and inertial (Earth) spatial realities.** 86 400 sec (one day) of passage of time is inertial (fractal) difference (delay) between Earth's dense mass within and empty solar space without. It is its eigenvalue or inertial overlap between positive/negative non-Euclidean curvatures, which is linear direction of time (Euclidean curvature) in between (surface of rotating Earth). 216 years of precessional time and/or 216 sec of time here and now (rotation) is singular quantum (inversion) of inertial space/time sequence of rotating Earth, which is fractal 25/24 reality of such inertial difference.

86 400/216=400=$(2x10)^2=20^2$ repetition of such singular 216-sec sequence, which is 3 min 36 sec, where 20 is common denominator of fractal 5/4 reality (eigenvalue or synchronicity of both, which is their inertion). 20 is difference between inertion and inversion of fractal 5/4, which is its instant double inertion (quantum change). Inertial 5/4 sequence repeated 20 times is equal (resonance) to its inversion (space and time or size and motion, particle and wave, attraction and repulsion etc.), when 2 waves resonate in inertial framework of each other (they add up to create double/inverse wave). 20-fold repetition is instant inertion/inversion=double inertion, while 400-fold repetition is double repetition or 2D fractal difference (eigenvalue of double/inverse fractal 4/5). 400 is $4x10^2$ or 2D fractal difference of inertial 4D space/time here and now. 400 is equal opposite (eigenvalue) of fractal 24/25 reality, which is inertial 0.96 difference, where 0.96 in space is 0.04 or $4/10^2$ in time and equally vice versa (1/0.96=1.041666, where 1.666 or 5/3 is fractal difference between double/inverse inertions and fractal 5/3 is light speed).

5184 years is 24x216, 5400=25x216 and 6480=30x216. 6480-5184=6x216= 1296 =3/2x864 and/or 1440-144, where 5184=4x1296 or 6x864. $6x86400/10^2$ =5184 years, which are 6 days of a week, while 3/2x864=1296 is 7th (inertial) day of a week, which is fractal difference (surface) between two equal and opposite passages of time (double/inverse weeks). 1296 years is relative/uncertain difference between constant passage of time here and now (6x864) and its precessional (accelerated) time (3/2), which is time of solar reality and/or constant gravity here and now. 1296 years of constant passage of time here and now (rotation) is 1 day in context of solar (relative) reality, which is instant quantum change at the end of 5184 years period. ¼ precessional period (5184 years here and now) is resonance between inertial space/time sequence (Earth) and its relative (solar) reality, when they are not equal opposites of fractal difference in between but same reality in context of inertial here and now (no inertion in between, where 0-inertion in time is infinite spatial inertion and equally vice versa). It is instant addition of past to the present sequence, which creates future sequence of passage of time. It is 1 day in context of instant density of empty solar reality and/or context of empty galactic (instead of solar) space, which is passage of time of solar (relative) reality.

Such 1-day quantum change is Earth in context of solar (accelerated) passage of time and not linear passage of time here and now (rotation), because in such instant Earth does not rotate in horizontal but angular direction (direction of angular tilt, which is not tilt but rotation through accumulation of its momentum). ¼ angular rotation is ½ reversal of rotation (swing) and ½ angular rotation is complete reversal of its rotation (reversal of east/west into west/east directions). ¼ period (5 184 years) is swing or reversible change of rotation, while ½ period (10 368 years) is 2x1/2 change, which is complete east/west and north/south reversal. Complete precessional cycle of 20 736 years (21 600 in context of orbital and 25 920 in context of precessional motion) is 360° curvature or one complete angular rotation. Angular rotation is linear direction of orbital or instant repulsion/attraction of Earth in direction of Sun. Linear direction, perpendicular on orbital rotation is orbital in time or orbital in context of relative spatial reality, where time is space and gravity (attraction/repulsion) is time. Relative spatial reality is solar reality in context of its rotation/orbital motion, which is space of galaxy or twice relative space here and now. Galaxy is time of solar reality, just as solar reality is time of Earth.

Rotation in solar context is linear direction in context of galaxy. Linear direction in solar context is curvature in context of galaxy and equally vice versa. Linear direction, perpendicular on round (spatial) orbital is orbital around core of galaxy (infinitely larger space and infinitely faster passage of time). Earth's orbital around Sun is curvature and its orbital around galaxy is infinitely eccentric curvature, which is linear direction here and now. Angular rotation of Earth is its orbital around galaxy. Singular (¼) quantum of such orbital is ¼ precessional motion. 1-day quantum change is 1 day in context of solar (relative) reality, when not 2, but all 3 directions of inertial space here and now are synchronized with solar reality. It is instant ¾ reality of uncertain 2/3 reality here and now, when passage of time is not passage but instant space and gravity is not changeless constant (singularity) but passage of time (linear direction). It is 1 day in context of galactic, instead solar reality, which is instant reaction here and now (relatively different space/time sequence of following period of passage of time). Galaxy is boundary of Earth's space/time reality in solar context—it is its ceiling and instant rebound in backward direction, which is momentum that creates following passage of time (double/inverse spatial discrepancy). Space/time is solar/Earth duality, while gravity (their eigenvalue) is their galactic double/inverse inertial context.

Such instant reversal of rotation in 1 day is of solar time is 1296 years difference in context of inertial passage of time here and now and/or 1080 years difference in context of orbital time. 1296 years is 5th (instant) quarter of fractal 4x1/4 (5184-year period), while 1080 years is 6th 1/5 (5400-year period) which is its inertial synodic/sidereal difference (overlap) that perpetuates passage of time or time wave created by Earth's rotational/orbital duality. 1296/1080/216 years is inertial relativity (overlap) between 3 different time curvatures (rotation/orbital/

precession), which is instant reconciliation in 5184-years intervals, which instantly adds 216 years to orbital motion and 1296 years to rotational motion (and/or vice versa), through instant quantum reversals of direction of its rotation and gradual (fractal) shift in its velocity, as well as fractal change of its orbital motion. Singular period of uninterrupted passage of time is 5184-year, which is linear direction between two equal and opposite singularities that are its quantum beginning and end (double/inverse inertial context of linear direction in between). Singularity in horizontal is linear direction in angular (vertical) perspective, just as round in horizontal is linear direction in angular perspective and equally vice versa. Singularity, which is boundary between two linear passages of time, is singularity only in context of its inversion (passage of time here and now). In its inertial context, it is vertical direction of its inverse horizontal direction. Singularity of linear passage of time is linear passage of time in angular direction. Such angular time is accelerated/decelerated passage of relative/uncertain (fractal) reality. Time here and now is horizontal and time of relative reality is vertical passage of time. Singular 1-day quantum change is double time—it is singularity of inverse time that is infinity of inertial time. It is 1296 years of constant passage of time of Earth's reality contracted into 1 day in context of its angular (solar) passage of time.

Each future quantum change expands (creates) time from inertial space, which is its past sequence, just as any past sequence contracts time into inertial space. Expanded time is expanded empty space of universe and contracted time is its dense mass within. Instantly contracted time is surface of dense mass, while gradually expanding time is expansion of universe. Mass is instant past time, while universe is gradual future time and equally vice versa—mass is gradual future space, while universe is instant past space. Earth (here and now) is inversion point between past spatial reality of universe and future spatial reality of micro-world, as well as between past time of micro-world and future time of universe, which is inertial 4x1/4 (double duality) in context of inertial ¼ here and now that is fractal 5/4 reality. Earth's motion in space is from without to within and in time from within to without, which is double/inverse motion of no motion in between. Here and now is double/inverse space that is no space at all and double/inverse time that is no time at all.

Perihelion orbital is 25 minutes discrepancy per year, which is additional quantum of time every 57.6 years. 20 minutes discrepancy (5/4) is perfect precessional motion here and now, which is additional quantum of time every 72 years. Difference between perfect and imperfect curvatures is their eccentricity in between, which is 5 min discrepancy (synodic/sidereal difference between precessional rotation here and now and perihelion orbital). 25 minutes x 30 km/sec orbital velocity is 45 000 km, which is 5 000 km more than one rotation or 40 000/5 000=4/5x10 fractal reality. 20 minutes of perfect precessional motion is 36 000 km ($360x10^2$, which is spatial reflection of time) or 4000 less than full rotation. It is 4000 km negative difference within, which is 5000 positive difference without or fractal 5/4 that is 5+4=9D. 5D within and 4D without are gravitational equal opposites (2x1/2) of inertial 1D here and now in between, which is positive inertion of one and negative inertion (inversion) of another. They are double/inverse fractals of singular (whole) 9D. Inertial difference in between is double/inverse (relative/uncertain) time of such spatial opposites and/or space in between such different times.

36 000km is 40 000/1.111 or space/time sequence that is 1/10 fractal difference. 360 is 1/10 fractal difference of 400 (double repetition or double fractal difference—see text on page 143), while 324 is 1/10 fractal difference of 360. 400 is singular 4D, 360 is its double/inverse duality and 324 is its inertial 3D sequence, which are 3 instant sequences or 3D spatial reality here and now (universe/ micro-world). 400 without is singular eigenvalue and 324 within is equal opposite of singular eigenvalue, which is its fractal 3/2. They are double/inverse fractals of inertial duality (360) in between. 400/360/324 sequence is double/inverse 40/36 difference in between, which is singular 4D fractal difference in between (here and now). 40/36 repetitive space/time sequences of inertial 4D are its relative/uncertain reality, while 400/324 repetitive sequences are its double relativity/ uncertainty, which are boundary of inertial 3D (where inertial 3D without is 4D within and vice versa, which is fractal ¾). 324/1.111=291.6 and 291.6/ 1.111=262.44 which are progressively smaller fractals of such instant reality here and now. Each successive fractal within is larger quantum reality without. Larger fractal sequence within is larger period of uninterrupted passage of time without (equally vice versa), which is limited by equally larger quantum spatial changes of inertial reality. In opposite direction—400x1.111=444.444 or instant infinity of singular fractals, which is instant gradual relativity without of fractal quantum reality within. Instant gradual relativity is exponentially increased irrationality of such sequences without, which is infinite diversification in inertial context here and now. Such relative infinity is real only in context of equal and opposite quantum boundaries, where spatial infinity is bounded by its quantum time and time infinity by its quantum space.

Synodic/sidereal difference of 5 min between rotational precession (20 min) and orbital precession (25 min) is additional quantum of time (1day) in periods of 288 years, which is singular ¼ quantum here and now in 288x90=25 920 years. Every complete 4x1/4 sequence (double duality) is singular quantum of larger 4x1/4, which is 16-fold repetitive sequence of singular ¼ quanta (and/or 20-fold repetition in context of fractal 5/4). Orbital plane is surface (2D) reality of singular (1D) orbital motion, which is double angular freedom of horizontal rotation. While ¼ precessional periods (6480 years) are fractal shifts or angular (vertical) rotation in horizontal direction of orbital plane, such larger periods are also fractal shifts in angular direction of orbital plane. Shift in horizontal direction of orbital plane are fractal quantum changes of 1D time, while shifts in angular direction are fractal quantum changes of 2D time (change of double/inverse times, where each change is only surface spatial change or change of 2 out of 3 spatial directions). 25 920-year period (20 736 years in context of inertial passage of time here and now) is singular ¼ quantum of such change, which is only ½ (reversible) swing in angular direction of orbital plane. It is complete 360° curvature of angular rotation in orbital plane in context of its 4x1/4 fractal changes and reversible ½ in angular direction (90° swing in angular direction and 360° curvature in its plane, which is 450° or 5/4 of a curvature). Each successive larger quantum period of time is larger accumulation of inertial difference, which is relatively larger momentum that creates larger angular difference.

2x25 920=51 840 years is ½ (irreversible) rotation in angular direction, which is 3/2 change of 360° in orbital plane and 180° in its angular direction, which is 540° change. 2x51 840=103 680-year period is complete 4x1/4 wave (360° curvature), which is one complete angular rotation following complete horizontal rotation in orbital plane, which is double curvature of 720°. It is 8th quantum change (4 quantum changes of singular precessional period of 25920 years in orbital plane, in context of 4 larger quantum changes of double precessional period in angular direction of orbital plane), whose motion is in "8" shape, just as 6th quantum change is in "6" shape, 7th quantum change in "7" shape or 1st quantum change in "1" shape and 2nd change in shape of number "2" etc. Such complete double time curvature is complete inverse spatial curvature and/or vice versa—such complete double spatial curvature is complete inverse time curvature. Singular precessional period (25 920) is 3D boundary that is boundary of spatial sequences in its time. Double precessional period (103 680) is 4D boundary that is boundary of its time sequences.

Illustration 9 - Quantum 2x1/2 rotational (time) sequences

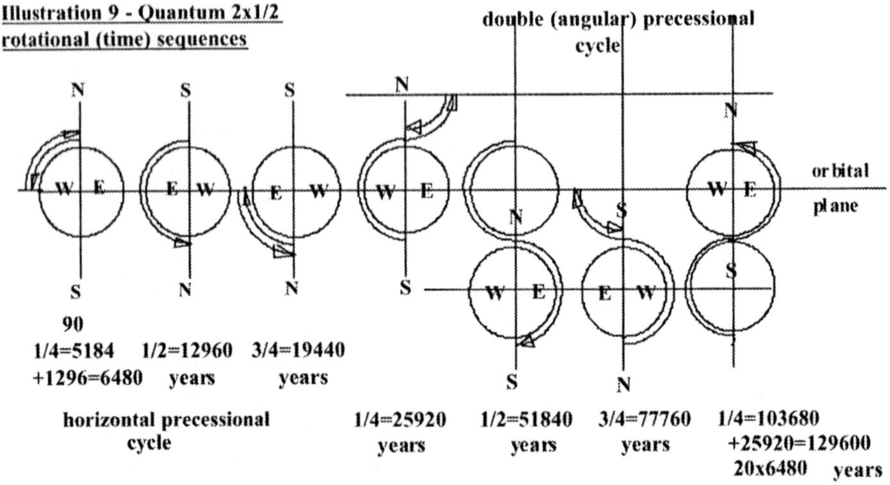

25 920-year period is singular quantum of relative (solar) space that is singular quantum of passage of time here and now, while 103 680-year period is singular quantum of twice relative (galactic) space that is singular quantum of changeless gravity here and now. Such period of time is gravitational framework of space/time sequencing, which is instant (double) quantum change. Relative difference of such larger 5/4 reality is ¼ of 103 680 or 25 920 years, which is period of 129 600 years in context of precessional (relative) time. Just as 5184-year period is instant addition of 1296-years in its angular direction (plane of ecliptic), which is 6480 years, so is 103 680 instant addition of 25 920 years in its 3rd direction (angular direction of plane of ecliptic), which is 129 600 years of its total relative time. **129 600** in context of inertial precessional (relative) time is 5/4x**103 680** years of rotational time and/or **108 000** of orbital time (6/5). It is not only instant change of time (motion) but also instant change of space (size). It is singular quantum of expansion/contraction of massive Earth, which is instant reaction of equal and opposite gradual expansion of empty space of the universe. It is instant reversal of gradual spatial expansion, which is its equal and opposite contraction (contraction of empty space without is expansion of dense mass within).

Just as 4x6480=25 920, which is singular ¼ of larger 103 680 (129 600) year period, so is 129 600 only ¼ of yet larger period of uninterrupted passage of (faster) time, because reality is infinite 2x1/2 curvature in context of larger/smaller 2x1/2 multiplication/division of inertial here and now. Horizontal and angular directions of ecliptic are 2D spatial freedom of inertial direction of rotational motion, just as surface of Earth is two-dimensional spatial freedom of rotation and orbital motion (vertical tilt). 2D space is 3D time, just as 2D time is 3D space. Two simultaneous motions are boundary of spatial size, just as double size is boundary of singular motion (i.e. rotation). Horizontal and angular curvatures of rotational motion are boundaries of rotational and orbital time as singular precessional eigenvalue. It is boundary of spatial double/inverse motion, which is singular ¼ quantum of its motion in time (relative 4D motion), which is in-depth motion or motion within/without inertial surface in between. Motion within/without is double (accelerated) instant motion or instant change of attraction/repulsion, which is instant expansion/contraction of inertial surface size. It is instant change of double/inverse density of mass within (i.e. de-densification) and space without (equal and opposite densification).

129 600-year period is singular ¼ period of such change in depth, which is only ½ fractal (reversible) change. 129 600=25x5184 and/or 24x5400 (fractal 5/6 and 6/5), which is 20x6480 (fractal 4/5, which is their eigenvalue) It is instant ½ expansion followed by instant ½ contraction, because of insufficient gathered momentum, same as ¼ precessional change, which is only ½ reversible 90° swing. It is 2¼ curvature or 720+90=810° (9^2x10), which is fractal 9/8 reality (9th quantum change in shape of number 9). It is ¼ curvature in depth of surface sphere or ¼ curvature of its density. Only twice same period of time (2x103 680= 207 360, 2x108 000=216 000 and/or 2x129 600=259 200 years) is ½ curvature that is singular quantum of irreversible change of density (900° curvature that is 9x10^2 or fractal 10/9=1.111). 2x1/2 inversion of such ½ difference is period of 414 720 years of rotational and 432 000 years of orbital time in context of inertial precessional (relative) time of 518 400 years (fractal 11/10=1.1, which is limitation of infinite 1.111 repetitive curvature of past sequence in 3rd direction to only its double/inverse surface duality). Such period of uninterrupted passage of time is complete 3D curvature (1080°) or singular quantum of inertial (spatial) 3D sequence.

It is boundary of inertial 3D (space/time/gravity or size/motion/density), which is singular (whole) gravitational quantum that is only ¼ equal and opposite antigravitational quantum, which is 4 times such period (1440° or 4D curvature, which is curvature of all 3 spatial directions and direction of its time, which is depth of spatial reality). It is boundary of inertial spatial reality, which is boundary of its inertial mass/energy or inertial density of its micro-world (atom). It is singular E/m sequence within of infinity of its fractal space/time sequences without (singular quantum of change of inertial half-life). 4D curvature is again only ¼ of 5D curvature, which is gravitational sequence of inertial mass/energy (inversion of equal and opposite gravitational sequence of space/time, which is infinite attraction of space without as opposed to singular attraction of mass within and/or vice versa). Each sequence of larger (faster) uninterrupted passage of time is created by its preceding quantum change, which expands singular fractal difference into infinite linear passage of time, which is angular (90°) inversion of past inertial space/time sequence. 90° inversion creates future fractal sequence that is future passage of time of inertial space here and now. It converts past spatial fractal difference into future time difference (future passage of time) and past time difference (past passage of time) into future spatial difference.

5D curvature is 1800°, 6D is 2160°, 8D is 2880°, 9D is 3240° and 10 D curvature is 3600° or 1/10 fractal difference (1.111). 11D is 3960°, which is 40° less than 4000°, which is complete inversion of inertion (4D here and now) that is double inertion. Such period of time is infinite passage of time of inertial space/time sequence of massive Earth. It is period of time between singular point within (an atom) to infinite size of solar reality without, which is double gravitational sequence or eigenvalue between dense mass within and its empty space without (double/inverse density). It is absolute period of time of Earth in its solar (relative) context, which is infinity of fractal quantum changes of density in between that vary its size/density from singular (atom) to infinite (solar reality). Here and now is ½ way in between such equal gravitational opposites, because micro-world is its past time, just as universe is its future time. It is ½ way in between solar life, because instant half-life of solar (relative) reality without is gradual half-life of its dense (uncertain) mass within. All the mass here and now is half-life of spatial reality, just as half-life of mass is all the space of empty universe, because dense mass within and empty space without are double/inverse times of inertial surface between mass and space here and now (Earth), which is their eigenvalue (overlap).

Here and now is always central or ½-way point of all its spatial and/or time directions. It is ½-way point between infinitely expanded empty space of universe and infinitely contracted space of dense mass within (Earth), as well as ½-way point between its infinitely expanded passage of future time and infinitely contracted passage of past time. Infinite space and time of universe are double/inverse infinity of singular here and now, just as here and now is double/inverse infinity of space and time of its micro-world within (atom). Equal and opposite infinities of space and time of universe/micro-world are double/inverse overlap of inertial here and now. They are reaction of self-interference of inertial here and now, which is its only reality. Universe/micro-world are infinite creation in space and time of inertial here and now and equally vice versa, because there is no difference between here and now and all its infinite space and time of universe/micro-world.

Precessional rotation is 4x1/4 (double duality) of inertial 2x1/2 duality of orbital motion (vertical tilt). Just as vertical motion expands singular points of inversion of its past (rotational) sequence into passage of time (spring/autumn between winter/summer), so precessional motion expands singular points of solstice/equinox into passage of time, which is time or boundary of constant passage of time here and now (duality of rotation/orbital). Only double (surface) time is constant passage of time in between, which is double/inverse gradual change of one in context of another. Only singular fractal difference is inertial passage of time. Its double/inverse curvatures are 2D fractal difference, which are equal and opposite spatial reality of time (relative/uncertain space). Time is only fractal between two spatial realities and/or vice versa—space is fractal between times. Whole difference is no difference at all, because it is complete curvature or complete cycle of gradual changes. Only rotation and orbital motion are instant spatial reality here and now, which is gradual change of time of one day and/or one year. Precessional motion is spatial reality only in singular ¼ quanta, which are instant quantum inversions of time here and now and not its gradual change, because it is singular quantum of accelerated time, which is constant passage of time of relative spatial reality.

¼ of precessional motion (5184, 5 400 and/or 6 480 years) is smallest (singular) quantum of infinite fractal relativity of time, while 20x5184=103 680, 20x5 400=108 000 and/or 20x6 480=129 600 years is its double (2x1/2) quantum (twice-relative space, which is singular quantum of galactic passage of time). Double quantum expands past singular time sequence by adding it to inertial sequence. 4x5 400=21 600 is whole quantum of precessional time and/or 1/5th fractal of larger time (slower motion = faster relative passage of time), which is 108 000 years. It is again only ¼ of larger 432 000-year quantum period, which is only 1/5th of larger 2 160 000 year period of faster uninterrupted passage of time (slower motion) etc. Each larger period of uninterrupted (faster) passage of time is proportionally larger quantum change of inertial space/time sequence here and now, because each larger period is inertial framework of its smaller (past) sequence within. Each larger quantum change generates longer period of uninterrupted passage of time, which is longer duration of its spatial effect here and now. Universe is largest quantum change that generates largest passage of time, which is constant passage of time between its double/inverse changes of inertion (complete expansion without, which is complete contraction within). It is passage of time between singularity of Planck length and light speed motion, which is infinite size of expanding universe and/or its black hole in opposite direction.

Precessional rotation is passage of time in quantum ¼ periods. Double precession is precession of orbital motion (difference between orbital of perihelion and precessional rotation). It is precessional motion of vertical tilt in context of spatial Earth, which is not real, unless it is singular ¼ quantum, which is instant change from fractal tilt to full angular rotation. Only infinite number (5184) of fractal tilts is one quantum of angular rotation, when momentum (tension) created by such small fractal difference is equal to singular inertion (¼ quantum) of rotation (time). Every action is equal and opposite reaction. Action that is smaller than singular quantum of inertion is not instant but delayed reaction, which is passage of time in between. Time is gradual accumulation of such fractal difference (action), which is instant reaction in period of time. Smaller fractal action is larger period of time between action and its reaction, while larger action is smaller period of time. Infinitely small inertial discrepancy is infinitely long passage of time before reaction and infinitely large discrepancy is no time at all but instant spatial reality here and now. Only infinite number of successive twisting reaches breaking point and only infinite mass is critical mass, when accumulated inertion is equal to its inversion (equality without and within or perfect duality of no inertion in between). Only resonance is real, when two equal opposites resonate within each other, which is instant singularity of two equal and opposite waves. Only inertionless reality on surface is infinite inertion in depth and equally vice versa.

Vertical tilt is additional quantum of time, which expands singular points of time of its past sequence (sunset/sunrise of rotation) into linear passage of time of fractal spring/autumn between whole quanta of summer/winter. Precessional rotation is additional quantum of time that expands singular points of solstice/equinox into constant passage of time, which is inertial time framework without of its past sequence of seasonal changes within (vertical tilt/orbital motion). Expanded winter/summer solstices are alternating ice and warm ages, while expanded equinoxes are ½-fractal (gradual) changes in between. 4x1/4 periods of precessional passage of time are double seasons, created by double solstice/equinox, which are its singular points of instant quantum changes in between. 4 seasons of 1 year are inertial (singular) framework of infinite number of days (4x90=360), while 4 seasons of precessional motion, which are 4 quarters of solar (time) rotation, are inertial framework of infinite number of years (4x6 480=25 920). Infinite number of different days is singular year and infinite number of different years is singular period of constant passage of time, which is one precessional cycle. Orbital time is boundary of rotational time and precessional time is boundary of constant passage of time, which is instant reality (eigenvalue) of both.

4 quantum periods of time are only ¼ of larger wave, which is 16 quantum periods of double precession. Such 16 periods are relatively different quantum time frameworks of its constant passage of time within, before repetition of complete sequence, which is relatively different space/time reality. Orbital motion (change of seasons) is gradual expansion of past singular points of inversion into complete seasons (passage of time), because vertical tilt is gradual motion. It is slower motion than rotation, which is faster time. Faster time in context of inertial here and now, which is slower passage of time is additional (expanded) time. Expanded time is fractal difference between double/inverse (linear/accelerated) passages of time. Rotational/orbital time difference (singular quantum of relativity) is equal opposite of rotation/size—it is inversion of rotational velocity, which is singular quantum of 216 years discrepancy per ¼ precessional period (864 years in context of one precessional cycle, which is equal opposite of 86400 sec of one rotational cycle). 216 years is fractal 24/25 difference between such double/ inverse passages of time (5184 vs. 5400 years). Such instant eigenvalue is instant change of fractal spatial sequence here and now into gradual passage of time of next sequence. Precession is not gradual but instant expansion of time, which is relative spatial sequence that is inertial context of passage of time within. It is instant spatial change that causes gradual passage of time. Instant spatial change reconciliates accumulated difference between inertial and accelerated (relative) passage of time by creating additional period of time. Just as sunrise is a dawn of a new day, so is quantum change of time a dawn of a new age.

CHAPTER 9

RELATIVE/ UNCERTAIN (FRACTAL) DIFFERENCE

Inertion is fractal difference between action and its reaction, which is singular quantum of such duality. It is double inertion of action and/or inverse inertion of its reaction. Inertion is eigenvalue of such duality—it is time of one and space of another, which is gravity in between. Inertion is limitation of equal and opposite infinities. It expands action by simultaneously contracting reaction and/or vice versa. Inertion is overlap between double/inverse curvatures, which is their singular double/inverse quantum. Inertion is common denominator in between two fractals. There is no larger or smaller space and/or larger/smaller time than inertial quantum, which is totality of all the space in time curvature and/or totality of all time in equal and opposite spatial curvature, where spatial curvature is infinity of time curvatures, just as time curvature is infinity of fractal spatial curvatures. Action/reaction are inversely proportional, which is their eigenvalue. Eigenvalue (inertion) is inverse proportionality or double/inverse reality in between two equal opposites, which limits action of one by expanding infinitely reaction of another and equally vice versa. Inertion is singular fractal difference of both—

action and reaction—it is infinite reality of one and singular quantum of another, where singular in time is infinite in space and equally vice versa. It is singular positive quantum of one (i.e. space) and equal and opposite singular negative (inverse) quantum of another (i.e. time).

Inertion between action and reaction is simultaneous reality (resonance) of both, which are its instant two equal opposites of space and time (1=2x1/2). In context of exclusivity of inertial action or inertial reaction (where action and/or reaction are not consequences of inertion but are changeless reality), its inertion (singular quantum) is its equal opposite, which is 1D fractal difference (10). Action is singular fractal difference of inertion and double fractal difference (10^2) of its reaction. Inertial space here and now is double fractal difference of its relative/uncertain spatial realities, while constant passage of time in between is their inertion (eigenvalue), which is delay between such spatial action and reaction, which are double/inverse fractal difference. Time is singular fractal boundary between two infinite spatial realities and equally vice versa—space is singular boundary between two infinite spatial realities. Time is reduction of infinite to singular space and space is reduction of infinite to singular time. Each successive wave of accelerated/decelerated passage of time is further reduction of successive spatial infinities and equally vice versa—each successive sequence of inertial sizes is reduction of infinite periods of time to singular number. Density sequences are reduction of infinity of both—space and time, because they are space/time eigenvalue (inertion). An atom is singular density (quantum) of inertial space, which is reduction of its infinity to singular number. Fractal difference is difference between infinity and singularity (C), while double fractal difference is double singular/infinite distance or difference (C^2).

Singularity/infinity are equal and opposite distortions of inertial here and now, which is neither by being both. Contraction of spatial infinity into time singularity is equal and opposite expansion of spatial singularity into time infinity. Infinity/singularity are not infinite/singular realities—they are only equal and opposite boundaries of inertial sequence in between. They are instant Doppler effect or instant double/inverse infinite fractal 2x1/2 refractions of white light into different colors, where relative reality of such colors, as well as their boundaries, are exclusive function of inertial perspective. Spatial infinity of universe is Doppler effect frozen in singular time, while Doppler effect of its mass within (expansion of universe) is its gradual reaction in infinite time here and now, because mass within is space/time inversion of empty space without—one is instant reality of no time, while other is equal and opposite gradual reality in time, which is 2D/3D relative/uncertain (fractal) difference (1.111 difference between diagonal of a square and diagonal of a cube, or difference between round and flat size—circle/square circumference). Reality is singular something, which can be interpreted in infinite number of relatively different ways, all of them equally real. Singularity/infinity are consequence of double/inverse non-Euclidean curvatures (2D fractal difference or double infinity/singularity) or relative difference between round and flat curvatures (cube and a sphere, which is 1D fractal difference or infinite 1.111 repetitive sequence). Infinity and singularity are consequence of difference between attraction and equal and opposite repulsion (double/inverse curvatures). Attraction in inertial perspective from within is always equal and opposite repulsion in perspective from without, because convex curvature within is concave curvature without.

Increase of infinitely small size within is equal and opposite decrease of infinitely large size without, which is their equal and opposite expansion/contraction. Eigenvalue (½-way) is no difference between such sizes. Size/motion is only relative difference/distance. Such difference is real only in context of inertial point of observation. Two equal and opposite (double/inverse) non-Euclidean curvatures, which are 2D fractal difference, meet ½-way in between as singular Euclidean curvature, which is +/-1D difference of both. 1D fractal difference in between is equal and opposite inertion of both. Inertion or eigenvalue is only fractal ½ in between two wholes (2D fractal difference), which is double repetitive 1.111 sequence or infinite asymmetrical 1.234567...differentiation. Any inertial reality is always fractal ½ or inertion between its two equal opposites without and within. Time is always fractal singularity between two linear spatial directions and equally vice versa—space is singular point between two linear directions of constant passage of time. Space and time, which are inertial duality that is physical reality is double/inverse fractal reality of 2x(2x1/2)=4x1/4 or singular ¼ quantum, which is fractal ¾ without and/or 5/4 within and equally vice versa. Its eigenvalue is $(3x5)/4^2$ =15/16 or positive 1/16 that is negative 1/15 difference. 16-fold repetitive sequence is boundary of inertial ¼ reality, where ¼ is eigenvalue between 1/3 and 1/5, whose boundary is 15-fold repetitive sequence. 15-fold repetition of fractal 5/3 is its resonance in context of 4x1/4, which is instant addition of 16th part that is their synodic/sidereal (relative) difference.

4x1/4, which is ¾ and/or 5/4 is 1D fractal difference, which is only fractal ½ in between two wholes (quanta). It is time of space and/or space of time, which is only surface boundary or overlap between two equal opposites. It is Euclidean (flat) curvature in between two equal and opposite non-Euclidean curvatures. 4x1/4 is time wave of spatial reality (particle) and/or spatial wave of time (inverse particle). 4x1/4 is relative motion between two different sizes and/or relative size between two different inertial motions. Fractals 1/3 and 1/5 (3D and 5D) are action/reaction of inertial 4D in between (¼). 3D and 5D are double/inverse precessional curvatures of double helix in context of inertial 4D in between (15/16). 3D and 4D are eigenvalue in 12-fold repetition of inertial sequence in between, while 4D and 5D are eigenvalue in 20-fold repetition. Eigenvalue of both is 12x20=240 (synodic/sidereal difference between 5D solar and 4D Earth's reality). 5D is inertion of double/inverse 4D/6D, which are its double/inverse precessional curvatures. 4D/6D are $(6x4)/5^2$=24/25 fractal reality, which is repetitive 24-sequence followed by instant (quantum) addition of its angular 25th part. In context of inertial 4D here and now (singular ¼ quantum), its 1D fractal difference (5D) is repetitive 20-sequence (time), which is instant 1/20 difference here and now (space) while its 2D fractal difference (6D) is 24-repetitive sequence (time) that is instant 1/24 difference (space) here and now.

Inversion is spatial reflection or instant space of equal and opposite passage of time (equally vice versa), which is square of such spatial sequence. I.e. 72^2=72/(1/72)=5184 and/or 1440^2=2 073 600 years. 72 and 1440 are spatial fractals of orbital motion (5 ¼ day difference) reflecting equal and opposite periods of passage of time. 5184 is singular quantum of constant passage of time or singular quantum of spatial reality of time and 2 073 600 is singular quantum of spatial reality of gravity. Only 5184 years is density, which is not passage of time but instant space (singular quantum) and only 2 073 600 is density, which is not constant force of attraction/repulsion but singular quantum of spatial reality. Simultaneous multiplication and division is a square (inertion and inversion are double inertion). Fractal 1/10 spatial difference repeated 10 times is inertial here and now (1), which is $10/(1/10)=10^2$=100 fractal difference in between. 10^2 is double (2D) fractal difference between inertial and relative/uncertain sequence. It is difference between inertial and relative and/or uncertain space (spatial reality of time, which is 1D fractal difference of its constant passage). 10^2-fractal difference is spatial reflection of constant passage of time and/or time reflection of its spatial (relative) reality. I.e. 25 920, 6480 and 129 600-year periods are reflected as 5 days, 1 day and ¼ day of orbital distance ($25920x10^2$, $6480x10^2$ and $103680x10^2$ km). Eigenvalue between year of passage of time and spatial distance is time velocity of 100 km/year and/or 0.01 year per km, which is 300 000 sec/km or light speed difference between space and time.

40 000 km circumference of Earth is $400x10^2$, where 400 is 1/10 fractal difference (1.111) from angular (time) curvature of 360°. Orbital motion of 360 days (time) is reflected as 40 000 km of spatial size. Likewise $400=4x10^2$ or 2D fractal difference from inertial here and now (4D), which is 20-fold repetition of

1/20 difference (20/(1/20)=20^2 =400), just as 360 is 6/(1/6)=6^2 or 36x10 (10 is fractal difference between 360 and 400). Diameter of Earth 12 000=120x10^2 km is ½ reflection of synodic/sidereal 240-sec difference, while its radius is ¼ difference. Space/time reality of Earth, which is its singular rotational velocity of 0.463 km/sec and/or 2.16 sec/km is 2.16x10^2x10^2=21 600 years of precessional period, which is singular quantum of spatial reality of time, whereas 216 is 6^3, which is fractal reality in between 6^2x10=360 days and $6^2/(1/6^2)=6^4$=1296 years of fractal 5/4 (relative) difference between constant passage of time (5184 years of rotational time) and its spatial reality (relative space, 6480 years). 1296 years is singular ¼ that repeated 4 times is inertial period of passage of time (5184 years), which instantly adds 5th quarter, which is inertial space/time difference (eigenvalue) of such duality. 6^4 is inertial 6D repeated in all 4 directions of inertial here and now and/or inertial 4D repeated (curved) in all 6 directions of fractal 3/2.

It is 6x4=24 or fractal 6/4, which is eigenvalue (inertion) between 24x10=240 sec of synodic/sidereal (orbital/rotational) difference and 24/10=2.4 fractal difference between orbital and precessional motion. 240=216x1.111 or 1/10 fractal difference of inertial 216 sequence. Its derivates (fractal differences or relative/uncertain sequences) are 4x6^3=4x216=864 or 1/10^2 of one day, which is fractal ¾ of inertial 6D that is eigenvalue in 12-fold repetition (864x12=10 368 or ½ precessional cycle), 12x6^2=432 that is 24-fold repetitive sequence (432x24) and 24x6=144-fold repetitive sequence (10368/144=72, which is spatial singularity of such infinite passage of time and/or 144^2=20 736 or complete precessional cycle). 24^2=576 or 1/10 fractal difference of 57.6° of perihelion orbital, just as 60-57.6 is 2.4 and 2.4^2=5.76 or inverse 1/10 fractal difference. 576=2x288=4x144=8x72=16x36=32x18=64x9, which is 6 fold 2x1/2 multiplication/division of singular 3^2=9, which is inertial 3/2 here and now (rotating Earth). 6-fold harmonics or 6-fold curvature, which is 6 quantum points of inversion in between (singular mass/energy of infinite space/time) is boundary of fractal 3/2 reality (6 planets/electrons within/without 4 double/inverse planets of Bode's sequencing). 576x9=5184=9^2x8^2 is double/inverse fractal 9/8 ($3^2/2^3$) that is 1/9 and/or 1/8 fractal difference, where 1/9=0.111 infinite repetitive sequencing of inertial here and now, which is its infinite boundary, just as equal and opposite 90° is inversion of 1.111x10^2. Fractal 9/8 that repeats in 9x8=72-fold sequences is double/inverse (space/time) infinity of singular here and now.

Fractal 1/10 of 400 is 360, while fractal 1/10 of 360 is 324=18^2 or $(3x6)^2$. 324=3/2x216, where 216=2x108=4x54=8x27, which are 1/10^2 fractal differences of 1, ½ and ¼ orbital precessional period, while 8x27=8x3^3 or (9x8)x3=3x72, which is triple repetition (rotation) of fractal 9/8. 216x2=432 and 432x2=864 or

$1/10^2$ time of one rotational period. Such sequencing is 5-fold repetition or fractal 5D in between 4D and 6D (324 and 576-sequences) 576/432 and/or 432/324 are successive fractal 4/3, which is equal opposite of 216x3/2. 16x324=5184 and 20x324=6480, which is 24x216=5184 and/or 30x216=6480. Fractal 5/4 difference of 1296 years without is 6x216 or 4x324 (fractal 3/2) within. 2x576=1152, which is $144=12^2$ fractal difference from 1296=9x144 (fractal 8/9).

Space/time is infinite 2x1/2 multiplication/division of inertial 3D, which is infinite repetitive curvature of its singular quantum 2x3=6. Singular quantum of passage of time is 5184=**6**x864 and 864=**6**x144, while 144=**6**x24 and 24=**6**x4, which is 4 times 6-fold repetition of inertial 4D here and now (Earth), while 6480=**6**x1080, 1080=**6**x180, 180=**6**x30 and 30=**6**x5 is 4 times 6-fold repetition of inertial solar 5D. 5184 is 4D sequence, while 6480=648x10 is 3D sequence, which is 1/10 fractal (relative) difference It is inertial 6x4=24 in context of fractal 5/4 solar/Earth duality, which ¼ difference within that is 1/5 difference without or 20-fold repetitive curvature, which is its eigenvalue. 24/20 is inertial fractal 5/6. Fractal 5/6 is boundary of inertial here and now. 5400 is $54x10^2$ or 2D sequence that is double fractal difference of inertial 4D here and now. 5400=**6**x900, 900=**6**x150 and 150=**6**x25 or $6x5^2$. It is only triple 6-fold repetition of twice larger fractal difference (5^2), because inertial 4D is double/inverse 5D within/without or 5^2-5=20-fold repetition of fractal 4/5 in between. **Space/time is quantum inertion between fractal 4/5=0.8 and 24/25=0.96 differences. It is singular 0.96/0.8= 1.2 or fractal 6/5, which is singularity between dual infinity and/or infinity between dual singularity.**

Quantum reality is limitation of infinite, because infinity is not real—it is only perspective (curvature) of singular reality here and now and equally vice versa—singularity is only perspective of infinite here and now, which are meaningless dual (fractal) descriptions of inertial 3D, because 3D is not exclusivity of two equal opposites but inclusivity of all equal opposites, where inclusivity in space is exclusivity in time and equally vice versa. Reality is supersymmetry (eigenvalue) of all its fractal comparisons, which is inertial here and now, because it is always comparison of here and now with itself (self-interference). Here and now is simultaneous infinity and singularity, as well as its quantum boundaries. Infinite empty space of universe in context of singular mass within is singular mass within in context of larger space without, because convex within is concave without and equally vice versa. Infinite emptiness within in context of equal and opposite infinite passage of time is singular instant density of no time without. Infinity is singular fractal difference—it is singular square without, which is inversion of infinite linear repetition within. It is singular horizontal principal of infinity of angular subordinates (Balmer) I.e. $3^2=9$ is singular equal opposite of infinite linear repetition in opposite direction, which is 9x2=18x2=36x2=72x2=144x2=288x2=576.

0 is singularity between double/inverse (positive/negative) 1's. 1 is also singularity between two equal and opposite 0's. Infinity of positive numbers, larger than 1, is only inverse singularity (0), just as singularity (0) in opposite direction from 1 is equal and opposite infinity of larger numbers in its angular depth. Left infinity is right singularity and right singularity is left infinity, because left and right are mirror image of each other—one is flat reflection of 3D reality of another, where 3D in space is surface in time and equally vice versa, which is fractal 3/2 duality of 1/10 (1.111) difference. Infinity of whole numbers in space is infinity of inverse fractals in time and equally vice versa (spatial infinity = time singularity). Infinity and singularity in two equal and opposite directions of 1 in between, which is their eigenvalue, are double/inverse reflections of such point in between, which is inertial infinity of one (0) and inertial singularity of another (infinity of larger numbers). 1 is angular ½ of such horizontal double/inverse singularity/infinity (2x1/2). Difference between infinity and its equal and opposite singularity is only relative/uncertain. It is exclusive spatial and/or time difference. In context of singular space/time, which is inertial 1 in between, there is no difference between them—they are only its double/inverse reflections (where double in time is inverse in space etc). 1 is inertial framework of both, singularity and infinity. They are exclusive reality only in its context. Different inertion is relatively different infinity/singularity (expanded in one and contracted in opposite direction). **Square of infinity without is its singularity within, just as inversion of singularity (derivation of exponent) is its equal and opposite infinity.**

Boundary of inertion is its inversion and boundary of inversion is its inertion, which are double boundary of double inertion in between. There is no larger/smaller (faster/slower) reality that is real here and now than double inversion of its inertion, which is its singular (1/4) quantum. Singular quantum (double/inverse inertion) is light speed difference of inertial here and now. Smaller/larger size/motion is only spatial or time fractal of its quantum, which is gradually accumulated to singular quantum only in context of passage of time, which is its instant reality. Larger than infinite (smaller than singular and/or more than improbable) spatial difference is passage of time in between and equally vice versa—larger than infinite time difference is fractal spatial reality in between. Quantum reality is 2x1/2 multiplication/division of inertial here and now, which is preceded by equal and opposite 2x1/2 multiplication/division of its double/inverse boundaries. It is 2x1/2 pattern of numerical sequencing, because numerical abstraction is inertial framework of infinitely differentiated reality here and now. Numerical abstraction is double/inverse or infinitely larger/smaller reality that is inertion (eigenvalue) of physical reality here and now. Numerical abstraction is singularity of infinity of fractal, relatively different theories. It is theory of theories or double/inverse theory, which is not theory at all by being all of them at the same time. Abstraction is inertial framework of infinity of relatively different theories, which are real only in its inertial framework. Numerical abstraction is not theory—it is reality of all uncertain/relative theories, because it is their supersymmetry (eigenvalue). All relatively different theories resonate in inertial context of singular numerical abstraction. Singular numerical pattern is simultaneous infinity of relatively different applications, which are expanded into different theories on surface, while in depth they are contracted into singular point.

Abstraction is infinitely larger/smaller physical reality, which is singular cause of infinity of physical effects here and now. Physical reality is double/inverse or space/time infinite expansion of numerical abstraction, which is its quintessence, just as abstraction is its equal and opposite contraction. Abstraction is form of physical reality here and now, which is its content. It is reality that creates reality here and now. It is infinitely larger singular reality without of infinity of different forms within. Mathematical abstraction is double/inverse reality of inertial here and now, which is 1D fractal difference, while abstraction of abstraction is no pattern of numerical sequences at all but only singular 1, which is common denominator of infinity of different numerical patterns. Singularity of all abstractions is double fractal difference, which is equal opposite of all relative physical (space/time) sequences. Reduction of infinity through double fractal difference to singularity is reduction of space and time to singular point of no difference right here and now.

Reality is infinite double/inverse curvature of inertial here and now in between. It is double/inverse wave of a particle and/or double/inverse particle of a wave. **Wave is singular reality (inertion), which is interaction or interference between two (double/inverse) particles, just as particle is synchronicity (eigenvalue) between two equal and opposite waves.** Inertion (here and now) of its double/inverse space and time, and/or space/time of inertion in between, is one out of infinite number of sequences of its double/inverse precessional curvature, where one fractal reality follows another. Every double/inverse duality is double infinity of singular inertion and every inertion is dual infinity of singular space/time (and equally vice versa). Dual action/reaction is singular reaction of inertion, which is its equal and opposite ½. Action and reaction are ¼ each of inertial ½, which is infinite 2x1/2 double/inverse spiraling (precessional) curvature. Here and now is only one of its fractal inertial sequences, bounded by its double/inverse space/time boundaries. Infinite precessional (fractal) curvature is infinity of fractal realities, where every fractal is only one out of infinite number of equal possibilities. It is infinite chain action/reaction.

Each action is infinity of double/inverse reactions, which are singular action of successive sequence of double/inverse infinite reactions. One action is one equal and opposite reaction here and now. In context of 1D fractal difference, it is two; in context of double fractal difference it is three, and in context of infinitely different fractal reality, it is infinite number of reactions. Reality is infinite double/inverse (anti-parallel) causal curvature, where singular possibility here and now, echoes in infinity of successive 2x1/2 fractals (waves). Action/reaction is space/time or physical reality. It is not only exclusive spatial or time reality. Reaction transcends exclusivity of inertial framework of its action and equally vice versa. It is reality of time and space, and lack of instant spatial reaction here and now is infinitely larger reaction in time. Larger uninterrupted period of time between action and its reaction indicates larger size/motion (momentum or inertion) of such reaction. **Action/reaction or causality is only law and it surpasses all inertial limitations.** Action/reaction is reality itself, which is infinite momentum (infinite in space is singular in time and equally vice versa—it is eigenvalue of all and nothing). Only numerical abstraction transcends all inertial (space/time) limitations.

Synodic/sidereal difference of 236 sec/day, which is 4 sec less than 4 min (i.e. 4th side of 4th shell), is singular numerical abstraction of space/time reality here and now. Such 3D numerical sequence is in its depth infinity of all fractal space/time theories. It is space/time line of symmetry (eigenvalue), which is total double/inverse description of both. It completely describes Earth's reality within (4D) and its equal and opposite solar reality (5D) without. 236 sec is inertion that perpetuates time wave between such two particles, as well as its double/inverse boundary in space and time. It is constant difference between two variables—relative and uncertain reality without/within. It is singular inertion of all precessional time waves (orbital/rotational duality) and sizes of its double/inverse particles (Sun/Earth and all massive particles in between, which are its 6-fold harmonics). Numerical sequences are not accidental and inconsequential numerals, but complete definition that transcends any fractal theory. Numerical abstraction does not differentiate between relative/uncertain forms (space, time, gravity, size, motion, energy, mass, etc) of one reality, because such differentiation is not real in its context. It does not discriminate between large and small, past and future, real and unreal, because any surface duality is unreal in its singular depth. If any fractal interpretation does not reflect causal numerical abstraction—it is not correct, because numerical abstraction has no equal opposite—it is not a theory, but reality of all theories. It is the end of all theoretical discussion.

Fractal 3/2 of 240 is 360 and their eigenvalue (resonance) is **86400 (1 day)=240x360.** $864x10^2$ is 3D inertion between its double/inverse fractal (surface) dualities (36 and 24). It is time of double/inverse (relative/uncertain) spaces. 864 is delay between solar action and its reaction here and now, which is constant fractal difference or constant passage of time. It is singular wave (time) in between two particles (space). 86 400 sec here and now is 240 sec in context of inertial orbital of 360 days and/or vice versa. 86400 sec (rotation) and 360 days (orbital) are two equal opposites of inertial 240 sec difference. 240 sec is perfect duality (no eccentricity), while 236 sec is its 3D sequence or 4 sec eccentricity. 235.6 is 4D sequence of double eccentricity (reflection of 23 hrs 56 min), which is 4.4 sec difference (fractal limitations of infinite repetitive curvature of inertial 4D here and now: $2x10^2+2^2x10-4/10-4/10^2-...=235.555...$). 240 is perfect inertionless fractal of action/reaction, without its inertion in between. It is perfect duality of orbital/rotational motion without its inertial size (Earth =4D) in between. Rotation/orbital motion of Earth in between is therefore 3D inertial sequence of perfect surface duality, which is 240-4=236. Earth in between is imperfection (eccentricity) of its double/inverse perfect curvatures and equally vice versa.

Earth (size) is fractal (relative/uncertain) difference, which is inertion that perpetuates such double/inverse motion. 236/4=59 and 240/4=60, which is 1/60 fractal difference or 6x10 double/inverse repetition of inertial sequence, which is $60^2=3600$ that is ½ of 7200. 1/10 fractal difference of 7200 is 720, which is 6480 years, (time boundary of spatial inertion in between). 3/2x3600=5400 of orbital precessional time. 6480 is 1/10 fractal difference of 7200, while 5184 is double fractal difference of 6480 (6480-2x648=5184, which is fractal 5/4 or 1.25 fractal difference that is 3D limitation of infinite 1.23456...). 7200 years is inertial context of its singular/double fractal difference, which are 5184 and/or 6480-years periods of time. 1/60 in context of synodic/sidereal difference precedes 60° curvature of perihelion orbital, where 60° is 2/3 of 90° here and now (and/or 5/6 of 72° of precessional rotation). 240 and 360 are fractal 3/2 in context of inertial ½ difference, which is gravitational context of such space/time (2/4=6/3=1/2). Earth and solar reality are different only in fractal space/time context. In gravitational context, they are no difference (½).

864 is gravitational sequence of dual space/time sequences (36/24). It is galactic eigenvalue of solar/Earth duality and/or vice versa—it is here and now, which

is eigenvalue (inertion) between solar and galactic realities. Space/time reality of Earth within is overlap between 1D/2D fractal realities without, which are its solar/galactic space/times, where solar space is galactic time and solar time is galactic space. Time here and now is fractal solar space, while gravity is fractal galactic space. Time is overlap between solar and Earth's spaces and gravity is overlap (symmetry or resonance) between Earth's and galactic spaces, where Earth's space is overlap between solar and galactic times. Earth is time wave of solar/galactic spatial interaction (and/or spatial wave of their interaction in time). **8-6-4=2^3-2x3-2^2** sequence is perfect description of space/time here and now as overlap between solar and galactic double/inverse fractal realities. It is one fractal (2x3=6) between two equal and opposite multiples (exponents). It is double 3D (**2**x3), which is inertion of its double/inverse 2D ($2^3/2^2$) or double time between two equal and opposite spatial realities. It is double derivate (2x3) of two equal and opposite exponential (infinite) realities. 2^3=8 and 2^2=4 are boundaries (infinity/singularity) of inertial 2x3=6 in between. They are infinite and singular space without/within (universe/micro-world) of constant passage of time that is instant space here and now. 4/8=1/2 are relativity/uncertainty of inertial here and now (6) in between, which are equal and opposite 1D fractal difference (+/- non-Euclidean curvatures).

Here and now is quantum eigenvalue or line of symmetry (overlap) between its equal opposites (space/time). Space and time here and now are double fractal of galactic and singular fractal of solar reality (double/singular infinity or double/singular relativity/uncertainty). 8-6-4 is double 4-3-2 sequence, because only ½ of reality is instant here and now, while another ½ is its gradual change (time). 4D is inertial context of 3D that is context of 2D (equally vice versa), where 4D/2D are relative/uncertain sequences or spatial expansion/contraction (double/inverse time) of inertial 3D sequence in between. 432 is fractal 3/4 without (motion) that is equal and opposite fractal 2/3 within (size) of inertial here and now in between (1, which is their difference). 2x3=6 and 3x4=12, while their eigenvalue is 12x6=72 (5 additional days) and 72^2=5184 years that is singular quantum of such reality. Resonance (eigenvalue) of fractal 4/5 that succeeds ¾ is 4x5=20 and eigenvalue of 72x20=1440 (1/4 additional day), where 1440^2=2 073 600 years that is limitation of inertial gravity here and now. Resonance with following 5/6 is 30x1440=43200 (double precessional period), which is 10^2 (double fractal difference) of inertial 432. It is boundary of space in time and/or time in space or total time of inertial Earth's spatial sequence (distance between surface of atom and here and now and/or inverse distance between surface of solar reality and here and now, which is complete curvature of expansion/contraction (double/inverse time or relative and uncertain spatial reality).

Reality is infinite repetitive sequencing of fractal 6 here and now. It is 6x1.111= 6.666, of which only ½ is instant here and now (3x1.111=3.333), while another ½ is its time (gradual change of instant reality, which is its equal and opposite reaction). Orbital period is 3 out of infinite number of such repetitive sequences. It is 6^2x10^1+6x10^0-6x10^{-1}+...=365.4 days per year, which is 365.25 in context of inertial 0.15 or 3/2x10^{-1} difference that is reality here and now. Period of one year is double/inverse multiple in context of one fractal in between or double (positive/negative) fractal difference, which is no difference here and now. It is 3D sequence of infinite inertion/inversion=double inertion Fibonacci sequencing. Fractals are infinity of different possibilities of singular multiple and equally vice versa. Double fractal difference is two simultaneous possibilities of one multiple, both of them equally real. Another possibility, which is same description, is: 6^2x10^1+6^1x10^0-6^0=365 days, which is inertial ¼ difference (0.25 is fractal 5/2x10^{-1}, where fractal 5/2 is inertion (quantum) of fractal 3/2—it is inertial gravitational wave of its fractal 3/2 space/time. 5/2 is double/inverse or 4x1/4 of inertial 2x1/2 wave. Negative 0.25 difference is inversion of positive 0.4 difference, because 0.25=1/4 is equal opposite of 4 in context of inertial 1/10 difference—one is fractal 3/2, while another is fractal 5/2 difference.

Fractal definitions are superficial theories, while numerical sequences are real—they incorporate all infinity of different theories into singular context. Singular abstraction is inertial framework of infinity of different theories. 8-6-4 sequence, which is eigenvalue (resonance) of 2-4 and 3-6 sequences, is singular definition of space/time and their gravity. 1D fractal difference of such sequence is inversion of one of its directions, which is its double inertion. Inversion of 864 is 648 and inversion of 648 is 486. Triple rotation is 3D sequence of inertion/inversion = double inertion. **864x10^2** (seconds) within is period of one day, while **648x10** (years) without is its boundary, which is one quantum period of time. It is 1D or space/time fractal difference in context of inertial orbital motion in between. Rotation within (seconds) is fractal reaction precessional motion (years) and equally vice versa. Precession without is multiplication, while rotation within is fragmentation of inertial orbital time in between (where multiplication in time is fragmentation in space and/or vice versa). Quantum period of 6480 years and rotational period of 86 400 seconds are inertion and its inversion—one is singular/infinite boundary of another, while here and now (spatial Earth) is ½ way in between (eigenvalue). Rotating Earth is real only in context of its quantum period, which is its inertial background (singular action of infinity of fractal reactions here and now). It is boundary of its time in (relative) space and/or boundary of space in time.

One quantum period of time (¼ precession) is infinite time of universe limited by inertial fractal curvature here and now. It is beginning/end of double/inverse constant passages of time, as well as gradual spatial changes. Infinite time of universe is not real, because all its time here and now is contracted within ¼ precessional period. Infinite time of universe is infinite repetition of such singular time sequence. ¼ is instant De Broglie' quantum leap, which is instant energy jump of massive particle that emits a photon. Time beyond its ¼ boundary is not time but instant space. Time beyond its singular quantum is not only one but infinite number of different spatial possibilities, all equally real (Everett's many worlds). Earth here and now is only one space/time sequence out of infinite number of relative sequences, which are simultaneous in context of inertial (gravitational) solar reality. Infinite numbers of relatively different spatial possibilities that are inversely different passages of time are instant density of no time difference in context of inertial solar reality (just as infinity of relatively different solar sequences are singular galactic sequence and infinity of galactic space/time sequences are singular universe). Passage of time here and now is exclusive reality of inertial spatial sequence (Earth). In context of solar reality—all infinity of different time sequences (infinite time here and now) are its instant spatial density (atom). Accelerated passage of time of solar reality (slower motion) is also its exclusive reality, which is meaningless in galactic context, just as twice-accelerated galactic time is unreal in context of universe. Time is instant spatial reality of infinitely larger space, because within precedes without in spatial direction, just as without precedes within in direction of time.

Passage of time (25 920 years) is instant rotational (round) size of Earth within (40 000 km) and linear spatial distance between here and now and core of galaxy (25 920 light years), because solar/galactic and solar/Earth spaces are double/inverse (rotational/linear) light speed difference. Galaxy and Earth are relative/uncertain space (universe/atom) of inertial solar space in between. Size of galaxy without and/or size of Earth within are double/inverse spaces of passage of time, which is inertial (fractal) solar space in between. Size of galaxy without and size of Earth within are double/inverse infinite reactions of singular passage of time here and now (tg and ctg). Such double/inverse sizes and their equal and opposite densities (attraction/repulsion) are singular time overlap in between curved in convex and equal and opposite concave curvatures. Earth is spatial curvature of time and galaxy is time curvature of space (+/- non-Euclidean curvatures) of linear direction of time in between (Euclidean curvature), which is yet another non-Euclidean curvature in 3rd direction, which is solar size. Galaxy is time (size of relative solar space) inverted inside out, just as time is inertial size (mass) here and now inverted inside out. Dense mass of Earth inverted inside out is empty space of solar reality, which is its equal and opposite passage of time (densification of empty space).

Dense mass of Earth is line of symmetry between empty space and passage of time. Only empty space is passage of time in future direction, while denser mass than inertial mass here and now is inverse (negative) passage of time, which is passage in past direction. Time is fractal reality, which is instant density of solar space. Time is not passage but instant space of solar reality. It is its instant dense mass within. Solar mass inverted inside out is twice-empty space of galaxy in context of accelerated (squared) passage of time. Precessional motion is infinitely small circumference of Earth (2D) and/or infinitely large radius of galaxy (1D), which is 2Π difference (double curvature, which is 2x360=720 or 72x10). Such double/inverse sizes are double/inverse infinity in context of singular time. Each is 1D fractal difference of passage of time in between, while both are double fractal difference—one is attraction/repulsion (gravity/antigravity) of another, because convex curvature within is concave curvature without.

Only double fractal difference that is space/space and/or time/time difference are two equal opposites (two quanta) which repel/attract each other (negative/positive polarities). Singular fractal difference, which is double/inverse passage of time between two spaces, is only neutral fractal in between (Euclidean curvature between two non-Euclidean curvatures). It is only their line of symmetry or fractal surface in between two opposites. Time is double/inverse eigenvalue of both—relative space without and uncertain space within, which does not affect either, unless ¼ quantum, when time is not time but space, therefore double/inverse attractive/repulsive reality of both, which is instant (singular) 4x1/4 of two equal and opposite dualities (2x1/2). Such ¼ quantum is instant reversal of changeless gravitational reality of both. It is instant repulsion of Earth that is instant attraction of galaxy and/or vice versa—instant attraction of Earth that is instant repulsion of galaxy. It is instant change of gravitational sequence, which is changeless reality of universe. Such change is instant change of passage of time within that is instant change of spatial reality of universe without (instant contraction that is reaction of its gradual expansion).

Numerical sequences are complete and unreserved description of any size/motion and their density. They are singularity of all fractal descriptions, which are only its partial derivates. Triple 8-6-4 sequence is complete description of physical (space/time) reality. It is one imperfect fractal between two perfect wholes and/or vice versa. Only imperfection is double perfection and only perfection is double imperfection. Two imperfect sequences of perfect 8-6-4 are 240-sec synodic/sidereal difference and 360-days year and equally vice versa—8-6-4 is imperfect sequence of perfect synodic/sidereal difference of 236 sec and/or perfect 365.25-days year. 240 and 236-sec difference are equally two imperfect curvatures of perfect 4 sec difference in between or two perfect curvatures of 4-sec eccentricity in between, because reality is fractal. It is reality of small quantum steps in between large gradual changes and/or large quantum steps in between small gradual changes. Quantum is boundary of infinite gradual change, because it is inversion of inertion that is double inertion. Quantum is inversion of infinity into singularity or space into time. It is end of one and beginning of next sequence. There is no gradual reality beyond its quantum boundaries, because any gradual reality beyond is in inverse direction, which is not real here and now.

Only quantum is real here and now. There is no difference between infinity and singularity—in inertial context here and now, they are same reality. Infinitely large is boundary of size, just as infinitely fast is boundary of motion—it is its limit, which instantly converts such infinity into next singularity. Constant angle of curvature is parabolae here and now, while in distance it is exponential (spiraling) round curvature. Constant angle of round curvature here and now is infinitely distorted in distance of space and time (tangent of such curvature), which is infinite parabolae that ends in singularity/infinity. Distortion of reality is proportional with its spatial distance, where distortion is double/inverse. It is simultaneous horizontal distortion of space in time and angular distortion of time in space. Infinity/singularity is perspective and not reality. Universe and atom are not real—they are twice-infinite distortions of same reality as here and now. Perspective of universe and micro-world is inversion (equal opposite) of their reality. They are observed through convex and concave curvatures, which is their complete double/inverse distortion. They are infinite distortion of singular reality and/or singular distortion of infinite reality, which is neither. Infinity is 90° curvature, which is horizon of inertial perspective. Any reality beyond is not real here and now, because it is around the corner (quantum boundary).

Any difference is relative—infinity of one is singularity of another and imperfection of one is perfection of another. 360 days of repetition of 240-sec synodic/sidereal difference creates additional quantum of time (86 400 sec), which is vertical (angular) tilt of rotation/size. 240 sec difference in context of orbital motion is angular tilt of rotation and 240 sec in context of rotation is its orbital motion. 4 min (240 sec) instant spatial difference is singular inertion (quantum) of infinite passage of time (rotation and orbital motion/tilt). It is quantum boundary that limits infinite time. It is instant spatial curvature of time and 1 min is its singular ¼ quantum. 4 sec discrepancy between perfect 240 and 236-sec difference is curvature within curvature or double/inverse curvature, which is double/inverse 1/60 boundary of time and its singular1/4 quantum is 1 sec. It is double fractal difference. Just as 86400/240=360 days, so is 86400/4=21600 years, because day/year are 1D fractal (relative) difference of time (orbital/rotational relative difference). 21600 years (**6^3x10^2**) and 360 days (**6^2x10**) are double/inverse curvatures of same inertial time (**6**—here and now). 1 year is boundary of infinite number of days and 1 precessional period is boundary of infinite number of years.

A year, day, precessional period or all infinite time of universe, as well as singular spin of electron, is only one time in different quantum boundaries (curvatures). Different periods of time are relatively different densities of only one time, just as different sizes are different density of only one space, where size in space is period of time in time and size in time is period of time in space. There is only one reality in infinity of different densities, which are its double/inverse quantum boundaries. Smaller instant (spatial) difference of time is larger period of passage of time and larger difference is smaller period of passage. 240 sec in space is 1 year and/or 1 day in time (Fibonacci sequencing—there is two singularities on beginning of infinity and/or vice versa) and 4 sec in space is ¼ precessional time, while 1 sec is complete precessional period of passage of time. Infinitely small fragment of time of the universe (spin of electron) is infinite time here and now. No spatial time difference (changeless universe) is infinite passage of time here and now and equally vice versa. 1 sec of time is static spatial wave (Wheeler-DeWitt formulation). It is infinite time of universe in context of only 1 sec and/or vice versa. 1 day is all the time of the universe apart from right now (1 sec), 236 sec is all the time of the universe, apart from time of 1 year and 1 sec is all the time of the universe apart from period of 25 920 years.

Reaction of spatial motion is equal and opposite time motion, which is its inversion. Faster spatial motion is slower time motion and equally vice versa. Spatial motion is motion of spatial size, while time motion is motion of period of time. Time motion is passage of time here and now. Faster motion is slower passage and slower motion is faster passage (larger time density). Reaction of rotation is orbital motion (angular tilt), which is time motion of spatial rotational motion. Reaction of orbital motion is precessional rotation, which is its time motion. Only double fractal difference is difference between equals (space/space); therefore orbital motion is only ½ fractal in between (space/time difference). Orbital motion (vertical) tilt is only fractal passage of time between rotation and precessional rotation. Orbital motion is not motion but instant size between double/inverse rotational and precessional motions. Orbital motion is instant surface between two rotating spheres (inertion). Orbital motion is double/inverse boundary of infinitely fast rotation within and infinitely slow precessional rotation without (which is infinitely fast time motion). Precessional (spatial) motion is reaction of equal and opposite time motion—it is discrepancy (overlap) between spatial and time motions. Precession of rotation is its ¼ quantum boundary and so is precession of orbital motion (angular tilt) its quantum boundary. Precession of rotation is boundary of such passage of time. Precession of orbital motion reflects as precession of angular motion (tilt), which is discrepancy between rotational precession and perihelion orbital.

Angular tilt (change of seasons) is equal opposite of rotation. Precessional rotation is singular gradual 90° (1/4) motion followed by instant singularly small shift, while angular tilt is infinite number of gradual motions, which are only ¼ of 90°, followed by instant ¼ (90°) shift, which is infinitely large shift or 90° instant angular rotation of horizontal rotation (upside down turning of Earth in ½ intervals, and only fractal ¼ swing in ¼ intervals, because one is positive and another one is negative discrepancy of perfect 90° curvature). Precessional rotation is passage of time here and now, while precessional angular rotation (change of seasons) is decelerated passage of time, which is boundary of passage of time (double time). Precessional rotation is boundary of time of one day and angular precessional rotation is boundary of time of one year. Precessional orbital motion, which is precessional angular rotation of Earth, is time of constant passage of time—it is its quantum boundary, which is real here and now only in ¼ quanta. Precession of angular motion is not real here and now, unless full ¼ quantum. Any change less than ¼ is only gradual accumulation of tension within (inside) massive Earth, which is not real on its surface, unless ¼ quantum. Gradual increase of tension in angular direction is caused by discrepancy (synodic/sidereal difference) between horizontal rotation (precessional rotation) and angular tilt, which is not instant reaction, because it is only fractal change.

Just as rotation is gradually pushed in angular direction by synodic/sidereal difference between rotation and orbital motion, so is angular shift. Such motion is not instant but quantum motion in ¼ intervals, because angular shift is fractal in between rotation/precessional rotation. Gradual shift of rotation in north/south direction is only increase of tension (momentum) in equal and opposite shift of angular tilt. It is instant shift only when such momentum is equal to its equal and opposite inertion (rotation). Only infinite number of inverse quanta (angular tilt) is singular inertial quantum (1/4 rotation in angular direction). Precessional rotation is gradual ¼ rotation, followed by relatively small shift (its small spatial discrepancy, which is large time difference), while angular ¼ shift of angular change of seasons is instant ¼ rotation of rotation, preceded by infinite number of fractal ¼ motions (tilts). Rotation is limited by its ¼ precessional periods, which are its boundary (infinity of quanta), while ¼ quantum changes are only singular ¼ quantum of equal and opposite angular rotation of Earth, where angular rotation is linear solar (attraction/repulsion) or galactic round orbital motion. Round orbital is time orbital and linear orbital is gravitational orbital. Gravitational orbital (pulsating expulsion/contraction) in solar context is round (time) orbital in galactic context, because galaxy is infinitely larger than solar reality and round motion around galaxy is infinitely eccentric curvature, which is linear motion in solar context. Such eccentricity is real only in solar context, while in galactic context; it is perfect curvature, because of 90° angular time difference (galaxy is infinitely slower spatial motion that is infinitely faster motion in time; therefore infinitely faster passage of time, where linear direction in solar context is infinitely accelerated into round curvature).

Angular tilt (orbital motion) and rotation are resonant only in singular points of solstice/equinox. Solstice/equinox are points where inertion of rotation is equal to inverse inertion of angular tilt and vice versa, which is instant change of direction of angular motion. Solstice/equinox are double/inverse 4x1/4 eigenvalues between horizontal and angular motion. In context of precessional motion, infinity of such instant points is gradual (uninterrupted) curvature, while periods of time in between (seasons) are not real. Periods of time between solstice/equinox are exclusive reality here and now. Each point of solstice/equinox is (1/288)° of precessional motion (1° per 72 years), which is its singular quantum. (1/288)° of precessional motion is 10000 km/5400 years/4=0.463 meters, which is spatial reaction of rotational velocity (time), where 1/0463=2.16 ($216/10^2$, which is singular quantum of time). Precessional motion is space of time—it is constant change (velocity) of size reflected as infinitely small fractal of its constant size. ¼ of size (10 000 km) divided by its velocity (0.463 km/sec) is 21 600 years of precessional period, which is boundary of such inertial space/time sequence.

90° of precessional motion is full 360° curvature of gradual changes of season, because there is 4 solstice/equinoxes in a year (every 4x1/4 wave is only ¼ of infinitely larger/smaller 4x1/4 wave). Solstice/equinox are only points where rotation and orbital motion are simultaneous reality (inertion/inversion=double inertion). They are only points real in context of precessional motion, which is their time (double time or time of time). Quantum change is possible only in solstice/equinox points, when double/inverse momentum between rotation and angular motion are synchronized (eigenvalue) in inertial context of their precessional motion (passage of time). Any gradual change of seasons in between such points is not precessional reality but only its gradual in-depth accumulation of tension. Only ¼ of gradual precessional east/west curvature is singular (instant) ¼ quantum of angular north/south rotation. Quantum is instant reality (no time, which is double/inverse or accelerated time of its fractal/relative reality).

Angular tilt (space) is +/-22.5°, which is ¼ of its north/south direction, just as 1 year (time) is 4 seasons, because space is equal opposite (inversion) of time and vice versa. Time of one precessional period is time of time of one year, which is its instant space. It is also 4x1/4 periods that are double/inverse time reflections of inertial time of one year. 4x1/4 precessional periods are expanded 4x1/4 singular points of solstice/equinox, just as 4 seasons are expanded singular sunset/sunrise, because any slower motion is faster (accelerated) time that expands singularity into linear direction of passage of time, which is inertial framework without of its past sequence of passage of time within. 4x1/4 precessional periods are double/inverse seasons (expanded seasons of warm/ice ages in ½ periods, interrupted by fractal ¼ changes in between, which are its gradual reversal/advance). 4 quantum points between 4 precessional periods are further expanded in context of double precessional wave (129 000 years), which are gradual curvature of rotation in angular direction of ecliptic. 4 quantum points between such double precessional periods are yet again expanded into 4 times larger time wave (518 400 and/or 432 000 years), which is not only curvature of rotational motion, but also gradual shift of orbital motion. 4x1/4 quantum periods of singular precessional motion (5184 years intervals) are shifts of rotation in horizontal direction of ecliptic and full precessional periods (20736/21600/25920 years), which are 4x1/4 of larger time wave (103680/108000/129600 years), are its shifts in angular direction of ecliptic (double curvature in shape of "8"). Such quantum shifts are exclusive horizontal/vertical change of rotational motion, which does not affect orbital motion.

4x108000=432000-year (518400) wave is change of orbital motion in horizontal direction of ecliptic, which is triple rotational curvature (curvature in direction of ecliptic, followed by additional curvature in its angular (vertical) direction, followed by its reversal, which is inverse angular curvature into previous plane of ecliptic. Such triple rotational curvature creates difference between initial and final (3rd) rotational curvature, which are in same horizontal plane of ecliptic through instant accumulation of triple double/inverse difference (curvature in "∞" shape with additional round curvature in between and above, which is curvature of rotation in angular direction of ecliptic. Such curvature in angular direction is difference between two different horizontal curvatures, which is shift of orbital motion). 4x432000=1 728 000 (and/or 4x518 400=2 073 600 years are not only shifts in horizontal, but also angular direction of ecliptic, which is 4-fold rotational curvature. 2 073 600-years is boundary of time changes (changes of orbital/rotational motion). Double/inverse quantum changes of rotation are singular quantum of orbital changes, while double/inverse orbital changes are singular quantum of spatial change, which does not affect only motion, but also size of Earth.

2 073 600=1440^2 (¼ additional day of a year) is boundary of double/inverse passages of time or absolute (4D) time boundary. Such quantum is singular gravitational change of size, which is its gradual contraction/expansion or change of inertial density/gravity. Time is space of relative (solar) reality and gravity is its time, which is space of twice-relative reality (galaxy). Gravity is time of time, which is double/inverse space here and now, and such time quantum is its singular change. 72^2=5184 years of time (inversion of additional 5 days) is singular quantum of solar spatial reality and 1440^2=2 073 600 years (inversion of ¼ day) is singular quantum of galactic spatial reality. Any passage of time less than 5184 year (6480) is not real in solar context—it is only its instant spatial density. Any passage of time less than 2 073 600 (2 592 000) years is not real in galactic context, because it is its instant spatial density. Such periods are singular quantum of change of solar/galactic reality. 2 073 600 is $1/10^2$ of solar spatial orbital around core of galaxy, which is time reaction here and now of such spatial reality without (2D fractal difference). Round non-Euclidean orbital curvature of solar reality (space) is linear (Euclidean) passage of uninterrupted time here and now. Such change is not only change of time (orbital/rotational) motion—it is also change of size.

Change of size of Earth is change of motion of solar reality. Quantum gravitational change is quantum time change of solar reality that affects all planets of solar system. 2 073 600=400x5184 periods, where 400=360x1.111 and/or 20^2. 20-fold repetition is eigenvalue of fractal 4/5, while its double/inverse repetition (400) is eigenvalue of double/inverse (dual) 4/5. It is eigenvalue of fractal spatial and time quantum changes or eigenvalue (resonance) of both—double/inverse rotational and orbital changes, which are singular 4x1/4 gravitational wave with frequency 2 073 600 years (20 736 x10^2 or inversion of precessional time, which is gravity here and now). It is singular quantum of solar quantum change (change of rotation), which is infinitely larger change in context of its orbiting planets. It is their instant contraction, which precedes gradual expansion of universe without. Gradual expansion of empty space without is equal and opposite quantum contraction of its uncertain mass within. 20-fold repetition of singular

5184-years rotational quantum changes is singular quantum of orbital change (103680) and its 20-fold repetition is singular quantum of gravitational change of size.

20 fold repetition of such sequence is 41 472 000-year period (43.2 and/or 51.84 million years in context of orbital/rotational times, which is 64.8 mill or $6480 \times 10^2 \times 10^2$ in solar context), is singular reversal of gradual (fractal) 20-fold repetition of inertial sequence, which is singular quantum of expansion as reaction of 20 smaller fractal contractions. 43.2 million years (64.8 in solar context, which is fractal 3/2x43.2 or 21.6 mill years inertial difference) is singular quantum of expansion. 20-fold repetition of such sequence is 864-mill year sequence (1.296 billion in solar context), which is singular quantum that does not affect only surface spatial density, but also its density in-depth. It is not only expansion of spatial size in context of universe but singular quantum of nuclear change of its mass/energy within (quantum of double/inverse fission). It is singular quantum of change of inertial density of mass/energy (atom), which is singular reaction within of infinity of gradual space/time reactions without. It is singular quantum of half-life of Earth. 20-fold repetition of such sequence is complete $E/m=C^2$ change, which is inversion of mass into energy and energy into mass. It is complete inversion of dense mass into empty space (expansion) and empty space into dense mass (contraction), which is quantum conversion between dense mass and dense energy or quantum change between massive reality within and energetic (solar) reality without (quantum of double/inverse fusion). It is complete half-life cycle of Earth in context of solar reality (inertial 3D), which is beginning of cycle in context of galaxy (4D). It is end of time in context of inertial space here and now and beginning of time in context of accelerated passage of time, where passage of time here and now is contracted into instant space, while singular gravity is expanded into linear passage of time.

CHAPTER 10

QUANTUM LEAP (DE BROGLIE)

Space is reaction of time and time is equal and opposite spatial reaction, while both are double/inverse reaction of inertial here and now in between, which is their eigenvalue (overlap). Instant spatial reality reflects as inverse gradual change that is its passage of time. Instant space is singular quantum of time and all the time is singular quantum of infinitely larger (relative) space. Time is fractal in between spatial sequences and equally vice versa—it is their double/inverse eigenvalue or flat curvature between two round curvatures, which does not affect either, unless ¼ quantum. It is double/inverse gradual/instant change of relative space without (passage of time) and uncertain space within (instant time), which is density of mass. Time is fractal ½ sphere between two spatial spheres within and without and equally vice versa. Duality of fractal spatial and time spheres is instant dual sphere, which is singular gravitational sphere. Gravitational sphere is 3/2 time and/or 3/2 space. Two gravitational spheres (attraction/repulsion) are 6/4 or fractal $(6x4)/5^2=24/25$ in context of here and now. Time and space are double/inverse overlap between two gravitational spheres. Each successive sphere is instant double sequence of its past inertion/inversion (action/reaction duality), which is infinite fractal curvature of double helix. Only double fractal difference or inclusivity of both equal opposites is yet another sequence. Singular quantum of space and/or time is ¼=0.25, while singular quantum of double fractal difference (gravity) is 1/25=0.04, because 1/25 is ¼ of 10^2.

Time is double inertion of space, which is its singular quantum (inversion), while gravity is double time inertion, which is its singular quantum. Previous inertion is inversion or singular quantum of next sequence. Each successive sequence starts as inversion of infinity into singularity, which is its starting point in time. Accumulation of such singularity in time is yet another inversion of infinity into singular point of next sequence. It is infinite fractal curvature (double helix), where only double curvature or double fractal difference is complete change, while fractal change in between is its passage of time. Only all the time is instant quantum leap from one into next sequence, where infinity of such quantum leaps is instant (in-built) density of space that is singular massive particle. Any inertion here and now is only surface of infinity of quantum leaps in its depth, which are its instant spatial reality. It is genetic code of such inertial sequence, while time of such sequence is its next spatial sequence (and gravity yet another future sequence). **Time is infinity of all future spatial sequences, just as space is past of all time sequences.**

Each inertial sequence here and now is double/inverse reaction of its future/past sequences. Inertial sequence is eigenvalue of all past and future sequences. It is simultaneous infinity of both. Spatial reality of inertial sequence is inverse reflection of its time, just as its time is inverse reflection of its space, because one is singular surface of infinite depth of another. Angular tilt is ¼ north/south distance, which is 2 500 km out of total 10 000 km in each direction, which is spatial reflection of orbital motion (fractal ¾). 2 500 km of spatial distance is ¼ discrepancy (overlap) between orbital and rotational motions, which is their time. 2x1/2 (both sides, which are inertion/inversion=double inertion) is 5 000 out of total distance of 20 000 km. 5 000 is ¼ of 20 000, which is fractal 4/5 spatial reaction of additional 5 ¼ day (time), which repeats in quanta of 20. It is reflection of precessional reality (time) in context of inertial space. Spatial reality of angular tilt is perfect spatial shadow of its time, just as precessional motion (time) is perfect shadow of its space (page 165). Just as Earth tilts in space creating 4 different spatial seasons, it also tilts in time, creating 4 different seasons of time, which are 4 different spatial seasons of solar reality (5).

Size of Earth is double/inverse reality of its time. 400×10^2=40 000 km is 360°x1.111 of its angular curvature. 360° is double/inverse time of spatial surface in between (convex/concave angular curvatures). Double time is its orbital curvature of 360 days, while its inverse time is instant spatial 360° curvature, which is reaction of rotation. Rotation is instant 360° curvature in context of no time (round size of Earth), while orbital is equal and opposite 360° gradual curvature in time. Rotation/orbital motions are double/inverse angular curvatures of horizontal (flat) here and now in between. They are instant/gradual or spatial and time angular curvatures of inertial sequence in between. 360° is instant space of one and gradual time of another. Singularity in context of rotation is infinity in context of orbital and equally vice versa, where infinite in time is singular in space. 1° of spatial change in context of orbital motion is 1° of time change in context of rotation and equally vice versa. Multiplication in context of one is fragmentation in context of another, because one is double/inverse context of inertial reality of another in between. Smaller fragment of spatial orbital is inversely larger fragment of time in context of spatial (rotating) Earth and equally vice versa. Smallest spatial fractal of orbital (singular quantum) is ¼ day, which is 21600 sec, where 21600 sec in space is 21600 years in time (and/or vice versa). It is singular spatial quantum of time that is 4-fold infinity of time quanta.

21600 sec of rotation is $6480x10^2$ km of orbital motion, which is 6480 years or ¼ of precessional motion (singular quantum), because singular quantum of time is 4-fold infinity of spatial quanta. ¼ quantum of spatial reality is whole quantum of time and equally vice versa, which is singular 4x1/4 sin wave of 360° curvature. Each complete 4x1/4 wave is only ¼ (90°) of infinitely larger 4x1/4 wave. $6480x10^2$ km is 1/1440th or 1/4x360th part of orbital, which is again 4-fold infinity, because reality is always 4-fold curvature or instant 3-fold curvature of its space and ¼ gradual curvature, which is its fractal time. 1440 is singular ¼ quantum of a curvature. Singular quantum of 4-fold curvature (4D), which is double or space/time curvature is 4x1440=5760, which is $10^2x57.6$ of perihelion orbital. Gravitational (6D) curvature is 6x1440=8640, which is 1/10 of rotational time. 9D curvature, which is 3D curvature of inertial 6D curvature (space of time and/or time of space) is 12960 years of ½ precessional period, which is 1/ 10th part of double precessional period of 129 600 years. 18x1440=25920, which is 20/18=10/9=1.111 of full 20-fold repetition, which is 28800 years. 28800 is circumference of round curvature, while 4x6480=25920 is circumference of a square within (diameter of circle=diagonal of square). 25x1440=36000 or $360x10^2$, where 25/100= ¼ singular quantum of larger curvature that is 360x1440=518400 of triple quantum period that is 10^2 fractal difference of 5184-year period. Supersymmetry of double helix is infinite. 360° curvature is one day in context of rotation and one year in context of orbital motion, where day is instant space (round Earth) of orbital time, just as orbital is instant space in context of precessional rotation, which is double 360° curvature (2x360=720 or 72 days per 1°, which is 1° per 72 years in context of rotation).

Synodic/sidereal difference of 236 sec/day is 111.111 km of additional motion in space per day (111.111x360 days=40 000 km, where each day of gradual shift of such difference is 1° of Earth's east/west angular curvature). Synodic/sidereal difference is time limit (curvature) of inertial space. Surface of Earth is instant (uncertain) time. Synodic/sidereal fractal difference is underlying constancy of Earth's curvature (constant angle of its curvature, which determines its size—and motion). It is its inertial framework (derivate of exponent and/or vice versa). 400x240 sec of perfect difference is 96000, which is 10^6 of 0.96 or 6D (6-fold curvature or 6-fold repetition) of fractal 24/25 (gravity). 400 (size) x 5184 (time) =2 073 600 (gravity), which is boundary of both (angular or finite curvature of infinite horizontal curvature), because it is 1440^2 that is $20736x10^2$ or boundary of precessional time (accelerated time of relative solar reality). It is full curvature or contraction of infinite horizontal curvature, which is linear passage of precessional time, just as $4x5184=20736=144^2$ is full curvature (boundary) of inertial passage of time here and now. 20736=21600-864, where 864 is $1/10^2$ of 86400 (one day or instant size of Earth in time, which is inertion or singular quantum of such infinite sequence). 21600 years is inertial period of time that includes its in-built inertion—Earth (see illustration 8), which perpetuates its own passage of time (infinite wave), while 20736 years is inertionless time on any point on its surface. 21 600 years is 20 736 years of rotational time in context of orbital motion (fractal 25/24).

20736 years is passage of time on surface of Earth (perfect wave), which is at the same time 21600 years in context of its orbital motion, which includes inertion of such particle in its wave. 72^2 (double/inverse orbital/rotational discrepancy of 5 days per year) is 5184 years of passage of time of such spatial sequence. 72 years per 1° of precessional motion is 18 years per ¼ of such motion, which is full curvature in context of angular motion of Earth (4 seasons per year). 18^2=324, which is 360/1.111 or boundary of 360° curvature (1/10 discrepancy, which is additional curvature, just as 400 is its double/inverse boundary). It is singular quantum of angular motion (seasons), which is 20x324=6480-year period that is its full curvature. Such complete angular curvature is only ¼ of horizontal (rotational) curvature. 400 is expansion of inertial 360° sequence in time without, while 324 is equal and opposite contraction in space within. 400 is relative, while 324 is its uncertain reality. 40000 km, which is 360° (20x18) spatial (instant) curvature, is only 324° (18^2) in time, because of 1/10 fractal space/time difference. Horizontal/angular curvatures are 1/10 discrepancy—they are 360/324 within and/or 360/400 without. 360/324 is double/inverse reality of inertial 400, which is singular 4D sequence, while 360 is its duality (2D) and 324 is its

3D sequence (double/inverse contraction or double curvature, which is an empty sphere or inertial size).

5184 years is 1/5 of precessional motion, while 6480-years is full curvature of angular motion (tilt), which is only ¼ precessional motion (fractal 5/4). They are eigenvalue in 20-fold repetition in time (i.e. metonic cycle, which is 19-fold repetition of inertial sequence—19+1=20). Spatial discrepancy generates time, just as time discrepancy generates space. 1/10 discrepancy between wholes (quanta) is 1/20 discrepancy of their fractal duality (and/or 1/40 discrepancy of 4x1/4 division/multiplication). 5184 years is singular 1/5 time quantum of solar 5D, while 6480 is singular ¼ spatial quantum of Earth's 4D. 5184 is 6-fold repetition of inertial $864x10^2$ sequence, while 6480 is 7.5-fold repetition, which is 3/2 discrepancy of 1296 years or additional 1.5 days per 6-day week (7th day, which is 3/2 time; therefore 4 weeks x 7.5=30 days of a month and/or 30x12 months = 360 days of a year). $864x10^2$ is double/inverse inertion (singular quantum) of inertial 4D here and now and/or inertial 5D solar reality. There are no different realities but only one in infinity of different curvatures, where every curvature is relatively different singular limitation of infinite and/or vice versa. $864x10^2$ sec is all the time of universe contracted into 40 000 km spatial curvature (and/or expanded time of an atom). Such time is passage of time of one day, which is instant (spatial) reality in context of its orbital motion. It is infinite relativity in context of rotation and infinite uncertainty in context of orbital motion.

Such time is infinity of different times that are real in equally different points on surface. Time of one latitude/longitude is different from time of another longitude/latitude. Curvature of dense Earth is infinite (total) relativity of time, just as is uncertainty of its space (reaction of equal and opposite time curvature of empty space). All relatively different fractals of time in context of inversely different points of space are instant reality in context of orbital motion. Rotational time is instant (uncertain) sequence in context of orbital, because it is everywhere at the same time. All its relatively different possibilities (Everett's many worlds) are simultaneous and instant reality of inertial space/time sequence of orbiting Earth. Rotation is instant (singular) time of time of orbital motion, which is not time but space. It is its quantum (quantum is inversion of inertion, which is space of time and/or time of space).

864 is singular 1/6 quantum of galactic 6D reality (inertial density within that is inertial velocity without). 1/6 of galactic reality is 1/30 of relative solar reality and/or 1/24 of inertial reality here and now (surface of Earth). Solar reality is empty space without Earth, which is passage of time, while galactic reality is dense mass within, which is instant time. Center of Earth is equal opposite (reac-

tion) of core of galaxy without. Surface of Earth is their eigenvalue in between. Any change of galactic reality is equal and opposite (inverse) change of dense mass within, just as change of solar reality is change of empty space without, which is change of constant passage of time. Change of density, which is instant (double/inverse) time, is twice larger change of passage of time. Eigenvalue of galactic/solar reality (empty space/dense mass) here and now is 24x30 or 1/720 discrepancy, which is double curvature or 1/10 fractal difference of 72. Fractal difference between 30 and 24 is 6 (here and now). 24x6=144 (inertion) and 30x6=180 (inversion), while 180x144=25920, which is their singular quantum boundary (double inertion). Fractal difference between 180 and 144 is 36. 144x36=5184 (inertion) and 180x36=6480 (inversion), while 144x180=25 920 (double inertion, which is eigenvalue of such fractal 4/5 difference of 36^2=1296 years discrepancy).

Surface of Earth is inertial overlap between inverse galactic reality, which is its instant dense mass within (instant linear attraction or density) and empty space of solar reality without, which is its constant passage of time (gradual linear expansion or de-densification). Surface of Earth is inertial 4D in between double/inverse 5D without and/or 6D within, which is duality of fractal 4/5 and 5/6, which is singular 24/25 here and now (surface). Passage of time in 4D context is unreal (more than improbable) in 5D solar context and twice unreal (twice uncertain) in galactic context. Only quantum periods of ¼ precessional motion (6480) are quanta (space) of solar passage of time and ¼ quantum periods of double precession (25920) are quanta of passage of time in galactic context. Solar passage of time is 4x1/4 quanta that are singular ¼ quantum of galactic passage of time. 4x1/4 periods of precessional motion are instant eigenvalue between solar 5D and Earth's 4D (angular rotation in orbital plane), while periods of full precessional curvature are also eigenvalue of 6D galactic reality (angular rotation in angular direction of orbital plane). Eigenvalue of galactic time is synchronicity with solar space and not only its fractal time, while larger quantum periods of time are also eigenvalue of galactic space, which is resonance with passage of time of universe (gravitational changes) and resonance with spatial reality of universe (E/m or nuclear changes).

Synodic/sidereal difference is 111.111 km per day (1°), which is compensated by gradual angular tilt. Angular tilt is 4x22.5° (4 seasons), which is 90° (¼) total curvature in context of one year, just as 360days/4 is 90 days per season (division in context of one is multiplication in context of another). It is 4/(1/4)=16-fold difference. 111.111km x360 days is 40 000 km and 40 000/16=2 500 km of such angular motion in north/south directions (likewise -240 sec x 360 days=86 400 sec and 86 400/16=5400 years of ¼ quantum period of time). Angular tilt is quadruple motion. It is double motion on both sides of sphere (double space) and in both directions (double time). It is 4x1/4 wave of inertial 2x1/2 (rotation), which is its singular quantum. It is 2x2x45=180° of total motion in course of one year, which is ½ of rotational curvature. If angular motion is 2x1/2 of rotation than precession is 4x1/4 and if angular motion is 4x1/4 than precession is 8x1/8 (etc). Precessional motion is double fractal difference (10^2), while angular rotation is singular fractal difference. Angular rotation is fractal 5/4 or inertial ¼ and/or 1/5 difference, which is eigenvalue in 20-fold repetitions (metonic cycle inclusive of inertial sequence). Precessional rotation is fractal 25/24 or 1/25 and/or 1/24 difference, while eigenvalue of both is 24x25=6x10^2, which is reflected as 6x10=60 or 1/6 fractal difference between precessional motion in context of rotational and orbital curvatures, which is perihelion (precessional) orbital (360/4=90 or singular ¼ quantum here and now is 360/6=60 or singular 1/6 quantum in context of orbital, which is 3/2 difference).

Synodic/sidereal difference is spatial eigenvalue of rotation/orbital motion (double/inverse times). 240 sec and/or 111.111 km difference between orbital and rotation is perfect 360-day curvature, which is uninterrupted passage of time between rotation and its angular tilt. 4 sec between 240 sec of 360-day year and 236 of 365.25-days year is 1/60 difference, which is discrepancy (relativity/uncertainty) between orbital and rotational times. 4 sec is 1.85 km per day, which is inertial difference (eccentricity) between 360 and 365 days, which is precessional motion. 40 000 km/1.85km and/or 86 400 sec/4 sec is 21 600 days in context of rotation and/or 21 600 years in context of orbital motion. 1.85 is 180/10^2 in context of its solar inertion (5). Inversion of inertial 180 is (0.555...)/10^2; therefore inertionless 1.8 is inertial 1.85. 1.8 km per day is **648** km per year, which is spatial fractal reaction of singular quantum of time. **648** km in context of 1 year of rotation is **648 000** km in ¼ day of orbital motion, where 360° curvature in context of orbital motion (year) is only 90° of rotation (day). 360° of inertion is 90° of inversion, which is singular 324° curvature (eigenvalue) in between (360x90=324x10^2). Inertion/inversion=double inertion is 360x90=324, because 1/90=0.0111...324° is 360° curvature that is 1/10 fractal difference.

1D fractal difference is 90°. 360° is 4-fold curvature or inertial 4D. 360° is double/inverse duality (4D), which is 4x10^2=400. 360/(4x4)=22.5° of angular tilt (change of seasons). ¾ curvature is inertial (instant) space in context of ¼=90° fractal difference, which is its time (angular curvature of horizontal space). 90° angular curvature is time of inertial space and/or space of time, which is one strand of double helix (repetition of singular here and now (1) in infinity of relatively different curvatures). 1.111 is 90° curvature (1/0.9=1.111, where 0.9=90/10^2), which is singular constant of relativity/uncertainty. It is double/ inverse 1D fractal difference, which is curvature of flat and/or flattening of round reality (1D difference between square and a cube and/or square and a round curvature). 90° angular curvature of inertial 360° is 324° and equally vice versa. Each 90° curvature is reduction (contraction) of inertial 360° for 1/10, where contraction in time is expansion in space and equally vice versa. Contraction of 360 is 324, while its expansion is 360+36=396, just as contraction of 400 is 360, which is 4° inertial difference in between (here and now). 360-36=324, while 324+32.4=356.4, which is 3.6° difference in between. 3.6 and 4 are double/ inverse fractal difference of inertial here and now, which is singular difference of 3.6x4=**14.4**. 4D curvature of inertial 14.4 is 4x14.4=57.6 years of perihelion (precessional) orbital, while ½ is 7.2 or 1/10 fractal difference of its inverse precessional rotation, which is two-fold curvature of spatial orbital (72= 2x360/10). Precessional rotation is 2x1/2 orbital, while precessional orbital motion is 8x1/8 precessional rotation (1D fractal difference between).

4-fold curvature (4D) of inertial 360° is **360-4x36=360-144=216**, which is its singular space/time quantum (3/2x144=216 and 3/2x216=324). 216/ 360=0.6, which is fractal 3/5 or light speed difference between such curvatures (3x10^5). Fractal 5/3=1.666..., which is infinite repetition of relative/uncertain (fractal) 6-fold sequences of inertial here and now (1). Inverse rotational velocity or time (2.16) is light speed difference from its size, just as is difference between one and 21600 years (difference between orbital and precessional motion), because linear passage of time in context of inertial 3D here and now is accelerated/decelerated time in context of relative solar reality (4D). Such accelerated/ decelerated time is real here and now only in ¼ quanta (inversion of inertial 4D). 4D curvature of inertial 400 is 400-160=240, which is synodic/sidereal difference between solar and Earth's reality. 240 sec is spatial reaction of time (400), just as 400x10^2 size of Earth is inverse spatial reaction of 240 sec of time. 400/240=fractal **13**/20 and 160/240= **7**/20, which are inertial 2x3+1=**7** and/or inverse 2x6+1=**13** or two equal gravitational opposites of inertial 20-fold spatial and/or time sequence, which is eigenvalue of fractal 4/5 (see illustration 8). 240 and 160 are fractal 3/2, just as are 360 and 240 (1D fractal difference). Curvature of synodic/sidereal difference is 360°x240=86400 sec of 1 day, while its inverse curvature is 240x160= 38400 km, which is 40000 km-4x400 and/or 40 000-160x10. 38400 km is inertionless and 40 000 is inertial size of Earth. Size of Earth within is passage of time without in context of synodic/sidereal difference in between. 240-24=216. 240 sec in context of orbital motion is 216 sec in context of rotation, because 240/216=1.111 and/or 216/240=0.9 or 90°/10^2 (curvature of flat 240 is round 216 and/or vice versa).

In context of Earth's 360° curvature, solar reality is 400° and in context of solar 360°, Earth is 324°. 360° of orbital is 324° of rotation and 360° of rotation is 400° of orbital and equally vice versa in context of time (400/324=1.111^2=1.23456...). 324/360/400 are spatial, time and gravitational curvatures, which is double/inverse fractal curvature of double helix. 324/360 and 360/400 are double/inverse 1D fractal difference or difference between instant space and its passage of time; while 324/400 is double fractal difference, which is difference between two relative spatial realities and/or two (double/inverse or positive/negative) passages of time. 1D is space/time fractal difference or difference between Euclidean and non-Euclidean curvatures, while 2D is gravitational difference, which is difference between convex/concave non-Euclidean curvature. Overlap between Euclidean and non-Euclidean (round and linear) curvatures is singular point of here and now in between, while overlap between non-Euclidean curvatures is convex/concave surface, which is inertial framework of such singular space/time point, which is its center. It is infinite space and inversely infinite passage of time of such here and now, which are double/inverse forces of attraction/repulsion and/or double/inverse densities without/within (space/mass and/or time/energy).

360°=4x90, where 4 is double fractal difference of relative 400°, while 90 is its curvature (1.111), which is 1/10 fractal difference (equal opposite is 400x9=3600 or 60^2). **4**x90=**400-40**, which is fractal 360° curvature between double/inverse 324°/400°. 360 is curvature of 400 and 324 is curvature of 360. **Curvature is ¼ fractal difference (90°).** Two equal and opposite 360° curvatures are one 324° in between and equally vice versa in opposite direction, just as 360° is singular curvature between two 400° (and vice versa). 2 horizontal 360° curvatures (double inertion) is one 324° positive (inertion) and one 400° negative (inversion) angular curvatures in between, because 324 within is 400 without. 324, 360 and 400 are 3 angular curvatures of inertial 3D, which are inertial space between double/inverse times and/or vice versa. Two of one is fractal second/third in between (2x1/2 is fractal 3/2). 360 is fractal 3/2 of 400 (singular fractal difference, which is ½ and/or 1/3 double/inverse difference that is inertial 6-fold curvature of Bode's periodicity). 324 is fractal 3/2 of 3/2, which is fractal 6/4 (1/4 and/or 1/6 difference without of double 1/5 difference within, which is 6/5x4/5=24/25). 324° curvature is time of inertial 360° space and/or space of inertial 360° time. It is 1/25=0.04 fractal difference (or ¼ of 10^2) of inertial ¼=0.25 within/without. 324/400 are double/inverse (relative/uncertain) eigenvalues (resonance) of fractal

360 in between. 360° of orbital and 360° of rotation is singular 324° in between (precession), which is angular 3rd half of 2x1/2 horizontal 360°. Eigenvalue of double/inverse 360° curvatures (2x360=720) is 324, where 720/324=2.222... which is infinity of repetitive fractal (relative) or dual (space/time) sequences of inertial duality here and now (see illustration 10).

90 is singular eigenvalue of double inertion = inertion/inversion. 5x90=450=500-50, 6x90=540=600-60, while 80 is double, 70 is triple, 60 is quadruple eigenvalue or 4-fold curvature of inertial difference etc—i.e. 5x80=400=500-2x 50, 6x80=480=600-2x60, 5x70=350=500-3x50, 6x70=420=600-3x60, 5x60=300=500-4x50, 6x60=360=600-4x60 etc, which is infinite fractal curvature of double helix. Larger number of fractal differences (eigenvalues) is proportionally larger number of simultaneous fractal curvatures. Decrease of acceleration is inverse increase of linear direction and equally vice versa (increase of inertion is equal and opposite decrease of inversion, which is its 1D fractal difference). Smaller inertial difference is larger inversion and equally vice versa. Smaller spatial discrepancy is larger period of time and smaller time discrepancy is larger spatial size (larger spatial discrepancy is smaller time in between). 9 is equal opposite (fractal) of 8, just as 8 is equal opposite of 7 and 7 is equal opposite of 6, which is infinite addition of past sequence to the present, which generates future (double) sequence.

324° is singular eccentric curvature (precessional rotation) between 360° rotation and angular tilt (orbital motion), because 324° is 90° angular curvature of horizontal 360° curvature (4x1/4=1). Inertial 360° Earth's curvature is instant 400° in space (instant size) and gradual 324° in time (motion) and equally vice versa. It is instant (inertial) 40 000 km of spatial curvature and/or 324° of double/inverse time curvature (2x360=720 and 720-72=648=2x324, which is 1/10 of precessional ¼ quantum or 1/40 period). Instant 40 000 km in space is 2x10=20-fold repetition of 324-year sequence in time, which is its fractal 4/5 time sequence (4x5=20). One day is 360 times synodic/sidereal difference (240 sec—see illustration 10) and one orbital is 360 days (2x360=720), while precessional motion is 324x20=6480 years, which is 20x360=7200 and 7200-720 or 360x18=6480 (6480=7200-720=8100-2x810). 90° angular curvature of horizontal 2D curvature (360° circle) is 2x1/4 or ½ of a sphere. 90° is inertial ½ quantum of 180° or double curvature, which is a sphere (360° is four-fold angular curvature or 4D sphere, which is 3D sphere and its angular density within, which is rotation without). Double curvature of singular curvature is 7200-720=6480-2x648=5184 (and/or 64 800-2x6480=51 840). Double curvature or double fractal difference is boundary of inertial 3D. 1D fractal difference (360/324 and 400/360) is curvature of flat reality and/or flattening of round curvature, while 2D fractal difference (324/400) are double/inverse curvatures, where curvature in space is flat time, just as curvature in time is flat space.

Illustration 10 -Superposition of orbital/rotational motions

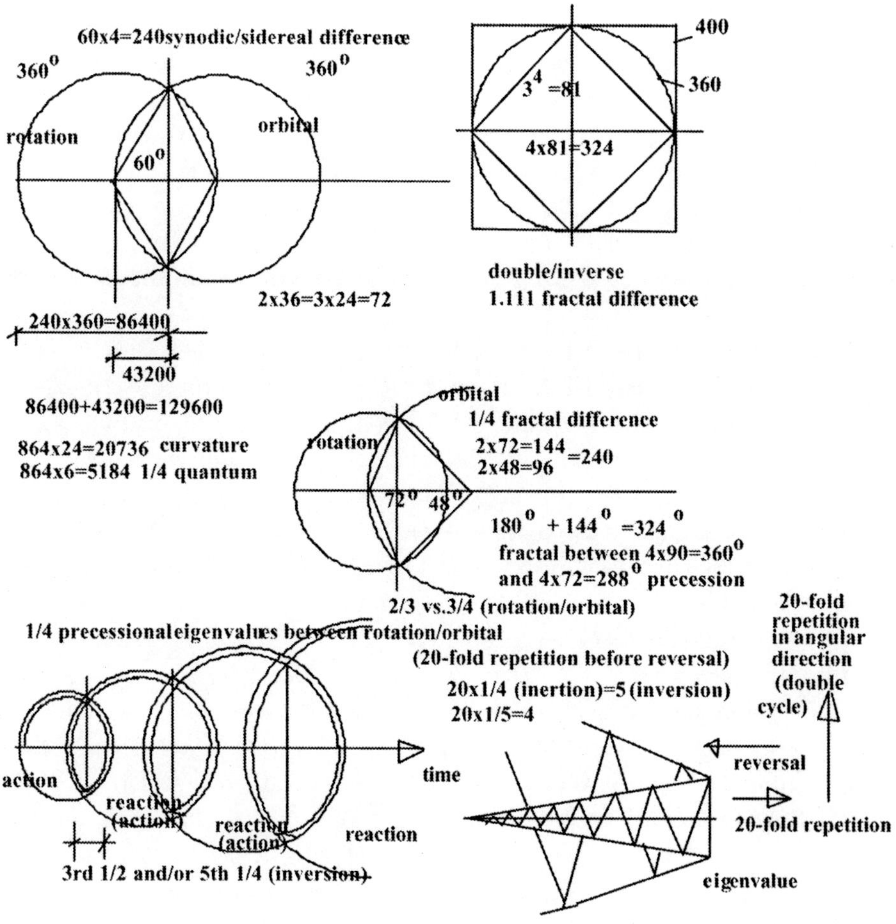

Rotation and orbital motion are 2/3 and ¾ reality of 1D fractal difference in between that is double/inverse angular 90° curvature (relativity of one is uncertainty of another, where relative space is uncertain time and equally vice versa). Rotation/orbital are action/reaction of each other (where spatial action is reaction in time, just as action in time is spatial reaction, because they are instant duality and any difference between action/reaction is relative/uncertain—they are instant or singular eigenvalue of both here and now). 2/3 without is 3/2 within, which is instant (spatial) reality in context of its time. Instant 2D space in context of its relative space (3D) without is instant 3D space in context of its uncertain 2D space within and equally vice versa in context of time. Eigenvalue of both is 2/3+3/2=13/6= **2.1666** sequence that is infinite 6-fold repetition of duality in context of its singularity. 3D sequence (3D limitation of infinite) is 2.16 of inverse Earth's velocity and/or singular 216-sec/year sequence. **It is infinite echoing of successive 6-fold sequences that are 1D fractal difference in context of inertial 6^3=216.** 2/3 is 1/3 difference in one and ½ difference in opposite directions of convex/concave curvatures, which is 6-fold eigenvalue or singularity of such duality, which is 1/6 fractal difference. 1/6 fractal difference is fractal 5/6 and/or 6/5, which is again singular 24/25-fractal curvature of inertial 1/24 and/or 1/25 differences that are singular 1/600 or double fractal difference (10^2) of 6D two steps before. **Inversion of inertial 2/3 is fractal 5/6 and its double inertion is fractal 24/25.**

Reaction of fractal size/rotation (2/3) is its orbital, which is fractal ¾. Their eigenvalues (4x1/4 of inertial 2x12) are 2/3+3/4=17/12=1.41666 (4D), 2/3+4/3=8/3 =2.666 (2D, which is infinite 6-fold repetition of inertial duality), 3/2+3/4=9/4=2.111 (2D, which is infinity of singular sequences of inertial duality) and 3/2+4/3=17/6=2.8333 (3D sequence). ¾ in space is 4/3 in time and equally vice versa, which is singular space/time ¾+4/3=25/12=2.08333 (12/25=0.48 or 48/10^2, where 48=6x8 or fractal 6D in context of triple duality 2^3), which is singular ½ quantum of fractal 25/24=1.041666 (24/25=0.96). ¾ and 4/3 are 1/12 fractal difference between double/inverse curvatures, which is fractal 11/12 and/or 12/11 that is singular 143/144 or 1/144 inertial difference, which is 144 repetitive sequence within and without, which is **144^2=20 736** years, which is limitation (double curvature or double fractal difference) of inertial ¾ rotational/orbital duality (time). 20 736 is inertionless period here and now, which does not account for inertion of Earth (rotation) in context of its orbital motion, which is 3/2 difference. 3/2x144=216 and its 4-fold curvature is 216x4=6x144=864 (inertion of time here and now, which is 1 day); therefore 20 736+864=**21 600** years of orbital time. Rotational/orbital times are 20 736/21600=0.96 or 24/25 fractality in context of inertial precessional quantum. Rotational sequence is fractal 1/25 difference of its orbital motion, which is equal and opposite reaction of 1/25 inertial difference of rotating Earth in context of perihelion orbital (57.6 instead of inertionless 60, which is 2.4 or 1/25 inertial difference, where eigenvalue of 2.4

and 60 is 144 and eigenvalue of 25 and 144 is 3600 or 360x10, while 25x60=150 or $3/2x10^2$ and 150-144=6, which is inertial 6D here and now in between).

Reaction of orbital ¾ is fractal 4/5 of precessional motion, which is singular inertial framework of both—rotation and orbital motion. 4/5 without and 5/4 within are 41/20=2+1/20, which is 20-fold fractal repetition that is singular sequence. 20-fold repetition is 90° angular curvature, which inverts 2D into 3D and/or vice versa (inversion of inertion that is double inertion). 2+1/20=2.05 or $2+5/10^2$, which is space/time duality (2) here and now in context of solar (5D) time, which is 5x2=10-fold repetitive sequence that is singular ½ quantum of inertial 5/4. 20/41=0.525, which is 25^2-4x25 and $25=5^2$, while double sequence is 1.05 or $1+5/10^2$. Fractal 4/5 is ¼ difference without that is 1/5 difference within and/or vice versa, which is 20-fold fractal curvature (fractal 19/20 or 20 +/ -1D fractal difference, which is 20/18 or 20x18=360. +/-1D fractal difference is 1/21 and 1/19, where 21x19=399 or 400-1 and +/-2D fractal difference is 22x18=396=360+36, which is inertial 3D here and now in between). Inertial 3D sequence, which is 2D fractal difference in between is 2/3x3/4x4/5=24/60=2/5. Each fractal is duality, which is 8-fold sequence or 4 inertial and 4 inverse sequences, i.e.2/3x4/3x4/5=0.7111, 3/2x3/4 x4/5=0.9 etc).

Reaction of orbital ¾ (inertion) is angular tilt of inertial size/rotation, which is fractal 5/4 (inversion) and/or orbital of orbital (perihelion precessional orbital), which is fractal 5/6 in equal and opposite direction (double inertion). Angular motion is 2500 km north/south in context of inertial 10 000 km (northern/ southern hemisphere), which is 4x1/4 in context of its additional variable ¼ difference (10 000 + its gradual 2500 km variability that is 12 500/10 000=5/4). Angular ¼ difference is 22.5° north/south inclination, which is angular curvature or time of time of rotation (which is in its turn time of inertial size). Angular tilt is singularity (quantum) of infinite inertion of rotating size—it limits (curves) its infinity. Rotation is 4x90°=360° inertial curvature, while orbital motion is its ¼ inversion, which is ¼x90°=22.5° Angular motion is inversion of inertial rotation, which is its double inertion without. Infinite number of rotations (spatial quanta) is singular angular tilt, which is equal and opposite singular time quantum (infinity without of singularity within). Angular tilt is inertial framework in which rotation is real. Angular motion is relativity of rotation, just as rotation is uncertainty of angular motion. Angular motion varies rotation, just as rotation varies angular motion. They are double/inverse times of each other in context of inertial spatial reality in between (here and now).

They are two equal opposites in context of inertial synodic/sidereal difference, which is spatial limit of both—rotational and orbital time. Synodic/sidereal difference is inverse spatial reality of Earth's size (2x1/2), while rotation/orbital motions are double/inverse time in between (2x1/2), which is double 4x1/4 duality. Sizes of Earth and/or synodic/sidereal difference are double/inverse times of inertial (rotational/orbital) time in between, which is its instant spatial reality. Just as rotation is instant (space) in context of orbital, so is orbital instant space of precessional time (inertial orbital includes rotation, just as rotation includes size and size includes density/gravity, because any future sequence is duality of past/present sequences). Size limits (curves) infinite density, rotation curves infinite size, orbital limits infinite number of rotations and precession limits infinite number of orbitals, because each successive curvature is inversion (time) of previous inertial sequence. Singular quantum of curvature that is real here and now is ¼. Each succeeding motion (reaction) is 90° angular curvature of its previous motion (action). Instant spatial limitation (synodic/sidereal difference) of such double/inverse times is equal and opposite time limitation of their instant spatial reality!

Angular motion of rotation is reaction of orbital and reaction of angular motion of rotation is equal and opposite angular motion of orbital, which is its precessional curvature (time). It is precession of orbital motion without, which is reaction of fractal 5/4 angular motion within. It is fractal 5/6 orbital of perihelion, which is its gradual motion in time (angular orbital of orbital). Horizontal eccentricity of orbital is angular eccentricity of rotation and equally vice versa. Angular motion of Earth within and orbital precession without are double/inverse reactions of inertial orbital motion in between. They are its spatial and time reactions, which are slow angular tilt of spatial Earth (seasons) and twice as slow precessional orbital motion. Angular tilt is curvature or time of rotation, just as orbital precession is curvature or time (limitation) of orbital motion. One is real only in inertial context of its time, which is its inversion within that is double inertion without. It is infinite echoing in back and forth directions of inertial here and now in a zigzag pattern of Brownian motion (see illustration 10), where each successive sequence is infinitely larger space in context of equally slower motion that is infinitely faster passage of time (and/or passage of time here and now is infinitely slower). Each successive sequence is infinitely emptier space and infinitely larger period of (denser) time. Each successive sequence is acceleration of time and deceleration of space (larger space), because it is curvature of flat direction and equal and opposite flattening of round curvature of previous sequence (equally vice versa in direction of micro-world, which is inversion of universe that is its double inertion without).

Symmetry in context of one sequence is asymmetry in context of another. Angular tilt within and perihelion motion without are two symmetrical 1D reactions of orbital in between. Such symmetry in context of orbital is asymmetry in context of rotation. While angular motion is its 1D fractal difference, perihelion orbital is its 2D fractal difference. Each space/time or size/motion duality (2x1/2) is only ½ of its larger/smaller duality (relative/uncertain 4x1/4), which is fractal 3/2. Fractal 3/2 in context of 2x1/2 wave is fractal ¾ in context of 4x1/4 wave, which is fractal 5/4, because 4x1/4 is only ¼ of larger double duality. Symmetry of 2x1/2 wave is asymmetry of its angular 4x1/4 wave, which is its curvature (90° difference). Waves without/within are double/inverse waves of inertial wave in between, which is both waves at the same time (eigenvalue); therefore not wave at all but particle and equally vice versa. Symmetry of wave without is inverse symmetry of wave within. Reaction of angular motion, which is fractal 5/4, is precessional rotation, which is 4/5. Angular motion is instant space of precessional rotation, which is its time. While angular motion is 5/4 of 90°, where 90° is its instant space and remaining ¼ (22.5°) is its variability (time), precessional rotation is 4/5x90= 72°. One is positive and another is negative 5th quarter, because they are double/inverse times and one is negation (limitation or boundary) of another. Precession is relativity of angular motion, just as angular motion is uncertainty or instant space of precessional time.

72° (4/5) of precessional motion is limitation (curvature) of angular tilt (5/4). It is time of inertial passage of time here and now in which such time is not passage, but instant space (relative sequence). 72° of precessional motion is infinite period of passage of time here and now, which is singular quantum of its spatial (instant) reality. Such infinite passage of time here and now is singular quantum of relative (solar) space. Singular quantum of time here and now (Earth) is inversion of 4D, which is ¼, while singular quantum of solar time (5D) is 1/5. ¼ is quantum of relative space and 1/5 is singular quantum of its time, which is accelerated time (gravity) here and now. Singular ¼ difference within (angular motion) is singular 1/5 difference without (precession). ¼ and 1/5 are double/inverse differences between convex/concave curvatures of time without and within. Precessional rotation is consequently 5x1/5 wave or 5x72=360°, where 72° is 4/5 of 90°. 1/5 is negative (inverse) fractal discrepancy of inertial 4x1/4 wave between precessional and orbital motion (1296 years relative discrepancy between 6480 and 5184 years), just as ¼ is positive (inertial) discrepancy between rotation and orbital motion. ¼ within is 1/5 without, which is singular eigenvalue in 20-fold repetition. Double/inverse reactions of 4/5 precessional rotation (time) in between are its reaction in context of dense Earth within and empty space of solar reality without, which are equal and opposite fractal 5/6. Precessional motion is time (eigenvalue or line of symmetry) between its double/inverse spatial realities (uncertain Earth within and relative galaxy without). Symmetry in context of solar reality is asymmetry in Earth's context. In Earth's inertial perspective—solar reality is singular and galaxy is double relative reality or its time and time of its time (its boundary or curvature).

Consequence is that fractal 5/6 reaction here and now is instant spatial density, while 5/6 in context of solar empty space is gradual motion in time, which is orbital of orbital (perihelion orbital). Reaction of precessional rotation is time of orbital motion and/or instant spatial curvature here and now, which are equal opposite horizontal/angular curvatures (90° difference). Fractal 5/6 in context of empty solar space is its rotation in context of galactic time, which is 6D. Singular quantum of 6D is 1/6 or 360/6=60°. Galactic time is hexagon (6x60=360), solar time is pentagon (5x72=360) and time here and now is a square of 4 linear directions or 4 different periods of constant passage of time in context of one precessional cycle (4x90=360). 5 precessional cycles (5x72=360) are 5 linear directions of accelerated time of relative (solar) reality, which is singular quantum of twice accelerated galactic time, which is 6 relatively different time curvatures (6x60=360). Galactic and solar times are fractal 60/90=2/3 and 72/90=4/5 of inertial time here and now. 2/3 in equal and opposite direction is 24/25.

Rotation is 4x90=360° east/west spatial curvature, while angular motion is its inversion (singular quantum) that is double inertion, which is 1/4x90=22.5° curvature in both directions (south/north). 22.5° is constant angular inclination, which is spatial curvature (time) of orbital time. Precessional motion (time), which is fractal 4/5, reflects as spatial 5/6. 5/6 in context of 5/4 of angular motion is singular 24/25, which is 1/25x90°=3.6° curvature. 22.5° is 2x1/2 north/south angular tilt of rotation (4x1/4 change of seasons, because 2x1/2 in space is 4x1/4 in time and equally vice versa), while 3.6° is east/west angular tilt of angular tilt, which is precessional rotation. It is 4x1/4 or 4 quadrants of a circle on both poles, which is total 8x1/8. Angular tilt is 4x22.5=90° and its tilt (wobble) is 8x3.6=28.8° (2x14.4). While rotation is singular east/west motion, angular motion is double north and south motion on opposite sides of a sphere that repeats in both directions in course of one year (singular time quantum). Precessional motion is diagonal motion in between or equal and opposite wobble on opposite sides of sphere and on both poles (south/east and south/west on north pole and north/east and north/west on south pole). Precession repeats in 4 different directions (4 quadrants of a curvature) in course of singular time quantum of one precessional cycle.

Curvature of 360° rotation is instant spatial 22.5° angular inclination of its horizontal ecliptic. Only surface of Earth rotates, while its angular inclination is constant; therefore one orbital is one inside out (inverse) rotation of such constant inclination, which is equal and opposite time rotation of spatial rotation. Earth's rotation in solar context (plane of ecliptic) is no rotation in context of galaxy (direction of angular inclination). Just as only one side of lunar surface faces Earth, so Earth faces galaxy only with one (polar) side, because Moon and galaxy are equal opposites (inversion of inertion or uncertain reality within of relative reality without). Moon within and galaxy without are double/inverse reaction of solar reality in inertial context of Earth in between (and/or vice versa). Space of solar reality is double/inverse time of Moon/galaxy and equally vice versa. 3.6° angular curvature of precessional motion is 10^2 or double fractal difference of 360° curvature, which is 1/25 of 90° in between. It is inversion of its inertion or its singular quantum. 3.6° of inertion in context of spatial curvature (size) is 240 sec in context of time curvature (motion). 3.6° is 3/2 spatial reaction of time of synodic/sidereal $2.4x10^2$ sec difference. Precessional motion is 2x1/2 angular tilt and/or 4x1/4 rotation, which is inertial 2x3.6=7.2° of 180° and/or 4x3.6=14.4° of 360° curvature (1/25 difference). 90° is singular quantum of time and 3.6° is its singular spatial quantum within, which is its inertion. 25x3.6=90° is inertial curvature of inertionless 90-3.6=**86.4**° angular curvature (24/25 time), where inertionless in time is inertial in space and equally vice versa. 3.6° is positive/negative fractal difference between inertial spatial and time curvatures—it is additional angular curvature of one, which is equal and opposite flattening of another. It flattens spatial reality (obliquity) and expands its time.

86.4° is inertionless 90° or ¼ singular quantum of time here and now, just as **57.6**° is inertionless 60° or 1/6 singular quantum of galactic time in context of inertial solar time (72°) in between. 86.4=90-3.6 and 57.6=60-2.4, where 3.6 is 1/25 of 90, just as 2.4 is 1/25 of 60, which are 2D fractal difference of 360 days and/or 240 sec. 72° is inertial quantum of time (precessional rotation), while **86.4-72=14.4** and **72-57.6=14.4** are its double/inverse fractal differences (eigenvalues). **Inertial 72° in between is 4/5x90, which is 5/6x86.4 and/or 5/4x57.6, which is 6/5x60, because it is singular 4x1/4 eigenvalue of equal opposites of Earth's space/time within and galactic space/time without.** 14.4 is negative angular difference between rotation/precession (Earth's/solar time) and/or positive difference between orbital/precession (solar/galactic time). 14.4 is relative discrepancy of one and uncertain discrepancy of another, where relative in space is uncertain in time and vice versa. Inertion/inversion=double inertion is 2x14.4=28.8, which is 3D fractal difference of inertial precessional curvature of

28 800 years, which is 28800-2880=25920 years (1.111 or 1/10 difference). 4x6480=25920 is a square (quantum) in context of curvature (relativity), whose circumference is 28800 and radius=$\sqrt{2}$x6480. 25920 in context of solar time is in its turn 25920-2x2592=20736 years here and now (2x1.111= 2.222 or 1/5 difference, which is singular quantum of solar time).

25920-year period is 5x5184 years (72^2), 4x6480 (72x90) and/or 6x4320 (72x60), which is fractal 3/2 discrepancy between 3 and 2 zodiacs (precession/orbital discrepancy). 90x60=5400-year period of 2.5 or 5/2 zodiacs, which is ¼ quantum period in context of orbital time (period inclusive of Earth's inertion, which is 2.4 discrepancy between 60 and 57.6). Positive inertion in context of orbital is negative inertion in context of rotation, which is 5400-1/24x5400 and/or 5400x0.96=5184. 5184 years is singular eigenvalue here and now (rotational time) of both 5400-years period (orbital) and 6480-year precessional period. **5184 is 57.6x90 and/or 86.4x60**, which are inertial time in context of inertionless space and/or inertial space in context of inertionless time. It is perfect/imperfect curvature that is twice perfect curvature (**72^2= 5184**). 90-72=18 and 72-12=60, which is 18/12 or fractal 3/2, which is singular ½ quantum of 36/24 duality (1D fractal difference) that is singular 14.4 eigenvalue (2D fractal difference—18-3.6 and/or 12+2.4=14.4). Such 14.4° discrepancy between inertial period of time here and now, which is 5184 years, is 216-years slower time than orbital motion and/or 1296 years slower time than precessional rotation (and/or 5184=6/5x4320 which is 864-years faster time than orbital precessional motion). Such difference is not real here and now, unless singular ¼ quantum, which is instant reaction of accumulated difference. 6480=6/5x5400 (where 5400=5/4x4320) and/or 5/4x5184, which is 1/5 difference between precession and orbital motion that is also ¼ difference between precession and rotation or eigenvalue of ¼ and 1/5 differences. **5184 is singular 4x1/4 eigenvalue that reconciles ¼ fractal difference between rotation and orbital motion, as well as equal and opposite ¼ difference between rotation and precessional rotation (which is 1/5 eigenvalue between precessional rotation and precessional orbital).**

Size/motion is relative/uncertain (space/time) duality. It is not exclusivity of one, but inclusivity of both. There is no difference between Earth's spatial rotation (its rotation in empty space) and/or equal and opposite spatial rotation of empty space (around Earth). Rotation of dense mass in empty space is equal and opposite rotation of empty space without, which is reaction or time rotation of Earth. It is both at the same time, because reality is not exclusive space (size) or time (motion), but inclusive space/time, which is singular gravity. Instant size in space is gradual motion in time and gradual motion in space is instant size in time (relative space). Reaction of spatial orbital of 360 days is equal and opposite orbital in opposite direction of empty solar space (which is time). 360 days in one direction is additional 5 days in opposite direction, caused by such drag. 360° of spatial orbital is 5° of equal and opposite orbital of empty solar space in opposite direction, which is ¼ quantum of rotation in 6480 years. I.e. 360/5=72 days of rotation that is additional day of equal and opposite time, which is 5 days per year or ¼ orbital in 18 years (saros cycle). 18 years is ¼ or singular quantum of rotation (90° or ¼ difference), which is 4x18=72 years per 1° precessional rotation. 18x360°=6480 years is ¼ of precessional motion, which is instant addition of singular ¼ quantum of solar time (relative space).

Both—spatial and time motions (360+5 =365 days) are likewise only singular spatial motion in context of empty space of galaxy, which is its time. 365 days of spatial orbital motion is additional ¼ day in equal and opposite direction, which is rotation of empty galactic space (double time or gravity). 5 additional days is relativity (time boundary) of inertial time here and now (rotation), while ¼ day is double relativity, which is time boundary of solar time (gravity here and now). It is additional ¼ day per year, which is singular quantum in 4 years (1440 days); therefore ¼ orbital in 360 years. Rotation is 4x1/4 singular quantum of orbital motion, which is 360x360 (90°x1440 years, which is complete orbital) or 360^2=129 600-year period (Milankovich cycle) of additional ¼ quantum of galactic time, which is real here and now (twice relative space or relative time, which is gravity). 6480-year period and 129600-year period are equal and opposite (time) motions of inertial spatial motion here and now. Spatial motion is not motion but instant size of its time motion (6480-year ¼ period or 25 920-year cycle), just as such time motion is not motion but instant size of its gravitational motion (129 600-year ¼ period or 518 400-year 4x1/4 cycle). Such fractal equal and opposite (time) motions are not real here and now, unless ¼ quantum (De Broglie). Gradual precessional and double precessional motions (6480 and 20x6480=129600) are only gradual increase of tension, which is not real unless ¼ quantum. 6480-year period is gradual increase in tension of relative space (time), while 129 600-year period is gradual increase in depth of twice relative space (gravity). Motion is not time, unless singular ¼ quantum. Only ¼ of such precessional motions are instant time (singular quantum) here and now.

Space/time (size/motion) is relative (fractal) differentiation, which has only subjective meaning. There is no exclusive reality, only infinite 2x1/2 fractal multiplication/division, which is infinity of relatively different fractal space/time spheres, one within the other. Here and now is inertial spatial sphere (size) of double/inverse infinity of relative/uncertain time spheres (motions) within and without. Sphere that is 1D fractal difference without/within is positive/negative passage of time of inertial spatial sphere in between which is its relative/uncertain spatial reality (universe/micro-world). Such spheres are synchronized only on surface and not in depth (fractal 2/3). 2D fractal difference is only linear synchronization, which is gravity here and now, while 3D fractal difference is synchronicity of singular point and 4D is beyond event horizon (horizons of space). Each sphere is 4x1/4 of its previous or succeeding space/time sphere. Such spheres are double/inverse motions of inertial motion here and now. Only ¼ of such inverse/double motion is instant (quantum) reality here and now (see illustration 11).

Illustration 11 -Relative/uncertain (quantum) time difference

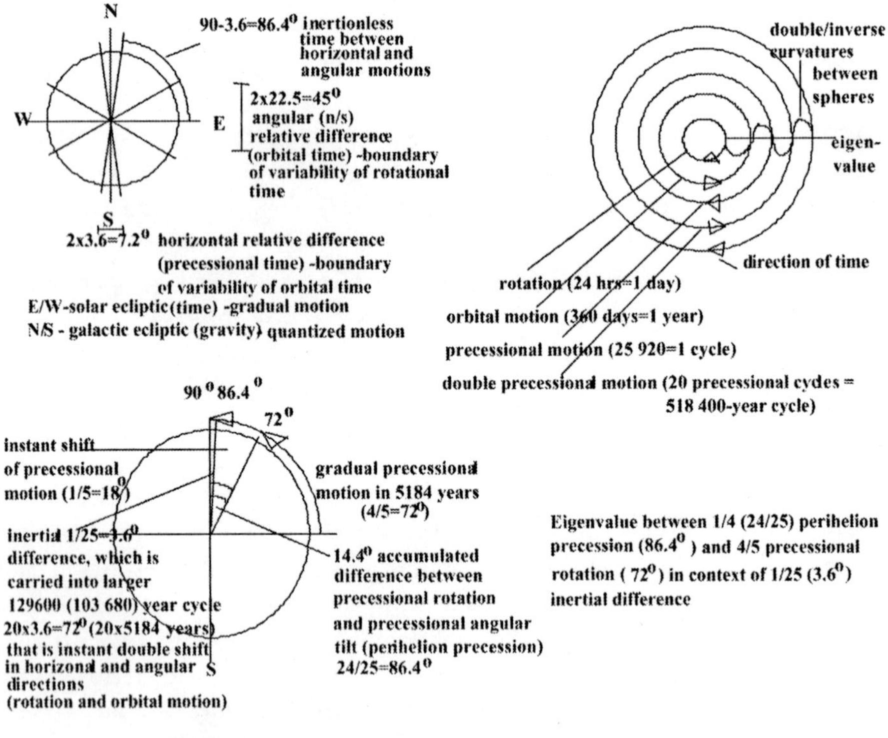

Constant vertical inclination (space) of 22.5° is inertion of inverse orbital motion (time). Spatial vertical inclination in context of time (gradual spatial change) is constant orbital motion, just as constant plane of ecliptic (equator) is constant spatial rotation. Horizontal coordinate system (plane of ecliptic) is solar coordinate (90°), while its angular inclination (1/4x90=22.5°) is galactic coordinate and one is time of another's instant (spatial) reality. Plane of ecliptic and angular inclination of polar axe are time and gravitational coordinates of inertial space in between (Earth). Angular inclination (polar axe) is direction of galactic orbital, just as plane of ecliptic is direction of solar orbital. Solar orbital is spatial (gradual) motion, while galactic orbital is time (quantized) motion, which is real only in ¼ intervals of precessional motion. Solar orbital is uninterrupted east/west motion, while galactic orbital is quantized north/south motion (relative/uncertain difference).

Rotation (space) in context of orbital motion (time) is its gradual angular north/south tilt (seasons), which is cause of inverse east/west tilt (1/25x90=3.6°) or precessional wobble. Precessional wobble is gradual rotation of angular inclination and therefore also rotation of angular tilt (change of seasons), which is an effect of angular inclination. Such gradual rotation of angular tilt is not real, unless ¼ quantum. Less than ¼ quantum is only increase of tension in depth of surface reality of angular tilt. Consequently, ¼ precessional motion is boundary (limitation) of orbital motion and angular tilt (change of seasons). ¼ precessional wobble of polar axe is singular time quantum of orbital time (change of seasons), which is infinite time and singular quantum of relative space. It is singular ¼ quantum of galactic (north/south) orbital. Just as orbital (angular tilt) is boundary of infinite repetitive rotation, so is precessional rotation boundary of infinite number of orbitals. It is boundary of both, because it is boundary of constant passage of time, which is instant inversion of infinity into singularity or 90° curvature (1.111, which is change of space/time sequences). 24x216, which is 72^2 or 18x288 (14.4x360 etc) is 5184-years, which is boundary of orbital and therefore also boundary of rotational time. It is boundary of infinite period of constant passage of time, which is instant inversion of infinite time into singular quantum of (relative) spatial reality.

Earth's precessional rotation is 1° in 72 years, while equal and opposite perihelion (orbital) precession is 1° in 57.6 years (60 years of inertionless time, which does not account for size of such orbiting particle, which is its 1/25 discrepancy). 72/60=6/5 faster precessional orbital than precessional rotation (slower passage of time) or 1/6 discrepancy without, which is 1/5 discrepancy within. While precessional rotation accomplishes 1/5 of its motion (72°), perihelion (orbital) precession, which is gradual increase in north/south tension, accomplishes 86.4° (1/4 less 1/25 inertial difference). Inertial 86.4° is inertionless 90°, which is singular ¼ quantum of angular (north/south) motion. 1/5 of horizontal motion (precessional rotation) is ¼ of angular motion (perihelion precession, which is orbital precession and therefore precessional curvature of angular tilt or change of seasons). ¼ of perihelion precession is complete (360°) curvature of gradual changes of seasons, because one season is ¼ of orbital motion. 24/25 parts of inertial precessional orbital, which is ¼ of inertionless time is instant eigenvalue between 4/5 and 5/6 fractal realities of precessional rotation and its precessional angular curvature (orbital).

Precessional motion is reaction of discrepancy between rotation and orbital motion, while double precessional motion is result of discrepancy between perihelion (orbital) and rotational precessions. Perihelion and rotational precessions are inverse (equal and opposite) curvatures—one is horizontal, while another one is angular curvature. Such 1D fractal difference is infinite precessional curvature of exponentially smaller fractal differences between convex/concave curvatures that are exponentially larger periods of time between its quantum changes (beginning/end of such exponentially larger periods of uninterrupted passage of time). Larger period of uninterrupted time is larger duration of inertial spatial sequence in between. Perihelion and rotational precession are fractal 4/5 reality, where 1/5 discrepancy of one is ¼ discrepancy of another, which is eigenvalue in 20-fold repetitive sequences (inertionless metonic cycle 19+1=20). Inertial discrepancy of such 20-fold repetition is 1/24 within that is 1/25 without (fractal 24/25). Such 1/24 discrepancy repeated 20 times is 20/24 or 5/6 discrepancy without, which is 4/5 discrepancy within (20/25), which is yet another quantum leap of inertion/inversion= double inertion. Frequency of such changes is always 20-fold repetition of inertial sequence.

Orbital (perihelion) precession is inversion (reversal) of angular tilt or north/south directions, while precessional rotation is inversion (reversal) of rotation or east/west directions. Perihelion precession and rotational precession are gradual increases of tension created by duality between rotation and orbital motion, which is instant reaction only in ¼ periods. While accumulated tension of angular tilt (orbital motion) is full ¼ quantum in period of 5184 years (57.6x90°), curvature of precessional rotation is only 4/5 of ¼ (72x72°), which is its instant 1/5 shift indirectly through angular rotation. 72° of precessional rotation is 4x1/4 or 4x18° wave, which is instant 5th quarter of 18°. Accumulated 90° tension in north/south (angular) direction is instant ¼ angular rotation, which is singular quantum of Earth's galactic orbital motion. Equinox ¼ change is insufficient momentum, because inertion of 90 is 3.6°, which is only 86.4° angular rotation (less than 90°). It is only reversible swing, while solstice ½ change is 21.6+86.4=108° (more than 90°), which is complete ½ irreversible change of north/south and east/west directions (21.6° is inertionless 22.5° angular inclination, where 0.9° difference is 1/25 inertion of such wave. Inertion between angular tilt and its precessional motion is 86.4-21.6=64.8°, which is spatial reflection of time). Such instant 18° shift of precessional motion is achieved through singular ¼ and/or ½ quantum of angular rotation.

Singular ¼ quantum of angular rotation (equinoxial swing) is 86.4° advance and subsequent 90° reversal (and/or vice versa), which is 3.6° inertial difference in between. 3.6° difference displaces equinox point for 1/6th of a season (21.6/3.6) or 15 days (15 days is inertionless 14.4+0.6, where 0.6 is 25/24 fractal difference). Solstice (½) change is upside down toppling of Earth or ½ angular rotation, which is double angular advance of 90°+86.4° or 108+72=180°-3.6°, which is again 3.6° discrepancy or displacement of solstice point for 1/6 (15 days). Positive 3.6° equinoxial displacement is followed by negative 3.6° solstice displacement and equally vice versa, which are 4 relatively different quantum changes of one precessional cycle. Positive 3.6° equinoxial displacement shortens spring/autumn for 15 days, while negative 3.6° solstice displacement prolongs summer/winter seasons. Prolonged summer/winter are cyclical warm/ice ages in ½ precessional intervals, with partial changes ½ way in between. Equinoxial changes in between further strengthen such warm and/or cold periods on surface, while in depth they reduce their momentum, preparing for equal and opposite change.

Only ½ of a sphere is instant reality here and now, while another ½ is depth of such surface. Ice age on surface is gradual change toward warm period in its depth and equally vice versa. Only when gradual accumulated change in depth is singular ¼ quantum it causes instant change on surface. Full precessional cycle is only ¼ of larger 103680-year cycle, which is additional ¼ reversible swing followed by complete ½ angular rotation bellow/above ecliptic at the end of each precessional cycle, which inverts positive/negative 3.6° difference. Consequently, subsequent precessional period is prolonged spring/autumn and/or shortened summer/winter. Yet another precessional cycle is again prolonged summer/winter but through inverse 3.6° displacement, which is followed again by prolonged spring/autumn through inverse 3.6° displacement. In every case, 3.6° displacement in orbital (rotational) plane is equal and opposite displacement in angular direction, because 22.5° (21.6°) angular inclination between solar and galactic ecliptic is their constant eigenvalue. Shift of solstice/equinox points is also 3.6° shift of polar axe. 4 precessional periods are shift of polar axe in 4x90° directions.

In every ¼ quantum change, precessional rotation is instantly advanced for 1/5 (18°, which is equal opposite of angular tilt, i.e. 2500 km), either through reversible swing or through ½ angular rotation, which is instant advancement from 5184 to 6480 years of ¼ precessional period. ¼ precessional period is 5184=18x288 years here and now, which is 18x360=6480 years of precessional motion, where 288=360-2x36 or 360/1.111=324 and 324/1.111=291.6, which is 288 in context of inertial 3.6° difference. 1296 years is relative (fractal) difference between rotation and precessional motion or time discrepancy between space/time (Earth/solar) motions. ¼ quantum change is 1296 years of constant passage of time contracted into 1 day, because 1 day in context of solar (relative) reality is 1296 years in context of inertial Earth's reality. It is instant reversal of 5184 years of gradual spatial change in period of 1 day. Gradual passage of time between seasons is not real in context of relative (solar) reality. Only its instant points of inversion, which are gradual and uninterrupted passage of time of relative (solar) reality, are real in context of relative (solar) spatial sequence. Solstice/equinox points are singular ¼ quanta of passage of time of solar reality (1 orbital=4 solstice/equinox points). Only solstice/equinox are eigenvalues between rotation and orbital motion and/or orbital and precessional motion, which are real in context of fractal solar reality; therefore only 25 920-fold repetition of such points (6480 years) is singular spatial quantum of relative reality. One precessional cycle of 25920 years, which is whole quantum of solar spatial reality, is $4x25920=360^2$ =103680 seasons, because seasons within are periods of double precession without in context of inertial precessional motion in between (one precedes another).

Quantum change is always solstice/equinox change because only such points are real in context of solar reality. It expands such singular points of time within into linear passage of time without, where such expanded time is inertial context of slower passage of time between such singular points (change of seasons). Equinox/solstice quantum change is instant change **when perihelion orbital point is ¼ season or 14.4 days apart from solstice/equinox**, which is 15 days less 1/25 inertial difference (0.6). 14.4 is 14 days 9 hours and 36 minutes discrepancy or same period of time as between repeating solar and lunar eclipses. It is instant advancement of precessional rotation for 18° (14.4+3.6 inertion), where 18° is 1/20 of 360°, which is 6/5 hrs, while 14.4° is 25/24 hrs. Consequently, such precessional shift is also **instant reversal of gradual expansion of time (deceleration of rotation) in previous 5184-year period**. It is instant increase of rotation that decreases passage of time for 1 hour (reduction from 24 to 23 hours and/or from expanded 25 hours back to initial 24-hour cycle, because every day is longer than the previous day, which is accumulated difference of 1 hour in period of 5184 years—see text on pages 188/9). It is instant inversion of gradually deteriorated momentum, which is not real, unless singular 1/24 quantum (1 hour discrepancy). 1/24 quantum is 4/5 fractal discrepancy within that is 5/6 discrepancy without or 5184-year period, while 20x5184=103680 years, which is same 20/24 or 5/6 discrepancy, therefore also instant quantum change (103680/129600).

103680-year change is instant reversal of additional 1 hr discrepancy, created by its smaller 1D quantum changes, because 1D changes are gradual change in context of singular 2D change. 2D quantum change is correction of correction or correction of discrepancy within discrepancy. Each quantum change reconciles 4/5 discrepancy, while fractal difference is carried into next cycle, because it is its inertion. 4/5 discrepancy is ¼ difference repeated 5 times and/or 1/5 difference repeated 4 times (eigenvalue in 20-fold repetition). Inertial ¼ difference, which is carried over into yet larger cycle is always 24/25 discrepancy. Repeated 4 times in 4x1/4 (one full cycle), it is 1/5 of 5 times larger difference, which is reconciled in 5 cycles that is one infinitely larger cycle. At the same time, it is 1/5 difference reconciled in 4 cycles, because ¼ in space is 1/5 in time and equally vice versa. Each quantum change is ¼ spatial correction, which is 1/24 time correction and/or vice versa (remaining +/-1/24 difference without/within is 24/(1/24)= 576, which is 600-6x4 and/or 4x144). Each 5184-year quantum change is reconciliation of 1 hour (gradual slowing of rotation) in context of its 24/25 discrepancy, which is 2.4/2.5 min or 144/150 sec discrepancy. Repeated 4 times, which is whole cycle (5184x4=20736); it is accumulation of 9.6 min or 576 sec difference. 5 cycles is 48 min or 2880 sec, which is 4/5 of an hour difference and therefore instant change in 20x5184=103680 years, where 103680 is 4/5 of 129 600 or instant addition of 25920 years of precessional (instant) time.

Likewise—5x2.4=12 and 4x12=48, which is 5x5184=25920 year period repeated 4 times (it is 5x2.5=12.5 and 4x12.5=50 and/or 4x2.5=10 and 5x10=50, which is 5/6 of one hour, because 5/6 within is 4/5 without). 103680-year period is inverse 1 hr correction from 20-fold inertial 1 hr correction. 5184-year change accelerates decelerated rotation, and 103680-year decelerates accelerated rotation. While 5184-year change increases deteriorated rotation, 103680-year change decreases such increased rotation. 5184-year period is boundary of gradual spatial change or constant passage of time, which is its instant inversion. 103680-year period is boundary of such 5184-year changes, which is its instant inversion. Such inverse correction is instant shift of precessional motion from 90° back to 72°, because of accumulated 24/25 difference that remains after each quantum reconciliation. Inverse quantum change is double spatial change, which is not angular rotation only in plane of ecliptic, but also additional rotation bellow/above ecliptic, which is change of orbital and not only change of rotation-see illustration 9. Likewise—inertial 24/25 discrepancy of 3.6°, created by 18° advancement of precessional rotation, is accumulation of 20x3.6=72° in 103 680 years period, which is instant quantum change in opposite direction. 103 680-year quantum change is reversal of 20x5184 year smaller quantum changes, just as each 5184-year change is reversal of gradual change in context of passage of time.

1/24 increase of rotation in ¼ periods is 5184+216 =5400 years, where 216 years is discrepancy (relativity) between rotation and orbital motion. It is 1/24 decrease in passage of time created by rotation and/or 1/25 increase in passage of time created by orbital motion (slower motion is faster passage of time and/or vice versa). Every 24 years of constant passage of time generates 1 year of relative discrepancy, which is real only in ¼ quantum periods that are 9/10 or 216 out of 240 accumulation (240-24 or 240/1.111). 24x216 (inertion/inversion) is 5184 years of rotation that is 5400 years of orbital motion and/or 6480 years of precessional motion. Gradual 216-year discrepancy between rotation and orbital motion that builds up in 5184-year period is instantly reversed through such quantum change, where instant reversal of gradual accumulation is positive 1/25 fractal difference, just as inverse gradual reversal of quantum change is negative 1/24 difference and/or vice versa, because positive in time is negative in space, just as 1/24 in time is 1/25 in space. It is perpetuation of 24x25=600 or $6x10^2$ inertial difference (here and now), in space and time, which is infinity of different spatial sequences in exponentially larger periods of time (20-fold repetition of non-inertial metonic cycle).

Orbital and rotational times are same passage of time only in context of inertial ¼ precessional period, while in larger periods, they are fractal 24/25 difference, just as precessional and rotational times are same only in periods less than 5184-years, while in larger periods, they are 4/5 discrepancy. 5184-year period, which is instant addition of 1296 years of angular time is singular quantum of 4x1/4 precessional period that is 20 736 (25 920) years. Accumulated 24/25 difference is 216 years per ¼ period and/or 14.4 days per year (perihelion discrepancy), which is 1 hr decrease in rotation per day. Instant acceleration of Earth's rotation in ¼ periods of 1 hr corrects such accumulated discrepancy, which is not real, unless ¼ quantum, which is instant reaction. 1hr per day x 360 days per year 14.4 days advance of perihelion motion, which is thus synchronized with precessional motion (synchronization of gradually built-up relative gap between rotation and orbital motion in previous 5184 year period of passage of time).

Such complete precessional period, is in its turn, singular ¼ period of 103680-year time wave, and 103680-year wave is ¼ of 414720-year wave, which is again only ¼ of larger 4x1/4 wave. Complete 4x1/4 cycle is also additional ¼ change in angular direction. 25920-year period is also reversible swing in angular direction of ecliptic, while 2x25920 =51840-years is complete ½ angular rotation bellow/above plane of ecliptic and 4x1/4= 103680-year period is also reversible swing in 3rd direction etc. 5184-year period is instant addition of 1296-years (5th quarter) of uncertain time (relative space) of precessional motion (and/or 216 years discrepancy between rotation and orbital motion, which is ½ zodiac or 1080 years between rotation and precessional motion). 1296 years discrepancy is not time, but space. 103680-year period is likewise, instant addition of 25920, which is total of 129600-years. Every 20-fold repetition is instant addition of ¼, which is relative space/uncertain time of inertial passage of time in between. Such exponentially fractal times are not real here and now, unless their gradual accumulation is singular ¼ quantum. Larger fractal difference is smaller discrepancy; therefore larger period of time in which accumulated difference is ¼. At the same time, such larger period of uninterrupted passage of time is equally larger quantum change of spatial sequence in between, because larger inertion without is smaller inversion within. Inertion/inversion are two equal opposites, while their eigenvalue is double inertion ($C/(1/C)=C^2$), because double helix multiplies through division and divides through multiplication.

There is only one reality in infinity of relatively different curvatures. Difference is only on surface, while in depth it is always same inertial here and now. Just as space is inversion of time, or universe inverse reflection of its micro-world, so is Earth inversion of solar reality and Moon is inversion of Earth's reality (where Moon is equal opposite within of galaxy without). Earth orbits Sun, while Moon orbits Earth. Sun/Moon are space/time duality of inertial Earth in between—one is inversion of inertion of another. Spatial reality (size) in context of solar reality is time (motion) in context of lunar reality and equally vice versa. Moon is uncertainty within of relative solar space without and/or relativity within of uncertain (quantum) solar time without. Moon is ¼ Earth's size, because it is 5/4 fractal difference between Earth and Sun, just as Sun is 111 times larger than Earth, because it is its infinite difference or 90° curvature (109 times larger is 111 times larger inclusive of inertial duality here and now). Moon is flattening (4x1/4 square) of round Earth, while Sun is rounding of flat Earth. Inertial discrepancy (synodic/sidereal difference) between solar/Earth duality is space/time here and now, which is gradual spatial change in context of constant passage of time, interrupted by quantum changes in ¼ periods of its subsequent precessional motions.

Moon within is reaction of solar reality without in context of inertial Earth in between, therefore Moon's motion around Earth is inversely proportional with Earth's motion around Sun. Every point of space/time is in perfect balance with all other points, therefore each of such points perfectly reflects here and now (as well as any other, because sequences are only relatively different—i.e. difference is only on surface, not in depth). Inertion of Earth/solar duality is inversion of Earth/lunar duality, which is double inertion here and now. Moon/Earth distance is non-inertial 1 light second, while Earth/Sun distance is 8 light minutes, which is 8x60=480 or 48x10 difference. 48 is inertial sequence (singular quantum) of lunar time, just as 864 is inertial sequence of Earth's time. 864=18x48, i.e. Earth is instant 18-fold repetition (saros) of lunar time, which is inversion of 2^2=4 and 2^3=8 sequences that adds 2x3=6 fractal sequence in between. There is no space/time reality of Earth without Moon, just as there is no such reality without Sun. Moon is time (boundary) of solar spatial reality, just as Sun is time of lunar spatial reality. Solar and lunar perspectives are inversely proportional curvatures of space/time, just as are universe and its micro-world.

Lunar orbital cycle is 29.5 days, while its inertion (synodic/sidereal difference of its orbital) is 1/29.5=48.8 min, which is perfect 48 min in context of inertial 0.8=4/5 fractal difference. Discrepancy of inertial 24 hrs here and now is 48 min of synodic/sidereal difference of lunar orbital, which is again 0.8 min=48 sec discrepancy or infinite 48 within 48 repetitive sequence. It is infinite 2x1/2 repetition of inertial 24 hrs here and now. 48/0.8=60 is difference within difference, which is same inertial 1/60 discrepancy between 4min/4sec of synodic/sidereal difference and its inertion (time of space and/or space of time, which is curvature of flat reality or limitation of infinity). 1/60 is discrepancy between orbital and rotational motions, which is their precessional curvature. Lunar orbital motion is equal opposite of precessional (perihelion) motion. Lunar reality is 480/8, while solar is 240/4=60, which is 2x1/2 multiplication/division. 1440/48=30 is difference between rotational time (1 day) and inertial time of lunar orbital, where 30 is ½ of 60. Lunar orbital (tidal cycle) is instant 30-day month (space), which is reaction of gradual perihelion precession of Earth's orbital (time). While Earth is 23hr 56 min of rotation and 24 hrs of orbital or—4 min synodic/sidereal difference, lunar orbital is 24hr 48 min or +48 min difference. 48+4=52 weeks of one year (in context of 5/4 days discrepancy, which is inertion of such duality), while 48/4=12 months (52/12=4.333 weeks per month). 52x7=364-day year and 12x30=360-day year, which is 4 days discrepancy that is inertial 4D here and now between such equal opposites (360/4=90° curvature, while 360x4=1440 etc).

Synodic/sidereal difference of 48 min is 48x60=2880 sec, which is 12x240 sec or 12-fold repetition (months) of Earth's synodic/sidereal difference. 2880/1.111=25920/10, which is 1D fractal difference or time (boundary) of such inertial time. 25 920 years is boundary of inertial 2880-sec sequence, which is its singular spatial quantum. 86400 sec/4 min =21600 years of precessional cycle and

86400/48= 1800, which is 1/12 of such cycle (86400=1440x60 and 60=48+12). 2880 sec is 4/5 of 1hr, 1/30 of 1 day, 1/900 of 1 month and 1/10800 of 1 year. It is 10800-fold repetition, which is boundary of orbital motion (1/2 precession) and/or 900-fold repetition of months, which is their boundary (900/12=75, which is 72 years in context of inertial 24/25 difference or 1° of precessional curvature) and 30-fold repetition of days, which is their boundary (1 month). 1 inertial month is 30 days, while 30x0.96=28.8 is non-inertial month. 90 orbitals (¼ curvature) is 2700 days x 0.96 is 2592 days (7.2 years), which is 1/10 fractal inversion of precessional cycle (precessional cycle is 360° or 4-fold curvature of inertial 72 years. Inertial 24/25 difference of 72 is 3 years, which is its inversion or quantum. 3x90°=270 is singular inverse curvature, which is 2D fractal difference or 27000 years. 27000x24/25=25920. It is $3x86.4x10^2$ and/or $3x90x10^2x24/25$, which is inertial 3.6° discrepancy).

360 orbitals (full curvature) is 10 800x0.96= 10368 days (28.8 years) and 4-fold curvature is 1440x30=43200 or 43200x0.96=41472 days etc. Gradual lunar orbital motion is perfect reflection of quantized precessional motion, because such motion is equal opposite of precessional perihelion precession. Lunar spatial orbital (gradual change) is time of precessional motion (boundary of gradual change, which is instant quantum change). Precessional motion (in years) without and lunar orbital motion (in days) within are action/reaction of each other in context of inertial 360° curvature in between, which is their eigenvalue. Lunar orbital (size/motion) is ½ gravitational eigenvalue between Earth's rotation within and its orbital motion without; therefore it is equal opposite (space) of precessional motion (time). Time of precessional motion is instant space of lunar orbital (tidal 30-day cycle) and instant space of precessional motion is time of lunar orbital. Precession without and lunar orbital within are angular 2x1/2 of horizontal 2x1/2 rotation/orbital motion, which is singular 4x1/4 wave.

Precessional motion is action/reaction of 30-day tidal cycle. 5400 years of ¼ precession x 12 tidal cycles per year = 64 800 tidal cycles per quantum period of time (1/10 fractal difference), which is 25x2592 cycles or 2592/12=216 years. 12 tidal cycles are perfect 360-days year, while additional 5 days are 5400x5=27000 days (75 years =900 tidal cycles), which is 1.111 fractal difference of 25920 days (72 years or 864 tidal cycles). 72-year period, which is 1° of precessional curvature and/or 864 tidal cycles is inertial discrepancy, which is inversion of inertial here and now in period of 64 800 tidal cycles. Such instant change, which is addition of relative 1080 years (5400+1080=6480) is instant reality of 1080x12=12 960 tidal cycles (1/5 period). Consequence of infinite repetition of tidal cycles is gradual change of relative size/motion (Earth), which is instant reaction only in ¼ quantum periods.

Earth/Moon are duality (conserved momentum of singular inertial system) and gradual change of size/motion of one is equal and opposite change of another. Tidal cycle between Moon and Earth is gradual change of Earth's size and motion, which reflects in increased distance between them. Such gradual increase in distance is 4 m/100 years (0.04) or 259.2 m per ¼ precession and/or 5.184 km per 103680-year period (which is 103.68 km per 2 073 600 years of gravitational cycle). Because of tidal drag there is gradual deceleration of Earth's rotation, which is acceleration in its passage of time. Such gradual change of motion (time) is not real unless singular ¼ quantum, which is 216 (6^3) years relative discrepancy between rotation and orbital motion. Gradual deceleration of rotation is equal and opposite latitudinal (east/west) expansion or gradual increase in equatorial bulge, which is built-up of tension in longitudinal (north/south) directions. Such increased obliquity (spatial eccentricity) is not real (does not cause equal and opposite reaction) unless singular quantum, which is ¼ or 90° of angular perihelion precession, which is instant angular shift (1/4 and/or ½ quanta of angular rotation, which is rotation of rotation or double/inverse rotation). Because of 24/25-fractal inertion, such shift is not 90° but 86.4°, which is 3.6°-north/south displacement of accumulated bulge. It is instant reversal of gradually increased bulge and/or decreased rotation.

Gradual increase of bulge is increase of size, which causes equal and opposite decrease in rotation. Quantum change is instant (spatial) reaction of such gradual (time) action, which is instant correction of accumulated space/time imbalance. It is exclusive Earth's reality, because in solar (relative) context, passage of time between quantum changes is not real (it is its instant spatial density or inertial mass). Rotation/orbital (passage of time) is not real in solar context. Only Earth's angular (north/south) rotation, which is not round orbital but linear direction is real in solar context. Angular (north/south) rotation, which is linear direction perpendicular on its orbital, is Earth's double/inverse orbital round core of galaxy. In other words—Earth's reality in solar context (passage of time) is not solar reality. Only Earth's galactic orbital is passage of time in solar context, because context of solar reality is galaxy, just as context of Earth's reality is solar reality. Angular (north/south) rotation is quantized reality of infinitely long periods of time in between only in inertial context here and now (surface of Earth). In solar context, such angular rotation, which is gradual orbital in context of galaxy is its constant passage of time, which is double or accelerated passage of time here and now.

Passage of time of solar reality is changeless gravity here and now, which is changing only in quantum steps of infinitely long passage of Earth's time. ¼ precessional change is singular quantum of galactic orbital—it is singular quantum of solar spatial reality and/or singular quantum of its passage of time. Just as passage of time here and now is not real in solar context, so passage of time of relative (solar) reality is not real in galactic context. Orbital round core of galaxy and/or angular Earth's rotation are exclusive solar reality. In context of galaxy, such time is instant density (mass). Only singular quantum of expansion/contraction of massive Earth is singular quantum of passage of time of galaxy and/or singular spatial quantum. Time of galaxy is expansion of universe—i.e. it is equal and opposite antigravity of gravity here and now. Because of fractal asymmetry, gravity within mass is instant, while antigravity without (space) is gradual. Constant passage of time of galactic reality is inversion of such gravity/antigravity—it is inertial (non-expanding) space without (same passage of time as here and now), while gradually expanding/contracting massive reality within (Earth) is its instant reality. Infinite gradual expansion of massive Earth is instant density of no time in context of galaxy. Infinite passage of time of Earth's reality (complete expansion of universe) is instant galactic reality. Solar reality is instant and Earth's reality is double/inverse instant massive density within. Solar reality is an atom and Earth is its orbiting electron, which emits a photon in ¼ intervals of precessional motion (De Broglie). Such singular quantum of energy is singular quantum change of space/time here and now.

Tidal drag causes decrease of rotation (time of 1 day) of 1.666...millisecond per century, which is 360x1.666=0.6 sec in decrease of velocity per year or 60 sec=1min accumulated time difference per century. ¼ precessional period of 5184 years is accumulated difference of 51.84 min (0.864 hrs), which is inertial period of non-inertial 54 min=0.9 hrs (24/25 fractal discrepancy) and 54 min=60-6 or 60/1.111 or 1/10 fractal difference. Tidal drag causes 1 hr discrepancy in rotational motion in period of 5184 years, which is fractal 24/25 difference of time. 5184 years is singular quantum of time, which is 360 days/24=15 days accumulated discrepancy (where 15=14.4+0.6, which is its 24/25 discrepancy) between rotation and orbital motion. It is 15 (14.4) days or 1/6 seasonal discrepancy between orbital (perihelion) and rotational precessional motion, which is instant quantum correction. 4 precessional quantum changes are 4x15=60-2.4 (24/25 discrepancy) =57.6 days, which is 2/3 or singular quantum of double quantum change, which are 25920-years periods. It is accumulation of inertial 3 days difference remaining after singular quantum correction (3x4=12), which is 60+12=72 days or 4/5 discrepancy, which requires correction of correction that is not only angular rotation in plane of ecliptic, but also bellow/above its plane.

Infinite number of corrections in plane of ecliptic is passage of time of solar reality, while corrections bellow/above plane of ecliptic (gradual angular rotation of plane of ecliptic) is passage of time in context of galaxy, while passage of time of solar reality is its instant surface spatial density and passage of time here and now is in-depth density of surface density. **15 days of accumulated discrepancy is total of 15x5184=77760 days or 77760/360=216 years** (2592 lunar tidal cycles). 5184 yeas is accumulation of relative difference between orbital and rotational motion of 15 (14.4) days per year, which is 24/25 discrepancy of total of 360 days, as well as accumulation of total time difference of 216 years per precessional period, which is its 24/25 discrepancy. Lunar tidal cycles are time or boundary of inertial passage of time here and now. They are cause/effect of precessional motion and its equal and opposite discrepancy (inertion/inversion= double inertion). Tidal waves are discrepancy between solar reality without (rotation/orbital motion) and lunar reality within (lunar orbital motion). They are time here and now between double/inverse spatial realities of Sun and Moon, where Moon is equal opposite within of galaxy without.

Solar/lunar duality is 18-year repetitive **saros cycle**, which is 18 years required for 90° rotation (singular quantum) of empty solar space, which is equal opposite of Earth's orbital motion (5 days discrepancy, which is 5° motion of empty solar space per year or 18 years per ¼ curvature). Its double/inverse reactions are instant 18° precessional corrections, as well as lunar nodal cycles of 18.6 years. 18.6 years of nodal cycle is 18 + inertial 0.6 year discrepancy (18/24=0.75, where 0.75 is 5/4 of 0.6, which is precessional/perihelion 72/57.6 difference increased by its 24/25 inertion). 18/0.6=30 day tidal cycle, which is its inertion in between; 18+10.8=28.8=4x7.2 and 18-10.8=7.2=1/4x28.8, while 18x0.6=3x3.6=10.8, which is 10.8 days discrepancy of saros cycle (18.03 years, where 0.3 years is 10.8 days, which is ½ of 216/10 and/or 3/2x7.2 etc). 360° curvature (4x1/4) of 18-year cycle is 18x360=6480, which is 5/4 of 18x288=5184 and 6/5 of 300x18=5400. 30 (tidal cycle) x5184–25920 (5/4 inertial discrepancy) = 129600 years of double/inverse precessional period. Inertion of 18-fold cycle is its 5/4 difference of 3.6, which is non-inertial 14.4 period. 3.6° angular curvature of linear 14.4 period and/or 14.4° angular curvature of linear 3.6 is 51.84 or $1/10^2$ of quantum period of time (precessional time discrepancy). 3.6° inverse curvature of 5184 is 5184/3.6=1440 years of Sothic cycle, which is 1440+21.6=1461.6 years (21.6=3/2x14.4 or 2D fractal difference).

Sothic cycle is 1° curvature of double/inverse precessional period (90° x1440=129600 years). 3.6° inertion of such cycle is 3.6x1440=5184-year quantum period of time. 3/2 Sothic cycle is 1 zodiac of 2160 years, while ¼ precessional period of 6480 years is 4.5 or 9/2 cycles. ½ Sothic cycle is 720 years (1 ecliptic sequence), which is 1D fractal difference of 72 years, or 1° of precessional curvature, while 360° curvature (25920 years) of precessional motion is 18-fold repetition (saros) of Sothic cycle (18x360), which is 36-fold repetition of ecliptic sequence. Sothic cycle is 4x20=80-fold repetition of saros (80x18) in context of inertial 21.6 discrepancy (6/5x18=21.6). It is 4-fold curvature of inertial 4D here and now (4x360). Sothic cycle is duration (time) of saros, just as saros is duration of Sothic cycle. They are inertion/inversion=double inertion here and now. 5184-year period is 200x18 (saros) x 1.44 years ($1/10^3$ of Sothic cycle), where 200 is 10-fold repetition of 20-fold cycle, where 20 fold cycle is eigenvalue of fractal 5/4 and 10-fold repetition is eigenvalue of fractal 5/2. It is 360°x10x1.44, where 1.44 years is 518.4 days (14.4 years =5184 days). Lunar reaction of such solar reality is Moon's monthly ascension through its nodes, which are spaced by 1.44°. Such nodal regression is equal opposite of Earth's precessional motion. 360° curvature of nodal regression is 518.4 months, which is 43.2 years (30x1.44 years).

Solar and lunar space/time sequences are inertion/inversion of each other. 235 synodic months of 19-year **Metonic cycle** are 10x24 less 5, which is 5D inertion of solar reality that is its singular quantum (240/5=48 and 235/5=47, which is 47/48 fractal discrepancy), just as synodic/sidereal difference of 236 sec is 240 less 4, which is 4D inertion of Earth's reality, which is singular quantum of solar reality (240/4=60 and 236/4=59, which is 59/60 fractal difference). Solar/lunar sequences are 60/48 =5/4 fractal reality or double/inverse time waves of inertial particle (Earth) in between. It is 60-48 or 1/12 fractal difference in between, which are 12 months of one year. 60+48=108, which is 90 +18 or full curvature in context of 1/5 fractal difference. 108/12=9 years, where 576x9=5184, 600x9=5400 and 720x9=6184 or 9^2=81 (difference between Earth's and lunar mass), while 81^2-81=6480 years. Likewise: 4x81=36x9=324 and 324x20=6480 etc. Inertial 1/48 discrepancy (i.e. space) is ½ of 1/24 discrepancy (i.e. time), which is fractal 5/2=2.5x1/60 discrepancy (gravity) and/or vice versa. 240 is 2D fractal difference of inertion of perihelion precession (2.4). 240^2=57600 and 57600/5184 is 11.111-year cycle of solar activity (time), which is 1D fractal difference (eigenvalue) of its inverse spatial reality without (10/9 planets = 1.111, where inertial difference between 9 and 10 planets orbiting empty space without of dense Sun within is Earth).

CHAPTER 11

QUANTUM COSMOLOGY

Relativity and quantum uncertainty are double/inverse fractality. Space and time are inertial here and now turned inside out in directions within and without. There is no definition of physical reality, without its inertial background, which is its singular eigenvalue. Only comparative reality is real and space and time are double/inverse comparison with inertial here and now. Space/time difference is singular/infinite or light speed difference, which is infinite motion of singular size and/or singular motion of infinite size. Passage of time is only fractal overlap between two relative spatial sequences (without and within), just as space is fractal overlap between two relative passages of time. **Infinite period of passage of time here and now is singular quantum of infinitely larger instant space in context of no time and equally vice versa—no passage of time here and now is infinitely long period of time of infinitely smaller space.** Space/time is double/inverse perspective, where spatial and time distances are distorted proportionally with distance of observation, because space in time context and/or time in spatial context are double/inverse (convex/concave) curvatures. Singular quantum of space is time and singular quantum of time is space, because one is boundary of another. Space and time are double/inverse non-Euclidean curvatures; therefore they are equal and opposite singular limitations (curvatures) of infinite reality of another.

Singularly small and/or infinitely large space is double/inverse boundary or horizon of inertial time here and now, which is always ¼ (90°) curvature, just as singular instant and/or infinite period of time is double/inverse boundary (90° curvature) of inertial space (here and now). Size of the universe and its micro-world, as well as its passage of time are real exclusively in perspective of inertial here and now as frame of reference. In context of relativity/uncertainty, such infinite/singular spatial or time reality is meaningless. It is not reality, but its perspective, just as infinitely small object in distance is not really small, but only distorted in curvature of perspective. Relativity and uncertainty are double/inverse curvatures of inertial perspective here and now. They are only its double/inverse or space/time applications. Singularity/infinity, just as in any perspective are only consequence of curvature and boundary of curvature is +/-90°, which are its uncertain perspective within and relative perspective without. Opposite side of sphere is always beyond horizons of such double/inverse perspectives, because it is equal opposite of inertial here and now (gravity of antigravity and/or vice versa). Reality is always only here and now, while any distant/different reality is same as here and now, only in perspective of double/inverse (space/time) curvatures.

Space is real only in comparison with its equal and opposite time and time is real only in comparison with its space. Exclusive spatial reality without its time context, just as exclusive time, without its equal and opposite spatial context, is not real, but infinite distortion. Double helix has no beginning/end—only its fractal sequences are limited by its double/inverse (space/time) curvatures. Each sequence is only one out of infinite number of sequences, all equally real (many worlds). One spatial sequence is infinity of time sequences (infinite passage of time) and one time sequence is infinity of spatial sequences (different instant densities/sizes). Sequencing is quantum reality, while its content is gradual relativity (where quantum space is relative time and quantum time is relative space or gradual relative motion is quantum change of size and gradual change of size is quantized motion etc). Mass within is instant spatial reality of infinity of different past time sequences, while empty space without is gradual passage of time, which is infinity of future time sequences, each sequence being quantum period of time. Mass within is instant past time and space without is passage of time, which is instant energy of such mass.

There is no difference between Planck length sized space here and now or infinite spatial universe contracted to singularity in direction of past time—they are same time in double/inverse perspectives (angular difference 90°). They are double/inverse time beginnings of inertial spatial sequence here and now, just as infinite universe and singular black hole are its double/inverse ends. Such 4 points are 90° ends (boundaries) of convex/concave curvatures of inertial 5th point in between (here and now), which is their singular overlap. Such universe/ micro-world is exclusive reality of inertial here and now, while any different inertial point is relatively different curvatures and therefore relatively different sizes/ motions and densities of its universe/micro-world, which are also different beginnings/end (only relatively different, because boundary is always 90°). Universe has singular beginning in time only in perspective of inertial here and now, just as spatial horizon of observation is exclusively determined by position of point of observation. While it is different horizon and different reality within its boundaries, curvature that determines horizon is always 90° (light speed difference). Earth in perspective here and now is flat surface determined by ½ curvature of empty space without (sky), while in inverse (infinitely large) perspective—it is not flat but round curvature within in context of flat spatial reality of universe without.

Curvature of flat Earth is equal and opposite de-curving of its spatial inertial context, where de-curving of space is equal and opposite curving of time, which is only linear direction (passage of time) in context of flat Earth here and now. Curvature of massive Earth within is equal and opposite curvature of empty space without, which is also curvature of infinite linear passage of time. Curvature of infinite passage of time is its limitation within boundaries of finite, which is spatial reality of passage of time (relativity). Larger spatial reality without is larger limitation of its time, which is smaller angular curvature or relatively denser time in context of relatively emptier space. Infinitely large spatial size is singular time curvature. Such reality is real only in perspective from inside such curvature, while from outside, it is vice versa—it is infinitely small and dense space. Convex curvature in perspective from within is at the same time concave curvature from without; therefore force of gravity within (contraction) is antigravity (expansion) without and equally vice versa (where without in time is within in space and without in space is within in time, which is 4x1/4 wave). Direction of time generates same causative (double/inverse) spatial curvature—larger distance of past time is smaller and larger distance of future time is larger spatial curvature (smaller distance of future time is likewise smaller space—i.e. black hole). Spatial beginning/end in time and/or time beginning/end in space are double/inverse boundaries of perspective, which are 90° curvatures. They are exclusive reality of inertial here and now, without meaning in such larger space/time context.

They are exclusive boundaries of inertial quantum sequence, which are real only in context of ¼ precessional time, which is instant change of inertial sequence here and now. Infinity of such inertial quantum changes is gradual change of reality of universe and its micro-world, which is gradual shifting of its relative beginnings/ends (they rotate). It is passage through different many-worlds sequences or gradual passage through infinity of different possibilities of inertial sequence here and now. Instant quantum change is instant change of space/time reality, which is instant change of universe and its micro-world. Such different inertial sequences are only relatively different, because boundary of perspective is always inertial ¼ (90°). Point of observation is always focal point of its 4x1/4 double/inverse (space/time) perspectives. Here and now is always singular overlap between double/inverse (convex/concave) curvatures. It is always ½ way between all equal opposites in spatial and time directions. Space/time reality is a sphere with here and now as center of its gravity (and/or antigravity, because spatial gravity is antigravity of time and equally vice versa). Point of observation is ½ way between infinitely large and infinitely small and/or between infinite past and equally infinite future. It is right at the edge of galaxy between infinite empty space without and dense core of galaxy within (1/2 way between infinity and singularity), which is permanent perspective, regardless of relative position. It is ½ way between infinite density and infinite emptiness; ½ between past and future (i.e. half-life of solar reality); ½ way between something and nothing.

Beginning of universe (15 bil. years) is therefore only 90° perspective of same infinite universe as here and now (and so is its end). At the beginning, universe was already 15 bill years old, because singular point of space in perspective of 15 bill years of past time is contraction of 15 bil years of its previous time. There is universe within universe, beyond boundaries of its beginning or ends, in space (Planck length) or time, infinitely many times. Singularity of space in distance of time is consequence of inertial perspective, which is double/inverse curvature, which expands space and contracts time in one and contracts space and expands time in another direction, because every action is equal and opposite reaction in inertial context which is no change. Past direction of time is inversion of universe, which is without, into universe, which is within. It is change of perspective from universe as infinite empty space to universe as infinitely dense mass of micro-world. Such universe at the beginning (or before) was same universe as here and now (15 bill years old), just as universe here and now is only singular point of infinitely dense (and hot) mass within in perspective from without, which is direction of future (infinitely faster) time. In other words—15 bill years from now (infinitely expanded universe), present universe will be its singular, infinitely dense (singular) beginning, and its stars and galaxies contracted beyond singularity of Planck length. There is no reaction, without equal and opposite action—expansion in space is equal and opposite contraction in time, which is infinite spiraling of double helix, where every infinite/singular point is instant addition of inversion to its inertion, which creates double inertion.

Expansion of universe is passage of time and its red shift is same reality as here and now in angular perspective, which is increased energetic activity of such mass (i.e. quasars etc). Red shift is static or time Doppler effect. Mass embedded in distant (curved) space is not observed in horizontal, but increasingly angular perspective. It is increasingly less static surface reality of mass and increasingly more its nuclear depth (energy). Mass in distance of space is increasingly energy or mass in context of exponentially accelerated passage of time, which is decelerated rotation. Linear expansion is reaction of exponentially decreased spatial rotation and/or decreased gravity, which is increased antigravity (repulsion). Linear and rotational directions are double/inverse time directions of passage of time. Faster rotation is slower linear motion and equally vice versa, because one is equal and opposite angular curvature of another (90° difference), where linear spatial direction is curvature in time, just as round spatial rotation is linear time direction (light speed difference). Linear spatial expansion is rotation of universe (empty space) in time (which is relative or infinitely larger space without inertial space here and now), which is equal opposite of spatial rotation of dense mass. Linear expansion of universe is reaction of passage of time here and now, which is increasingly less passage of time and increasingly more spatial reality of time.

Spatial expansion is equal opposite of passage of time here and now, which is passage of time in opposite direction or anti-rotation. It is gradual precessional motion (infinitely slow motion in opposite direction), or inverse rotation of empty space without, which is reaction of inertial rotation of dense mass within, which is not real here and now, unless singular ¼ quantum. Expansion of universe, which is inverse rotation of empty space here and now, is contraction of time, which is equal opposite of spatial expansion. Spatial expansion is gradual time curvature of space, while time contraction is gradual spatial curvature of time. Spatial expansion is gradual relativity in direction without, while time contraction is equal and opposite quantum uncertainty within. Space is gradual relativity without and instant uncertainty within, while time is gradual uncertainty within and instant relativity without. Mass is instant density, while its empty space is gradual densification in quantum steps, because they are fractal equal opposites and what is instant in spatial context is infinite passage of time in context of time, just as instant in time is gradual passage of time in spatial context. Mass is instant past time and gradual future space, while empty space of universe is gradual unfolding in quantum steps of future time, which is past space.

There is no instant reaction here and now of gradual expansion of universe. It is instant inversion only in ¼ quantum steps of precessional motion, which is instant reversal of spatial expansion, through equal and opposite time contraction, which is singular quantum of expansion of mass within. It is instant fractal shift of beginning and end of universe in direction of its future time, because it is singular quantum of rotation in time of empty space, which shifts boundaries (90°) of inertial perspective. It is shift of inertion here and now, which is equal and opposite shift of physical reality of universe/micro-world. It is instant passage from one to another space/time possibility of inertial here and now, which is singular quantum of space as passage of time, because space is infinitely faster passage of time (infinitely slower rotation) than time, just as its density (gravity) is twice infinitely faster passage of time. Such accelerated and twice-accelerated passages of time are real only in ¼ quantum steps, which are infinite and twice-infinite uninterrupted passage of linear time (5184 and 103 680 years).

15-bill years is angular spatial distance, as well as linear time distance between Planck length and here and now. It is infinitively small size (micro-world) within infinitely large universe (infinitely fast passage of time, which is infinitely slow motion or infinitely small time), as well as infinitely large space (universe) of infinitely small time (infinitely slow passage of time). 15-bill years of time distance is equal and opposite reaction of 150-mill km spatial distance of Earth's orbital, which is fractal 10^2 difference (1 year=100 km in context of orbital motion is 1km=100 years in context of equal and opposite linear expansion of empty space, because rotation is attraction, while linear motion is repulsion, which is double light speed difference or $10^2/(1/10^2)=10^4$, which is fractal of orbital velocity in context of inertial light speed. 10^4 is 4-fold curvature of inertial 4D here and now). **Time distance without (size of universe) and spatial distance within (size of orbital) are double/inverse fractal 3/2**=1.5 realities—one is double/inverse boundary (limitation) of infinite reality of another. Orbital within is inverse time (boundary) of spatial universe without, just as universe is double time of spatial orbital within. Universe is singular time quantum of orbital and orbital is singular spatial quantum of universe (and/or vice versa). Singular quantum of spatial reality of time and/or time reality of space, which is quantum change of such changeless reality, is ¼ precessional period or 5184 years. It is 4x1/4 cycle of 25920 years of constant passage of time in context of orbital (solar reality), which is spatial linear distance (size) in context of universe (instant distance from core of galaxy).

Passage of time (motion) in context of solar reality is instant spatial size of no time in context of galaxy. Eigenvalue between passage of time here and now (solar reality) and instant spatial reality of galaxy (size) is light speed difference, which is their overlap. I.e. time of orbital x inertial gravity/density = light speed, which is their eigenvalue (space in between). 86400 sec x 360 days =31104x10^3 sec, which is orbital time and/or 31104x10 km/sec, where 31104= 32400x0.96 (which is fractal 24/25 discrepancy) or 6/5x25920 (32400 is positive and 20736 is negative 5/4 discrepancy of inertial 25920, just as 31104 is positive and 21600 is negative 5/6 discrepancy, which are space of time and/or time of space or 4 directions of inertial 4x1/4 wave). 324/1.111=291.6, which is 3.6 discrepancy (inertion here and now) from perfect (non-inertial) 4x72=2x144=288° (360=100x3.6, while 288=80x3.6, which is fractal 5/4 or space/time difference). 288° is ½ angular curvature of perihelion precessional motion (½x57.6x10=288). 288+12=300 and 300+24 =324, which is fractal ½ difference between solar/ Earth's reality (12/24), just as 24/48 is fractal ½ difference between Earth/lunar reality. 12/24 and 24/48 are two equal and opposite ½'s of inertial ½ in between, which is 360° curvature here and now (double/inverse multiples of 6, which are 12-fold multiples). 300 is singular eigenvalue between equal and opposite 288° and 324° curvatures, where 324=360-36 and 288=360-2x36 (just as 216=360-4x36, which is infinite 2x1/2 multiplication/division of double helix—1.111). 324° is singular and 288° is double strand of double helix, while 300 is their line of symmetry (3x10^2).

311 040 km/sec is inertial distortion (discrepancy or curvature) of non-inertial 300 000 km/sec light speed, caused by gravity here and now (size of particle in context of wave or motion). 300 000 km/sec light speed is eigenvalue between 288° and 324° curvatures, where 324° curvature is time of spatial 360° curvature and 288° is time of time or gravity of space (double curvature). Light speed is eigenvalue (line of symmetry) between constant passage of time without and changeless gravity within, because it is equal opposite (energy) of mass here and now (Earth). Light speed is double/inverse curvature of inertial here and now. It is round time of flat space and/or flat time of round space. Light speed is eigenvalue between space and time—it is instant size of one and motion (passage of time) of another. Light speed is 1 year of passage of time in context of round orbital motion and no time but instant 1/25920 fractal of linear spatial distance (quantum) in context of equal and opposite galactic orbital. 25 920 years of passage of time (number of orbitals) is instant space of galaxy (25 920 light years distance), because orbital time in context of gravity is light speed and 25920 orbitals are 25920 years of light speed (25920=20-fold repetition of 1296=6^4 or 6x1=6x6=36x6=216x6=1296).

Orbital here and now (1 year) is singular ¼ of galactic size, while 4 orbitals are its whole quantum. Spatial size of galaxy is 6480 time quanta (6480x4=25920 years). 25920 years is time (linear spatial distance, which is rotation in time) between core of galaxy (black hole) and here and now. It is ½-time between singular space within and infinite space without or one 2x1/2 sequence of multiplication/division of double helix. It is time period between two subsequent passages through a black hole and each 6480-years period is fractal ¼ adjustment of such distance. End of each ¼ precessional period is instant ¼ contraction of galactic space (quantum reaction within of equal and opposite gradual expansion without). 4 of such quantum steps are distance between edge of galaxy and its core. It is double spiraling motion of galaxy in its depth, which is real here and now only in ¼ steps. It is spiraling of solar reality towards black hole within and/or equal and opposite spiraling of core of galaxy in direction without, which is again fractal 4/5 reality. Fractal difference between such double/inverse (within/without) spiraling is double precessional motion, which is 4x25920=103680 year period (1/2000 or $1/20x10^2$ of its spatial orbital, which is singular ¼ quantum, while whole quantum is 1/500 or $1/5x10^2$, which is 500x414720 periods, where 414720=24/25x432000 or 4/5x5184000). Spiraling is double/inverse curvature, while rotation and orbital motions are 2 equal and opposite curvatures.

Spiraling is singular eigenvalue of two equal and opposite non-Euclidean curvatures. It is space/time or singular gravitational reality of two equal and opposite exclusive spatial motions (where one is spatial motion in time, while other is time motion in space, which is relative/uncertain description of inertial reality here and now). Rotation is spatial reality, while spiraling is in time, because space is only surface of time (where space and time difference is always relative/uncertain). Time is double/inverse space, which is instant duality of two equal and opposite spatial realities. It is simultaneous horizontal and angular rotation, which is spiraling. Just as time is context of spatial motion, so is gravity context of time motion. It is spiraling only in context of inertial spatial perspective within (inside), while in context of time, which is inertial here and now of such gravitational reality (without or outside), it is two equal and opposite motions, which are rotation and orbital motion separated by passage of time in between. It is same reality as here and now, where rotation and orbital are separated by passage of time, while in perspective of solar reality, where passage of time here and now is not real, such two motions are gradual spiraling away from solar center (and/or spiraling of Sun in opposite direction, which is fractal 4/5). Galaxy is inside out solar reality, where space is inverse passage of time and time is instant double space. It is inversion between rotation and orbital—its orbital is round curvature without and its rotation is linear direction within (which is round rotation in time).

Earth's and solar rotation are 90° angular (time) difference, which is flattening of one out of 3 spatial directions, therefore its 3D rotation is projected only as double linear motion on its surface, which is inertial 3D (space) here and now. 3D solar reality is space and time; therefore exclusive spatial reflection is flattening of curvature, which is equal and opposite rounding of linear direction in time. Solar rotation is linear distance between core of galaxy and infinite empty space in opposite direction. It is not rotation on surface (spatial reality) but rotation in depth (time). Likewise—rotation of galaxy is on surface (space), while its orbital motion is in 3rd direction (gravity) between surface reality here and now (space) and its depth (time), which is only surface space of solar reality, because Earth's reality is only one out of 4 linear directions (4 simultaneous possibilities) of surface solar reality. Galactic orbital is beyond space/time boundaries, because it is their eigenvalue, which is changeless gravity here and now. Its singular ¼ quantum is 432000-year period.

Infinite space of the universe (15 bill years) is ¼ of its orbital. Beginning of universe in time is spatial distance from here and now and limb of its orbital in one direction. Its beginning and end, which is ½ of its orbital is distance between here and now and light speed expansion of infinite universe without, and light speed rotation of its black hole within. ¼ of its orbital, which is infinite universe and all its time, is space/time without, while another ¼ is its density within (mass/energy). ½ of galactic orbital is universe and its micro-world, because its time is density—it is motion between surface of atom within and surface of infinite universe without. 15 bill years of past time, which is density of an atom is 15 bill years of infinitely empty space of universe, which is its future time. Such different densities are only surface reality, which is no difference in its depth. Atom-sized universe is same universe as here and now, just as universe here and now is size of an atom in context of future 15 bill years (difference between rotational light speed of an atom and linear light speed of expanding universe, which is double light speed difference here and now in between). Distance between surface of atom and its nucleus is yet another ¼ of galactic orbital (total ¾), while distance between surface of nucleus and Planck length sized center within is yet another ¼ (complete orbital).

Distance between surface of atom and surface of nucleus within, as well as distance between surface of nucleus and its Planck-sized center is yet another double/inverse light speed difference (C^4), which accounts for discrepancy in envisaged uniform expansion of universe (its accelerated expansion in the beginning). Planck length is infinity of galactic orbitals within, which is in-built genetic code of its outer sequences, which are its equal opposite. Every inertial sequence here and now is real only in context of its double/inverse inertion within/without, which is fractal 4/5 difference, while such fractal (infinitely smaller/larger) sequence is real only in context of its double/inverse inertion, which is 24/25 double fractal difference here and now. It is infinite curvature, where singular (inertial) quantum on surface is infinite number of inverse quanta in depth. Each of infinite number of quanta is yet another infinity of inverse quanta in depth (Balmer) etc. Singular quantum is inertion and each inertion is infinity of inverse inertions within, which is infinite curvature of double helix. Double helix is two equal and opposite gradual curvatures (double/inverse or space/time infinities), interrupted by periodic quantum changes (singularity), where such two curvatures intersect each other.

One orbital of galaxy is 60 bill years (which is distance of density between Planck length sized and infinite universe), while its spatial reality here and now (universe) is ¼ of that period. Its orbital is gradual expansion of universe (time curvature or relativity), which is instant reaction here and now in ¼ periods (spatial curvature or quantum uncertainty). Solar rotation is time (double space) in context of gravity, which is its time, while galactic orbital, which is equal and opposite direction of solar rotation (same scalar, different vector), is gravity (double time or attraction) in context of antigravity (expansion), which is its time (therefore universe is rotation in antigravity, which is double gravity, in context of singular gravity/antigravity, which is their eigenvalue etc). Gravity is time of solar reality and space of galaxy. It is linear direction in depth of surface space (equal opposite of rotation, which is its reaction). 25920 years is period of one rotation in time of solar reality, which is linear distance between here and now and core of galaxy. It is singular ¼ gravitational quantum, which is infinite passage of time here and now and singular spatial quantum of solar reality. It is infinite passage of time here and now, instant surface space of solar reality and twice-instant space of galaxy, which is density within of empty space without (mass).

Each ¼ precessional change is instant ¼ contraction/expansion. It is instant ¼ contraction of galactic space and/or instant expansion of space within (mass), which is instant expansion of time without and contraction of time within (reaction). It is instant incorporation of past period of infinite passage of time into its singular spatial sequence within and simultaneous incorporation of singular gravitational sequence without, which is infinite period of its future passage of time (5184 years). It is instant addition of one space/time sequence to genetic code of inertial here and now in context of its unfolding subsequent sequence, which is one pulse of galaxy. Pulsations of galaxies are increased with spatial distance, which is increased angular curvature or accelerated passage of time (passage of time is tangent function of such curvature). Such pulsation in infinite spatial distance, which is 90° time curvature is not quantum change between infinitely long passages of time in between, but gradual, infinitely faster uninterrupted series of changes, which is increased energetic and decreased massive reality of galaxies at the edge of universe. Each pulse of distant galaxies is infinite passage of time of 6480 years here and now contracted into no time at all. Pulsation of distant galaxies is infinitely fast double/inverse spiraling, which is real exclusively in inertial perspective here and now. It is galaxy in perspective of its micro-world (atomic) reality or galaxy at 90° angle, where space is time and time is space (mass is energy and energy is mass). Pulsation of galaxy is depth (time) here and now and its surface (space) on edge of universe, because space here is time, and time here is space at its infinite/singular (double/inverse) boundaries.

Pulsation of galaxy is quantum reality, which is passage of time in between quantum changes, only in context of inertial here and now, which is its uncertain reality within. In context of solar reality (without), where passage of time here and now is not real, pulsation is its gradual change or uninterrupted passage of time, which affects different sizes/motions (planets) in different time intervals, because of their relatively different passages of times. Different passages of time of different sizes, which are relative/uncertain reality of planets, are instant spatial density of solar reality. Instant solar spatial density is relatively different space/time duality of its orbiting planets—i.e. larger size of orbital is slower motion and faster passage of time, while smaller size is faster motion and slower passage of time (Kepler). 5184 years between pulses is exclusive reality of Earth's inertial passage of time. Larger/smaller sizes/motions are relatively different periods between pulses, which are only surface (spatial) reality here and now. Such relatively different periods between pulses in context of inertial passage of time here and now (Earth), are always 5184 years in context of inertial passage of time of such different planets. Size/motion and distance of a planet determines speed of its passage of time and therefore periodicity between pulses, which are not pulses, but uninterrupted passage of time of solar reality, which is light speed difference from its orbiting planets within.

Pulsation of galaxies from perspective of orbiting planets is not pulsation but double/inverse rotation of solar reality. It is its rotation and orbital motion and constant passage of time in between. In context of solar reality—empty universe outside of galaxy is its pulsating galaxy, where such pulses are twice-infinitely long periods of passage of time between double/inverse quantum changes here and now, which are not exclusive Earth's reality, but simultaneous reality of all orbiting planets. Equal and opposite pulsations of universe, which are its instant expansion as consequence of gradual contraction of galaxy and/or vice versa, are smaller spatial discrepancy within (here and now), which are larger periods of time without (20-fold multiples). Pulsations of galaxy are gradual change of space/time sequences, while pulsation of universe, which are space/time changes of solar reality, are E/m changes here and now (equally vice versa). Pulsation of universe, which is galaxy of solar reality, affects stars of different sizes/motions (densities) at different time intervals, because of their relatively different rates of passage of time. Such different intervals are again only surface reality, because in inertial context of each star, it is same time interval.

Infinite number of pulses of galaxy is one pulse of universe, and infinity of pulses of universe is singular pulse of double/inverse universe, which is beyond space/time boundaries of inertial here and now. All different rates of pulsation of universe, because of different passages of time of different stars, are instant surface (spatial) reality of galaxy, just as all different pulsations of double/inverse universe in context of relatively different passages of time of its fractal, relatively different galaxies within, is instant surface reality of universe, which is beyond boundary of infinite and/or singular space (event horizon). Here and now is 3D, while universe is 4D, because time is spatial direction of universe, while 3D space here and now is one of its surfaces. 5D is double/inverse or twice relative universe, which is additional curvature of gravity as spatial direction, while time is its surface and 3D space here and now is one of its linear directions. 5D universe is instant surface space of all different 4D universes, which are relatively different passages of time. They are instant density of its mass within. 4D universe is also instant spatial surface of different 3D realities, which are different passages of time that are its instant density (equally in direction of micro-world). Universe here and now is only one possibility or one surface of 4D universe, which is infinite number of equally real 3D universes. Likewise—infinite number of relatively different 4D universes are instant density of singular 5D universe.

Infinite number of pulses of galaxy, which is infinite number of quantum changes here and now (infinity in 4/5D context is limited to 20x20=400-fold sequence, i.e. 20 in each direction of fractal duality), is singular pulse of universe and infinite (20x20=400) number of pulses of universe is singular pulse of double/inverse (5D) universe etc. Pulse (time) of larger spatial reality is smaller spatial discrepancy here and now, which is larger period of uninterrupted passage of time. Infinite number of galactic pulsation (400) is infinite number of different time sequences of space here and now, which are 20 different sequences of rotation in context of 20 different orbital sequences. It is singular spatial quantum or singular spatial sequence of time (2160000/2073600 years or 432000 of singular ¼ quantum, which is singular pulse of universe that affects all planets of solar reality, because it is singular solar quantum change of rotation, which is instant change of size/density of all its planets. 400 such sequences is singular pulse of 5D universe (864 mill years), which is singular E/m change or singular sequence that affects reality of an atom. Such change affects all stars of galaxy simultaneously, because it is singular quantum of change of their size. 864 mill is 6/5 of 720 mill (non-inertial time) and 720x20=14.4 bill years, which is 24/25x15 bill years of space/time reality of universe. It is 20 different rotational sequences of galaxy that are its singular orbital sequence.

15 bill years that is space and time of universe is exclusive Earth's reality. It is only one out of twice-infinite number of different possibilities, all simultaneously real, although in any inertial perspective, it is always 15 bill years—only difference is content (relative size/motion) of such space/time perspective. 15 bill years is not time of infinite space, but instant spatial density (mass) of universe, which is only singular point within of its infinite universe without, which is yet another dense point within yet larger 15 bill-year universe without (equally in direction of micro-world within, where here and now is infinite expanding universe in context of 15 bill years of passage of time). 15-bill years of universe without are reaction of 150-mill km of orbital (spatial) distance within, which is constant passage of time in between, in period of 25920 years (solar reality). 25920 years of passage of time in context of Earth's orbital is instant 25920 light years distance of solar orbital.

Each it ¼ precessional change is shift in position of Earth from edge of galaxy to ¾ distance toward its nucleus, which is equal and opposite shift of nucleus in direction without (4/5 fractal difference). Spatial shift within is time shift without and equally vice versa, which is constant position of Earth ½-way between dense core within and empty universe without. Space of galaxy is gravity, which is instant reality of space and time, and because change of one is equal and opposite change of another, Earth's position at the edge of galaxy is relative/uncertain (totally meaningless) description of its position. Quantum precessional change is instant contraction of universe in direction within and equal and opposite expansion of galaxy without, which is instant reversal of gradual recession of stars. It is instant fractal (quantum) change of changeless sky here and now, which is reaction of equal and opposite change of density of massive Earth within. It is simultaneous equal and opposite change without (empty space) and within (dense mass), which is no change on surface, because only infinity of such double/inverse fractal changes are its singular ¼ quantum.

Fractal difference between space/time and time/gravity is light speed, while double fractal difference between space and gravity is double light speed. Space (size) of galaxy and time (motion) here and now are space/time or light speed difference (C=d/t, which is fractal 5/4), while space of galaxy and space here and now are double light speed difference (C^2=E/m, which is fractal 24/25). Linear passage of time here and now is round spatial motion, while round passage of time of solar reality is its linear spatial motion here and now. Light speed is space/time eigenvalue (line of symmetry) between space and time (C=d/t)—it is its mass/energy. Light speed is infinite motion of energy and singular size of mass, where infinity of different sizes without is singular motion within, just as infinity of different motions without is singular size within and equally vice versa (fractal asymmetry between singular and infinite).

Singular quantum of passage of time in context of light speed is 25920/300000= 0.0864 or 86400×10^{-6} (year x sec/km), which is 10^{-6} (year x day)/km or 6 fold curvature of inertial here and now (10). Light speed is difference between time here and now (rotation) and its boundary—precessional time (relative space). It is orbital in between, because year x day is orbital time; therefore precessional time in context of light speed is 0.00036 day/km and/or 2777.777 km/day, which is 31.104 sec/km (non-inertial 30 sec/km, which is equal opposite of orbital velocity). 40000/2777.77= 14.4 days, where 40000 km is size of Earth. **Light speed is accumulated difference of 14.4 days between orbital and rotational motion, which is inertial precessional period of 25 920 (and/or 103680) years.** 21600/300000=0.072 is 5/6 of inertial time of rotation, or orbital/rotational discrepancy (precessional curvature). It is 0.0003 days per km or 3333.33 km per day, which is 40000/3333.33=12 days or inertial 2.4 days discrepancy, which is 1D fractal difference ($12^2/10=6 \times 2.4=14.4$) between orbital motion of Earth and equal and opposite rotation of empty space in opposite direction. Such double/inverse rotation is Sun/Earth orbital duality without and/or Moon/Earth orbital duality within in context of singular galaxy in between, which is their 4x1/4 (5/4) eigenvalue.

Light speed is space inverted inside out, which is its time and equally vice versa. It is time of space, space of time and gravity of both—space and time. Universe/micro-world is infinity of relatively different double/inverse curvatures of same constant, where any difference between sizes and motions is only surface of one reality, which is always double/inverse light speed (here and now). Galaxy is solar system inverted inside out—it is time reaction of its space and/or spatial reaction of its time, just as universe is galaxy inverted inside out, where time of fractal reality within is space of equal and opposite fractal reality without, because singularity of one is infinity of another. Solar system is inverted Earth and Earth is inverted singular here and now, which is inversion of an atom, which is inversion of nucleus, which is inversion of singularity of Planck length at its center (where spatial inversion without is time inversion within). It is total of 8 fractal sequences in between 2 gravitational sequences within/without, which are its singular/infinite surface boundaries within/without (Bode's periodicity, which is fractal 3/2 in context of gravitational 5/2 or 10 planets/electrons that are 10 inversion points of 3D space/time wave). Size and motion are double/inverse variables of constant light speed. Increase of size is equal and opposite decrease of time and equally vice versa. Infinite increase of one is singularity of another, which is instant quantum change of inertial sequence, when accumulated difference between rotation and orbital motion is 14.4 days, which are different 4x1/4 quantum periods of time (4-fold multiples of singular ¼ quantum). Size and motion are double/inverse relativity of inertial here and now, which is their only constant (double/inverse light speed).

Fractalization means that one always precedes another (double/inverse spiraling). It is infinite self-interference or repetition of inertial eigenvalue in different 3D sequences, which is gradual E/m change, which is no change at all, unless it is total change (whole number). Duality of one inertial wave is two equal and opposite spatial and time waves. Space is surface of time and time is surface of space—they are equal and opposite fractal 2/3 curvatures of double helix. Only time = space (C=1) is eigenvalue of fractal 2/3 reality, which is an instant quantum change of E/m and equally vice versa—only mass = energy (C^2=1, where 1 in space is 0 in time and vice versa) is quantum change of space/time. Such instant space/time change is singular gravitational quantum, which evens the odds. Each instant change is one E/m frequency (spectral line), which is instant release of accumulated tension (energy) in preceding period. There is no difference between micro-world and its universe—they are inversely proportional. They are both quantized relative reality and/or relative quantization, which is infinity of spatial waves in context of one time wave and/or infinity of time waves in context of one spatial wave. Static or no wave (Wheeler-DeWitt) is infinity of inverse waves, because **any particle of time is spatial wave and any spatial particle is time wave**—i.e. any size is infinity of motions and any motion is infinity of sizes; any multiple is infinity of fragments and any fragment is infinity of multiples (multiple in time is fragment in space). It is infinite double helix, where inertion inverts into double inertion through infinite precessional curvature of double helix, and what is instant in space is gradual in time, just as instant in time is gradual in space. Space is infinite time potential and time is infinite spatial potential, just as mass is infinite potential of energy and energy is potential of mass. Double helix is infinite 4x1/4 potential, where direction of its curvature is determined exclusively by inertion of singular choice of point of observation. **Point of observation is the only vector of scalar double helix and it is the only form of its content.**

3D space coordinate system is only surface of 3D space/time coordinate system, where time is equal and opposite angular in depth surface coordinate system. 3D space/time is in its turn only surface of gravitational 3D coordinate system, while 3D space is only one of its linear directions. There are 2 (double/inverse) directions of possible changes (curvatures) of any inertial direction of size and/or motion in between. Double/inverse changes of rotation are its angular curvature in horizontal and angular direction of orbital plane. Changes in orbital plane are 1D spatial changes, while changes in angular direction of orbital plane are 2D (double/inverse) spatial changes, which are also 1D time changes. 1D changes are fractal rotational changes, while 2D are multiple rotational changes that are fractal orbital changes. They are exclusive reaction of inertial passage of time here and now (Earth) in context of solar passage of time (pulsation of galaxy). Each successive change in twice larger period of time is twice larger spatial change. 3D quantum change is change of 3 spatial directions, which is 2D (double/inverse) time change or linear gravitational change. It is multiple orbital change and fractal change of size, while 4D is double/inverse gravitational change or change of two directions of linear gravity, which is multiple change of size and fractal change of its density.

It is expansion or contraction of such sphere, where expansion in space is contraction in time and equally vice versa. 3D/4D changes are not exclusive Earth's reality but solar quantum changes that affect all orbiting planets (pulsation of universe). They are quantum changes of solar rotation/orbital or changes in horizontal and angular directions of its orbital plane around core of galaxy. 5D/6D quantum changes are reaction of pulsation of double/inverse (5D) universe, which are galactic quantum changes of its rotation and orbital that affects all its stars. It is change of motion of galaxy, change of sizes of its stars and E/m or density change of its planets. 5D is multiple change of spatial density (mass) and fractal change of time density (energy), which is simultaneous multiple expansion in one direction and fractal contraction in opposite (angular) direction). It is multiple gravitational and fractal antigravitational change. Singular quantum of 6D change is complete inside out inversion between dense mass within and empty space without (distance between Planck length and surface of light speed expanding universe, which is 4 times all the time of the universe or 1 galactic orbital), while 5D change is exclusive expansion and/or contraction, which is distance from here and now and surface of expanding universe or ¼ galactic orbital, which is 15 bill years).

6D quantum change is whole galactic orbital that is 4 pulses of 5D universe, while 5D change is ¼ orbital, which is 1 pulse. 4D change is 4 pulses of 4D universe that are one solar orbital and 3D change is 1 pulse that is ¼ orbital. 2D/1D changes are 4 and 1 pulse of galaxy that are one and ¼ solar rotation. It is always one step behind in order to gain momentum for 2 steps ahead, which is infinite 2x1/2 fractal curvature that adds past sequence to the present to generate future sequence. Each whole quantum is only ¼ of larger quantum. Because of surface fractality, space is always ahead of time and time is ahead of gravity and equally vice versa in direction of time, which is infinite double/inverse Fibonacci spiraling, where gradual time curvature is interrupted by spatial curvature and equally vice versa. Double/inverse interruptions are reality, which is 2D fractal difference. It is not exclusive relative or uncertain reality, but both at the same time (twice relative reality, which is not time of space but space of space). Quantum changes are incorporations of additional quantum of time, created through precessional motion. It is gradual expansion of inertion here and now, which happens only in periodic quantum leaps. Larger spatial change is larger period of time and larger time change is larger space.

Infinity of singular quantum changes is reversals between rotation and vertical motion (tilt). I.e. constant rotation in short period of time (1/4 precession) is rapid succession of inverse directions of rotation in long period of time (infinity of precessional periods). Such rapid succession of reversals is east/west swinging, which is time reflection of its spatial seasonal north/south swinging (tilt). At the same time, series of north/south inversions is only constant rotation in one (north/south) direction. Even such north/south rotation in context of yet larger time is only swinging, and such swinging (tilt) in context of yet larger time is rotation, etc. Rotation in context of 1D time is 2x1/2 tilt in context of 2D time and yet another 4x1/4 rotation in context of 3D time. Spatial rotation is therefore only tilt in time and spatial tilt is rotation in time. It is two time possibilities of one space and/or two spatial possibilities of one time. Infinite number of inversions is infinite number of possibilities of inertial here and now. It is sphere within sphere and what is fast rotation in context of one is only slow swinging in context of another sphere. In context of 3rd sphere in between, which is double/inverse reaction of such horizontal/angular rotating spheres, such simultaneous dual rotation (time) is not rotation at all but constant size (space). Such constant size is rotating in context of yet larger sphere around its axe, where speed of rotation is exponentially increased from poles toward equator, which is ½ way between poles. Such difference (overlap) in speed of rotation is consequence of inequality of its 2 constituent rotations, which is inertial size in between (light speed difference from both). Its horizontal rotation is infinitely faster than its vertical rotation (or vice versa), therefore such rotation is an exponential gradual change of speed between poles and equator.

Earth's time is real only in context of solar (precessional) time and solar time is real only in context of galactic time. One is space, another one is time and 3rd one is their gravity—they are 3 directions of inertial 3D or 3 spheres, one within the other. Each larger spatial sphere is slower motion (faster passage of time), where such double/inverse motions are synchronized only periodically (eigenvalue). Rotation of Earth is reaction of Earth's gravity, while its orbital is reaction of solar gravity (rotation/linear motion). Orbital of solar reality is reaction of galactic gravity. Inertial overlap between two motions is passage of time. Motion (time) is reaction of gravity. Galactic gravity that causes solar orbital, which is its passage of time, is only fractal of Earth's reality, therefore real only when accumulation of such fractals is singular ¼ quantum of Earth's reality. Orbital motion is singular eigenvalue of two linear motions—one vertical and one horizontal—it is 2D reaction of 1D linear solar gravity. Solar gravity is 1D, while galactic gravity is 0D. Reaction of solar gravity is round orbital that is limited passage of time, while reaction of 0D galactic gravity is linear direction, which is infinite passage of time in context of equally infinite space. Solar reality attracts—it curves space in concave curvature, while galactic gravity is its equal opposite—it repels space, by curving it in convex curvature. Solar reality is space curved within, which is limitation of infinite passage of time, while galaxy is space curved without, which is infinite passage of time of equally infinite empty space. Singular cause is infinity of effects—it is infinite number of possibilities, all of which are equally real. Infinite space/time (universe/micro-world) is reaction of 0D galactic gravity. Linear force of galactic gravity is not real here and now, because it does not create motion (time)—it is exclusive time of solar reality—only Sun orbits galaxy, while planets orbit Sun. **Galactic gravity (attraction) is real here and now only when Earth is synchronized with solar reality (resonance).** Earth is synchronized with solar reality (eigenvalue) only when additional time created by it is incorporated into inertial time here and now—i.e. only in ¼ intervals. Galaxy is repulsive reality in periods of constant passage of time and attractive reality in context of quantum changes, when Earth's time is resonant with solar time (i.e. Earth's reality is not 3D but 4D).

Earth and Sun are synchronized only on surface—3rd direction is their relative/uncertain direction in between, which is their overlap or passage of time between two spatial spheres (infinite number of possibilities). They are 3/2 (double/inverse) fractals of each other. Synchronization (eigenvalue) of 3 out of 4 fractal directions is at the same time fractal 2/3 synchronization of Earth/galaxy (from 1/3). They are two equal and opposite effects of same cause. It means that in such instant, galaxy is not 1D but 2D eigenvalue with here and now. It is an instant when Earth's time is solar time and Earth's center of orbital motion is not Sun, but core of galaxy. It is instant reconciliation of fractal difference between Earth's and solar reality, when they are not equal opposites but same inertial space/time, which is context of both. It is end of passage of Earth's time and beginning of passage of solar time, which is light speed difference. It is also reduction of double to singular fractal difference between galactic and Earth's reality, when reaction of galactic gravity is not infinite empty space, but same concave curvature (spatial orbital) caused previously by solar attraction. It is inversion from repulsive to attractive reality of galaxy, when reaction of galactic gravity is not infinite repulsion of empty space in context of infinite passage of time, but attraction that is limited passage of time of its dense reality within.

It is reduction of infinite 3D space here and now to only a compressed surface, while previous passage of time is instant depth of such surface or its 3rd direction. It is across the board 1D shift, which is reduction of inertial 3D space on surface of flat Earth (sky) to singularity and solar space, which is context of round Earth, to only infinitely small surface between here and now and galactic space (sky), while galactic space is not infinitely distant empty space, but infinitely close, replacing solar space. It is instant incorporation of galactic spatial reality into reality here and now, when background inertion is not inertion but action/reaction and action/reaction (duality) is inertion. It is an instant, when universe is not expanding reality, but galaxy spiraling and contracting in direction within, while double/inverse (5D) universe, which was previously beyond space/time (light speed) boundaries replaces expanding universe. Solar reality is not fractal difference or time, but right here and now, while galaxy is not double fractal difference or gravity, but time. **It is moment when universe instantly stops expanding and it contracts in direction within—toward Earth**.

Galactic 0D attraction in context of gradual spatial change, which is constant passage of time, is infinite repulsion of empty space without (infinite passage of time) and singular contraction of dense mass within (no passage of time, but instant density). It is infinite 3D space here and now, which instantly changes in course of quantum change, when 3D is reduced to only 2D freedom of choice, because Earth's clock stops and Earth's spatial reality is measured in context of solar, infinitely faster passage of time (light speed difference). Such additional solar time is not in horizontal, but angular direction, which is 3rd half of fractal 3/2 and/or 5th quarter of fractal 5/4. **Such event is light speed motion of Earth or its passage through wormhole of time. It is passage through a black hole, because accumulated difference between orbital and rotational time is light speed (fractal) difference.** In 4D context of quantum changes, stars of galaxy are not light years away, but instant reality of no time difference. It is space/time inversion, when space is time and time is space and when galaxy is not 25 920, but only **0.0864 light years or 81.648 x 10^{10}** km away (25 920/C, just as time of rotation x light speed =86400x3x10^5=25920x10^6 km and 25920x10^6 x360=933120000 x10^4, which is size of orbital). It is only 1/50th part of distance from Sun to nearest star and/or 54 432 times further away from Earth than Sun, where 54 432 = 2 x 27 216 and 27 216-1296=25 920, which is 1/10^2 of 129 600 or 1/20 fractal difference.

It is inversion of space into time and time into space, when light radiated by Sun is not light but darkness and Sun is not a star (energy) but massive planet in orbital around galaxy. It is period when Earth does not orbit in space, but in time, and when its space is not space, but time, while its time is not time, but space. There is no coincidence in perfect curvature of double helix. Reality is numbers that have meaning. It is infinite 2x1/2 fragmentation/multiplication of inertial here and now, and their infinite recombinations in constant flux. Every form of physical reality is only particular numerical sequence, which is its **genetic code**. Such numerical sequences are infinitely simple spatial sequences in context of time (universe) and infinitely complex time sequences in context of space (micro-world), where simplicity in space is complexity in time and equally vice versa.

In a moment, when Earth and Sun are same inertial space/time (eigenvalue)—center of galaxy exchanges place with Sun as focus of its orbital motion. In such instant gravitational solar vacuum, galactic gravity is not 0D, but 1D, and Earth orbits galaxy instead of Sun. It orbits galaxy, while absorbing additionally created solar (precessional) time. While absorbing such time, it cannot orbit Sun, because such additional (precessional) time is solar time and Sun can rotate only in context of galaxy, it cannot rotate in itself, without background of galactic reality. Likewise—Earth can rotate only in time and while it absorbs additional time, which will fuel it through next period, it can be real only in context of larger—

solar time—and solar time is relative or infinitely faster passage of time. In other words—Sun and Earth are same inertial reality here and now in context of their larger galactic reality. They are not equal opposites but only equals and Sun is not center of Earth's orbital—it is same reality as any other planet of solar system, while they both orbit galaxy. Round orbital around galaxy, which is infinitely larger space without and denser mass within is infinitely eccentric solar orbit, which is linear motion vertical on its spatial orbital. Orbital around galaxy in context of infinitely larger gravity is instant repulsion/attraction, which is linear motion toward or away from Sun.

Rotation in context of solar reality is linear motion in context of galaxy and equally vice versa. Earth gradually stops its orbital around Sun, because solar gravity is instantly superseded by inertion of galactic gravity (difference between solar/galactic gravity). It either moves away or toward the Sun in spiraling curvature of its inertion, depending if Earth is positioned between Sun and galaxy or vice versa—Sun is positioned between Earth and galaxy. Attraction follows repulsion and equally vice versa—one quantum change is linear motion away, and another one is toward Sun, which are double/inverse oscillations of orbital, where one step behind is followed by double step ahead—it is one small (1/2) movement toward Sun, followed by larger (2) movement away. In context of solar time, planets do not orbit Sun—they are spiraling away from it, just as they are straightforward linear expulsion in context of galactic time. In solar context, planets are spiraling electrons (spiraling = orbital + infinity of spectral changes), while in galactic context; they are photons of light (energy) emitted by an atom (solar reality).

Quantum change is galactic eclipse, when Sun and core of galaxy are in syzygy. Solar/lunar eclipse is spatial reality (Moon between Earth and Sun or vice versa), while galactic eclipse (Sun between Earth and galaxy and/or vice versa) is time or gravitational eclipse. Solar eclipse obscures light (energy) and galactic eclipse obscures or increases gravity, depending if galaxy is in conjunction or opposition. In course of spatial (solar) eclipse—light is obscured by smaller mass within (Moon), while during time (galactic) eclipse, gravity of galaxy, which is only source of solar light, is either obscured by Sun, causing instant cessation of light emission, or Earth is between Sun and core of galaxy, which is instant accelerated solar flaring. Moon is completely covered by umbra during lunar eclipse, because it is smaller than Earth, and so is Earth covered by gravitational umbra during galactic eclipse when Sun and galaxy are in conjunction. When they are in opposition, Earth in between is focus of solar flaring, intensified by gravitational pull of galaxy. It is magnifying effect of gravitational lensing (accelerated time), parallel to passing of ray of light through magnifying glass, which focuses its intensity into one point (Earth). Equinox quantum changes are conjunction and solstice changes are opposition.

Angular tilt (change of seasons) and resulting precessional motion of Earth are nodes and their gradual regression in context of inertial solar reality. Such double/inverse motion is gradual regression of solstice/equinox points in orbital context. Just as solar/lunar eclipse occurs only when lunar orbital crosses nodes in point of conjunction, so galactic eclipse occurs only when point of equinox/solstice crosses (inverts) with precessional motion, which is inverse or galactic orbital. Only when points of solstice/equinox are regressed for ¼ orbital and orbital (perihelion) motion advanced for ¼ seasonal change (angular tilt) are they perfect duality (nodes crossing path of orbital motion). Rotation and orbital motion are double/inverse resonance (synchronicity) only in solstice/equinox points, which are its nodal points. Galactic eclipse occurs only when such points are regressed for ¼ orbital curvature and when perihelion/precessional discrepancy is 1/6 seasonal change (14.4 days), which is once every 6480 years. One precessional period is 4 points when Sun and core of galaxy are in syzygy and aligned in galactic plane of ecliptic.

Moon and galaxy are angular equal opposites of horizontal Earth/Sun duality (4x1/4 wave). Moon is inversion (uncertainty) within of galaxy (relativity) without. It is gravitational balance or inertion (eigenvalue) of Earth/Sun duality within, just as is galaxy without. Size of Moon is reaction of time (motion) of galaxy and motion of Moon is reaction of size of galaxy. Moon in its orbital is fractal difference or overlap between Earth and solar space/time realities. When Moon is in conjunction and/or opposition during lunar/solar eclipse, such gravitational balance (fractal difference) is reduced to 0; therefore Earth and Sun are perfect duality, which is instant change (eclipse). During eclipse, Sun and Moon do not balance each other, but one increases gravitational pull of another. Infinite number of lunar/solar (spatial) eclipses is one instant galactic (time) eclipse. Moon in conjunction obscures solar light (energy) by its size (mass), while Sun in conjunction obscures galactic mass (gravity) by its energy and equally vice versa when they are in opposition. Galactic eclipse is positive or negative gravitational or time obscuration, while solar/lunar eclipse is spatial obscuration, because they are light speed difference.

Solar attraction is galactic repulsion and equally vice versa. Attraction of planets in context of solar reality is their repulsion in context of galaxy, because they are positive/negative non-Euclidean curvatures, which are light speed difference in between. Galaxy attracts stars and stars attract planets. When galaxy attracts planets (quantum change), they are repelled by solar reality and equally vice versa—when they are repelled by galaxy, they are attracted by solar reality (passage of time). Only opposites attract, while equals repel. Double difference is infinitely large difference, which is no difference at all. Only an odd fractal difference is attractive, while even difference is repulsive. Earth quantum changes (1D/2D) are changes in context of repulsive solar and attractive galactic reality. Solar quantum changes, which are gravitational (3D/4D) quantum changes here and now, are changes in context of repulsive galactic reality and attraction of infinite empty space of universe without, because galaxy and Earth are synchronized in all three spatial directions and 2 out of 3 gravitational directions.

CHAPTER 12

CONTINUITY

Each successive quantum change is instant inversion of inertion into double inertion, which is creation of inertial space/time sequence, which is inertial size in context of its passage of time, which is boundary of its duration. Fibonacci sequencing of double helix is infinite number of such fractals, which are always recombined in different order, ensuring there is no two same sequences (irrationality of Π/φ). Only when all possible permutations (sequences or harmonics) of one fractal difference are exhausted, it initiates new sequence of fragmentations, by adding past sequence to the present in order to create future (different) sequence. Each new sequence carries an imprint of all infinity of previous sequences in its genetic code. Earth here and now is final result of infinity of such gradual and instant changes, where largest imprints are recent ones, while smallest imprints are sequences long past Where gradual in space is instant in time and equally vice versa). Boundary of such imprints is boundary of time or triple inversion of time in past direction.

Inertial context of Earth's rotation is its orbital, which is its boundary or time (curvature), just as boundary or time of orbital motion is precessional motion and time of precessional motion is double precessional motion etc. Orbital motion is fractal reaction of equal and opposite rotation of empty solar space, just as precessional motion is reaction of its equal and opposite orbital, while double precessional motion is fractal reaction of rotation of twice-empty space of galaxy and triple precessional motion is reaction of its orbital etc. It is infinity of motions within motion, where each preceding motion is instant size of succeeding motion and twice preceding motion is its gravity. There is no difference between motion and size, because double/inverse motion is size and double/inverse size is motion. It is infinity of relative/uncertain spheres, where spatial sphere (size) is overlap between double/inverse time spheres (motions) and time sphere is overlap between double/inverse spatial spheres, which are double/inverse singular fractal difference (light speed). Eigenvalue of such double/inverse fractal differences is double fractal difference between two spatial and/or time spheres, which is singular gravitational sphere (dense mass within and empty space without). Inertial size here and now is singular cause of infinity of motions within and without (universe/micro-world), just as its equal and opposite motion is cause of infinity of sizes, one preceding another. It is infinity of gradual spatial waves in context of instant time wave here and now and infinity of gradual time waves in context of its equal and opposite instant spatial wave (and/or infinity of gradual gravitational waves of instant space/time here and now).

Earth's quantum changes in ¼ intervals of precessional motion are 0-gravity in one spatial direction, which are pulsation of Earth in north/south direction as well as horizontal pulsation of its orbital plane. 4x1/4 changes are changes in 4 directions of orbital plane (east/west and north/south). Double quantum changes are fractal pulsation in angular direction of orbital plane. Solar double/inverse quantum changes (instant expansion/contraction), which are 0-gravity in 2 spatial directions, are horizontal and angular pulsations of entire solar system, which are 3D change of orbital plane and size of Earth. They are 90° (reversible) and 180° (irreversible) angular rotations of plane of ecliptic, which are instant angular rotation of empty solar space. They are instant reversals of rotation of all orbiting planets, which is singular reversal of their orbiting motions from left to right and/or vice versa, which is complete inversion of time direction of solar reality. 4x1/4 such smaller changes is ¼ of larger quantum change, which is not only angular rotation, but also ¼ (reversible) and/or ½ (irreversible) rotation of solar ecliptic (empty solar space) above/below galactic ecliptic or plane of solar orbital motion round core of galaxy. It is 180° of spatial angular motion of Sun, preceded by double/inverse angular motion of its orbiting planets. It is not singular (½) angular rotation of surface (orbiting planets) around its singular center within, but double/inverse angular rotation of surface (positive/negative 180° curvature, which is one 4x1/4 wave) and singular equal and opposite angular rotation of center of gravity (i.e. Sun) around its surface (½ of a wave, which is 2x1/2 of solar reality that is at the same time 4x1/4 space/time multiplication/division of its orbiting planets), which is its equal opposite, just as are double quantum changes of Earth's reality (see illustration 12).

Illustration 12 Space/time matrix of double/inverse curvature

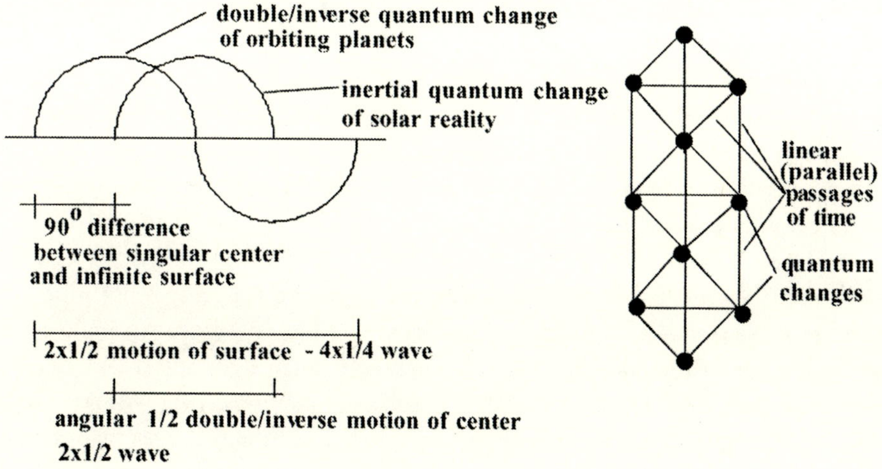

Matrix of double/inverse (90°) sucessive quantum (precessional) changes

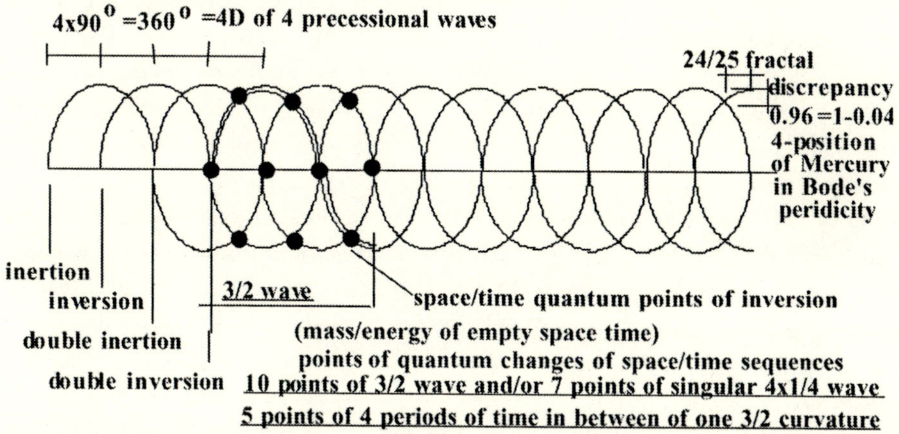

Surface without is reaction of singular point within—it is infinite reaction of singular action. Singular point of gravity within is reaction of infinite surface without, which is its equal and opposite antigravity (equally vice versa). Point within is singular space of its infinite time, which is its surface without and equally vice versa. Singular point of gravity (center of rotating mass or empty space) is singular mass, while space between such point and its surface is its infinite universe, and rotation of surface is its infinite expansion. Earth is universe of its singular and infinitely denser point within (light speed difference), while its surface is its gradual expansion, because gradual accumulation of tension in depth, which is not real on surface, unless quantum, is time or gradual spatial change of such point within. Exponentially denser mass within Earth's surface is at the same time exponentially increased empty space in inverse perspective from its center, because density within/without are fractal light speed difference and what is dense mass from outside is empty space from inside, just as universe here and now is empty space, while in context of its surface, it is dense mass within.

Infinite surface and singular point within are space/time reality. There is no difference between Earth, atom, nucleus, galaxy, universe or any other sphere. It is overlap between two equal opposites—space within and space without, which are light speed difference of density in between. Such light speed difference is fractal passage of time of two (relative without and uncertain within) spatial spheres. Passage of time of inner (singular) sphere is infinitely slower than passage of time of outer (infinite) sphere, because they are light speed difference. Passage of time of inner sphere is instant density of outer sphere. Distance between singular center of sphere and its surface is time curvature of empty space in between—it is linear direction or instant space in context of concave curvature without and round curvature or passage of time in context of convex curvature within. Rotational velocity on surface is reduced to singular velocity in its depth, where no spatial velocity is light speed velocity in time and equally vice versa. Difference in rotational velocity is equal and opposite change of density in depth of surface sphere. Change of motion/density is change in rate of passage of time. Surface rotational velocity is 0.463 km/sec or 2.16 sec/km, which is $2.16/C=72x10^{-7}$ (sec/km)^2, where 10^{-7} is fractal difference between surface and its core (2x3+1), which is orbital time of Earth that is $\Pi x10^7$ (curvature without is linear direction within) and 72 (sec/km)^2 is time discrepancy between double/inverse rotational motions (Earth's orbital around Sun, which is rotation of surface in one direction and equal and opposite rotation of empty space within in opposite direction).

Discrepancy of any relative/uncertain size/motion without and within is always 1/72, because 72=24x3 or 1/24 discrepancy of fractal 3D without/within. Likewise—C/0.463= $6480x10^2$, just as 25920/40000=0.648 year/km or $6480x10^{-4}$, which is inversion (time) of 1.54321 km/year precessional motion, where 40000-size of Earth and 25920-precessional interval. 6480 is curvature or boundary; therefore time limit of such inertial (spatial) sequence (463 m/sec rotation). **Time, which is limitation of space, is discrepancy between inertial spatial sequence and light speed, just as space, which is limitation of time, is discrepancy between light speed and time.** $648x10^2/360$ days in a year is 1800 years or $18x10^2$, where 18 year is ¼ orbital period of inverse rotation of empty space or ¼ of 72 years that are 1° of precessional motion (also period of nodal regression), while 4x1800=7200 and 7200/1.111=6480 years. Orbital time is $\Pi x10^7$ sec and/or $32400x24/25=31104x10^3$ sec, which is 0.03 or $3x10^{-2}$ discrepancy. Orbital time/light speed = 103.68 (sec^2/km) or $103680x10^{-3}$, which is singular period of double/inverse precessional interval and/or $10368x10^{-2}$, which is ½ of inertial precessional interval.

Quantum period of time is discrepancy between orbital time and light speed, which is boundary of such inertial sequence of passage of time. Discrepancy between orbital time and light speed is 24/25=0.96 discrepancy, which is inertial gravity here and now (1/103.68=9.645 m/sec^2, which is 0.15 or $3/2\ x10^{-1}$ discrepancy). 9.6 m/sec^2 is inertial gravity that includes here and now (particle of perfect wave and/or wave of perfect particle), which is 10x0.96 or $10\text{-}4x10^{-2}$, where $4x10^{-2}$ is inertion of inertionless 10 m/sec^2 (inertionless in space is inertial in time and equally vice versa). 9.8m/sec^2 gravity is ½ way between such equal opposites—(10+9.6)/2=9.8 m/sec^2. Gravity is overlap between time and light sped, which is inverse overlap between space and light speed. It is inertial space/time, which is double light speed. Singular gravity on surface is double/inverse space and time without and within, which is infinite time of empty space without and no time of dense mass within.

Dense mass within is inversion of empty space without—it is universe turned inside out, where time is not constant passage but instant density. Universe is within, just as is without Earth (as above so bellow)—it is instant dense universe within and gradual empty universe without. Time of inner universe is space, just as space of outer universe is time. Only difference between such double/inverse spatial realities is inertial here and now, which is surface reality of Earth. Surface reality of Earth is only cause of infinity of universes within and without, which is its double/inverse echo. There is only one reality in every point of space/time, which is always right here and now. **Quantum change of space/time is at the same time quantum change of its inertion, which is right here and now.** Such change is incorporation of inertial here and now and its double/inverse (space/time) boundaries, in order to generate double sequence of here and now in context of its inverse space/time. It is addition of one genetic sequence in evolution of double/inverse curvature, which is infinite change of no change.

Consequently—rotation of gravitational center around its surface is fractal equal opposite of rotation of surface around its gravitational center (2D quantum change). It is inversion of inertion that creates double inertion. They are expansion of Earth without that is equal and opposite contraction of empty space within or gradual inversion between empty space without into dense mass within and change of dense (timeless) mass within into empty space of the universe without, which generates passage of time. It is time motion, which is angular motion of horizontal space. It is rotational or circular motion within/without or double/inverse motion in depth of spatial reality, in which dense and small reality within gradually shifts in direction without, and at the same time empty and large reality without gradually curves in direction within. Infinity of such gradual solar 2D quantum changes, which are fractal expansion/contraction of its orbiting planets are singular quantum of galactic change, which is not gradual but instant expansion/contraction in depth of orbiting planets. Each of solar 2D quantum change is one fractal of one quantum of half-life of its orbiting planets. 20 solar changes are 20 fractals that are singular quantum of half-life of its planets and 20 quanta, which are 20 quantum changes of galaxy is their absolute time boundary in context of solar reality, from inception on its surface to repulsion beyond its surface boundaries.

Time from inception to repulsion, which is inception in outer solar sphere that is galaxy, is period between two 2D galactic quantum changes (double/inverse fractal changes of galactic density). Such time period is complete space/time inversion between surface of an atom and surface of solar reality. Here and now is always ½ of such period, which is ½ of contracted time within, which is its instant dense mass, and ½ is its expanded time without, which is infinite empty space of solar reality. Here and now is always singular center of universe without and infinite universe (surface) of singular center within. ½-time of universe is frozen in instant density of its mass and ½ is expanded into infinite passage of time of empty space. Here and now is boundary between dense mass and empty space, which is double/inverse curvature. It is curvature or limitation of infinite passage of time into quantum periods, which are quanta of spatial (future) reality of time of dense mass within. Only infinitely empty space is gradual uninterrupted passage of time, while its boundary, which is surface of mass within, is not infinite but quantum periods of time, where each quantum period is one curvature. Each quantum period of time is one quantum of spatial reality of time or one quantum of relative reality. It is creation of one, infinitely small massive particle of empty space, which is passage of time or relative reality. Infinity of such quantum curvatures of time is gradual densification of empty space without, which is simultaneous gradual de-densification of mass within.

10 quanta of galactic quantum changes are past frozen in dense mass within and 10 quanta are future time of orbiting planets in context of solar reality. 10 quantum changes are complete inversion of dense mass within into empty space without and empty space without into dense mass within, which is not only complete space/time change (324°), but double/inverse space/time change or complete E/m change (288°), which is double fractal or double light speed difference. Such complete change is inversion of space into time and time into space, which is inversion from mass to energy, just as planets are created by inversion of energy into mass. Galactic quantum change is fractal change of density of solar reality, which is instant reversal of previous period of gradual change, which is gradual fusion of energy (waves) into mass (particles). Dense Sun is inversion of its empty space without—while it attracts dense mass of orbiting planets without; it repels dense mass within, which gradually accumulates on surface. Quantum change is instant inversion between singularity and infinity or instant crossing of surface between dense mass within and empty space without. It is instant change between space without as passage of time within and passage of time without as instant space within. Such instant change is repulsion of gradually accumulated mass within into empty space without, which is inception of massive planet on solar surface.

Such planet is gradually repelled through quantum changes, which are equal and opposite attraction of galaxy. Each quantum change is one fractal step in repulsion of planet from solar surface toward infinitely empty space of galaxy. Solar space is density within of empty galactic space without and gradual repulsion of orbital is gradual shift through solar density towards its surface without, which is inversion of singular center (Sun) within. Infinite space and time of orbiting planets is instant solar density of no time (double/inverse space), where solar reality in context of galaxy is not Sun, but entire rotating solar system, while Sun is only singular center of its gravity within. Empty space of solar system is dense mass rotating in context of galactic empty space, which is twice-empty space here and now. Overlap or inertion (friction) between rotation of solar reality in context of galaxy and inverse rotation of Earth in context of solar reality is gradual precessional curvature here and now, which is constant passage of time between double/inverse spaces. Gradual repulsion of orbiting planet is spiraling curvature between surface of Sun and surface of solar reality in context of galaxy. Spiraling curvature is duality of double/inverse rotations—horizontal (east/west) rotation here and now, which is round spatial orbital in context of passage of time and its gradual angular (north/south) rotation, which is angular depth (time) of such horizontal rotation.

Angular quantum rotation is time of spatial horizontal rotation. It is linear repulsion away from Sun, which is gradual motion through depth of solar density (mass), toward its surface. It is time motion or motion through relative space, which is dense solar reality in context of galaxy. It is space only here and now, while in context of galaxy it is dense mass within space. Simultaneously in context of galactic, infinitely faster passage of time, which is empty space of universe, it is not spiraling but linear light speed motion of infinitely small massless particle (photon). Rotation is Earth's reality and spiraling is solar reality, which is fractal reality between rotation and linear motion, which is galactic reality. Earth is singular quantum or inertion of solar reality, just as solar reality is quantum or inertion of galactic reality and galaxy is inertion of universe, where each successive spatial sphere is infinitely faster passage of time (infinitely slower motion). Passage of time between spheres is light speed difference. Round in context of one is spiraling in context of another and yet linear motion in context of double fractal difference. Spiraling is fractal reality between rotation and linear motion. Orbital motion is equal opposite of gradual rotation of empty space, which is rotation of dense solar reality in galactic context.

Earth, which is infinitely denser particle within empty solar space, is its inertion (infinitely heavier), which causes friction. Such friction is rotation of Earth in opposite direction of solar rotation. Orbital is not motion—it is size of solar reality. It is solar reality or empty space that rotates in opposite direction of orbital, while its planets are stationary. Size of orbital is double/inverse motion—it is infinitely larger rotation of empty space and infinitely slower reaction caused by friction, which is gradual precession (perihelion). Orbital motion of planets around Sun, which is center of gravity of solar reality and/or inverse rotation of empty space of solar reality, while planets are stationary, are double/inverse perspective of same reality. While planets orbit Sun in context of Earth, in context of solar reality—planets are stationary, while solar space rotates in opposite direction. Just as Sun does not orbit Earth as it appears, so Earth does not orbit Sun as it appears—it is rotation of empty solar space in context of galaxy and equal and opposite rotation of Earth in context of solar reality. There is no exclusive description, because reality is fractal—it is duality of two equal opposites. Only one exclusive description, without its equal opposite is infinite distortion of reality. Earth is stationary (it does not orbit but only rotates) inside rotating

solar space. Orbital motion of Earth is an illusion, just as is solar orbital around Earth (where illusion is totally relative/uncertain description or complete distortion of reality)! Just as apparent orbital of Sun around Earth is caused by Earth's rotation, so is its apparent orbital around Sun caused by equal and opposite rotation of solar empty space (solar system) in context of galaxy.

Orbital motion of Earth around Sun is meaningless simplification without its equal opposite, which is rotation of empty solar space in opposite direction. Every action is equal and opposite reaction. Exclusivity is always causeless (one-sided) description, which is infinite distortion, because reality is inclusivity of action and reaction, cause and effect. Action without reaction is only ½, while whole reality is 2x1/2 in context of yet another, angular ½, which is only ½ of yet larger, angular 2x1/2, which is infinite fractal double/inverse curvature of double helix. Rotation of Earth is fractal overlap between such two equal opposites. As solar space rotates in one direction, it rotates Earth, which is its inertial (infinitely heavy) particle within, in equal and opposite direction. Increase of distance from core of solar reality (Sun), towards its surface in context of galaxy without, is its decreased density. Decreased density is relatively larger space and relatively faster passage of time, which is decreased rotation of solar space and therefore reduced equal and opposite orbital motion of planets at larger distance (Keplers harmonics). Different orbital velocity of different planets is due to their relatively different inertions (different sizes/motions and/or densities of such planets in context of equally different density of empty space without, which results in different rotations of such planets and consequently different orbital velocity). Different planets are different relative weight discrepancy (gravity) between dense mass within and empty space without, which are different orbital motions.

Orbital motion is consequence of different rates of rotations of dense space within and empty space without, which are double/inverse densities or densities that are light speed difference. Light speed difference is not only difference between double/inverse motions, but also difference between double/inverse (infinite/singular) sizes. Increase in motion is simultaneous decrease in size and equally vice versa. Orbital motion is light speed difference between Earth's size (dense mass) within and solar size (empty space) without. Light speed is eigenvalue between size and motion and change of one is equal and opposite change of another (C=d/t). Light speed difference of motion indicates equal and opposite light speed difference of sizes—it is infinite size without in perspective of singular point within and/or singular point within in perspective of infinite size without, which are light speed difference of density. Rotation (spin) of an atom is reaction of Earth's rotation, just as Earth's rotation is reaction of solar rotation and solar rotation is reaction of galactic rotation etc. Each two rotating spheres without/within are light speed difference. Atom rotates because it is infinitely smaller size and rotation here and now in context of such small size is light speed motion, just as light speed motion of infinitely larger solar reality is only infinitely slow rotation here and now.

Rotational velocity of 463 m/sec (2.16 sec/m) is light speed motion in context of 648 000 larger size, because C/0.463 (2.16C) is 648000 or $6480x10^2$, where 6480 years is time (boundary) of such fractal space/time sequence of rotating Earth. $6480x10^2x400x10^2$ =25920x10 km, where $400x10^2$ is size of Earth. In context of sphere of empty space without, which is that size, Earth within spins with light speed (25920x10 km/86400 sec =C). It is sphere with radius of 1 light day (distance light travels in 1 day). Diameter of Earth's orbital is 8 light minutes, while 1440 min/8 min=180° curvature. Earth's distance from core of solar reality is 1° or 1 AU, while surface of such sphere is 180° or 180 AU, which is ½ of total E/m curvature, which is its space/time curvature in direction without or infinitely large size of empty space that is passage of time of 1 day. 180° is equal opposite of 18 years that are ¼ orbital of empty space (5 days discrepancy caused by their friction). 1° is inertion (24/25=0.96 discrepancy or gravity here and now) of such curvature of empty space. 180 AU is 4x1/4 distance of Pluto, which is 40 AU distance (40x0.96=38.4 AU), 8x1/8 Uranus' distance (20x0.96=19.2), 16x1/16 Saturn's distance (10x0.96=9.6), 32x1/32 Jupiter's distance (5x0.96=4.8) and 64x1/64 asteroids distance (2.5x0.96=2.4AU), which is progressive 2x1/2 spiraling of double helix (Neptune is fractal ½ between Uranus/Pluto, just as Venus is fractal between Earth/Mercury—see Bode's periodicity, illustration 13 and text).

Each 2x1/2 point of inversion is E/m of such empty space or orbiting/rotating massive planets, because such singular points are instant reality of both—space and time. They are points of instant time, where time is not passage of time, but instant double/inverse space within of empty space without, which is instant density (mass) of empty space or planet/electron in motion (particle, which is inertion of a wave). Pluto's orbital is ¼ or 90° curvature of empty space in direction without (fractal ¼ difference of inertial ¾ reality, which is time within of space without and/or space within of time without. Pluto is ¼ curvature or 6 light hours distance from core of solar reality (Sun), which is singular ¼ quantum of total curvature of 24 hours or 1 light day that is sphere of empty space, which is light speed difference from here and now (Earth). 40AU is perfect 6 hrs=360 min or ¼ day, which is inertial 0.96 discrepancy—38.4AU or 5.76 hours (Perihelion precession), which is 345.6 min, which is equal opposite of one eclipse year

(345.6 days). Pluto is 90° curvature, which is fractal ¼ difference. Such difference is boundary (corner) of inertial space here and now, while remaining ¾ are time of such space, which are around the corner or in depth of spatial reality here and now (beyond horizon).

Orbital position of Pluto is singular ¼ quantum of solar reality of empty space, while distance of 1 day (180AU) is its surface. Pluto's position is limb of such sphere (90° curvature or infinite distortion of inertial spatial reality here and now, which is function of tangent) and further ¾ of solar spatial reality are beyond its horizon, because they are time without of such inner space of solar system of 10 planets (incl. asteroids). 360° curvature of such distance, which are 360 days of one year is 64800 or 6480x10AU (which is 5184x10^2 min or 21600 hrs), while distance to nearest star is 4x360=1440° or 4D curvature, which is 4 light years or 2592000 or 25920x10^2 AU, which is equal opposite of distance Earth travels in its orbital in course of one day in km. 1 km in context of round orbital (solar reality) is 1AU in context of linear distance (galaxy), because they are light speed difference. Position of Pluto within is ¼ of 1 light day, which is surface of solar sphere that is infinitely slow motion and infinitely faster passage of time. In context of such infinitely slow rotational motion on surface, Earth's rotation within is light speed rotation of an atom. 4-fold full curvature of such inertion of 1 day is complete 4x1/4 inversion of solar reality without/within), which is its nearest star (from limited round curvature to unlimited linear distance, which is only curvature of yet larger fractal inversion). Distance to nearest star is 4 and distance to core of galaxy is 25920 light years, which is 25920/4=6480 years. **4 years without (distance to nearest star) is ¼ day within (distance to furthest planet), which is 1440-fold difference.** 18x1440=25920 years, which is time or boundary of space, as well as 25920 AU or km, which are spatial boundaries of time and where 18=20-2 or 20/1.111 or inertial 90° curvature (saros) of 4x5=20 (fractal 4/5).

Orbital motion of different planets is reaction of their different passages of time (which is larger or smaller difference than light speed, because only Earth's orbital time in context of its gravity is light speed motion). While empty solar space rotates, Earth is its infinitely heavy inertion, which remains stationary. It only rotates in opposite direction. 1 year of time is rotation of solar space in opposite direction of orbital, which is inversion of 30-km/sec orbital velocity (33.333 sec/m). It is 360-fold rotation of Earth in context of one equal and opposite rotation of empty solar space. Discrepancy between solar and Earth's rotation is additional 5 days, while ¼ day is discrepancy between Earth's and galactic rotation, where solar rotation is in time and galactic rotation is gravitational rotation here and now. Smaller fractal discrepancies are reaction of rotations of larger spatial spheres, which is infinite (irrational) curvature. Such 5 days of discrepancy is built-up of gradual precessional (inverse) rotation of Earth and ¼ day is gradual, infinitely slower built-up of double/inverse precessional rotation, which is precession of precession.

Precession of rotation is gradual accumulation of inertion (friction), which is not instant reaction, unless singular (¼) quantum—i.e. ¼ of rotation, which is ¼ of Earth's size. Only when accumulated tension is ¼ precession of rotation is such tension enough for singular quantum of motion of Earth in opposite direction. Accumulation of tension, through constant friction is gradual built-up of angular (north/south) force, which has no effect, unless ¼ (90°), which is instant push or toppling of rotation/size. Such angular motion is also gradual built-up of reaction in equal and opposite direction, which is instant push in angular direction of ecliptic, which is consequence of accumulation of ¼ day discrepancy. While precessional quantum changes are reaction of gradual accumulation of discrepancy between solar and Earth's rotation, double/inverse quantum changes are reaction of gradual accumulation of discrepancy between Earth's rotation and angular tilt of solar reality, which is reaction of its orbital in context of galaxy; therefore reaction of galactic rotation. There is no difference between Earth, solar, galactic or any other quantum changes of larger or smaller (emptier or denser) spatial sequences.

Larger space is faster time; therefore larger period of time between such changes and equally vice versa. Sizes smaller than here and now are infinitely fast quantum changes, which is constant flux of physical reality on surface of Earth. Sizes between here and now and size of an atom within are acceleration of such quantum changes, which are not quantum changes at all but gradual change of physical reality (evolution and growth of molecular life, which is infinitely fast spiraling of double helix—i.e. it is exclusive spatial double helix, because it is smaller than singular time quantum). Sizes smaller than atom are light speed difference, which is instant space/time inversion, where infinity of quantum changes are not periods of time in between, but instant spatial reality. An atom is instant reality of infinity of quantum changes, because it is inversion of inertial here and now in direction within. Earth in its depth is inverted universe and atom is its double inversion. It is singular quantum of inverse time, just as universe is singular quantum of double time in context of inertial here and now in between. Surface of an atom is singular/infinite point of inversion—it is refraction point between infinitely fast quantum changes and no change at all (mirror-image, where infinity reflects as singularity and inertion as inversion).

Gradual repulsion of orbiting planet is reaction of double/inverse interaction between solar and galactic gravity. Galactic attraction is gradual precessional curvature of solar reality, just as solar reality is gradual precessional curvature of Earth's reality. Consequently, a planet through singular and double/inverse quantum changes gradually spirals towards surface of solar reality, which is end of inertial solar sequence and beginning of galactic sequence of such planet. It is complete expansion of its mass in direction without and contraction of empty space in direction within, which is densification of passage of time into instant mass within or complete cycle of fission (complete half-life of dense massive planet), which is equal opposite of gradual fusion. Dense core of solar reality (Sun) is gradual fusion of energy into mass, while its outer sphere, which is empty solar space, is inversion of such fusion, which is gradual decay of mass. Gradual decay of mass in context of equal and opposite gradual fusion of energy into mass is double/inverse fusion/fission, which is fusion of mass and energy into singular gravity.

It is inception of a planet on surface of solar reality or its inception into galactic spatial sequence, which is infinitely emptier space without and infinitely denser core within. Consequently, it is infinitely denser planet, where mass and energy or dense mass and empty space (space and passage of time) are both its instant density. Such planet does not rotate around Sun, but core of galaxy, which is inversion between quantum and gradual reality or space and time. In context of solar reality, galactic gravity and galactic rotation/orbital are quantum changes in between infinitely long passages of time, while in context of galaxy; there is no such passage of time in between. Previous quantum changes are its gradual spatial change and its constant passage of time, because in context of galaxy, time is instant spatial density, while gravity/antigravity is its constant passage of time. Gradual angular shift of orbital, which is gradual spiraling of massive planet away from surface of dense solar core (Sun) towards its surface, through infinity of quantum changes and in context of infinite passage of time is in galactic context instant density of solar space and no time. Galactic spatial sequence is linear motion away from solar surface, which is same orbital round core of galaxy as solar reality. Such orbital is gradually shifted through quantum changes in context of galaxy.

Solar reality is not equal opposite of such planet in galactic orbital, but only its equal. It does not radiate light, just as it does not radiate light in context of quantum changes here and now. Solar reality is only a planet in context of galaxy, while its orbiting planets within are unreal, because their fractal passages of time are less than quantum of passage of time in context of galaxy. Sizes and motions of such planets are only its instant density. A planet is only temporary compressed even density of solar reality into singular point, just as electrons are infinite densification of space which is even density into singular point that rotates in context of de-densified empty space. Orbiting planets/electrons are infinite/singular density of their eigenvalue, which is space of even density (no difference), because they are only one out of infinite number of possibilities. Simultaneous reality of infinite number of different possibilities is even density, while only one possibility is always concentration of even density into singular points, which are rotating in remaining empty space.

Number of such singular points is determined by number of directions of such reality, which is in inertial context here and now, always 3 spatial (instant) directions and one gradual (time) direction. Larger number of directions (dimensions) is larger number of inversion points of such matrix, therefore smaller densification/de-densification, which is larger number of less dense singular massive points within and smaller empty space without (smaller period of passage of time), which is context of their motion. In inertial space/time (3D) context it is always 6 inertial points, while in exclusive spatial (or time) context, number of such points is 4 (4x1/4 wave). In context of both—number of such points is 10 (Bode's periodicity and periodicity of shells/electrons, which is double/inverse Bode's periodicity). Orbiting planets are instant density on surface, just as instant density on surface in perspective of singular core of such density within (light speed difference), which is double/inverse density, it is only empty space orbited by planets. Every solid mass on surface is empty space of double/inverse density of its core within, which is orbited by planets/electrons. It is infinity of spatial spheres, one within another, which are light speed difference from each other.

Fractal ½ reality in between spatial spheres is double/inverse passage of time and/or double/inverse density of empty space without and dense mass within. Such fractal reality is instant (singular) in one and gradual (infinite) reality in opposite direction. It is time flattened into instant spatial surface in one direction and expanded (curved) into infinite reality in opposite direction. It is gradual change of density in direction within, which is instant double/inverse density in direction without (infinitely dense mass in context of empty space), where inside in space is outside in time and equally vice versa. Likewise—it is gradual change of passage of time from within to without, which is instant space of no time on surface. Duality of gradual/instant or double/inverse spatial spheres and their fractal time sphere in between is gravitational 3D sphere, which are two equal opposites in their inertial context or linear 2x1/2 in context of their angular ½. Fractal passages of time between spheres are their spatial discrepancies, where largest discrepancy, which is smallest period of passage of time, is between two successive spheres. Each successive fractal passage of time is light speed difference from one before. Larger difference is smaller spatial discrepancy, which is larger period of passage of time. It is infinite quantum reality of spheres within a sphere.

Larger passage of time is larger duration of angular in-depth or time motion, which is motion through relative density from within to without, where within in space is without in time. There is no outside cause of all infinity of different quantum or gradual spheres—only cause is inertial here and now, because outside in space in inside in time. Here and now is singular wave, which is eigenvalue or synchronicity of double and inverse waves, therefore inertion (particle) of such wave and no wave at all. It is singular (inertial) cause of infinity of its spatial and time waves, which are its infinite self-interference, through different spatial and time (gradual and instant) oscillations, because inertial space in context of all its time and/or vice versa is one non-differentiated reality (Kepler). There is only one reality, which is infinity of different choices on surface, because it is no choice in its depth and equally vice versa.

Time is change, regardless if change is affecting space or time. Change of time is larger time, which is time of both—space and time. Time of smaller space is space of larger time and equally vice versa. Time of one day is 24 hrs and time of one year is 360 days. One day is space of time of one year, just as one hour is space of time of one day. Small discrepancy is inertial difference between double/inverse spatial spheres—it is infinitely small discrepancy here and now, which is infinitely larger angular spatial reality, where reality here and now is only its minute discrepancy. Discrepancy or difference, which is infinitely small or large (because infinitely small spatial discrepancy is infinitely large time discrepancy and/or vice versa), is always starting point of change of inertial spatial reality on surface, which is its time or boundary. Instant change is double/inverse spatial inertion, which is all infinity of its time without or within. Space is instant reality. Instant space here and now is all infinite time of smaller spatial sphere (micro-world), just as all infinite time here and now is instant space of larger spatial sphere (solar reality). Difference is always relative duality of two equal opposites, which is infinite double/inverse self-interaction of one reality. It is two surfaces in mutual spiraling curvature (double duality of 4x1/4 in curvature of its 3rd direction), where **within precedes without in space, and without precedes within in time.** Double/inverse precession of space/time is fractal 3/2 reality, where one is always surface (2D) of depth (3D) of another.

There is no spatial reality without its equal and opposite time and equally vice versa. One is real only in context of its equal opposite, which is its inversion or negation (affirmation is double/inverse negation). Only simultaneous reality of both is real—one without another are totally unreal, which is double/inverse infinite distortion of reality. Space/time is double/inverse comparative reality—there is no space without time and no time without space, which are negation of each other. Exclusivity of space or time (gravity or mass etc) without simultaneity of its negation (inversion) is not reality but its negation, because only action and reaction are real. There is no action without reaction and no reaction without action. Reality is double/inverse causative curvature or simultaneous duality of inertial and inverse waves (where inversion is double inertion). Double/inverse (anti-parallel) precessional or causative curvature is duality of space/time waves, which are simultaneous reality (eigenvalue) here and now. Space/time waves are always 1D fractal difference in between—if spatial waves are 2D, its time waves are 3D and equally vice versa, because one is surface of in-depth reality of another. 3D wave (sphere) has 6 and 2D wave (circle) has only 4 inversion points. Duality of both is 10 inversion points (see illustration 12). 2D and 3D waves are linear and accelerated waves, where acceleration in space is constant linear direction in time and equally vice versa, because one is derivate of exponential reality of another. Linear spatial waves are accelerated time waves and linear time waves are accelerated spatial waves. It is duality of positive/negative waves, which are exponentially accelerated in one and exponentially decelerated in opposite direction.

Time waves are decelerated in inverse and accelerated in double space and spatial waves are decelerated in double time and accelerated in inverse time. Such double/inverse waves are exponentially compressed in one and expanded in equal and opposite direction from inertial point here and now, which is their equal and opposite ½ eigenvalue. Fractal 3/2 reality is linear direction of one and exponential increase/decrease of another. Double/inverse exponential increase of one in inertial (linear) context of another is periodicity or repetition (saros) of one in context of another, which is its changeless inertion. Difference between accelerated and linear waves is light speed, which is difference between square and a cube (or circle and a square). It is infinite repetitive curvature of inertial here and now (1.111...) that is infinite fractal self-interference. Each sequence (1) is inertion or linear wave of infinite repetition of its double/inverse accelerated waves (Balmer—each principal is infinite number of subordinates). Infinity in context of 3D is 3x2=6-fold repetition or 6 max/min (inverse) points of 3D wave, which are 6 surfaces of a cube and/or 6 points of inversion (90°) on surface of a sphere. It is 2D wave rotated around its linear direction. In 2D context, infinity is limited to 2x2=4 points of inversion of 2D wave (4x1/4), which are 4 sides of a square and/or 4 points of inversion of circular curvature. Space/time is reality of both—

space and time, which is synchronicity of 3D/2D waves or 10-fold repetition of inertial here and now in space and time.

Fractal space/time reality of 3D/2D linear/accelerated (exponential/derivated) waves that are 1.111 repetitions is summary meaning of Kepler's laws. Cube of space that is square of time is harmonized repetition of inertial here and now (particle of its double/inverse waves), which is ½-way in between such double/inverse waves. It is space/time or exponential description of numerical abstraction of fractal 3/2 reality, which is Bode's $3x2^n$ periodicity. Bode's periodicity is harmonized exponential repetition of 2D time wave in context of its inertial 3D spatial wave. 6 outer planets (Earth through Uranus) are repetition or successive inversion (refraction) points of time wave in context of inertial 3D space, while 2 planets beyond (Neptune/Pluto) and 2 inner planets (Mercury/Venus) are double/inverse curvature or acceleration of such wave. They are additional spatial curvature of inertial time curvature of 2x3=6 planets. Space/time is fractal 1D difference; meaning one side is compressed, while its opposite side is expanded. Inner planets are instant compression, while outer planets are gradual expansion. 6 outer planets are gradual expansion of inertial here and now (Earth), followed by inverse compression of Neptune/Pluto, which are beyond boundary of inertial 3D space.

Mercury/Venus are instant compression in opposite direction. Mercury/Venus are instant spatial density of no time within of infinite empty space in context of passage of time of outer planets. Outer planets are time of spatial reality of 2 inner planets (and Neptune/Pluto are their inverse space). Mercury/Venus and Neptune/Pluto are double/inverse spatial spheres of fractal time sphere in between (6 planets). Mercury/Venus and Neptune/Pluto are contraction of 3D wave in between into 2D wave on its edges, which is inertial framework without of inverse reality within and/or equally vice versa. 2+6+2=10 planets of solar system is simultaneous duality of 3D and 2D space/time waves. Mercury within and Pluto without are further contraction of such 2D waves into singular 1D wave (see illustration 13). Mercury within and Pluto without are inertial 1D framework of double/inverse 2D within, which are Mercury/Venus duality within and 8 outer planets. 8 outer planets are further expanded into 6 planets, while inner planets are not, because reality within and without Earth are double/inverse (fractal) reality, where one is compressed 2D and its opposite is expanded 3D of fractal 2/3 in between (Earth). Outer planets are space of time of inner planets and equally vice versa. Earth is only point, which is simultaneous spatial and time reality (only Earth's orbital time x gravity = light speed). Inner/outer planets are eigenvalue of inertial Earth in between, exclusively as spatial or time reality (orbital time x gravity = larger/smaller than light speed).

Planets are only fractal eigenvalue of inertial here and now—they are exclusive spatial reality, which is exponential distortion of their passage of time and/or vice versa—they are synchronized time, which is exponential distortion of their spatial reality. Earth is only point of space and time, which is light speed difference. Rest of the planets in their spatial context, are progressive distortions of inertial passage of time here and now. They are not reality, but its distortion. Planets are inversion points where spatial and time waves (curving surfaces) refract, creating instant duality of space and time, which is instant mass of empty space. They are instant density (particles) of double/inverse empty waves. Acceleration/deceleration of such waves in context of each other is exponential Bode's periodicity—they are compressed in direction toward core of solar reality and expanded in direction of its surface (galaxy). In context of solar time, which is infinitely faster (contracted) passage of time than expanded time here and now, such 10 planets are only one out of an infinite number of possibilities. All infinity of different possibilities is infinity of different passages of time here and now, which are instant density of solar reality. Orbiting planets in solar context are everywhere at the same time, because they are its uncertain reality (instant mass within). They are massive points only as surface (time) projection of solar 3D (space).

Space is time wave and time is spatial wave. One is vibration from within towards without and another one is equal and opposite vibration from without toward within. One is double/inverse echo of another. They are double/inverse waves created by interactions of solar reality within and galactic reality without (anti-parallel strands of double helix), which are double/inverse spatial spheres of fractal time sphere in between. Time sphere in between are orbiting planets created at points of intersection between such accelerated/decelerated inward/outward (space/time) waves. They intersect ½-way (in gravitational context, which is simultaneous reality of space and time), creating larger (double) duality in between, which is mass/energy of empty space/time. Earth is inertial point or eigenvalue (½-way) between solar reality within and galaxy without. Inner/outer planets are equal and opposite interaction between Earth and solar reality within and/or Earth and galaxy without. Earth is singular point of inversion between Sun within and core of galaxy without, while planets without/within are further interaction between Earth and solar reality in one direction and Earth and galaxy in opposite direction, which is 2x1/2 multiplication/division of inertial 2x1/2 between Sun and galaxy that is Earth in between. Earth and solar reality intersect again in ½-way point, which is Venus. Venus is inertion of further duality, which creates Mercury ½-way in between. Mercury/Venus are 2 harmonics of 2x1/2 wave in direction within (Sun).

Galaxy without is infinitely larger spatial sphere than solar reality within (light speed difference). Earth/galaxy double/inverse waves are 3D waves without of 2D Earth/Sun waves within, which intersect at infinitely larger spatial distance that is orbital position of Uranus (inversion of polar axe). Uranus is ½-way between Earth and galaxy, which is inertial point of waves in direction without and within. Direction without is orbital position of Pluto, which is ½-way (gravitational equilibrium point) between Uranus and galaxy, while interaction between Pluto without and Uranus within is yet another ½-way point in between, which is orbital position of Neptune. Neptune is time reaction without of spatial reality of Uranus within and/or spatial reaction within of time of Pluto without. Larger spatial distance between Earth and Uranus than between Earth and Venus in opposite direction is additional 4 planets in between (Mars through Saturn). Galaxy is larger and more distant attraction without than solar reality within. Solar reality is larger attraction, because it is infinitely closer, while galactic reality is smaller attraction because it is infinitely further away (they are same size in spatial perspective as Sun/Moon here and now, therefore obscuration during eclipse is total). Such double/inverse (relative/uncertain) attraction of solar reality within and galaxy without is larger empty space without than empty space within, which is larger fractalization without than within, resulting in 2 planets within and 7 planets without. Outer planets are relative reality (time curvature) of uncertain reality (spatial curvature) of inner planets.

Once they intersect, such double/inverse waves create instant E/m of empty space/time (double duality), which is inertial point of double/inverse reflection of waves without/within, which intersects yet again ½-way in equal and opposite directions, creating another duality—and such waves intersect yet again, etc. Boundary of number of such harmonics is inversion of inertion, which is double inertion. Boundary of fractal 3/2, which is fractal ½ within and fractal 1/3 difference without (difference or overlap between convex/concave non-Euclidean curvatures without/within) is 2x3=6, while boundary of fractal ½ is 2x1=2, which are 2 inner planets and Neptune/Pluto beyond orbital position of Uranus. Boundary of fractal ¾ is 3x4=12, boundary of 4/5 is 4x5=20, boundary of fractal 5/6 is 5x6=30 harmonics etc. Only whole numbers are quanta, while its fractals are real only when accumulated difference is singular quantum. Empty space beyond orbit of Pluto and/or inversely empty space before orbit of Mercury are infinite densification of one and infinite de-densification of another wave, which is not eigenvalue here and now, which is simultaneous reality of both. Increased multiplication in space is increased fragmentation in time and only multiples (wholes) of both are real here and now (i.e. Bode's periodicity is periodicity of whole numbers, where exponent varies from 0 to 6).

Infinity of all different harmonics is instant density of solar reality. One side is always expanded and another is contracted, where expansion in space is contraction in time and equally vice versa. Such equal and opposite expansion/contraction in space and time is double/inverse causal property of precessional double helix. It is double/inverse densification of space in time and/or time in space, which is fractal repetition (saros) of one in context of another. It is repetition of time waves in context of inertial space, which are instant orbiting particles (E/m of empty space/time), as well as repetition of spatial waves in context of inertial time, which are quantum time periods of instant changes of such orbiting particles. Quantum periods of time are reaction of instant reality of orbiting massive particles—they are their time (boundary), which is gradual change of their instant spatial reality. Reality is acceleration/deceleration or exponential relativity/uncertainty of inertial here and now in between. Acceleration (double/inverse relativity/uncertainty) is an increase of double/inverse (space/time) possibilities, apart from inertial possibility here and now. It is broadening of horizons, where any inertial choice here and now is balanced against infinity of different choices—either gradually or in quantum leaps.

Two waves that are resonant, interfere and create double/inverse wave, therefore it is irrelevant if such duality is described as only one singular wave or infinity of its constituent waves. Planets are harmonics of fractal 5/2, which is eigenvalue in 5x2=10 whole multiples (½ of fractal 5/4). Such 10 multiples are instant resonant duality of space/time waves, where space=time, which is double space/time or mass/energy (particle of a wave and/or wave of a particle). They are fractal remainders of multiplication or overlapping corners between two curved surfaces. Every horizontal curvature creates an infinitely small angular curvature, because of inertion of size in its context. I.e.—sharp bend is always smaller contraction in direction within and larger expansion in direction without. Such contraction is increased density within and decreased density without, which is 2D overlap of linear reality and/or 3D overlap of 2D reality, where 3D in space is 2D in time and equally vice versa. Only such bends or folds are eigenvalue of their fractal (relative) reality. They are singular reality (curvature) of two equal and opposite directions, which is mass/energy of space/time. Such bend is double time of linear space and/or double space of linear time. Vehicle in motion has to slow down before bend, because gravity of such bend is twice gravity of straight road. Additional centrifugal force is double (accelerated) time of time of linear motion. It is additional density or mass of such bend.

Planets are bends or sharp loops of inertial wave, which is reaction of solar/galactic duality, with a difference that such bend is not bend of linear direction, but 4-fold bend of 3 spatial and 1 time direction, which is instant 5D reality. Energy (light speed) passing between such two centers is retained by gravitational loops as mass. Mass is energy held together by centrifugal force, because centrifugal force is repulsion of mass and attraction of energy. Infinite increase of centrifugal force is its instant reality in context of no time, when such force is simultaneously everywhere. Infinite increase is instant inversion, which is double inertion. Infinite centrifugal force of mass is singular quantum of its inversion, which is 90° curvature. It is singular quantum of attraction of energy, because mass here and now is energy in context of solar reality. Light speed of repulsive force is instant attraction. Mass is instant attraction of energy (light), just as energy is instant repulsion of mass. Such linear energy (light speed) in context of light speed difference between solar and galactic reality is curved by gravity of instant space/time bend into sphere in points where space=time (eigenvalue). **Gravity of mass is equal opposite of light speed motion. Gravity of any mass is constant, which is 1/C—it is double/inverse light speed or light speed motion in space and time. Gravity is constant of space/time and/or double/inverse variability of space in time and time in space.** Different mass is same gravitational constant in relatively different convex/concave (space/time) curvatures. Gravity is 3rd (angular) direction of inertial size and motion. Every massive object in its motion is same gravity, which is inversion of light speed motion. **Gravity of light speed is $C/(1/C) = C^2 = E/m$. Gravity of mass is energy and gravity of energy is mass. Gravity of space is time and gravity of time is space, while gravity of gravity is antigravity.**

Illustration 13 - Fractal 3/2 (Bode) periodicity

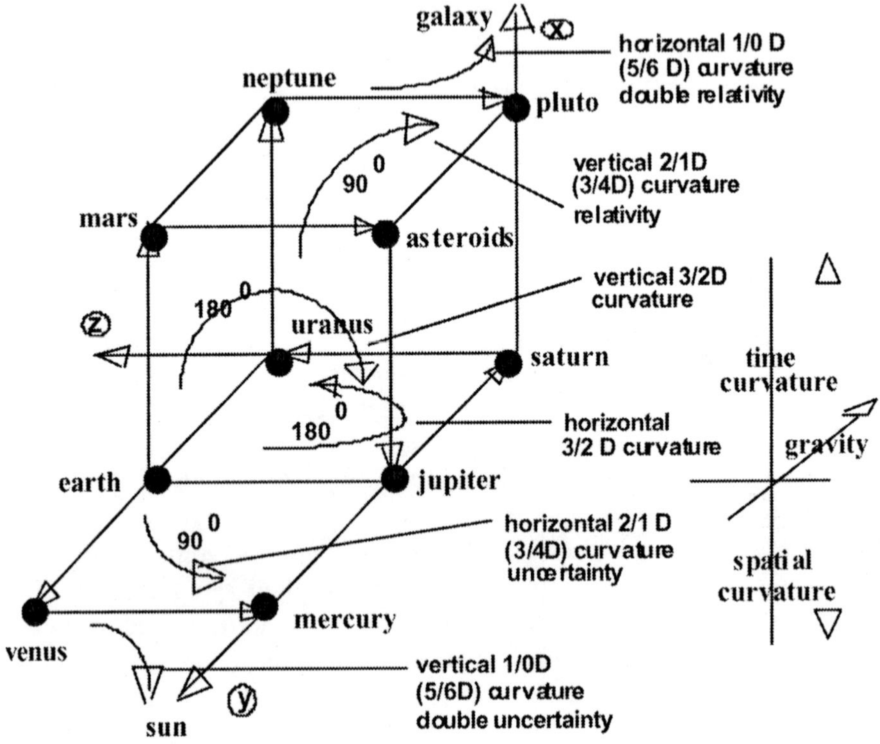

Reality is inclusivity of both equal opposites. It is simultaneous multiplication of one and fragmentation of another, where fragmentation in inverse context is multiplication in inertial context; hence double multiplication (and double fragmentation). Multiplication in space is fragmentation in time, where fragmentation in time is multiplication in space. Multiplication/fragmentation in context of instant duality is double multiplication and/or double fragmentation. Fractal 2/3 is 2x3=6 +1 (here and now) = 7, which are 7 planets and/or shells of an atom, as well as 7 quantum periods of time that are real in context of inertial 3/2 (space/time) reality here and now. 7 different sizes/motions (particles) and 7 quantum periods of time are space/time boundaries of inertial here and now, which is relative inertion (singular point) of such reality. Every quantum period of time is fractal shift of space and time in equal and opposite directions (acceleration/deceleration), which is shift of its boundaries (spatial size and time duration).

Different inertial space/time sequences are different boundaries, which are always 7 inertial shells within and/or without. Orbits of outer planets (incl.

Earth, which is its nucleus) are 7 shells of inertial here and now (Earth) in solar/galactic context (Neptune is fractal ½ between Uranus/Pluto). 7 shells without is 2+1=3 shells within. Bode's periodicity is $\pm 3x2^n$, where negative curvature is within and positive curvature is without. Venus is 10-3=7 and Mercury is 10-6=4 and/or 7-3=4, which is boundary of inertial 4D here and now (inertial ¼ difference of fractal 5/4 and/or 0.04 or $4/10^2$ difference of fractal 24/25=0.96). Fibonacci sequencing is double/inverse singularity of first two members (angular and horizontal singularity, which are double/inverse inertion of infinite repetition in 3rd direction). Consequently—distance between Earth/Venus and Venus/Mercury is linear repetition of ½ difference in opposite direction (3), while outer planets are exponential increase in distance. Outer planets are addition of orbital position of previous planet to exponentially larger $3x2^n$ repetition, where $n \leq 6$ (Mars, n=1 is 10+6= 16, asteroids, n=2 is 16+12=28, etc through Uranus, n=5 is 100+96=196. Pluto is n=6, therefore 196+192=388, while Neptune is fractal ½ in between. It is 196+100, which is twice previous position of Saturn and/or 3/2x196).

Gradual change of inertial wave is instant change of inverse wave and equally vice versa, because one is acceleration of linear direction of another. It is double/inverse acceleration (acceleration/deceleration) without/within of inertial particle in between, because one is flattening of round curvature of another and/or curving (singularity or limitation) of linear direction (infinity). Instant spatial exponential $3x2^n$ periodicity is gradual $3x2^n$ in time and vice versa. Instant spatial orbital of planets are their gradual change in time, which are quantum leaps of such inertial spatial sequences. **Precessional quantum changes are instant time changes in spatial context, just as orbiting planets are instant spatial changes in time context.** Inertion of time here and now is inversion of rotational velocity, which is 2.16 sec/km (6^3) that is $2.16x10^2$ =216 years in context of orbital and 2160 years (1 zodiac) in context of precessional motion. $2160x3x2^n$ is frequency of exponentially larger 2x1/2 quantum (instant) energy change of inertial mass (orbiting particle). For n=0, period is 6480 years (1/4 quantum), for n=1, period is 12960 (1/2 quantum) and n=2, period is 25920 years (whole quantum). N=6 is 414720 (432000 in context of orbital motion), which is singular ¼ gravitational quantum (change of size) or boundary (time) of quantum time changes (changes of motion).

Angular difference is time distance. Space/time reality, which is size and motion of orbiting planets, is inertial difference between solar and galactic double/inverse gravities (attraction/repulsion). Solar reality is spatial reality, which is distance of orbital, while galaxy is its time, which is angular inclination of direction of rotation in context of orbital plane. East/west are spatial (solar) directions and north/south (Pleiades) are time (galactic) directions, while their constant angular difference is inertial space/time sequence of orbiting/rotating planet. Size of planet is difference (overlap) between such double/inverse directions. Ecliptic is surface of galaxy, just direction of polar axe is surface of solar reality—they are double/inverse strand of double helix, which is 3D space/time here and now. Exponentially larger distance without is decreased solar and increased galactic gravity, which is gradual change of angular inclination between orbital plane (east/west) and rotation (north/south axe), which are double/inverse directions of solar and galactic gravity. Rotation/orbital are dual space/time reality, while non-differentiated motion, where orbital time is same as rotation is exclusive spatial or time reality, which is not different but same inertial space/time (gravitational) sequence as center of its orbital (i.e. Moon). Orbital position of Uranus (inverted rotation) is ¼ angular curvature, while position of Pluto is ½ curvature—Uranus is boundary of space in time, while Pluto is boundary of space/time in context of gravity. Pluto's orbital position without (40 AU) is inversion of Mercury's position within (4), which is 100-fold or 10^2 (double fractal) difference.

Chapter 13

TIME COORDINATES

Duration of quantum change is inversely proportional with duration of gradual spatial change in context of constant passage of time, which is its instant spatial reality. Quantum change is infinitely short period of time proportional with such infinitely small spatial sequence. It is singular quantum of double/inverse (relative) time. Singular quantum of rotation is its synodic/sidereal difference of 240 sec or 4 min, which is 1° of its curvature (360x240=86400 sec and 4x360=1440 min), while singular quantum of synodic/sidereal time difference is 4 sec (1/60 that is discrepancy between rotation and orbital motion). 4 min is singular spatial quantum, which is 1D fractal difference of rotational time, while 4 sec is singular quantum of its double/inverse time (2D fractal difference). 4 min is spatial discrepancy between day and night, which is time of sunset/sunrise. It is spatial discrepancy of inertial time, which is its relative/uncertain difference. Size of Sun, which is simultaneous end of one and beginning of another cycle, is spatial discrepancy of rotational time wave. It is inertial (singular) point that perpetuates such wave (3rd half and/or 5th quarter). Rotation is 360-fold repetition of such inertial size in one day (240 sec, which is 111.111 km on surface of Earth). Inertion of inertion, which is double inertion, is inversion of such 4 min spatial discrepancy, which is 4 sec.

Singular (1D) fractal difference is only ½ of space/time reality—it is exclusive passage of time or exclusive instant space without its equal opposite. It is only action without its reaction and/or vice versa. Singular difference is fractal passage of time between two spatial spheres and/or instant space between two inversely proportional passages of time. It is only overlap between convex/concave curvatures and not curvatures themselves. Only double fractal difference (double light speed) is whole reality, which is difference between two spatial and/or time spheres. Singular fractal difference is space/time (attraction), while double fractal difference is space/space and/or time/time difference (double/inverse repulsion). Singular fractal difference is passage of time, while double difference is its boundary (curvature or limitation of infinity). Double difference is singular quantum of spatial or instant reality of time. It is singular quantum of relative space. Only space of space is relatively different spatial sequence and only time of time is different time, while space of time and/or time of space is double/inverse gradual change, which is not different reality, unless singular ¼ quantum. 4 min is spatial boundary of rotational time, while 4 sec is its time boundary. 4 min is inertion of instant spatial reality (size), while 4 sec is inertion of passage of time here and now (motion). **1 day is 6^2x10=360x4-min spatial quanta and/or 6^3x10^2= 21600x4 sec time quanta. Singular ¼ quantum of spatial and time inertion of rotational time is 1 min and/or 1 sec. 360-fold repetition is spatial boundary within that is time boundary without (orbital time of 1 year), while 21600-fold repetition is double/inverse time boundary without and within. It is boundary (time) of orbital motion without and inverse boundary (time) of rotation.**

Relativity/uncertainty of rotational time is 4 sec discrepancy per day, which is not real, unless singular ¼ quantum. Singular ¼ quantum is 21600 =**6^3**x10^2 sec, 360 =**6^2**x10 min or **6** hrs, which is ¼ of rotation/size. 21600/360=60 years is singular ¼ quantum of such time in context of orbital motion (perihelion precession). Singular time discrepancy of rotational time (4 sec) is real in context of orbital motion only in 60-year quanta, which is singular time quantum (1°) of orbital motion. Only 90-fold repetition, which is ¼ orbital motion, is boundary of passage of time, which is discrepancy (overlap) between rotational and orbital time (90x60=5400 years is singular ¼ quantum period of time). Spatial discrepancy of 4 min per day is cause of gradual angular tilt (1440/4=360), which is 4 seasons (reaction of 4x1min), while 4 sec of its time discrepancy within, is boundary of gradual change of seasons (60 years, which is 1° of precessional curvature). Just as sunrise/sunset are singular spatial quantum of rotation, so are 4 points of solstice/equinox singular spatial quantum of orbital motion or angular change of seasons. Solstice/equinox points are fractal difference or overlap between different seasons. They are singular (inertial) point, which perpetuates such passage of time, just as sunset/sunrise are singular quantum that perpetuates rotational cycles.

Sunset/sunrise and/or solstice/equinox are singular points of equal and opposite linear passages of time, because they are 1D fractal difference, which is contraction/expansion of time (inversion between space and time). They are space/time difference, which is infinite perpetuation of passage of time, while 2D fractal difference, which is inertion of such singular points is their time boundary or curvature. Inversion of inertion is double inertion. Inversion of inertial singular point within is double inertion or time of time without. Singular spatial quantum of solstice/equinox, which perpetuates seasonal passage of time, is 1 day, which is 5th quarter or simultaneous (overlapping) end of one and beginning of another cycle. Singular spatial quantum of 1 day, which is 4 min, is simultaneously singular time quantum of seasonal passage of time (orbital motion), just as 4 sec is time of spatial 4 min in context of rotation. 4 min is relative/uncertain time difference between different seasons, just as 4 sec is relative time difference between day and night, because it is simultaneous reality of both. 90 days of one season is accumulation of 90x4 sec =360 sec or 6 min. per season. Space/time is fractal 3/2; therefore 4 out of 6 min is inertion of orbital motion, while remaining 2 min per season is perpetuation of infinity of larger cycles of time. Each solstice/equinox is 4 min discrepancy, which is 4x5=20 min relative discrepancy (incl. inertion) per year. Accumulated rotational discrepancy is 6x4=24 min per year, of which 20 min is balanced through angular tilt (fractal 20/24=5/6 or reconciliation of 5 out of 6 min discrepancy, while remaining 1 min is perpetuation of infinity of larger cycles of time).

4-sec/4-min are relative time discrepancies (2D fractal difference) of rotational and ¼ orbital motion (one season), while 4 min and/or 1 day are their relative spatial discrepancies (1D fractal difference). Inertial spatial discrepancy of ¼ precessional time is one season or 90 days (¼ orbital time), while its time discrepancy is 1 day, which is spatial difference between seasons (solstice/equinox). Spatial discrepancy of inertion is time discrepancy of inversion and equally vice versa. 1 day of spatial discrepancy per season (1D difference) is 1 day of time discrepancy per ¼ precessional period. ¼ quantum change of time is accumulation of fractal 4-sec (and/or 4-min) difference, which is 1-day discrepancy. 20 out of 24 min discrepancy per year are reconciled through gradual angular tilt, while remaining 4 min x 6480 years is 25920 min or 18 days discrepancy in context of orbital motion. 18 days x 4/5 =14.4 days, which is 6480 x 4/5 = 5184-year period in which accumulated difference is singular quantum, which is ¼ of 1/5 and/or 1/5 of ¼ (18=1/4x72, which is 1/5 of 360° and/or 1/5x90, which is ¼ of 360°

curvature). It is singular ¼ quantum of orbital motion (angular tilt) and/or or 1/5 of precessional rotation or singular 1/20 quantum of both. 14.4 days (inertial 15x0.96, because 5x18 and 6x15=90 are fractal 5/6 which is fractal 24/25 in double/inverse context) is accumulated discrepancy between precessional and orbital motion, which is singular quantum of relative space. Singular quantum of relative space is singular quantum of infinitely long passage of time, which is reconciled through instant quantum change that advances precessional rotation and regresses perihelion for 18 days through instant 1/24 change of rotational velocity. It is singular quantum of solar spatial reality in context of galaxy or singular quantum of galactic gravity (orbital), which is instant ¼ angular nod and/or ½ angular (north/south) rotation of horizontal east/west rotation, which is its instant spatial reality.

Such angular nodes and/or ½ rotations create instant momentum of following passage of time, which is exhausted by previous period through gradual slowing of rotation. Instant upside/down inversion creates new out of old passage of time, by inverting directions. Gradual sowing of rotation in space (i.e. horizontal direction) is its gradual increase in time (angular direction) and equally vice versa, which is instant change in ¼ intervals. Only duality of equal opposites is real, therefore negation in context of one is affirmation in context of another. Instant inversion is instant double negation that creates affirmation, because there is no difference between up and down—up in one is down in opposite direction and decrease of momentum in one is increase of momentum in opposite direction (gradual unwinding in one is winding up in opposite direction). ¼ and/or 1/5 double/inverse difference created through quantum space/time change is inertion of infinity of larger time cycles. **Duration of ¼ quantum change in 5184-years intervals is 1 day** or 1440 min, which is inertial 1600 min or 1.111 day ($1600=16x10^2$ or $(4x10)^2$, where 16 min is accumulated time difference of 4 seasonal changes in one year, without inertion, which is perpetuity of time waves and 4 min is discrepancy of one season. 1600=80x20 or 80-fold repetition of 20 min discrepancy and/or 4x400, which are double/inverse fractal differences).

Spatial discrepancy, which is singular spatial quantum of ¼ precessional motion, is one season, which is instant change in duration of 1 day. It is relative/uncertain discrepancy between two quantum periods of passage of time (overlapping end of one and beginning of another). Spatial discrepancy, which is singular spatial quantum of whole precessional period, which is in its turn only singular ¼ quantum of double precessional period created by discrepancy between precessional rotation and orbital (perihelion) precession, is one year. Singular time quantum of ¼ precessional curvature is 1 day, while singular quantum of whole precessional cycle, which is ¼ of 103680-year Milankovich cycle, is 5x1 day (just as 5x4=20 min discrepancy per year, because ¼ within is 1/5 without and/or vice versa). It is 4 days per season, including repetitive inertion of such wave, which is 1 day. Reconciliation of ¼ difference is remaining 1/5 difference within difference, while reconciliation of 1/5 difference is remaining ¼ difference within difference, which is double/inverse infinite fractal 4/5 curvature. End of each precessional period is 1 day of angular rotation in orbital plane, followed by additional 5 days of ½ rotation or ¼ nod bellow/above orbital plane, which creates 4 seasons of time of following precessional cycle. It is double/inverse change, which is 2x1/2 larger change or simultaneous change of 2 directions of time, which are not only gradual change of rotation, but also fractal changes of orbital motion. It is simultaneous change within and without, which is change of double and inverse waves (see illustration 12). **Duration of precessional 20736-year change is 6 days, which is creation of passage of time of its following 360° cycle. It is instant 6 days, which create its 7th day that is constant passage of time and gradual spatial change in period of 20736 years (where instant in space is gradual in time and equally vice versa).** 6 days is 8640 min, 518400 sec and/or 144 hrs. It is 5184x10^2 sec that creates 4x5184 years, 144 hrs that creates 4x14.4-days discrepancy and 8640 min that creates 86400 sec of one day, which is fractal reality in between.

Duration of ¼ changes is 5°=1 day, duration of precessional changes is 5^1+5°=6 days, while duration of quantum changes between double/inverse Milankovich cycles (103680 and/or 129600 years), which are 5-fold repetition of precessional cycles that instantly adds 1/5 and/or ¼ period of time is 5^2+5^1=25+5=30 days. It is 5 days of change of orbital/rotational motion, followed by 25 days of gradual change of size in north/south direction (expansion/contraction). Such quantum change is instant change between low and high-eccentricity cycles of orbital motion. High eccentricity is reversal of gradual change in past passage of time, which wears-off in following cycle, because eccentric in space is round in time and round in space is eccentric orbit in time. It is instant reversal of gradual east/west increase in obliquity of Earth through gradual accumulation of equatorial bulge. Equatorial bulge is gradually shifted in north/south directions in course of past 20 precessional changes. Each ¼ change is its positive/negative 3.6° tilt, which is total 72° accumulation in 103680-year period (singular ¼ quantum). Each 3.6° shift is only temporary balancing, because northward shift is followed by southward shift and equally vice versa. 3.6° shift is angular tilt within angular tilt (change of seasons), which is tilting of equatorial line as well as tilting of polar axe (1/10^2 tilting of space/time coordinates, while change of seasons is only ¼ gradual shift of spatial coordinates). Such fractal shifts of coordinates are double/inverse change of seasons (ice/warm ages), which are gradual rebalancing of mass in north/south directions, through alternating attraction of surface liquid in opposite directions.

North ice age is south warm age and equally vice versa, which is attraction of water in opposite polar directions in subsequent ½ precessional periods, where it is deposited into ice. North and south are two equal opposites—northern hemisphere is land, topped with sea, while south is sea topped with land in polar regions, which is double/inverse difference. Passage of time is gradual accumulation of such discrepancy, which is instant change in ¼ quantum periods. Milankovich cycle is singular ¼ quantum of 414720-year period (432 000 in context of orbital motion), which is singular gravitational quantum or singular quantum of change of size (expansion/contraction). Precessional changes are reversible changes. They are instant reversal of gradual changes in preceding period or gradual wobble in east/west directions. Double/inverse quantum changes are not reversible, but twice-larger shift, of which only ½ is reversible in following period, while another ½ is permanent change of space/time coordinate system. It is singular quantum of gradual inversion between north/south and east/west directions. **Duration of quantum change at the end/beginning of 103680-year Milankovich cycle, which creates its following passage of time, is 30 days or 1 month, which are 43200 min, 2592000 sec and/or 720 hrs.**

Singular gravitational quantum, which is period of 414720 years of passage of time in between is instant change of $5^3+5^2=150$ days or 5 months duration (216000 min, 3600hrs or 12960×10^3 sec). It is instant reversal of accumulated difference of previous 4x1/4 period, which is instant surface expansion/contraction in both—north/south and east/west directions. Just as equinox is singular spatial point (eigenvalue) between two equal and opposite solstices, which are its double/inverse reality, so are angular precessional shifts. They repeat twice in between (east/west direction) and only once in north and south directions (without and within), which is gradual accumulation of tension in such direction (increased obliquity). East/west accumulation of discrepancy precedes north/south direction, which is singular quantum of expansion in all 4 directions. 5-fold repetition is singular 2.592 mill-year period, which is $5^4+5^3=750 \times 0.96=$ 720 days or 2 years in duration. 4-fold repetition is singular 10.368 mill-year period, which is quantum change of $5^5+5^4=3750 \times 0.96=3600$ days or 10 years in between. 5-fold repetition is 51.84 mil-year period, which is quantum change of $5^6+5^5=18750 \times 0.96=18000$ days or 50 years duration. 4-fold repetition is $5^7+5^6=$ $93750 \times 0.96=90000$ days or 250-year quantum change in between two equal and opposite 207.36 mil-year periods of passage of time (one solar orbital round core of galaxy). 5x207.36=1.0368 bill-year period of passage of time, which is two equal and opposite quantum changes, which are its beginnings/ends, in duration of 1250 years. 4x1.0368=4.1472 bill-years of passage of time, which are quantum changes of 6250 years at its beginning and end.

6250x1.04=6500 years, which is inertionless ¼ period, where 1.04=24/25 is equal and opposite reaction (inversion) of 0.96=25/24 (it is +/-0.04 or total $0.08=2^3x10^{-2}$, which is inversion of fractal 3/2 here and now). 6500 is perfect duality of 3D sequence of 6480 years. 6500-20=6480, where 20 is inertion of repetition in such cycle. 6500/20=325, where **$325/324=(13x25)/18^2$**, which is 1D fractal difference of $24/25=(6x4)/5^2$. Fractal 24/25 is +/-1D symmetrical discrepancy (6D without and 4D within) of inertial 5D in between, which is their double/inverse eigenvalue ($5/(1/5)=5^2$). Fractal 325/324 is +7/-5D asymmetrical discrepancy (13D without and 25D within) of inertial 18D in between, which is their double/inverse eigenvalue. 18=3x6, while 13=2x6+1 and 25=4x6+1, which is double linear direction or double/inverse expansion of singular point, where singular point is previous symmetrical sequence of no discrepancy in between (+1-1=0 of 24/25 sequence is 1+1=2 of 325/324 sequence). Fractal 24/25 is eigenvalue of (4/5)/(5/6), which is 2D fractal difference, while 325/324 is eigenvalue of (13/18)/(18/25), which is 6-fold repetition of such difference or 12D fractal difference (7+5). It is 13-fold repetition of fractal 1/25 difference and/or 13.5-fold repetition of equal and opposite 1/24 difference, which is singular ½ quantum difference. 325/324 is singular (½) quantum of inertial 24/25 duality, just as 24/25 is singular quantum of fractal 4/5 reality. It is difference within difference or double repetition of fractal 2D difference, which is inertial 4D difference (10^4). 6500 is non-inertial period of 6480; just as 26000 years is non-inertial period of 25920, 130000 of 129600, 5200 of 5184 years etc.

2D fractal difference is space/space or time/time difference, which is difference between two spatial sequences without/within in context of double/inverse passages of time in between (and/or two passages of time in context of singular spatial reality in between). 2D fractal difference is exclusive difference between uncertain space (dense mass within) and inertial here and now or relative space (empty space of universe without) and here and now. Inertial here and now (Earth) is dense mass within of empty universe without and/or empty space without of dense mass within. Each sequence is +/-2D fractal difference from here and now, while 4D fractal difference is difference of both. It is gravity/gravity difference, which is difference between two equal and opposite expansions in context of inertial contraction in between and equally vice versa. 2D fractal difference is infinite spatial universe without, as well as singular mass of an atom within, while 4D is difference between surface of atom and infinite surface of universe (light speed expansion). 2D is double light speed and 4D is quadruple light speed difference. Infinite space and infinite time of universe are only ½ of 4D reality, which is equal opposite (mirror image) of its micro-world, which is another ½. Universe is gradual expansion (time) of instant contraction (space) of its mass within and equally vice versa—it is instant passage of time without of gradual spatial contraction of no time within.

Universe is ½ of time, which is its constant passage in context of empty space and micro-world is another ½, which is instant space of dense mass. Universe is gradual future and micro-world is instant past time, which is double/inverse direction of time—one is space of time of another and equally vice versa. One is sin and another one is cos wave, which are double/inverse curvatures of double helix, where one precedes another in one direction and vice versa in opposite direction (2-way street). 2D fractal difference is double helix and 4D fractal difference is double/inverse double helix. 2D fractal difference is space/time and 4D fractal difference is gravity/antigravity double helix, where each strand is yet another double helix of space/time. Infinite number of space/time sequences is only one strand, while equal and opposite (inverse) sequencing, which is anti-parallel with first, is second strand. One strand is expansion of empty space in direction without, while another strand is equal and opposite contraction of dense mass within. Dense mass is inversion of empty space; therefore it is equal and opposite expansion of inverse empty space (time) in opposite direction, because singular space is infinite time, just as infinite space is singular time. 4D fractal difference is double expansion (spatial expansion is contraction of time etc). It is past and future expansion, which is instant density of mass within and empty space of universe without, which is future density of mass (or density of future mass).

Each larger quantum change within is larger period of time without, which is infinite fractal 4/5 spiraling of double helix. 4.1472 bill-years in context of rotation is 4.32 bill in context of orbital and 5.184 in context of precessional motion. Such period of time is 10-fold curvature or 10 different quantum periods of time (Bode), which is spatial sequence of massive Earth in context of solar reality. It is all its past time. 4.32 bill-years period is inversion between space and time or gradual and quantum changes. Quantum change of such period, which is 6480 years, is complete negation (curvature) of constant passage of time here and now, which is same duration. It is infinite passage of time, which is gradual, infinitely slower spatial change of 4.32 bill-years contracted into 6480 years of its equal and opposite quantum change, which is its negation or instant reversal. It is change between time and space as passage of time, which is inversion of all the future time without (universe) into instant space within (mass) and change of instant space within into infinite passage of future time without. It is infinite expansion of mass into empty space without, which is simultaneous contraction of empty space without into dense mass within. Such change is change between universe and micro-world, where future universe is micro-world within, just as micro-world within (dense mass) is future universe. It is time distance between inception of Earth on surface of solar reality and here and now, which is in-depth (angular) distance of density between size of an atom and inertial size of Earth here and now. It is infinite relative change of size and density of inertial here and now (Earth) in one quantum leap of time, which is same duration as infinite passage of time here and now.

4.1472=64x10^4x6480-year periods, where 64=8^2 or 2^6, which is 6-fold curvature of inertial space/time duality here and now. It is also 8x10^5x5184, where 8=2^3, which is 2x1/2 or 1D fractal difference. 4.32 bill-years is quantum period of time preceded by quantum change that is same duration as inertial passage of time here and now, which is infinite increase of Earth's spatial size from singular (atom of micro-world) to infinite (orbiting planet of solar reality), which is equal and opposite decrease of its size in context of time. It is creation of infinite passage of time in context of spatial reality without, which is instant end of passage of time in context of inverse spatial reality of micro-world, where space is passage of time and time is instant space. Duration of such sequence is followed by another 6480-year quantum change, which is yet another infinite increase in size and equal and opposite decrease of relative density from inertial size here and now to infinite size of solar reality in context of galaxy. Next sequence is infinite increase of size between solar reality and galaxy, followed by increase between galaxy and universe etc, which is infinite anti-parallel fractal curvature of Fibonacci sequencing, where future evolutionary sequence is created by addition of present (i.e. space) to the past (i.e. time) sequence.

Spatial universe is ¾ and micro-world between surfaces of atom and here and now is remaining ¼ of 4D spatial reality, just as micro-world is ¾ and universe is ¼ of time, which is one double/inverse gravitational sequence. 3 spatial shells of universe (solar reality/galaxy/universe) are 3 inverted time shells of micro-world (atom/nucleus/Planck length singularity) in context of inertial (7th) shell in between, which is rotating Earth here and now (eigenvalue). Universe, which is 3 fractal space/time shells, one within the other (inversion of all 3 inertial directions), is 3x4.1472 bill-years in context of its inertial discrepancy here and now

(Earth). In context of orbital time, it is 4x4.32=17.28 and 17.28x5/6=14.4 bill years or non—inertial 15 bill years. 5x4.1472= 20.736 bill years, which is 10^2=100 orbitals of solar reality round core of galaxy (20.736=14.4^2/10 or 14.4x1.44). 3/4x20.736=3x5.184=15.552, where 15552x24/25x 325/324= 15552x0.96x1.0030864= 15552x0.962962…= 14.976 bill years, which is 0.024 discrepancy from 15 bill. 15000/24=625 is fractal **625/624=25^2/(24x26)**, which is symmetrical 6D fractal difference or difference within difference, which is singular quantum of inertial (24/25)/(25/26), as opposed to ½ quantum, which is 324/325.

Size/density of Earth here and now is spatial discrepancy between light speed and double light speed expansion, which is relative/uncertain time difference of spatial universe; therefore larger size of universe is smaller discrepancy with here and now in its context. Reality is always here and now, which is singular inertion (cause) of relative/uncertain space/time without and within. It is always rotating Earth here and now, regardless of its relative density/size. Earth is infinite universe without of its dense mass within, just as is infinitely dense mass within of its infinite universe without. It is infinite empty space and infinitely accelerated passage of time (infinitely slow motion) in context of inverted reality within (atom), which is its 2D fractal difference. Likewise—Earth in context of all its infinite passage of time is only one instant spin of infinitely dense atom in context of solar reality, which is inverse 2D fractal difference of same inertial reality as here and now. Passage of time between different spatial sequences is only comparative or totally uncertain/relative (fractal) difference. It is one reality in context of relatively different passages of time, which are only different surfaces of same content.

Perihelion precession is 1° per 57.6 years, which is 90° or ¼ in 5184 years. At the same time, precessional rotation of Earth is only 72°, which is 90/72=5/4, where ¼ discrepancy within is 1/5 without and vice versa. ¼ and 1/5 discrepancy are double/inverse curvatures of 4D within (Earth's dense mass) and 5D without (empty space of solar reality) in context of inertial here and now in between (surface of rotating Earth), which is their fractal reality. Accumulated discrepancy of 18°, which are 18 days in context of orbital, is ¼ of 72° (within) and 1/5° of 90 (without). 18° is common denominator of both. Discrepancy between perihelion and precessional rotation is not real, unless singular quantum, which is 18°. Discrepancy less than 18° is only increase of tension within, which has no effect on surface. It is singular ¼ quantum of 72°, which is instant surmounting or inversion of inertion. 18° is ¼ of 72° without, while 3.6° is its 1/5 within, which is inertial quantum of such discrepancy. 18° is 1D (space/time) and 3.6° is 2D fractal difference (gravity) or difference within difference of 72° on surface. 18° is time discrepancy of 72° (its singular quantum), while 3.6° is double/inverse time discrepancy, which is singular quantum of its relative spatial reality. 18° is singular time quantum and 3.6° is singular quantum of spatial reality of time.

18° discrepancy is instant advance of precessional curvature for 18-3.6=14.4°, from 72° to 86.4°. 14.4 is 1/5 quantum of 72° and/or 1/6 of 86.4° (fractal 5/6 within of 5/4 without). It advances precession only to 86.4°, because 90° is not multiple of 14.4 quantum (3.6° inertial discrepancy in between) 3.6° relative spatial discrepancy is accumulated to singular 14.4° time quantum in 4x5184 periods. Each 5184-year period is instant advance of precession for instant (virtual) 1296 years (14.4°) to 6480 years (86.4°). It is instant advance in context of spatial reality of time and not its constant passage. 4-fold repetition is 4x86.4=360-14.4=345.6° spatial curvature, which is 345.6 days in context of orbital motion (eclipse year), where 345.6=360x24/25 (24-fold repetition of inertial 14.4 quantum). 4 periods of 5184 years are accumulation of additional 14.4° to 57.6° and its discrepancy (double/inverse inertion) from 3.6° to 14.4°, which is total of 72° (because such relative spatial quantum repeated 4 times is singular quantum of time or instant inversion of inertion). It is accumulation of 4x1296 years ($1296=14.4x10^2/1.111$) of virtual time (relative spatial discrepancy) to singular quantum of passage of time, which is 5184 years. 4-fold repetition of inertial passage of time is creation of additional quantum of passage of time, through gradual accumulation of inertial discrepancy from ¼ to whole quantum. Precessional period of 4x6480 years is 5x5184 years, because of 4x1296 relative discrepancy between periods.

One precessional period is 5x5184 or 4x6480-5184=20736=4x5184 years. Quantum reality is non-commutative. 5th quantum period of time, which is reaction of inertial 4 periods is simultaneously end of one and beginning of another precessional cycle (see illustration 14). It is overlap or transition between two cycles. It is 5th singular point of 4x1/4 wave, which is at the same time end of one and beginning of another wave. 5th period of time is gradual reversal (reaction) of 4 instant advances in previous periods. It is singular gradual reaction of 4 instant fractal actions, because inertion of four fractal actions is singular quantum of relative space, while inertion of singular gradual reaction is its 4-fold accumulation, which is change from relative spatial reality into passage of time. 4x6480 =4x86.4=360x0.96=345.6°, while 5x86.4=432° curvature (432/360= fractal 6/5 reality without of fractal 4/5 within). 345.6° spatial curvature is only 288° time curvature, because constant passage of time is 4x72=288° (gradual precessional motion), while its spatial (relative) discrepancy is 4x14.4=57.6°. 360° of time curvature is 5x72° or 5 quantum periods of passage of time. 360° in time context is 432° in spatial context, because time in spatial context is not round, but spiraling curvature, which does not ends where it started. 360° of spatial curvature is incomplete 288° time curvature, because space precedes time (fractal difference). Space is 5/4 of time (¼ positive discrepancy without) and time is 4/5 of space (1/5 negative discrepancy within). End of time curvature in spatial context is shifted from its beginning for ¼ in angular direction, just as spatial curvature is shifted for 1/5 in opposite angular direction in context of time.

Illustration 14 - Double/inve rse 4/5 precessional repetition

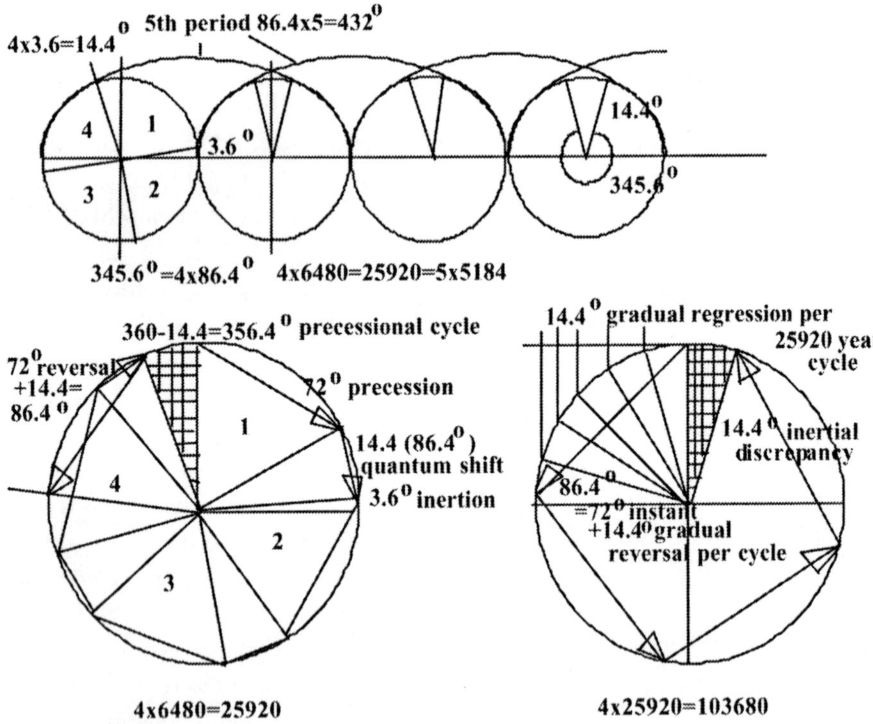

72° of precession is eigenvalue or symmetry between 90° and 86.4° without and 14.4° and 18° within (3.6° difference). 72° is 20-fold and 86.4° is 24-fold repetition of 3.6°. 72/90° is 4/5-curvature (18° discrepancy) within of 90/86.4° without (3.6° discrepancy), which is 24/25-double/inverse fractal reality. 72/90 is larger (inertial) 4/5 without of smaller 4/5 within (double/inverse inertion), which is 18/14.4°. 1/5 and ¼ of 3.6 are 0.9° and 0.72°, which are yet another smaller fractal 4/5 that is 2D fractal difference of 72/90. 72° is 1/5 and 90° is ¼ of 360°, which is larger fractal 4/5 without, etc. Fractal 4/5 is singular eigenvalue (resonance) of both in 20-fold repetition, where such repetition is 24/25 or double/inverse discrepancy. 20-fold repetition of inertion is 25-fold repetition of inversion. (129600-year period is 20x6480 and/or 25x5184). Precessional double/inverse space/time curvature is perpetuation of fractal difference within fractal difference, which is 20-fold repetition of singular 4/5-fractal difference without and double 24/25-fractal difference within. Fractal 4/5 difference (5184) of smaller period of time (25920) is 24/25 difference of larger period of time (129600-years), just as fractal 4/5 difference of such larger period (25920 years) is 24/25 difference of yet larger period (518400 years). 3.6° discrepancy of larger period is 14.4° discrepancy of smaller, 72° discrepancy of twice-smaller period of time (where 14.4 without is 18 within, just as 72 within is 90 without) and complete 360° curvature, which is passage of time and not its discrepancy, of trice-smaller period. Space/time is perpetuation of double/inverse difference, which is ¼ discrepancy of 1/5 and/or vice versa. 5/4 is space/time or 1D fractal difference, while double fractal difference (gravity) is simultaneous ¼ discrepancy of 1/5, which in its turn is discrepancy of 1/6. Singular eigenvalue of double fractal difference is 24/25.

It is relative/uncertain description of precessional period as 4x5184=20736, 4x6480 =25920 or 5x5184=25920 years, because 4-fold repetition within is 5-fold repetition without and equally vice versa, which is eigenvalue in 20-fold repetition. 20x5184= 103680 years of Milankovich cycle of low eccentricity orbital. 103680-year period is 4 precessional cycles of 5x5184=25920 years each or 5 precessional cycles of 20 236 years each. Singular ¼ quantum of such double/inverse precession is 5x5184=25920 years, while quantum change between two precessional cycles is at the end of 4th 5184-year period, while 5th period is its reaction, which is 1st period of next cycle. Consequently, singular quantum of double/inverse precession is 5/4 of 4x1/4 precessional cycle, where 5th quarter is angular transition between two horizontal cycles. 103680-year period is 4x25920 cycles and/or 5x20736-year cycles (4/5), because space/time is double/inverse spiraling curvature (see illustration 14), where symmetry of one is asymmetry of another and equally vice versa. Quantum change is always 4-fold repetition within, followed by 5-fold repetition without and/or vice versa. 5184-year period is singular quantum change, while 4x5184=20736 is double (angular) quantum change. Because 20736-year period is only 4/5 of singular ¼ quantum (25920) of double/inverse precession, only 5-fold repetition is larger quantum change at the end of 103680-year period. Space/time waves (harmonics) are ¼ or 90° difference from each other (illustration 12), which is ¼ discrepancy in their phases. ¼ phase discrepancy is 5th quarter in between two successive waves. Only 5/4 of 1/5 is 4/5 of ¼, which is larger/smaller fractal 4/5 reality.

Inertion (¼) or discrepancy (quantum) of 5184-years time waves is 1296 years, while inertion of 20736-year waves is 5184-year period (expansion of relative spatial difference of 1296 years into 5184-year period of passage of time through 4-fold repetition). 5th period is reversal of previous 4 instant (quantum) changes; therefore it is passage of time shifted in angular direction of 4 previous periods, which are passages of time in one horizontal direction. 5th period of 1st cycle is 1st period of 2nd cycle. It is inversion of 1st and inertion of 2nd cycle (relative discrepancy of 1st and uncertain discrepancy of 2nd cycle). Each 5184-year wave generates 1296-year relative discrepancy, which is instant 14.4° shift at the end of period, and each 20736-year wave generates 5184-year discrepancy of its 5th cycle. Quantum change at the end of 20736-year cycle is reversal of precessional motion, because it is singular ¼ quantum of double/inverse 103680-year period, which is precession of precession or gradual regression of precessional motion. Smaller quantum period of time is only ¼ of larger quantum period of time. 4-fold repetition of 5184-year period is 345.6° curvature, which is regressive accumulation of 14.4° discrepancy per 20736-year cycle. Each precessional cycle is gradual 14.4° regression of precessional motion (equal opposite of lunar nodal regression of 1.44° per cycle-see illustration 14). Such motion is not real, unless ¼ quantum. 5-fold repetition (5x20736) is accumulation of 5x14.4=72° regression, which is 4/5 of 90° (1/4) that is instant (quantum) inversion.

Quantum change at the end of 103680-year period is consequence of gradual 14.4° discrepancy at the end of each precessional cycle. It is whole quantum of double/inverse precession, which is ¼ of yet larger precessional cycle of 518400-years, which advances double precession in quantum leaps (5x14.4=72°, which is its instant 18-3.6=14.4° shift) Advance of double precession is regression of precession. Shift at the end of 5184-years advances precessional motion, while shift at the end of 20736-year period is singular ¼ quantum of double precession. Double precession is gradual regression of precessional motion, just as precession is gradual regression of rotation. Rotation and double/inverse precession are west/east rotations without and within of precessional east/west motion in between. Instant advance of double precession is regression of precession and advance of rotation, just as instant advance of precession (1296 years or 14.4° shift) is regression of rotation. Likewise—singular ¼ (1/5) quantum of 3D precession, which is 103680-year period, is 14.4° regression of double precession, advance of singular precession and reversal of rotation (see illustration 14). Each larger quantum change is gradual accumulation of smaller spatial discrepancy, which is larger quantum change here and now (rotation/size). Precession is time or boundary (curvature) of rotation, because it is its equal opposite (inverse rotation) and double precession is time or curvature of precession.

Each larger rotational motion is exponentially slower motion (faster passage of time) in opposite direction of its previous (inverse) motion and same direction as its twice-previous motion (2D fractal difference). Each larger motion is gradual regression or curvature of its previous motion and advance of one is regression of another, and at the same time it is twice larger advance of motion that is double fractal difference. Quantum interruptions are instant reversals, while double/inverse motions of inertial motion in between, which are fractal (light speed) difference is gradual reversal. Instant reversal of inverse curvature, which is gradual reversal is instant advance of inertial curvature, which is gradual advance and equally vice versa. Reversals are contraction and advances are expansion, where expansion in space is contraction in time. Reversals are fragmentation and advances are multiplication, where fragmentation in space is multiplication in time and equally vice versa. Curvature, less than 360° is fractal, while curvature larger than 360° is multiple. Fractal in context of larger precessional motion is multiple in context of smaller precessional motion (2D difference) and equally vice versa, because convex curvature of one is equal and opposite concave curvature of another (action/reaction).

Just as gradual accumulation through 4-fold/5-fold repetition of 14.4° and 3.6° is instant quantum reversal in 5184 and 20736-year periods, so is 20-fold repetition of smaller 0.72° discrepancy accumulation into singular 14.4° quantum, which is exponentially larger quantum change on surface (also—20x3.6=72°, 20x14.4=288° and 20x72=1440° or 5 full curvatures of precessional time). Quantum change at the end of 20736-year precessional cycle is reversal of 4x1296-years accumulated relative discrepancy, which is its 5th 5184-year period, while quantum change at the end of 103680-year period is reversal of 4x5184=20736 years accumulated difference through interpolation of 5th periods between 2 waves. 5th period of passage of time is relative discrepancy or inertion that perpetuates 20736-year time waves, just as 1296 years is spatial discrepancy between 5184-year waves (space of smaller wave is time of larger wave, which is 1D fractal difference). Just as accumulated 4x1296 year difference is 5th period of precessional cycle, so is accumulated 4x5184 year difference additional cycle of double/inverse precession. 103680 years is 5x20736-year cycles, which are 4x5184= 20736—years of accumulated difference, which is its additional 6th cycle without.

Double/inverse precession is fractal 5/6 without of fractal 5/4 within. Eigenvalue of both is fractal 24/25. 6th precessional cycle, generated by quantum change at the end of 103680-year interval is overlapping cycle between two waves—it is overlapping end of 1st cycle and beginning of 2nd cycle. 6th cycle is instant reaction of gradual accumulation in preceding 5 cycles; therefore direction of such time is yet again shifted in angular direction. 6th cycle, which follows quantum change at the end of 5th cycle, is high eccentricity orbital of Milankovich cycle. High eccentricity is reaction of singular ¼ quantum gravitational change, which is not only change of rotation but also change of orbital motion (2D time change or space and time change). 6th cycle is gradual reversal of instant quantum change, which is inversion of preceding 5 cycles, as well as inertion of succeeding 5 cycles of low-eccentricity. Additional 5th period of precession and additional 6th cycle of double/inverse precession are singular and double fractal difference of inertial 4D here and now. They are its double relativity/uncertainty or 6D, which is boundary of fractal 3/2 reality. 4 periods of precession within are 5 periods without and 5 precessional cycles within are 6 precessional cycles without. 6x20736= 124416, while 5x25920=129600, which is 5184 discrepancy in between which is 24/25 inertial discrepancy of such cycle (129600x0.96). 1296-year (¼ of 5184) discrepancy perpetuates precessional cycles and 5184-year (¼ of 20736) discrepancy perpetuates double/inverse precessional cycles. 4x5184=20736-year discrepancy perpetuates 518400-year waves.

103680=20x5184, 124416=24x5184 and 129600=25x5184. Inertial 103680-year cycle, which is 5x20736, generates additional 20736-year precessional cycle and 20736-year cycle generates additional 5184-year cycle (20+4+1=25 or 5^2 in between 4x6 within/without). 103680-year cycle, which is 4x25920, generates additional, 5th 25920-year period, which is its inertion. 5x5184=25920 and 5x25920=129600, while 5x129600= 648000 year cycle. 648000-year cycle is 5x103680=4x129600=518400, which is instant quantum change followed by overlapping 5th 129600-year cycle. 5th 129600-year cycle is 6th 103680-year cycle, which is 622080 years, followed by additional 25920-year cycle, which is its 24/25 (gravitational) inertion. 5x518400=4x648000=2.592 mill-year cycle, followed by additional overlapping 6th cycle, which is its end (inversion) and beginning (inertion) of next cycle (2.592+0.648=3.24 mill-years). 5x2.592=4x3.24=12.96+3.24= 16.2 mill-years cycle and 5x12.96=4x16.2=64.8 mill year cycle, followed by 5th 16.2 mill-year cycle, which is 81 mill-year time wave. 4x81=5x64.8=324 mill years, followed by 5th period is 405 mill years. 5x324=4x405=1.62 bill. years followed by additional 405 mill, which is 2.025 bill-years etc. It is perpetuity of 5-fold repetition, which is inertial framework of

4-fold repetition without and 6-fold repetition within (equally vice versa). 5-fold repetition is inertion (eigenvalue) between double/inverse 4-fold and 6-fold repetition, which are 4D/6D without/within inertial 5D in between. 5D is gravitational wave between double/inverse space/time waves without and within.

Double/inverse 5-fold repetition is $5/(1/5)=5^2=25$-fold repetition, which is fractal difference within of surface reality without, which is 5x4=20-fold repetition of fractal 4/5 reality. 20-fold repetition on surface is 25-fold repetition in depth and equally vice versa. Fractal 5/4 waves are space/time waves, where 4-fold curvature is boundary of double/inverse 5-fold curvatures and equally vice versa. It is total time of inertial spatial sequence and/or total space of inertial sequence of passage of time. They are space in time curvature and/or time in spatial curvature (negation of inertion in inverse context). They are quantum periods of time of inertial spatial sequence, which is also frequency of their quantum changes (inversion points of such 4x1/4 wave). Space/time waves are 4-fold repetition of 1/5 discrepancy followed by 5-fold repetition of ¼ discrepancy, which is 4-fold repetition without and 5-fold repetition within (and/or vice versa). 4x5184=20736 and 20736x5=103680-year wave, just as is 5x5184=25920 and 4x25920= 103680. Time waves of inertial spatial sequence and/or vice versa are 20-fold repetitive curvature, which is double fractal 5/2 reality (2x5x2). 20x103680 is 2073600-year wave, which repeated 20 times is 41.472 mill-year wave (1/5 solar orbital round core of galaxy) and its 20-fold repetition is 829.44 mill-year wave (4-fold orbital).

2073600 years is 400 $=4x10^2x5184$ years ($1/10^2$ solar orbital), which is double fractal difference of inertial passage of time here and now. One solar orbital of 207.36 mill years is yet another double fractal difference, which is 40000x5184 years, where 40000 is reaction of Earth's size here and now. 829.44x5=4.1472 bill years, which is instant past time (Earth's density or mass), while 4.1472x5 and/or 25x829.44 (4-fold repetition followed by inversion of inertion, which is its 5th period) is 20.736-bil years period. 1/5 of such period is 4.1472 and ¼ is 5.184 bill years, which are non-inertial and/or of inertial period of all past time. Past time is instant density of mass within (Earth), while remaining 4/5 is its infinite passage of time, which is instant emptiness of universe without. Past time is instant spatial density, which is particle or discrepancy of such time wave. Past time is size/density of mass (singular Earth), which is inertion within of its inversion without (infinite universe). Instant past time within (dense mass of rotating Earth) and instant future time without (empty space of universe) are double/inverse non-Euclidean curvatures of flat reality here and now, which is gradual passage of time in between. Mass within and empty space without are double/

inverse spatial sequences of passage of time in between. They are double/inverse (relative) spatial realities of time, which is passage only here and now.

Passage of time here and now is only fractal reality between two equal and opposite spatial sequences of time. It is passage only within its boundaries, while without such boundaries; it is instant double/inverse spatial density. Smallest boundary of passage of time is 5184-year period, while largest is 20.736 bill years. Passage of time here and now is double space without (universe) and inverse time within (mass). 4.1472 and/or 5.184 bill years is instant spatial sequence or massive particle within of its time wave without, which is 20.736-5.184=15.552 bill years. It is instant size of infinite and empty universe (15.552x24/25x325/324x625/624 = 15.552x0.96x1.030864x1.0016025 =15 bill years). Massive Earth (past time) is inertion of future time (universe), which is its inversion or change of instant density within into infinite passage of future time. Massive Earth (past time) is inertial ¼ discrepancy of 4x1/4 wave, where remaining ¾ are its infinite empty space and/or passage of time. ¼ spatial discrepancy (Earth) of 4x1/4 time wave is inversion of passage of time here and now, which is inertial ¼ discrepancy of 4x1/4 spatial reality of Earth. 15-bill years are passage of time only in context of all its past time, which is total massive density of Earth. Only its complete expansion into empty space without (which is dense mass of double/inverse spatial reality) is such passage of time. Fractal expansions or changes of density are proportionately smaller periods of time.

Expansion/contraction of Earth in context of empty universe (reaction of its expansion/contraction) is quantized reality of exponentially larger quantum steps, which is fractal 20-fold repetition in context of larger/smaller 25-fold repetition, which are 1D and 2D fractal difference of infinite fractal curvature of double helix, which is infinity of relatively different fractal repetitions, one in context of another. Smallest quantum of passage of time is inertial 5184-year period, which is 1/5 distance (radius) between here and now and core of galaxy or 1/10 of its size (diameter). Only such distance is passage of time, while any larger size of empty space (universe) is not passage of time, but instant empty space. Passage of time is real only in its spatial shell, which is 1/10 galaxy (1.111). Exponentially larger size of empty space is passage of time only in quantum steps, where each quantum step is light speed difference, therefore exponentially larger quantum change of density and its size/motion.

4x5=20-fold repetitions of inertial passage of time of previous sequence is instant inversion into double inertion or larger spatial sequence. It is instant de-densification of mass within and densification of empty space without, which is infinitely larger period of uninterrupted passage of time (expansion of massive density without and contraction of empty space of universe in direction within). 20-fold repetitions are successive (larger) time waves of inertial passage of time here and now (5184). At the same time, such time waves of inertial mass here and now (Earth) are instant spatial density (spatial waves) of double/inverse density of universe. 1D fractal difference of inertion is 2D fractal difference in context of inversion and equally vice versa. 20-fold repetitions are time waves of inertial space and/or spatial waves of inertial time. Space is 1D fractal difference of time and vice versa. 2D fractal difference is inversion of inertion—it is depth of surface reality. 1D is space/time and 2D is space/space or time/time fractal difference, which is gravitational difference. Gravity in context of inverse wave is time in context of inertial wave and equally vice versa. 4/5 is 1D and 24/25 is 2D fractal difference. 4/5 of inertial wave is 24/25 in context of inverse wave. Any wave is reality of both—it is 4/5-time wave of inertial reality, which is 1D fractal difference and at the same time, it is 24/25 wave of inverse reality, which is 2D fractal difference. 4/5 is past and 24/25 is twice-past sequence of inertial present sequence (where past in context of dense mass is future in context of empty space).

Future sequence is addition of past to the present sequence, where one is 4/5 and another is 24/25-fractal reality. Consequently—20-fold repetition is time and 25-fold repetition is gravity of inertial spatial sequence here and now. 25-fold repetition (5^2 or gravity, which is double or squared fractal difference) is 1/5 fractal difference from 20-fold repetition (fractal 4x5, which is passage of time). 5th additional (transitional) period between two cycles is difference between time and gravity. 5th period is double (gravitational) quantum change of previous 4x1/4 quantum changes of time. 4x1/4 changes are exclusive time changes (changes of motion), while 5th period is double/inverse change, which is spatial change (change of size/density). 4x1/4 time changes are instant (singular) space of larger change, which is its 5th period. 5th period is 1D fractal difference from its 4x1/4 wave. It is expansion of previous singular point into double/inverse (anti-parallel) linear direction. 25-fold repetition is gravitational framework without of its different time periods within. 20-fold repetitions are time and 25-fold repetitions are gravity of inertial space here and now.

4x1/4 wave is 4 double/inverse periods of 90°, where each 90° is 1D fractal difference. 4x1/4=360° singular wave is 4 fractal (different) waves, which are ¼ shift in phase (see illustration 12). 4x1/4 singular wave is 4 smaller waves, shifted 90° from each other. ¼ within is 1/5 without and 1/5 within is 1/6 without etc and equally vice versa, which is infinite fractal curvature. In context of inertial wave (here and now), its successive or preceding waves are exponentially expanded/contracted, because of exponential progression of difference between them. It is always same ¼ difference only between two inertial/inverse waves without and within, while exponentially larger difference is also exponentially larger fractality, where fractal in space is multiple in time and equally vice versa. Such successive space/time waves of inertial wave here and now are exponential multiples within and without. They are exponentially larger/smaller periods of time and/or larger/smaller sizes. Larger fractal difference between waves is exponentially smaller spatial discrepancy within, which are exponentially larger periods of time without and equally vice versa—they are exponentially larger space without and exponentially smaller periods of time within.

Inertion (gravity) of 4x1/4 wave is ¼ (overlapping begging/end between waves), which is 4/5-fractal difference. Inertion (¼) is perpetuity of additional 4x1/4 waves, which are shifted in phase for ¼. Inertial 4x1/4 wave, which includes its gravity, is 5 periods of such wave, which is inertionless wave. Just as inertion of 4x1/4 is ¼, so is inertion of 5x1/5 wave 1/5 and inertion of 6x1/6 is 1/6 etc (1/6 without is 1/5 within and 1/5 without is ¼ within etc, which is infinite dual differentiation of singular reality that is no difference). Singular inertion is infinity of relatively different fractal space/time waves. Different waves are all equally real (many worlds) but not here and now. Inertial passage of time is singular wave, while its relatively different fractal harmonics are relatively different passages of time. Such different passages of time are not time, but instant double/inverse space. They are instant density of mass within and/or inverse density of empty space without. Different passage of time is real here and now, only when its accumulated tension is singular ¼ quantum, which is instant change between spatial and time reality, which are ¼ (90°) difference. Only when discrepancy is perfect 90° are such waves resonant. Only resonant waves are same momentum (inertion), which is instant merging of two waves into larger singular wave.

Illustration 15 resonance of precessional waves

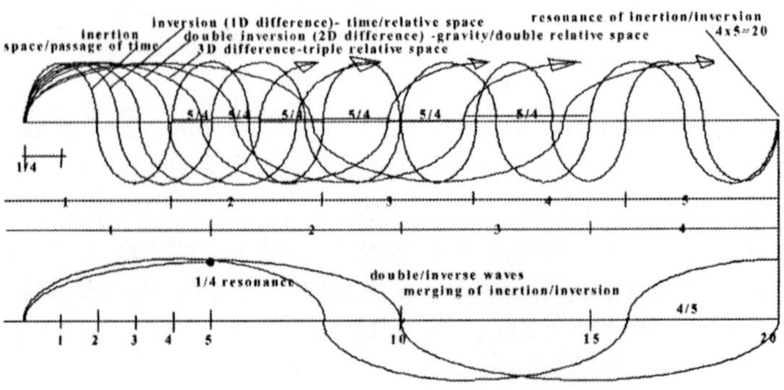

1296 years is ¼ discrepancy of 5184-year time wave and 1/5 of 6480-year spatial wave. They are resonant only in 5-fold repetition of 5184-year wave, which is 4-fold repetition of 6480-year wave, because it is 20x1296, which is their eigenvalue, when one precedes another (see illustration 15). It is inertial 4x1/4 of 6480-year wave, which is non-inertial 5x1/5 of 5184-year wave. It is instant merging of double/inverse space/time waves into singular (gravitational) 25920-year wave. Likewise—4x5184 =20736 is 1/5 and 5x5184=25920 is ¼ of 103680-year wave, which is their point of resonance (20x5184). 4x6480=25920 is 1/5 and 5x6480=32400 is ¼ of 129600-year wave (20x6480), which is gravitational wave of 103680-year time wave (1/4 discrepancy). 4x20736=82944 is 1/5 and 5x20736=103680 is ¼ of 414720-year wave, which is their resonance and instant merging into larger wave. 4x25920=103680 is 1/5 and 5x25920 =129600 is ¼ of 518400 year singular wave (20x25920). 414720 is time wave of 518400-year gravitational wave (overlapping 5th 103680-year period in between), just as 103680 is time of gravitational 129600-year wave and 20736 is time of 25920-year gravitational wave etc. At the same time, 518400-year wave is time of gravitational 648000-year wave; just as 25920 is time of gravitational 32400-year wave.

¼ is fractal difference between double/inverse waves, where ¼ within is 1/5 without (and 1/5 within is 1/6 without etc). Wave, which is singular fractal difference, is time and double fractal difference is gravity of inertial space, which is constant passage of time here and now. 5184x5/4=6480, which is resonance in 25920-year periods, while 6480x5/4=8100-year wave, which is resonance in 5x6480=4x8100=32400-year periods (1D fractal difference). 8100-year wave is double 5/4 reality, which is resonant with inertial 5184-year period in 129600-year periods ($(4/5)^2=16/25$; 16x8100=25x5184) It is 2D fractal difference, which is gravitational wave of inertial 5184-year time wave. 8100x5/4 =10125-year wave, which is resonant with inertial 5184 year wave in 648000-year periods (triple fractal difference, which is $(4/5)^3=64/125$ fractal reality or 64x10125 =125x5184). 10125x5/4=12656.25-year wave (i.e. only 2D fractal difference is real in 3D context, while larger difference are not multiples, but decimals or fractals, which are not real in its context). It is $(4/5)^4$ or 3240000-year wave and 12656.25x5/4 =15820.312, which is $(4/5)^5$ or 1024/3125 fractal reality of 16.2 mill-year singular wave etc.

Larger than 5184 years are precessional time waves, which are instant contraction of empty space of universe without into passage of time here and now (within), while smaller than 5184-year waves are inverse precessional waves, which are instant expansion of dense mass within into constant passage of time without. 5184x4/5 is 4147.2-year wave (1D difference), which is not real as passage of time, because it is fractal and not multiple (decimal division). It is instant density of mass, which is passage of time only in 20736-year periods (4x5184=5x4147.2). 4147.2x4/5=3317.76-year wave (2D difference) is $(4/5)^2$ or fractal 16/25, which is resonant with inertial passage of time in 82944-year periods (4/5 of 103680). 3317.76x4/5=2654.208-year wave is fractal $(4/5)^3$=64/125, which is 331776-year wave (4/5 of 414720-year wave). Such time wave, which is triple fractal difference, is instant density of mass in periods less than 331776 years, when accumulated discrepancy is equal to inertion of constant passage of time, which is instant inversion of dense mass within into passage of time without (expansion). 20736-year period is inversion of dense mass within into fractal passage of time on its surface, while 32400-year period is equal and opposite inversion of empty space without into passage of time within (32400 is 3/2 of 20736, i.e. 1.5x20736=31104, which is 0.96x32400).

25920-year change is fractal space/time inversion in between. 32400-year wave is reaction (time) of 20736-year wave (space) in context of inertial 25920-year wave in between. 20736-year change precedes 25920-year change, which precedes 32400-year change. Likewise—82944-year change, which is inversion of mass within into time without (action), precedes inertial 103680-year wave, which precedes 129600-year wave, which is inversion of empty space without into passage of time within (reaction). 331776-year change (inversion of mass into time, which is expansion) precedes inertial 414720-year change, which precedes 518400-year change (inversion of empty space into time, which is contraction). It is infinite double/inverse fractal curvature, which is instant infinity without/within of singular here and now in between. Here and now is infinity of fractal time waves within, which are its instant mass and inverse infinity of fractal spatial waves, which are its instant empty space (universe) without. Their resonance is constant passage of time in between, which is double/ inverse (relative/uncertain) boundary of both. Mass is time or curvature of empty space of universe, just as empty space is inverse time or boundary of its mass, because only double negation is affirmation in between. Such double/inverse curvatures are fractal reality of quantum steps of gradual reality in between.

Each 4x1/4 wave is ¼ of larger/smaller wave, which is infinite number of double/inverse waves without and within, which are preceding each other at 90°. Such even (symmetrical) precession is singular strand (infinite 1.111 repetition), while in context of both of its equal opposites, it is accelerated/decelerated precession, which is double helix (two strands, which are 1.111^2=1.23456...). Double helix is inversion of inertial symmetry, which is complete asymmetry, where asymmetry in space is symmetry in time and equally vice versa. Each wave is singular point of further double/inverse infinity, which is infinite fractal differentiation of waves. All waves are equally real, but not here and now. Waves that are real now are not real here (empty universe without) and waves that are real here and not real now (dense mass within). Such fractal waves, which are not here and now are instant double/inverse density of empty space without and dense mass within. They are not passage of time (waves), but double/inverse (relative/uncertain) spatial sequences of inertial spatial sequence in between (i.e. Earth). Each inertial period of time (here and now) is synchronicity of one pair of space/time waves, which is gradual change in time, where waves without gradually invert within and at the same time, waves within gradually invert without (anti-parallel space/time motion). Each period of time is one time wave, while all infinity of different fractal waves are instant space within (mass) and without (universe), which are passage of time only in different quantum periods of passage of time.

Each wave is 4-fold curvature, infinitely many times, which is infinite fractal differentiation in context of double/inverse (2D) fractal difference. It is infinity of different frequencies and different phases, which are resonant in relatively different quantum periods of time. Because each wave is double/inverse infinity of different waves, there is no quantum periods of time or two quantum changes that are same. Each quantum period of time and each quantum change is eigenvalue of relatively different fractal waves, which is infinite irrational curvature (Π/φ). Just as each day is different from all infinity of other days, so is every quantum period of time and every quantum change resonance of different infinity of waves without and within, which is relatively different sequence of inertial reality here and now. One numerical sequence on surface is double/inverse infinity in its depth. Numerical abstraction is reduction of double helix to repetitive singular strand, while its depth, which is inverse repetition, is infinitely different content of each sequence. Relative certainty of numerical sequencing on surface is complete uncertainty of its content, because increased certainty of one strand is simultaneous increased uncertainty of its inverse strand and equally vice versa. Increased certainty of time in spatial context is decreased spatial certainty in context of time and equally vice versa.

CHAPTER 14

RESONANCE

Reality is resonance between two equal opposites. What does not resonate here and now is not real, because it is not eigenvalue, but empty space (vacuum) with no content. Nothingness of empty space is time that is not resonant with inertial space here and now and/or space that is not resonant with inertial passage of time. It is infinity of different space/time possibilities that are not real here and now. Here and now is temporary spatial reality in time (and/or temporary time of space), because in context of all the time, all spatial sequences are real and equally vice versa. Here and now is singular time sequence that is resonant with infinity of spatial sequences and/or singular spatial sequence that resonates with all different passages of time, which are its exponentially different spatial reality. Space/time is inertial infinity in context of inverse singularity—one is infinite differentiation, while another one is singular (changeless) passage of time. One is real only in boundaries of another, because only inversion of inertion is resonance (reality). Only inversion is perfect mirror image that is resonant with inertion. Infinity is gradual change and its inverse singularity is change in quantum steps, where each step is infinite gradual change in between. Change of changeless reality (i.e. time) is real only when accumulated change is singular quantum. Singular time sequence (infinity of different space) that is real here and now is not infinity of time (infinitely long passage), because space/time is relative/uncertain (fractal) duality and infinity is only relative description of inverse singularity.

Time is not infinity but singularity of space. It does not perpetuate present spatial sequence—it limits it. Spatial reality here and now is real only as long as it resonates with its inverse time. Out of tune is an instant change of space/time sequence, which is relatively different space in context of inversely different passage of time. Instant change of fractal time is instant change of relative space/time or gravitational sequence. It is instant change of density, which inversely proportionally changes densities of empty space without and its dense mass within. Physical reality is double/inverse or space/time duality, which is singular gravitational sequence. Each gravitational sequence is resonance between equal opposites, where one is real only in context of its own negation. Only resonant contradiction between something and nothing is reality of singular gravitational sequence. Something that is real here and now is reaction of equal and opposite nothing, which is its reaction. Only nothing that is resonant with something is real nothing, while relatively different nothingness is real only in context of future/past time sequences, because only equal opposite of something is nothing. Difference between something/nothing is only relative, which is certain period of time. Certainty is boundary of relativity and constancy is boundary of uncertainty.

Time is boundary of space. Inertial space/time sequence of rotating Earth is its rotational velocity, which is 463 m/sec. It is fractal duality of its space and time, where one limits another. Reaction of inertial rotational sequence is orbital sequence, which is its inverse space/time sequence. It is sequence of inverse (orbital or solar) space in context of its time, which is orbital velocity of 30 km/sec. Rotational/orbital space/time is singular 4x1/4 wave, where one sequence is space, while another sequence is its time (boundary). Rotation and orbital motion are negation of each other, which is singular gravitational sequence of reality here and now. One is real only in context of another, which is **30/0.463=64.8 or 6480/10^2** (double fractal difference, which is different space and different time; therefore different gravity or different reality). 6480-fold repetition is boundary of such inertial space/time sequence. It is not reality beyond its limits. Boundary of present rotational/orbital sequence and/or size of Earth and size of its orbital is 6480 years. It is time of resonance of one within another, which is gradual accumulation of fractal difference in between, which is gradual shift out of tune, which is not real, unless singular quantum (gradual change of time in its depth, which is not real on surface).

6480x10=64800 (1D fractal difference) and 64800/3=21600, which is 4x1/4 curvature (4D reality) of inertial period of time (6480). 6480x10^2=648000 is its inversion, which is 7.5x86400 sec or time of one week, which is 6 days of 86400 sec (6x86400 =518400 or 5184x10^2) and its inertial 7th day, which is 3/2 time (1.5x86400 =129600). 7th day is boundary or inversion of inertial 6 days and equally vice versa. 6 days (5184x10^2) are duration (passage of time), while 7th day (1296x10^2) is its inversion or boundary (fractal difference in between). 6480/C (light speed) = 0.0216, which is 2.16/10^2 or inversion of 0.463 km/sec rotational velocity. In other words, Cx2.16 and/or **C/0.463 km/sec=648 000 km**, which is ¼ day of orbital motion or singular quantum of orbital space/time that is its singular boundary (limitation). 6480 years is time boundary of inertial space, just as 648000 km is spatial boundary of inertial passage of time. Likewise—**C/30 km/sec=10000 km** or ¼ size of Earth, which is its singular quantum. Rotational time limits spatial orbital, just as orbital time limits spatial size of rotation (Earth). 6480 years is time of rotation that is resonant with orbital space and time of orbital motion that is resonant with size here and now (Earth).

6480 years is ¼ of precessional motion, which is 1.54321 km/year velocity and **1.54321xC=462963 or 0.463x10^6** where 0.463 km/sec is rotational velocity. It is 10^6 km/year (25x40000 km) difference between precessional and rotational motion, just as there is 10^4 difference between orbital motion and light speed. 1.54321 km/year/0.463 km/sec=**3.333 sec/year relative discrepancy** between precessional and rotational time, which is 21600 sec or ¼ day accumulated time difference (singular quantum) in context of 6480 years (86400/1296=66.666=20-fold repetition of 3.333 in context of orbital motion, which is inversion of 0.015=3/2x10^{-2}). 6480x10^6 km discrepancy is 43.2 AU or total distance between center and boundary of solar system (Pluto's orbital 38.4AU=8/9x43.2). Accumulated space/time difference in 6480 years is all the space of solar reality, which is ¼ time discrepancy here and now (inversion of infinite space into singular time quantum or singular quantum of relative spatial reality).

6480 years is 90° space/time curvature between space and time, which is boundary of inertial resonant reality here and now (light speed difference). Rotation in context of light speed is C/0.463=2.16xC=6480x10^2 and precessional motion in context of light speed is 1.54321xC=0.463x10^6. **Light speed C is 6480 years x 0.463 km/sec x 10^2 and/or 0.463 km/sec/1.54321 km/year x10^6. Light speed is resonance between inertion and inversion (space and time).** 1.54321x0.463=0.72 and 2.16x3.333 =7.2, while 3.33/2.16= 1.54321, which is inversion of 0.648 or 6480/10^4. ¼ precessional period is light speed difference between rotation and orbital motion and/or space and time (equally vice versa—time is difference between light speed and space, and space is difference between light speed and time). Singular difference here and now is light speed difference in context of 6480 years of passage of time, which is boundary of such inertial space/time sequence of reality.

C/40000km (size of Earth) = 7.5 Hz of Schumann resonance (reaction of 7.5-days and/or its 4x1/4 multiplication/division, which is time of 1 month). 7.5 is inertial period of non-inertial 7.5x0.96=7.2, just as actual Schumann frequency is 7.8, which is 7.5x1.04 (equal opposite of 0.96, i.e. 7.8/7.5=1.04; 7.5/7.2=1.041666; 7.5/7.8=0.9615384 and 7.2/7.5=0.96, while 7.8/7.2=1.0833 or 1+5/6x10). 7.8 is double inversion (2D fractal difference of inertial 7.2), which is time or boundary of inertial space/time (rotating Earth). It is equal and opposite rotation of empty space without, which is reaction of rotation of dense mass within, which is 7.5x 86400 sec=648000 sec per sec or 2.16xC. Rotation (space) is real only in context of its negation (time), which is inverse rotation without. 7.5+/-0.3=7.2 and/or 7.8 is 0.6 or 6-fold difference (6D) in between. 7.8/0.6=13, 7.5/0.6=12.5 and 7.2/0.6=12 is fractal 13/12 (which is ½ of fractal 24/25) or 12x13=156-fold repetition ($3/2x10^2$ +6). Schumann resonance is eigenvalue of space/time duality here and now, which is equal opposite of additional 5 ¼ days of perfect 360-days year (72 years of one orbital of empty space as reaction of 72 orbitals of its dense mass within, which is singular ¼ quantum of 18 years). 1/7.2=0.13888 light sec per one inverse rotation of empty space without, while its 4-fold curvature (4x1/4) is 0.555 or 5x0.111 and 1/0.555=18/10 or 18 years of gradual orbital motion, which is instant 18 days in double/inverse context of rotation and/or 18° in spatial context of Earth (precessional quantum shift). 18 is saros, which is 18-fold curvature (18x360=6480) that is boundary of inertial space/time sequence of rotating/orbiting Earth. Saros is 65x324/325, where 65 is ½ of 130 or 13x10.

7.2xC=$21600x10^2$ km or 21600 years (100km=1 year) is limitation (time) of such resonant space/time reality. 0.6 discrepancy is time of time, which are 6 periods of double/inverse precession—6x21600x0.96=129600 years. 21600 (20736) years is 1st period of high eccentricity of Milankovich cycle, which is inertion (momentum) of its following 4x1/4 cycles. 1st period is cause of $2^{nd}/3^{rd}$ periods, which are 2x1/2 action, and $3^{rd}/4^{th}$ periods, which are its 2x1/2 reaction. 6th period, which is another high eccentricity period, is inversion of inertion (1st period) that is at the same time inertion of succeeding 6-fold cycle. 1st period is angular ½ of its $2^{nd}/3^{rd}$ and $4^{th}/5^{th}$ period, which are horizontal 2x1/2, while 6th period is its equal opposite. $1^{st}/6^{th}$ period are inertial 2x1/2 without of its inverse 4x1/4 within (1D fractal difference). Schumann resonance is instant spatial boundary of time, just as time is instant spatial boundary. Quantum change of reality is ¼ instant change of Schumann resonance (7.2+1.8=9 or

7.2+1.44=8.64), which is light speed change. Such change is relative/uncertain change of size/motion, which is relatively different resonant space/time (Schumann) frequency. Schumann frequency is double space/time resonance (4x1/4), which is singular gravitational sequence. It is temporary equality between space and time, which is constant light speed without of constant gravity within or constant density of mass/energy. Its change is instant double/inverse change of massive density within and density of empty space without (instant expansion/contraction). Change of density is instant change of dual (physical) reality, which is closing of one Fibonacci sequence in evolutionary chain of double helix (when gradual and reversible change in context of passage of time is irreversibly imprinted in its genetic code, which is inversion between time and space)

Space/time resonance is constant motion (light speed) without (empty space) and/or constant size within (dense mass), which is its gravity. Constant motion without and constant size within are singular fractal difference in between, which is passage of time without and instant or no time within. Light speed is double/inverse gravity (gravity of empty space without, which is energy within), just as gravity is double/inverse light speed (light speed of mass within, which is time without). Light speed without is reaction of gravity within, just as gravity without is reaction of light speed within. Change of one is equal and opposite change of another. Gravity is instant time within (dense mass) and light speed is instant space without (dense energy). Gravity is double fractal difference, which is one-dimensional freedom of change, while time is singular fractal difference, which is two-dimensional freedom and space is triple freedom of change. Only double spatial change is singular time change and only double time change is singular gravitational change.

Space/time is singular and space/gravity is double light speed difference, because space/gravity are double/inverse spatial sequences in context of fractal ½ time in between. Micro-world is inversion of reality without, which is positive/negative triple freedom or 6D in both directions. It is inertial 3D here and now, in its inverse 3D framework, suspended in nothingness, which is reality beyond its boundaries. Nothingness is double/inverse something, which is negation of reality here and now. Only inverse negation is total freedom of choice of inertial here and now, because only negation of inertion is totally inertionless reality. Inertion in space is inertionless in time and equally vice versa; therefore inertionless here and now is infinitely larger inertion, which changes only in infinitely larger quantum steps, because infinite number of different resonant space/time harmonics here and now are only singular quantum of such reality (space of inversion is space and time of inertion and equally vice versa). Infinite number of spatial waves resonates in context of singular time wave and infinite number of time waves resonates in context of singular gravitational wave, because only inertial infinity is eigenvalue of inverse singularity. Infinite number of different space/time realities, of which here and now is only one possibility, are simultaneous reality of inertial gravitational wave. Infinite numbers of physical realities, all equally real (only not here and now), are simultaneous reality of their inertial gravitational wave, because they are non-reality of each other. All relatively different passages of time are instant density of their gravitational wave; just as such gravitational wave is instant emptiness (nothingness) in context of their passages of time.

Infinity of different time waves (different passages of time) is successive 2x1/2 multiplication/division of inertial space here and now. Each successive wave is fractal (light speed) difference from previous wave, which is exponentially increased relativity or negation of inertial reality here and now. Infinity of time waves is infinite fractal relativity/uncertainty or negation of spatial reality here and now (just as infinity of spatial waves are negation of inertial time). Time waves are gradual, just as spatial waves are instant change of space. Spatial waves in context of space are instant, because they are double fractal difference, without fractal passage of time in between, just as time waves are instant in context of time, without fractal space in between. 1 day of quantum change is not time, but spatial wave—it is interruption of such time wave. Likewise—smaller time waves are interruption of spatial reality here and now. Such interruptions are equal and opposite ½ of ½ that is real here and now—they are surface of 3D reality of double helix, which is a "heartbeat" wave pattern, where spatial waves interrupt time waves and time waves interrupt spatial waves. Time is spatial non-reality and equally vice versa—space is nothingness of time.

Consequently—exclusive spatial or time waves are only ½ waves interrupted by ½ periods of nothingness (heartbeat), because its equal opposite is only its in-depth reality, which is not real on surface. Double helix is fractal, which is always only ½ of its reality in context of any inertial sequence. ½ of double helix is space of time and/or time of space, which is negation or inversion of any surface reality. Reality is interaction between space and time; therefore spatial reality is infinity of spatial waves in context of one time wave, just as time is one spatial wave, which consists of many time waves. Their interaction creates infinity of interruptions—from infinitely small to infinitely large. Infinitely small interruption is gradual change in infinitely small and fast fractal steps, while infinitely large interruption is not interruption or change at all—it is constant changeless reality, because such interruption is infinitely long period of time or infinitely large spatial size.

Infinitely large period of time of inertion is infinitely small size of inversion. **Space is size without and period of time within and time is period within and size without**. Space is infinitely long interruption of time, just as time is infinitely long interruption of space. Universe is delayed time reaction in space and micro-world is delayed spatial reaction in time. Universe is spatial distance between an action and its equal and opposite reaction in time. Such spatial distance is no distance at all in context of time—it is its instant action and reaction. There is entire spatial universe between action and reaction in time, just as there is infinite time between spatial action and reaction. "Invisible connection" between applied action and its distant spatial reaction in context of micro-world is same principle as invisible connection between two heartbeats or between two gravitational quantum changes. It is interrupted or delayed reaction. Interrupted time creates space in between and interrupted space creates time in between. Distance in space is no distance in time and equally vice versa. Spatial interruption is instant passage of time and time interruption is instant space. No spatial connection between action and reaction is simultaneous time, just as no time connection between action and reaction is space. Reality in between any action and reaction is its inertion, which is its double/inverse negation (singular space of infinite time and/or singular time of infinite space). Inertion is framework of both—it is singular cause of all infinity of resonant vibrations. It is only reality, which is repetitively fragmented and multiplied by such echo.

Inertion of space/time is their singular gravity. Infinite numbers of different space/time sequences are one gravitational reality in different relative forms. Gravity is instant double/inverse light (energy), which is mass here and now. It is linear (flat) light without and curved (rotational) light within. Light is also double gravity—spatial (gradually expanding) without and massive (instantly contracted) gravity within (double/inverse curvatures). It is $C/(1/C)=C^2=g$ and/or $g/(1/g)=g^2=C$, which is double/inverse reality or eigenvalue of two equal opposites—size (C^2 or g) and motion (C or g^2). Gravity of Earth is $g=C/t$, where $t=1/C$ (orbital time of 1 year)-i.e. time is instant fractal (space) of light speed and/or light speed is instant fractal (space) of time in context of inertial gravity. In context of inertial passage of time, gravity is inversion of constant light and equally vice versa—light is inversion of gravity. Time of orbital motion is singular quantum of gravity, just as time of rotation is singular quantum of orbital time, which is its gravity. Time of orbital motion, $t=C/g=31\ 104\ 000$ sec; $t/C=103.68$ ($103.68 \times 24/25 \times 325/324 \times 625/624=10^2$, i.e. 103.68 is inversion of 0.96 (gravity). In context of constant light speed, gravity and time are equal opposites and in context of constant time—gravity and light are equal opposites, while in context of constant gravity—time and light are equal opposites. Orbital time 31104/6=5184 and 5x5184=25920, which is inertial 5184 of fractal 5/6 (and/or 31104/24=1296, where 31104/32400 are fractal 24/25 or gravity without of space/time within). 31184-25920=5184 years and 32 400-25 920=6 480. Gravity, time and light speed are triangle, and change of any one is equal and opposite change of its two equal opposites (while gravity, light speed space and time are 4x1/4). Light (energy) is equal and opposite reaction of gravity. It is inverse gravity or antigravity (expansion) of empty space, because time is instant gravity of mass and gradual antigravity of space.

Gravity is curvature of light and light is inverse curvature of gravity. Gravity (mass) is boundary of light (energy), just as light is boundary of gravity (mass). Uninterrupted gravity here and now (dense Earth) is exponentially increased periods of interruptions in context of its inverse empty space, while uninterrupted light speed of empty space without is complete interruption in context of dense mass within. Solar and galactic empty space is gravitational resonance only in exponentially larger time periods and/or larger spatial distances, while expanding universe is not spatial resonance at all, but exclusive time resonance. Only all the time of the universe is spatial eigenvalue here and now (total expansion/inversion). Empty space is inversion of dense mass (here and now), and its stars and galaxies are inversion of such empty space. Stars are double/inverse density, which is infinitely larger density, because it is simultaneous density of mass and empty space (mass and energy), while core of galaxy is twice larger density than solar density, because it is density of space and time. Stars are double and core of galaxy is quadruple density (gravity), which is twice and 4-fold denser center of twice and 4-fold emptier (and larger) spatial shell without. It is 2x1/2 and 4x1/4 larger gravity within and larger light speed without. Increased spatial size is proportional with decreased time and equally vice versa—decreased spatial size is inversely proportional with decrease of its time.

Stars are only infinitely small time periods of spatial galaxy. They are compressed even density of galaxy, just as galaxies are compressed even density of universe, which is exclusive reality of inertial passage of time here and now and totally meaningless reality in context of solar reality or galaxy (likewise—massive planets are compressed even density of solar reality, which is light speed difference between their passages of time). Relative reality without is uncertain reality within and equally vice versa. Only solar reality, which is everywhere at the same time in its orbital, and galaxy, which is everywhere at the same time, is instant spatial reality. Empty space, which is real here and now, is not resonant with inertial solar and/or galactic reality. It is its instant dense mass in context of its passage of time, while passage of time here and now is not resonant with such infinitely larger spatial realities. Stars are space or density compacted in time—they are massive interruptions of uninterrupted reality, which is real exclusively here and now. It is same uninterrupted space as here and now, infinitely fragmented in time.

Total spatial reality of galaxy is space **and** time of its stars—it is all the stars and all their motions at the same time. Infinitely long period of one orbital of a star is not time in context of inertial (spatial) galaxy, just as planetary orbital is not time, but instant space in context of solar reality. Spatial reality of universe in uncertain perspective from within is infinite distortion of reality through inertial passage of time, which is exclusive reality here and now, without any meaning in context of spatial reality that is light speed difference. Earth is not resonant with solar/galactic spatial reality in context of its own inertial passage of time. It is synchronized with solar reality only during its 5184-year quantum change and with galactic reality only in course of its 103680-year change. 5184-year quantum change is instant resonance with solar spatial reality, when it is not empty space of orbiting planets but even density. It is resonance between double/inverse spatial realities or resonance of two different passages of time. Empty space of solar reality is galaxy, just as empty space of galaxy is universe. Time (duration) of quantum change is massive solar reality, when Sun is not energy (light), but mass (gravity) and galaxy is not gravity (black hole) but light (energy). Double quantum change is resonance between passage of time here and now and passage of time of galactic reality, when galaxy is not empty space, but dense mass.

Passage of time here and now is exclusive Earth's reality, which is meaningless in solar context. Solar reality is inertial framework of Earth's reality. It is its infinitely larger reality, which is beginning and end (curvature) of passage of time here and now. Such passage of time is meaningless beyond spatial boundaries of solar reality. Solar reality is reality of reality here and now, because it is simultaneously its affirmation and its negation. Space and time here and now are only space of solar reality. Passage of time here and now is only temporary reality, totally unreal in solar context. Only quantum changes of reality here and now is solar reality, where such quantum changes are only its constant passage of time. Interruptions (quantum changes) here and now are uninterrupted passage of time of solar space. Solar reality is infinitely larger freedom between what is real and what is not, because it is 4D of inertial 3D here and now (or 5D of 4D). It is light speed difference, where gravity here and now is not changeless singular constant but linear passage of time and therefore time is not linear passage but surface of spatial reality. Gravity, which absolutely determines size of spatial reality ad its passage of time, is eigenvalue of space/time duality. Gravity **is** physical reality, because it is both equal opposites of duality at the same time. Gravity is negation and affirmation of space and time. It is form and content of physical reality.

Gravity is reduction of 3D spatial choice to only one choice, which is not choice at all. It is all infinite physical reality. In context of solar reality, singular gravity (here and now) is expanded into linear passage of time, which is not changeless constant but constant change. Gravity of solar reality is double/inverse gravity, which is antigravity (expansion) here and now. Gravity of solar reality is repulsion, while attraction here and now is its passage of time (constant change). Solar reality, which is 1D fractal difference, is additional freedom of choice, which is not possibility here and now (in context of massive Earth). Infinite physical reality (space/time duality) here and now is infinitely expanded in solar context, because gravity in its context is not constant but constant change. It is not its singular boundary (limitation), but its infinite differentiation. Physical reality here and now is only flattened surface of such larger (solar) reality. Solar reality is relative space, which is spatial (instant) reality of time, where time is not passage but instant density (mass) of empty space. Solar reality is time, which is beginning and end (curvature or boundary) of inertial reality here and now. Time is inertial framework of spatial reality, because it is infinitely expanded (larger/smaller) spatial reality. In context of time, time here and now is absolutely unreal (more than uncertain).

In its turn, passage of time of solar reality (gravity here and now) is completely unreal in galactic context, because such passage of time (gravity here and now) is its instant spatial density (mass). Gravity is passage of time in solar context (all different gravitational sequences between singular contraction and infinite expansion are its inertial period of time) and instant space in context of galaxy (all infinite expansion/contraction, which is period of time in context of solar reality is singular spatial quantum of galaxy). Galaxy is double fractal difference from inertial here and now, which is twice-infinitely larger freedom choice between what is real and what is unreal. Gravity is 3D space of galaxy, which is triple freedom of choice, opposed to singular choice here and now. Physical reality here and now expanded in additional 2 directions is physical reality of galaxy, while physical reality here and now (all infinite space and time) is only one of its linear directions. Gravity here and now is singular freedom of choice; in context of solar reality it is double freedom (negation and affirmation), and in context of galaxy it is triple freedom (4x1/4 negation/affirmation). Solar reality is resonance between one reality (here and now) and its negation, which is infinity of different possibilities that are not real here and now.

Solar reality is infinity of different possibilities, of which only one is real here and now. It is eigenvalue between something and nothing (positive and negative), where simultaneous something and nothing is its double something. Galaxy is further 2x1/2 expansion of space/time (physical) reality, which is its 4x1/4 multiplication/division. It is resonance between double something or simultaneous something and nothing (here and now) and its negation, which is double nothing. Such double something and double nothing is quadruple something (where quadruple within is triple without and/or vice versa, because double duality is fractal ¾). In context of solar reality, inertial (singular) reality here and now (Earth) and infinity of its different negations are equally real. In context of galaxy, one reality here and now and double infinity of its different negations are instantly real. Solar reality is negation of time and galaxy is negation of space and time here and now. Universe is 3D negation, where inertial here and now is negated instantly in all 3 directions. It is negation of space, time and gravity that are real here and now. In context of universe singular gravity is 3D antigravity (expansion), where each possibility here and now is 8 different possibilities at the same time (solar reality is 2 and galaxy is 4 different possibilities of inertial here and now).

Time, as negation of space is infinitely larger physical reality and gravity as negation of time is yet larger reality. Physical reality is contradiction—it is duality between affirmation and simultaneous, equal and opposite negation. If affirmation is singular possibility, then negation is infinite number of relatively different possibilities and equally vice versa. Infinity is limited by singular inertion—i.e. only whole numbers or quanta of inertial here and now are real possibilities of negation, while the rest is empty space, which is a different possibility only in context of relatively different inertial here and now (different density or different size/motion, which are successive gravitational sequences in context of different quantum periods). Negation of inertion here and now is always its infinitely larger reality. Physical reality is resonance between affirmation and negation, which are two perspectives (curvatures) of one reality. Resonance between space and time (size/motion) is their singular gravity, which is their temporary overlap or infinitely larger space/time reality.

Gravity is singular cause of spatial reality (size), as well as its constant change (time, which is its boundary). It is simultaneous affirmation (i.e. space) and negation (i.e. time), which is double/inverse comparative reality because it is comparative with space and time. Space/time is duality, while gravity is 3D. Space and time are only double/inverse surfaces of inertial gravity, which is singular physical reality here and now. In context of solar (3D) reality, physical reality is not bound by inertial gravity of present space/time sequence, which is only its flat cross-section (2D). In context of galaxy, physical reality is twice free of such inertial gravity, which is only one of its linear directions and in context of universe, one reality here and now determined by gravity, is total (3D) freedom. Total freedom is total relativity of such universe without, which is total (3D) uncertainty of inertial reality here and now within. Reality here and now is only one, out of three-times infinite number of singular points, which comprise universe. Solar reality is resonance of all different spatial possibilities, of which here and now is only one (quantum period of time); galaxy is resonance of all space/time possibilities (infinity of relatively different passages of time and different spatial sizes) and universe is resonance of all gravitational possibilities (infinity of different E/m densities).

Each larger sphere of reality is singular wave, consisting of infinity of smaller waves, where two resonant waves combine into double/inverse (larger) wave. Resonance between inertial/inverse waves that creates double wave is passage of time here and now, while resonance between such larger (double) inertial and inverse waves is instant change of passage of time, which is instant equal and opposite change of spatial density (size). Passage of time is infinity of spatial waves, which gradually, through infinity of resonant changes (gradual spatial change), accumulate to singular quantum of its inertial (time) wave. When spatial inertion is equal to time inertion, it is resonance between double/inverse (space/time) waves, when one is not singularity of infinity of another, but two equals that combine into double/inverse wave, which is instant change of space/time sequence (different passage of time of inversely different spatial reality). Gradual spatial change (passage of time) is exclusive reality of singular space/time sequence (5184 years). Gradual time change (infinity of quantum changes) is gradual progression through depth of solar reality. Quantum time change is only fractal (½) between double/inverse spatial changes, which are changes of size/density (gravitational quantum change, which is double/inverse precessional period). Infinity of quantum time changes (20) is gradual overlap between two successive spatial sequences (relatively different size/density). Gradual change of gravity (density) is progression through depth of galactic reality, while gradual inverse change (anti-gravity) is progression (growth) through depth of universe (which is time or negation of spatial reality here and now).

Two resonant (double/inverse) waves create singular inertion of their interaction, where such inertion is a new wave. Reaction of singular "nothingness" is infinity of different multiplication/divisions or infinity of relatively different space/time vibrations—one within the other. Each wave is nothingness or inertion of infinity of new set (packets) of waves. Larger wave is slower time and smaller wave is faster time. Small wave repeats itself infinitely many times within one period of large wave and equally vice versa—in context of passage of time of smaller wave, such larger wave is infinitely long period of time. Different waves are different space/time possibilities that are equally real. Each wave is one of the same possibilities—i.e. each wave is equally real. Such different space/time waves are not eigenvalue, because each wave is same wave—i.e. it is same inertial "nothingness". Only relative ratios of two equal (fractal) opposites of such inertion differ, while inertion itself is always the same. Inertion is common denominator of all space/time frequencies—it is common multiple of such fragments and/or common fragments of such multiples. All space/time waves are same inertial gravity. Such inertial gravity is only time of its space and/or space of its time. In other words—such gravity is only fractal ½ of yet larger 3D reality, which is same infinity of different waves. Physical reality is fractal ½, because it is comparative reality of space and time, where time is gradual comparison between different spatial possibilities and space is instant comparison between different time possibilities. **Space (within) is instant past time, just as time (without) is gradual future space.**

Space/time waves are electrons in its gravitational shells and gravitational waves are infinitely larger/smaller space/time waves (fractal difference), which is infinity of waves within a wave. Infinity of different space/time possibilities of one gravitational reality is infinite number of subordinates of singular principal (Balmer). Infinite number of such principals are only subordinates of yet larger principal, which is infinite number of space/time shells (quanta) within a singular gravitational shell (inverse quantum) and infinite number of gravitational shells within singular spatial and/or time shell. It is infinity of spatial shells (quanta) within singular time shell (quantum) and subsequent infinity of time shells within singular double/inverse spatial shell, which is gravity here and now. Space and time are singular fractal difference (90° difference of 1 out of 3 directions), while gravity is double/inverse fractal difference (curvature of 2 out of 3 directions). Universe is one spatial shell of infinite number of time shells, which are different time possibilities of same spatial reality. All the mass of the universe (here and now) is only one time shell of inertial space, while infinity of different time shells (equally real but not here and now) is its empty space ("dark" mass of the universe, which is real only in context of gravity, but not space and time, because gravity transcends relative boundaries of space and time—i.e. it is future/past mass and energy).

Different time shells are different E/m density, which is exclusive gravitational reality (massless and motionless in context of inertial space/time shell here and now). Different time shells are different spheres of reality, which are real here but not now. E/m, which is real here (space), but not now ("dark" mass) is infinitely larger reality than here and now, because it is its infinite number of negations in time (fractal $9/8=3^2/2^3$, which is 9-fold without and/or 8-fold difference within). Each time sphere (singular quantum of 5184 years) is different possibility of same space (and/or different spatial possibility of the same time, which is spatial shell). Infinity of different passages of times (time spheres) is only one instant gravitational shell, which is light speed difference (i.e. only total or light speed change, which is change between singular and infinite size of universe and/or its density, is singular quantum of gravitational reality that is singular change between gravitational shells). Different gravitational shells are real neither here nor now, because they are beyond space/time boundaries of inertial reality here and now (beyond singular/infinite boundaries of space/time, which is reality beyond event horizon of black hole or Planck length, as well as beyond horizon of light speed expansion).

Different time shells are gradual reduction of certainty within, which is increased relativity without. Only infinite gradual change (increased uncertainty/relativity) is singular quantum of gravitational reality. Different gravitational shells are multiple uncertainty/relativity without of fractal (gradual) increase/decrease within. Double/inverse gravitational shells of inertial shell in between (here and now) are more than uncertain (negatively uncertain) reality within (beyond singular Planck length) and/or more than relative reality without (larger than light speed expanding universe). They are space/time negations of inertial here and now. Gradual decreases of inertial reality here and now (increased relativity/uncertainty) is space/time reality (passage of time), while its quantum changes, which are its interruptions, are instant multiple changes of such reality. Quantum changes of space/time, which are gravitational changes, are instant changes, because they transcend passage of time. Relative/uncertain reality is gradual reduction (negation) of reality in time and space, while different gravitational reality is complete (multiple) negation of inertial here and now. Just as any reality smaller than singular quantum (Planck length) is complete randomness and totally unreal, so is space/time (physical reality) here and now, which is singular quantum of double/inverse (relative) spatial reality (universe).

Space is time waves and time is spatial waves of singular (inertial) gravity. Mass of universe is infinity of spatial frequencies of same time, while mass of micro-world is infinity of time frequencies of same space. Spatial periods of an atom are time periods of the universe, and time periods of an atom are spatial periods of the universe. There is no future or past time, because future time of empty space of universe is past time of its dense mass and equally vice versa. It is instant gravitational reality. Only relative/uncertain difference between such past/future times is fractal ½ reality. Total reality, which is 2x1/2 (two equal opposites), is no difference. Future time is infinitely distant past time, just as past time is infinitely distant future time. Space/time is fractal ½ reality, which is 2x1/2 or duality between past and future times. All the future time there is, is time of one day, which is also time of one year and/or all the time of universe in context of its equal opposite, which is all its past time. They are only different spatial shells of only one time and/or different time shells of one space. Only time there is, is time right here and now, in context of infinity of different fractal curvatures (boundaries). Inertial period of 1 day is passage of time between all the future time expanded into instant empty space of solar reality without and all past time contracted into instant dense mass (Earth) within. It is resonant reality on surface of Earth, which is line of symmetry between size within and motion without (double/inverse densities). Inertial period of one year is passage of time between instant empty space of galaxy without and dense solar mass within, which is resonance on surface of such orbital.

Resonant duality here and now (passage of time) is angular ½ of horizontal 2x1/2, where passage of time in context of dense mass within is inverse (negative) passage of time in context of empty space without and equally vice versa. Rotation and orbital are positive/negative passages of time, which is inertial (singular) passage of time in between. Passage of time here and now is resonance between such equal and opposite smaller passages of time. It is time of precessional motion, which is gravity of rotation and orbital, which are its space and time (double/inverse passages of time). Orbital and rotation are double/inverse reactions of inertial spatial reality of Earth (dense mass within/empty space without). Precession is equal and opposite negation of both, because it is inversion without of spatial size within, while rotation/orbital is fractal reality in between (passage of time between two fractal spatial sequences). Spatial reality (Earth) within and precession without are two equal opposites and change of one is equal and opposite change of another. Rotation and orbital motion do not change inertial Earth, because they are gradual (fractal) increase of tension, which is infinite passage of time. Precession is boundary or limitation of such infinity. Infinite passage of time is real exclusively in context of rotation/orbital and not in context of precession, which is its negation. Period of 5184 years is 90° change (curvature) of such reality, which is instant inversion of past into future and future into past time. 5184 years is boundary of future time, which is instant past time (E/m density) of yet larger future time. Future/past is only relative determination of arrow of time, which is meaningless in context of solar reality, just as different spatial sizes and densities are meaningless in context of galaxy.

Time in context of spatial inertion is space in context of inertial time and equally vice versa, which is fractal passage of time between two spatial spheres and/or fractal spatial reality between two times. Space/time is relative/uncertain fractal 3/2 reality, which is $3x2^n$ and/or $2x3^n$ Bode's periodicity. Bode's periodicity are time frequencies of spatial Earth and/or spatial frequencies of Earth's time, which are double/inverse (space/time) resonant harmonics of inertial here and now. It is double/inverse curvature, which is ½ within and/or 1/3 without and which is resonant only in multiples of 6 (common denominator). In context of passage of time of solar reality, which is galactic space, it is $\mathbf{6^2/5=7.2}$, which is 4x1.8 and/or ¼ of 28.8, while in inverse context, which is gravity, it is $5^2/6=4.1666$ and/or $\mathbf{6/5^2=0.24}$, which is ¼ of 0.96 and/or 4x0.06 (7.2/0.24=30-fold repetition or ½ of 60 and 7.2/0.06=0.432, which is ½ time of rotation in context of 5D fractal difference etc). 7.2 and 0.24 are perfect duality without/within, which are 1D fractal difference in between (shift of decimal space). In context of solar space, which is passage of time here and now, it is $5^2/4=6.25$-fold repetition of fractal 0.16, which is 2/3 of 0.24 or repetition of exponentially larger quantum periods of time (5184x6.25=32400, 25920x6.25=162000 etc). 6.25 is 3D sequence of dual 0.16 sequence. In context of galactic passage of time, which is expanding universe, resonant reality is $7^2/6=8.1666$, which is infinite repetition (echo in time) of limited 3D here and now and in context of passage of time of universe, it is $8^2/7=9.14285$...or total randomness of inertial here and now.

Universe/micro-world is increased (compounded) 2x1/2 multiplication/division of inertial gravity here and now, where multiplication in space is division in time and only reality of both (resonance) is real here and now. It is fractal 2x1/2 resonance between inertion and its inversion, which are instant gravity (size) of dense mass within and instant light speed of empty space (motion) without. Light speed is double/inverse gravity and gravity is double/inverse light speed. Universe/micro-world is space/time infinity of fractal resonant harmonics between gravity and light speed. Gravity is density of light speed in space and light speed is density of gravity in time and equally vice versa. They are double/

inverse or positive/negative non-Euclidean curvatures of inertial here and now in between (surface of Earth). One is infinite curvature of singular curvature of another (inertial radius of singular curvature within, which is infinity of curvatures without created by its different tangents). Tangent/radius is double fractal difference of singular $2r\Pi/r = 2\Pi$ and/or $1/2\Pi$ difference (inertion). Tangent of light speed (miles per sec, where mile is 0.625=6.25/10 km x $(12x14)/13^2$ discrepancy) is tgC = $\sqrt{5}$ = 2.236, where 2 is inertial space/time duality (abstract inertial framework) and 0.236 is its sidereal/synodic fractal difference, which is space/time here and now. Synodic/sidereal difference is constant angle of curvature of light speed. 2.236 =tg36° x tg108°, where $36=(2x3)^2$ and $108=3x36=(2^2x3^3)$, which is 1D fractal difference (108° without and 36° within is 72° in between. 108-90=18° or ½ of 36° difference, just as is 90-72, which are inner/outer angles of pentagon—72+108=180°. 6x18=108 and 4x18=72, which is fractal 6/4=3/2 in context of inertial 5x18=90 in between, which is fractal 24/25. Likewise—108+36=144, which is 2x72 and/or ½ of 288, where 72x288=20736 years or boundary of such fractal reality, etc).

Sum of all fractal space/time variations is always 0 (nothing), and their difference is always 1 (something); therefore their multiplication/division is infinite differentiation of singular gravity/light speed. Each fractal difference is light speed difference, which is infinite/singular (double/inverse) difference: $C^2/(1/C)=C^3=CxC^2=$ (d/t)x(E/m)= (a/g)= const. 3D is triple fractal variability, which is **Ed=mt** in context of inertial gravity (C^3=const.); **gt=da** in context of inertial (changeless) $E/m=C^2$, and **Eg=ma** in context of inertial space/time (C=d/t=const), where g-gravity and a-antigravity (inverse gravity or gradual repulsion of instant attraction and/or vice versa). Line of symmetry of any duality is its inertion, which is symmetrical direction on surface and asymmetrical direction in its depth. Consequently, it is also $C^3/(1/C^2)=C^5$ etc. 3D is symmetry of **Egd=mta**, where Egd are 3 equal and opposite directions of mta, each being same 3D coordinate system, only in inverse perspectives. Time and repulsion of mass and/or space and attraction of energy are double/inverse (relative/uncertain) reality of constant E/m in between. In context of variable E/m, where $E/m=C^2=(d/t)^2$=const, $Et^2=md^2$, because space is double/inverse time ($\mathbf{d=t^2/(1/t)=t^3}$) and time is double/inverse space ($\mathbf{t=d^2/(1/d)=d^3}$). The only reason why such 3D is not empty (unreal) is inertial here and now, which is equality of all unequal space/time forms and reality of all fractal and temporary space/time sequences (singular whole number of double/inverse fractal infinities).

Difference of any closed (inertial) system is inertial here and now, regardless of its relative size/motion/force. It is same reality in all infinity of forms and no part of it can be taken out, because such part is all the reality there is. Atom is all reality of the universe there is, and so is a star, galaxy, drop of water or any other form whatsoever. There is no different or other reality without or within a drop of water, because it is all reality there is (and its equal and opposite negation). A drop of water is infinite number of fractal universes within and without, where every such universe is reality of same space and same time that are double/inverse here and now. 3D is inclusive (non-differentiated) reality. Only comparative reality (duality) is relative/uncertain, which is double/inverse infinity of forms, while 3D is a constant of no size/motion. Comparison is meaningful only in dual (space/time) context and meaningless in singular (gravitational) context. Exclusive duality of two equal opposites, without its inertial or inclusive context is only fractal ½ reality. Reality is equal opposite of non-reality—one is real only in inertial context of another. There is no non-reality without reality and equally vice versa—there is no reality without equal and opposite negation. They co-exist in double/inverse (anti-parallel) double helix, because there is no possibility of one without equal and opposite possibility of another.

Every possibility is real only in context of its equal and opposite impossibility, where difference between possible and impossible is only relative/uncertain. Unreal or impossible is meaningless description without its double/inverse inertion and equally vice versa. Only impossible possibility and/or possible impossibility are real (they resonate). Reality means nothing, without its equal and opposite negation, because negation is its infinitely larger reality. **Negation of inertial reality is its time**, which is its exponential increase. Double/inverse times are double negation that creates affirmation (inertial space) in between. Negation in space is affirmation in time and equally vice versa, which is double negation and double affirmation that creates quadruple affirmation in context of gravity, which is double/inverse space. Positive/negative space or time difference is square of such difference in gravitational context. Distinction between possible and impossible is only relative 2D perspective. There is no such distinction in context of inclusive 3D, because any reality that is real in space is unreal in time and equally vice versa. Boundary of exclusive reality is singular quantum period of time, which is 5184 years. Reality here and now is exponentially increased impossibility beyond such boundary and totally impossible beyond boundaries of spatial universe without and/or singularity of space within, which is its total time (negation). It is exclusive reality of inertial density (mass) here and now, and without such mass there is no such reality either.

Different possibilities of inertial mass are fractal change of $E/m=C^2$, which are double/inverse spatial possibilities of time (and/or time possibilities of space).

One mass is infinite number of relatively different passages of time of inertial size and/or different sizes of inertial passage of time. It is infinity of gradual space/time changes within its boundary, which is double light speed difference (just as infinity of gradual spatial changes is real within boundary of singular light speed difference, as per C=d/t). C^2 is singular gravitational quantum of reality, while infinity of quanta (C^3) is its infinite multiplication/division. C^2 is only one out of infinite number of possibilities, all equally real. $E/m=C^2$ is gradual (fractal) change of inertial possibility here and now, while its multiples are quantum changes of reality (inversions between possible and impossible or certain and uncertain). $E/m=C^2$ is distance/difference between two different (inertial and inverse) possibilities. It is difference between real and unreal. Change of reality is change of E/m. Gradual change is fractal E/m change (less than 1), while total change is complete E/m change (1). Only infinity of quanta of inertion is singular quantum of inversion and equally vice versa. Only twice larger space/time change is ½ mass/energy change and equally vice versa, where twice in space is ½ in time and vice versa. Double change of mass is ½ change of energy and double energy change is ½ change of mass (where ½ without is 2/3 within and 2/3 without is ¾ within etc).

Change of possible reality is time of spatial reality and/or energy of mass, which are fractal 2x1/2 reality of 4x1/4 quanta of change (i.e. 2(E/m)=½(d/t)=1/8(g/a) etc). Energy is infinity of fractal space/time changes and mass is inverse infinity of fractal time/space changes, where such double/inverse infinity of changes is singular C^2 (no change). $E/m= (d/t)^2$, where $d=t^3$ and $t=d^3$, therefore $E/m=t^4$ and/or $m/E=d^4$ (4D space without and/or 4D time within) and equally vice versa. $E/m=C^2$ is singular possibility only in 4D context. It is only possibility only in context of inertial space/time sequence or inertial density here and now. In context of all its time and/or all its space, it is only one out of infinite number possibilities, all equally real. In context of all its space and time, $\mathbf{E/m=C^{2x}}$, just as $\mathbf{d/t=C^x}$ (x=0,..6 in context of inertial 3D here and now, which is triple relative/uncertain curvature without and within, or 6D fractal difference between universe and its micro-world). In context of infinite double/inverse curvature of double helix, any space/time boundary is only temporary, which is infinite 2x1/2 multiplication/division of unbound $E/m= C^{2x}$ and/or $d/t=C^x$, because any inertial mass is double light speed difference of its equal and opposite energy and equally vice versa, infinite number of times. Mass is double/inverse energy, which is again double/inverse mass, which is again double/inverse energy etc. Infinite energy of singular mass here and now is infinite double light speed within double light speed difference, just as infinite time of spatial reality (and/or infinite space of singular time) is infinity of light speed differences, one within the other (where double light speed difference is E/m of space/time, triple light speed difference is its gravity/antigravity etc). Infinite energy (C^2) is singular quantum of mass, just as infinity of quanta of mass is singular quantum of energy.

Only force equal to inertion is duality of instant action/reaction. Only quantum of reality is possibility here and now. Only double infinity is singularity and double singularity is infinity. Only twice (square) infinite force of 3D is singular quantum of 4D and equally vice versa. Fractal difference is space/time or light speed difference (C). Only infinite 2D change is instant 3D reaction. Only double or C^2 change affects singular (C) change. Only E/m change is singular spatial or time (exclusive) change and only d/t change is singular mass or energy (exclusive) change. Only gravitational change is inclusive change that affects both—space and time (mass and energy). Only twice inertion is singular quantum of inversion and/or vice versa. I.e. only twice as large inertion (C^2) as here and now (C) is singular quantum of its change, and then such twice-larger inertion is only ½ of its size/motion (i.e. 2x1/2). Such double inertion is instant time and/or singular quantum of spatial reality. It is constant change (motion) of instant space and/or constant change of instant time (size). Denser and smaller mass within is larger and emptier space without, which is exponentially larger energy without of mass within and/or vice versa, because energy of spatial reality is mass of time and energy of time is mass of spatial reality. Only double 2D and/or 4D inertion is singular quantum of double/inverse 3D inertion in between.

Reality, which is x-fold fractal difference from inertial here and now, needs (x+1) times inertion to be real (eigenvalue) here and now. Only (x+1)-fold multiple of light speed difference is possibility of such impossible reality here and now and/or vice versa (possibility of inertial here and now to be real in context of (x+1)-fold fractal difference). Only x+1 multiple of C^2 is energy that synchronizes reality, which is x-fold light speed (fractal) difference—and then such reality is only 1/(x+1) in its size or duration (larger size is smaller duration and vice versa). Mass is double/inverse energy and vice versa. Mass is derivate of energy and energy is derivate of mass. Energy/mass is singular fractal difference, while double/inverse mass (and/or energy), which are subsequent relative/uncertain sequences are double fractal difference or double derivate. Energy and mass are fractal ½ (light speed) difference of each other, while energy/energy and mass/mass are whole (double light speed) difference. Energy (and/or mass) here and now is derivate of exponentially fragmented/multiplied reality without/within. If energy, which is x-fractal difference is E^x, then such energy here and now (C^2 = const.) is $\mathbf{E = (x+1)E^x}$. Such energy (and/or mass) is real only as derivate of its larger/smaller fractal reality, i.e. **$dE^{x+1}=(x+1)E^x$ and/or $dm^{x+1}=(x+1)m^x$, which is infinite curvature here and now $E=E^x\,(x+1)(x+2)(x+3)$....**

Derivate is curvature or underlying constancy of its variable surface reality, therefore such energy is real in inertial context here and now (C^2) only as complete curvature—i.e. only all of it at once (only if same inertion as here and now). There is no gradual variability of reality that is different dimensionality. Only quantum is same curvature (inertion) as here and now. Only quanta are resonating, while infinite number of its fractals in between is empty space (nothing). Fractals are only depth of surface reality (not real here and now). There is no partial (fractal) energy of such different reality—it either is or is not (only whole numbers). Any different (more than relative/uncertain) reality is real in context of inertial here and now only as quantum reality. Only E/m change here and now (no fractal difference, which is less than double light speed) is gradual reality. One mass is infinite number of quanta of energy (C) and one energy is infinite number of quanta of mass (1/C), while E/m is infinity of both, which is singular quantum (C^2) of double/inverse energy (or mass). Double/inverse energy of mass and/or double/inverse mass of energy is its gravity/antigravity.

Energy of mass is its density, while double/inverse (infinitely larger or exponential energy) is its inverse density. Infinitely larger energy of mass is energy of its inversion or negation, which is energy of empty space. **Empty space is infinitely larger energy of dense mass, just as dense mass is infinitely larger energy of empty space. Dense mass within is singular energy in context of its empty space without, which is its infinite multiplication. Empty space is instant infinite expansion of energy of dense mass, just as mass is instant singular contraction of energy of empty space.** Solar empty space is double and galaxy is triple multiplication of inertial energy of dense mass here and now. Mass of empty space is its singular energy, while its emptiness is double/inverse (squared) energy. Empty space between nucleus/electrons and/or planets and stars (and galaxies) is infinitely larger energy than mass itself, because E/m is gradual, while space/energy is double/inverse (quantum) change of reality. Empty space is double/inverse density of mass, which is double light speed difference. Energy of empty space is double light speed difference of total energy contained in mass (as per $E/m=C^2$), because mass is only one, while space is infinity of different possibilities. Energy of mass is only surface of energy that is empty space.

Mass is singular possibility of time, which is resonant with inertial here and now, while empty space is infinity of relatively different densities, which are not resonant and therefore not real in context of inertial passage of time. Change of inertial density is gradual E/m change, while change between different quantum densities of empty space are quantum leaps in periods of time in between. Mass is instant (spatial) reality of different packets (quanta) of energy, which are only fractals of infinitely larger energy that is real only in context of passage of time. Such different quanta of energy of empty space are instant spatial reality only in infinitely long periods of time in between (light speed difference). Energy of mass is spatial packets and energy of empty space is time packets. Infinity of spatial packets (energy of mass) is singular quantum of energy of empty space. Infinity of spatial quanta of energy is singular quantum of energy of time, which is infinitely larger and infinity of time quanta of energy is singular quantum of gravitational energy, which is its infinitely larger energy. Energy of empty solar space is double and energy of empty galactic space is quadruple energy, which is instant (spatial) reality only in quantum periods of passages of time in between.

Energy is double/inverse mass without and within and equally vice versa—mass is double/inverse energy. Energy is resonant surface between two inverse densities of empty space without and dense mass within, where dense in space is empty in time and empty in space is dense in time. Mass of empty space is energy and energy is mass of twice empty space. Mass without is energy within and energy without is mass within. Empty space here and now is dense mass in context of solar reality and double density in context of galaxy. Spatial reality here and now (passage of time) is not real in context of solar reality, because its space is time here and now. Solar reality is not only one, but all possibilities of time (all different fractal passages of time). It is all different possibilities of mass (different densities), which are real only in context of infinitely long period of time here and now. Energy of solar reality is not energy of singular possibility of mass, but energy of all different possibilities. Solar energy is infinitely larger energy, which is energy of empty space. Galactic energy is also all different possibilities of empty solar space, which is yet larger energy and energy of universe is all different possibilities of empty galactic space. Energy of time, which is negation of space, is infinitely larger than spatial energy, and energy of gravity, which is negation of both, is yet larger energy. **Energy of nothingness is twice as large energy as energy of something (mass).** Infinity of relatively different nothing is yet larger energy here and now, which is resonant only in infinitely long quantum periods of passage of time.

Mass here and now is fractal reality of energy of empty space or energy of time, which is resonant only in infinitely long periods of time, when gradually accumulated space/time change is its singular quantum. Only when mass within (Earth) is resonant with singular quantum of empty space without (solar and/or galactic reality), is such energy real here and now. Larger negation of inertial mass here and now is relatively larger size of empty space (universe), which is resonant (real) here and now in accordingly larger periods of time. Larger spatial distance is larger period of time between their synchronization, because they are relatively larger disproportion of their space/time waves. Larger empty space is relatively larger negation of inertial density here and now, which is larger/smaller wave. Different sizes of empty space (and their orbiting massive particles) are relatively different time waves, which are different time possibilities of empty space. Spatial distance is time difference. Increase of spatial distance is relatively slower motion and faster passage of time, which is not resonant here and now instantly, but only in quantum periods of time. Larger difference between such possibilities, i.e. larger difference between time wave of Earth's reality and such different possibilities, means larger spatial distance/size, and consequently larger periods of time between their synchronization.

Smaller time synchronization and/or larger spatial distance is larger angular distortion of such reality here and now. Infinite distance is its total angular distortion. Reality is only here and now, and any increase of spatial distance and/or time difference (density) is not here and now of such reality but its distortion. Reality, which is now but not here, is angular distortions of such distant (larger/smaller) space in context of inertial passage of time here and now, which is an exclusive reality, without any meaning in its larger (relative) context. Equally vice versa—reality, which is here but not now (different time possibilities of empty space, which are different densities), is inverse angular distortion of passage of time (and/or distortion of inverse passage of time). Increased spatial distance is decreased space/time periodicity of resonance of such reality here and now—i.e. it is increasingly less real here and now (dense mass) and increasingly more empty space, which is longer passage of time in between (discrepancy). Planets/stars/galaxies in context of its increasingly empty space are increasingly smaller spatial periods of such increasingly different space, which are increasingly larger periods of passage of time between their synchronization. Only inertial density is same space/time reality. Different density is relatively different passage of time. It is different possibility, which is not real here and now. Different reality is different time wave of inertial space and/or different spatial wave of inertial time (curvature), which is different relative density.

Only remaining density (overlap), which is equal to inertial density, is synchronized (real) with spatial reality here and now (Earth). Different realities are synchronized only on surface (2D), while they differ in depth (3D), which are their different passages of time. Increased spatial distance is larger angular distortion between densities and/or passages of time, which is increased empty space between such distant/different masses. Infinite difference (angular distance) is infinite empty space of no mass (i.e. expanding and universe beyond horizons. Orbitals of different planets are relatively different densities of empty solar space within and empty galactic space without, which are relatively different passages of time that are not resonant with passage of time here and now (Earth). Density is time curvature of space and density that differs from inertial density is not space/time, but exclusive spatial eigenvalue. Different densities of empty space (different planetary orbitals) are only ½ resonant. They are resonant with here and now (Earth's reality) only in context of space, but not time, because they are now but not here (different passages of time).

Here and now is line of symmetry between solar and galactic realities. Only inertial density of Earth's orbital/rotation is ½-way between solar and galactic reality (balance or resonance). Only Earth's time x gravity = light speed. Only Earth's time is inertial passage of time, because only Earth is here and now. Inner planets are relatively different resonance between solar/galactic realities, which are different passages of time (different spatial orbitals, which are different densities of empty space within/without). Inner planets are decelerated and outer planets are accelerated passages of time, which are inertial reality of their spatial orbitals (gravity x time ≠ light speed). They are not same space/time sequence as Earth's reality. Their passages of time are shifted in future/past direction of inertial time here and now. Likewise—in context of their passage of time, Earth's passage of time is not real. Earth's spatial reality in context of different planets is not same passage of time as here and now, but passage of time shifted in future/past direction (accelerated/decelerated passage of time). Orbiting planets, stars and galaxies are different passages of time, which are not real here and now, just as passage of time here and now is not real there. Space/time reality is only here and now, which is gravity x time = C. Distortion of such reality in context of inertial passage of time here and now, which is exclusive reality, results in increased discrepancy between passages of time and spatial reality of distant massive objects.

Distant massive objects are not same space/time or physical reality as here and now, but only same space, while their passages of time differ. Distant mass in perspective of inertial passage of time here and now, is only spatial shell of reality, without its content, which is passage of time. In context of their inertial passage of time, it is same reality as here and now, while here and now (Earth) is different reality (time x gravity ≠ light speed). Physical reality of Earth is not real in context of different massive objects, just as their physical reality is not real here, because they are only spatial and not time resonance. Different massive objects, are larger than $E/m=C^2$ difference, which is distance of empty space (nothingness) in between. E/m=C^2 is continuous curvature only in context of same density, which is no empty space (nothingness) in between. Only Earth is singular massive quantum of no interruption of empty space in between. Consequently, only Earth is singular quantum of passage of time, uninterrupted by its different possibilities (empty space). Different massive objects of universe/micro-world, which are distance of empty space in between are shifted passages of time, which is not real here and now instantly, but only in quantum periods of passage of time in between. Physical (space/time) reality of different spatial densities, which are different orbiting mass, is resonant only in context of their alignment during quantum change of time here and now. Quantum change, which is passage of time of solar reality and not inertial passage of time here and now (Earth), is only time when physical reality of different planets is synchronized with Earth's reality. Constant passage of time here and now is only their spatial (surface), but not time (in depth) eigenvalue.

Different densities are different space/time sequences, which are different gravity. Space and time are two equal opposites (variables) of inertial gravity, which are its double/inverse (resonant) curvatures (Π). Gravity is double/inverse or space/time curvature, which is $\Pi/(1/\Pi)=\Pi^2$ (9.86). Gravity is equal opposite of Π, which is space in time curvature and/or time in spatial curvature, which is instant overlapping mass (density) in between. Π is space/time constant of gravity (light speed). It is reality here and now, which is double helix. It is physical reality through its non-physicality and non-physical through its physicality. It is its simultaneous confirmation and negation, because it is simultaneously size (space) and motion (time) on surface, which is Π^2 (gravity/density) in its depth. Equal opposite of gravity is light speed. **Light speed and gravity are action/reaction in context of inertial Π (space/time), which is their non-physical abstraction (infinitely larger physical reality) and equally vice versa. Light speed is curvature (Π) of gravity and gravity is curvature (Π) of light speed.** Curvature (Π) is light speed of gravity and/or gravity of light speed (C/g=Π). There is no difference between gravity, light speed and space/time curvature (Π)—they are singular double helix, which is double/inverse curvature of infinity of different spheres of reality that are fractal passages of time in between different densities.

Spatial universe **is** micro-world of time, just as micro-world of space **is** universe of time. Round Earth is round space and flat time, and flat Earth is flat space and round time. Empty space is energy of mass within and mass within is energy of empty space without. Any reality that is here but not now and/or now but not here (interruption of even density with empty space in between) is double/inverse distortion, which is curvature of its spatial reality and inverse flattening of its time and/or vice versa. Different/distant or space/time reality is duality of round/flat or sphere and a cube (1.111 diagonal difference). Cube is two equal and opposite spheres and sphere is two equal and opposite cubes, which is fractal difference in between. Only here and now is synchronicity of both—sphere and a cube. Any distance without/within is fractality between cube and sphere. Sphere is equality (symmetry) of two inverted cubes and cube is symmetry of two inverted spheres; therefore in context of a cube it is not a sphere, but inverse cube, and equally vice versa. Sphere is surface of a cube and cube is surface (2D) of a sphere (3D). Cube/sphere is fractal 3/2 reality (equally vice versa). If space is cube (3D), its time is double/inverse sphere (2D) and equally vice versa (and/or square between double/inverse cubes and vice versa).

Exclusive sphere or a cube (round or flat) is an empty form without content, which is only fractal ½ of a whole. Cube is product of any two of its squares (surfaces). Such cube (2x2=4) is an empty cube. Only multiplication with a third surface (2x2x2=6) is its content (density). Third surface of a cube is time or density of empty spatial shell. Space/time is double cube (2x3=6), which is cube (3D) in context of one of its inertial surfaces (2D), which is its time—i.e. 6/4=3/2 space/time of double (asymmetrical or fractal) cube. Inertion is always an empty cube (or a sphere)—only relativity/uncertainty (fractal difference), which is its 3rd direction, is content that creates density within of emptiness without. Exclusive spatial or time cube is empty cube (product of 2 of its surfaces), which is not reality here and now that has content. There is no empty cube, because any cube is 3 and not only 2 sets of opposite surfaces! In other words, there is no inertion without its equal opposite (parallel inversion or negation, which is equally real, only not here and now, unless quantum). Spatial cube is time square and time cube is spatial square (or a sphere). Third surface of a cube (3rd half) is double/inverse time of empty space, which is its mass and energy. 3rd direction of a cube is gravity between space and time, which is content of double/inverse empty form.

Reality is absolute in any form. Relative reality is equally relative—i.e. it is its own negation, and so is uncertain reality. Relative reality is only relative description. Singular relativity/uncertainty within is infinite relativity of relativity and/or uncertainty of uncertainty without, because any reality is real only in background of its equal opposite. Reality on surface is real only in background of its own negation or time in its depth. Earth (inertion) is real only in background of its universe (inversion), and universe is real only in background of its Earth—one is negation or double/inverse (relative/uncertain) time of spatial reality of another. Inertial difference (overlap) between such equal opposites is gravity, which is inertion or equal reality of any relatively different space/time point. Earth is begging of universe, just as universe is end of Earth, where beginning in time is end in space and equally vice versa. It is double/inverse beginning, which is not begging but end and equally vice versa. **What did not begin cannot end, and what did not end cannot begin**! Earth has no beginning or end and still is—and so is universe or space or time or any of the infinite vibrations of singular inertion. Any relative curvature of space/time is infinite double/inverse (anti-parallel) repetition in space and time, through inertion of such singular duality, which is its perpetuity.

Beginning and end are only exclusive (fractal) surface duality (2D) of inertial 3D, which is inclusion of both—it has no beginning/end, because **it is** (it is an infinite beginning/end of inertion of any point of observation). Only what **is not** has beginning/end. End within is beginning without and equally vice versa. End of fragmentation is beginning of multiplication and end of multiplication is beginning of fragmentation, because fragmentation/multiplication are two sides of inertial 3rd side that is both at the same time. Multiplication in horizontal direction (numbers larger than 1) is fragmentation in vertical direction (less than 1), and yet a square in 3rd direction. End of space is beginning of time and end of time is beginning of space. End of space/time duality is beginning of gravitational reality and end of 3D is beginning of 4D, which is only relative 3D, etc. Reality begins and ends only here and now, which has no beginning/end and equally vice versa. Here and now is only singularity/infinity of its space/time and space/time is only singularity/infinity of such point. Double helix is triangulation of any surface duality, which is infinite double/inverse progressive fractal curvature.

Beginning in space is end in time and vice versa, which are double/inverse fractal curvature. One is negation of another. One fractal reflection is real only in context of its angular negation. One non-Euclidean curvature is real only in context of its inverse curvature, which is fractal overlap in between. Instant concave curvature within is gradual convex curvature without of inertial here and now in between. Inertial here and now is fractal 3/2, which is 1/3 without of ½ within, while time of inertial here and now is fractal 4/5. Fractal 4/5 is time of both—space and time here and now, which is their gravitational boundary. It is instant (quantum) boundary of gradual (fractal) reality and/or vice versa. I.e. 1 calorie is energy required for 1° in increase of temperature of water. Consequently, 100 calories increases temperature from 0° to 100°. At the same time it needs 80 calories to melt the ice, which is no increase in temperature at all, but only instant quantum change of reality, which is fractal 4/5 reality (80/100). 4/5 of energy between melting and boiling points are inertion of ice within. Ice, water and vapor are 3 space/time quanta of same inertion, which is double fractal difference (melting/boiling point) in between. It is 3 gradual (fractal) changes in between 2 instant (quantum) changes, which is fractal 3/2 inertion within of fractal 4/5 inversion without. Liquid water is double/inverse inertion between uncertain ice and relative vapor. It is asymmetrical symmetry of two equal opposites, where fractal change precedes quantum change and equally vice versa, which is resonance in packets of 4x5=20. 80 calories within is 100 calories without in context of inertial 20 calories in between, which is constant difference (inertion or overlap) between equal opposites. Water/ice/vapor is infinite 4/5 fractal curvature of inertial 20 calories, where one precedes another in anti-parallel curvature of double helix.

CHAPTER 15

INFINITE SINGULARITY (RENORMALIZATION)

Distortion is tangent of angular difference, which is eigenvalue of equal and opposite sin/cos curvatures. Sin/cos are double/inverse fractal (90°) difference or space/time (instant/gradual) waves, which are negation of each other. If sin is relativity, then cos is equal and opposite uncertainty of such angle (reality without/within) and their resultant distortion is inertion in between (tangent). Sin is negation of cos and cos is negation of sin, while their tangent is affirmation, which is double negation. Space/time reality is tangent of both waves. Tangent is overlap or discrepancy between equal and opposite waves, which is 1D fractal difference of both. Sin and cos are space and time, while tangent is their resonance, which is gravity in between (singularity of double/inverse infinities and/or infinity of double/inverse singularities). Relativity and uncertainty are fractal 2/3 within and fractal ¾ without, while their resultant tangent is infinite 1.111 repetitive curvature of double helix. Sin and cos are fractal 2/3 and ¾ curvatures without/within (where without in space is within in time etc), when curvature is 42° and 48° respectively, which is 6° difference in between. Tangent of 48° is 1.111, while tangent of 42° is 0.9, which is inverse 1.111. Double helix is double/inverse fractal curvature, which is duality of 2/3 and ¾ ($(2x4)/3^2=8/9$). It is inertial 6°

difference between 42° and 48° curvature, which is +/-3° discrepancy of inertial 45° in between. It is 3D space without and equal and opposite 3D time within inertial ½ in between, which is its here and now (singular point of double/inverse infinities).

Space/time reality here and now (universe/micro-world) is inertial 6° discrepancy between double/inverse curvatures, which is $6/90=0.0666=2/3\text{x}10^{-2}$ within and/or 90/6=15 =3/2x10 without. 3° curvature is space and equal and opposite 3° is time, while 6° is their gravity. Space and time are relative/uncertain 3° angular depth without/within inertial surface in between (here and now). Space and time of inertial here and now are 90/3=30-fold repetition, which is ½ (45°) of total reality that is 60-fold repetition of inertial difference (6x10). Sin/cos=tg is imperfect (irrational) ratio, same as Π/φ, because of infinite relativity/uncertainty of such curvature, where there is always further inertion or further precessional 3rd direction of such duality, and any attempt to determine it absolutely is liable to equal an opposite effect of uncertainty of such attempt. Every curvature is further 1/10 fractal discrepancy, which are its relativity/uncertainty (double/inverse discrepancies, which are space and time of inertial here and now). Inertial here and now is perpetual difference of any dual curvature, where infinitely large spatial difference is small difference in time and small spatial difference is large time. Tangent of 6° is 0.111=1/10x1.111, which is 0.1=sin and 0.99=cos (equal opposite of 0.01) that are double/inverse 1/10 difference.

3D is instant space without its context of passage of time and/or time without its spatial context, which are +/-3° fractal discrepancies. Space in context of linear passage of time, which is one out of 3 different fractal possibilities, is fractal 4D, which is +/-4° discrepancy of inertial (singular) ½ quantum in between. It is 41°/49° sin/cos curvatures, which is tangent of 1.15 and/or 0.87 (0.04 without and/or 0.03 discrepancy within of inertial 1.111), which are perfect 2/3 and ¾. Space in context of passage of time without and within (universe and its inversion—micro-world) are +/-4° angular depth of inertial surface, which is 90/4=22.5-fold repetition that is ½ of total 45° curvature (4/90=0.0444). 4D is inversion of 3D, which is inertial 10-fold difference in between. 3D curvature is flat 4D reality and equally vice versa. Inertial 3D curvature is inverse 4D curvature. 1D fractal difference between 3D and 4D (time of space) is sin 46°=0.72 or 72/10°, which is 0.03=3/10° difference from its cos curvature, while their tangent (gravity) is 1.04=25/24 (equal opposite of 0.96=24/25). Reality of both is 8° discrepancy, which is 90/8=11.25 (1D fractal difference of 9/8=1.125). Space/time difference (contraction/expansion) in both directions is +/-8° or total of 16° of angular depth, which is 53° and/or 37° without/within inertial ½ quantum in between (infinite 1.111 repetition). It is sin/cos, which are fractal 0.6=3/5 and 0.8=4/5, while their tangent (eigenvalue) is ¾ within (37°) and/or 4/3 without (53°). Sin/cos waves (6/10 and 8/10), which are space and time, are 1D fractal difference of previous 6° and 8° double/inverse discrepancies. They are double/inverse sin and cos waves, which are space of space and time of time. They are space and time (i.e. relative universe) of inertial fractal ¾ in between (orbiting Earth or space and time of inertial Earth), which is fractal 4/5 curvature. Duality of 4/5 and 3/5 (equal opposite) $(3x4)/5^2=12/25$ is exclusive ½ of fractal 24/25, which is discrepancy within discrepancy (double fractal difference or gravity within singular 4/5 fractal difference or time without).

If sin is time curvature of space (relativity), cos is its reaction—it is equal and opposite spatial curvature of time (uncertainty). Only both curvatures (space and its time) are physical reality here and now in between. Only double 2/3 in context of inverse ¾ is inertial ½ in between ((2/3)/(4/3) and/or (2/3)x(3/4)=6/12= ½). 1.111 (10/9) is constant of double helix. It is here and now (tg 45°), which is underlying linear constancy of round curvature (Π). 1.111 x Π =3.5 or 7/2 period, which is larger (galactic) period than solar 5/2 and Earth's 3/2 periods. Π is angular curvature (time), while 1.111. is its flattening (space or gravity) reflected as 11.1-year solar cycle. Numbers are abstraction of infinitely larger physical reality. They are undecoded language of the universe. Just as there is no accident or coincidence, there is also no accidental numbers. Each sequence of numbers is perfect description of its physical reality, which is its source (beginning/end). Numbers speak volumes. Sidereal/synodic difference of 236 sec are not random numbers—it is $2x10^2$ (inertion) + $3x10^1$ (inversion) = 6x10° (double inertion 2x3). Triple sequence is 3D; double sequence is 2D (perfect 240 of 236); 4D sequence is double duality, which is 2x2=4 without and/or 2/4= ½ within etc. In context of imperfect perfection and/or asymmetrical symmetry there is no room for coincidence. Inherited system of measurements is perfect description of reality. Reality is determined absolutely through inertion here and now. Only change of such inertion changes entire reality of universe/micro-world.

236/86400 (1 day) = 0.0027314. Non-inertial sidereal/synodic difference is 240 sec and 240/86400 = 0.00277777...(singular duality (2) here and now, which is reaction of infinite repetition of galactic reality). Additional 314 sequence is curvature (Π) or boundary of such infinite reality, which is inertial $3^3=27$ in between. Space/time is **almost** perfect (non-inertial) curvature, because here and now is its perpetual discrepancy (inertion), which exponentially increases/decreases its imperfection (eccentricity) and where increase in space is decrease in time and equally vice versa. Only 2D is perfect symmetry, while 3D is its imperfection and equally vice versa. 360 days of perfect year is perfect duality (non-inertial symmetry, i.e. 3/6=½). It is idealized non-inertial curvature that does not account for its inertion (Earth), which is inclusivity (gravity) of such exclusive space/time duality (3rd odd direction, which is precessional curvature of symmetrical even curvatures of equal opposites). 360 days is non-reality or boundary of inertial reality here and now, which is 365.25 days. Additional 5 days in context of time and/or additional 4 sec in spatial context (240-236) is fractal 4/5 (time) without of spatial 3/2 within.

360/1.1111=324, which is fractal 1/10 difference in between. 1/324 = 0.0030**864** (inversion) is singular quantum (boundary), which is time within of inertial space without (likewise: 26000/25920 and/or 130000/129600 is 1.0030864, which is 30-fold repetition of inertial 864). Such spherical/eccentric curvatures are consequence of perfect/imperfect discrepancy (5 ¼ day difference per year, which is 4 sec of synodic/sidereal time). **4** sec difference here and now is 4x20=**8**0 years difference per precessional period (26000-25920) and/or 80x5=**4**00 years per double/inverse precession (130000-129600) or 400x20=**8**000 years per triple precession, etc. Such imperfection is result of Earth's inertion or reality here and now, which is prism of any distant/different reality. Any 3D imperfection of perfect duality is exponentially increased/decreased inertial reality here and now, which is 3rd direction of any action/reaction. Imperfection is perpetuity of perfection and equally vice versa, where perfect in space is imperfect in time etc.

1/360=0.0027777=240/86400, 1/365.25=0.0027**378** and 236/86400=0.0027**314**. Double difference between synodic/sidereal difference (236) and orbital time (365.25) is 378-314= 64=2^6, which is fractal 6/4=3/2 or 6x4=24 (24 is 1/10 of synodic/sidereal 240 sec difference). 6+4=10, which is 1st/2nd orbital shell of 10 massive particles of empty space. Derivates of 2^6 are 6x5x4x3x2x2=2x720 or 1440=12^2. Sequence 314 (Π) is formulation of 3D (space) without and 4D (time) within, which is 1D fractal difference in between. It is fractal ¾ or 3x4=12-fold curvature without and/or 3+1+4=8-fold curvature within, which is 12/8=3/2 in between (here and now). 314 is Earth in context of solar (4D) reality, while 378 sequence is its negation (boundary) in opposite direction, which is 3D here and now as reaction 7/8 duality (time and gravity of empty space of universe). 0.0027**777**-0.0027**314**=0.0000**463** is rotational velocity of 463 m/sec (2.16 sec/m), because rotation (time) is 3rd direction of 360/240 spatial duality.

24=6x4 and 36=6^2 are double/inverse (3/2) duality, which is singular 3D in between (24x36=864). It is 36+24=60 or 6+4=10 resonant harmonics in context of inertial 6D (3+2=5, which is 5 harmonics without and 5 harmonics within). 3+6=9 and 2+4=6, which is 9/6=3/2, where division is inversion of addition, just as multiplication is inverse curvature of subtraction (6-3=3 and 4-2=2, therefore again 3/2 or 2x3=6). Likewise, (36x24)/(36+24)=**1.44** or $(12/10)^2$ wave, where 12=36-24=3x4. (36/24)x(36-24) =**1.8** or 3x6/10 wave and (36x24)/(36-24) =**7.2** (6x12/10) wave. 7.2/1.44=5 and 7.2/1.8=4 or fractal 4/5, while 1.8/1.44 = fractal 4/3 or rotational/orbital time of Earth. Addition/subtraction is derivate of multiplication/division, just as multiplication/division is derivate of exponential multiplication/division. They are inertion/inversion=double inertion of fractal 3/2 double helix.

<u>Illustration 16 -Inertion/inversion/double inertion (fractal 3/2 Fibonacci sequencing)</u>

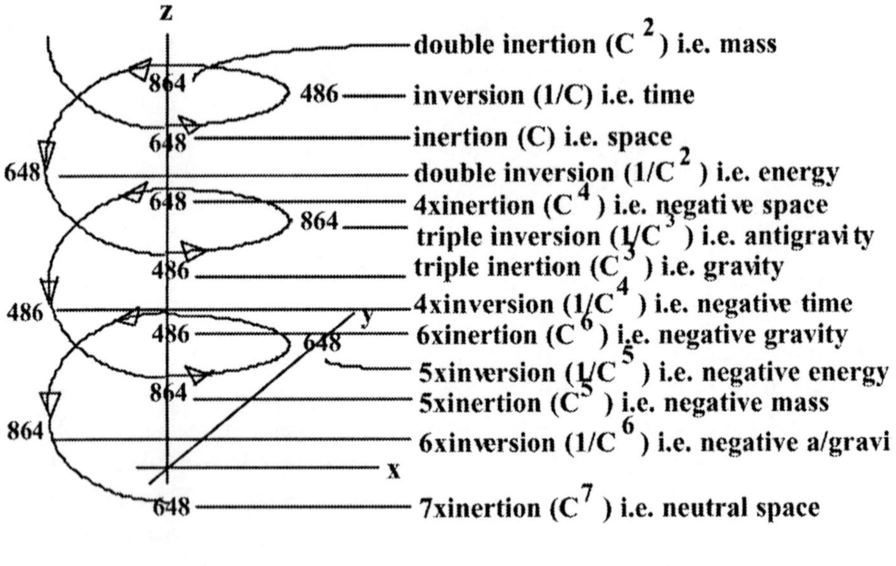

3 horizontal/2 angular curvatures= 6 eccentric curvatures

864×10^2 is time of rotational cycle, 648x10 is time of precessional cycle, while 648/864= ¾ (and so is 486/648). 864/486=1.777, which is uncertain/relative (fractal) reality between two multiples (whole numbers). It is angular (in depth) curvature of surface symmetry, which is inertial here and now (1) in context of infinite repetition of galactic wave (0.777). 3rd direction is always angular asymmetry of surface (spatial) symmetry and/or vice versa. See illustration 16—it is

864 (i.e. time) in horizontal curvature, inverted into 648 (i.e. space) in its vertical (angular) curvature, which is 486 (i.e. gravity) in resultant negative horizontal curvature (time = negative gravity and/or gravity = negative time in context of inertial space). Same 3D curvature repeats itself in inverse order, therefore 360° or 2^2=4 equal and opposite curvatures of double helix (each ¼=90°) is one loop of double helix. Each ¼ is relatively different fractal (different permutation of density), where ¾ within are 3/2 without. Double curvature is 2x3=6 fractal curvatures or 6 relatively different surface densities (3 positive + 3 negative). Double curvature is only fractal 2/3 of triple curvature 3^2=9, which are 3 inertial (neutral) densities between positive/negative densities without/within. 9-fold curvature is again only 2/3 fractal of 3x4=12-fold curvature, which is again 2/3 fractal of 4^2=16, which is fractal of 4x5=20, 5^2=25, 5x6=30, 6^2=36 etc.

8=2x4, 6=2x3, 12=3x4, 2^2=4, 3^2=9 etc are relatively different space/time waves of inertial here and now (2), where squares are times of fractal spaces in between (and/or vice versa). Any 3 numbers that are inertion/inversion (φ) = double inertion are fractal Fibonacci sequence of physical reality, which is addition of past (small) to the present in order to generate future (large) sequence (number). 6 successive sequences are 6 different surfaces of a cube or 6 different shells of inertial 3D here and now. It is space/time, mass/energy and attractive/repulsive forces. Time is inversion of inertial space and their double inertion (resonance) is mass (double negation = affirmation). Inversion of mass is energy and their duality is gravity. Inversion of gravity is antigravity, and their resonance is negative space (beyond event horizons of singularity/infinity, which is negatively probable spatial reality). Inversion of negative space is negative time and their duality is negative (inverse mass). Etc. Inversion of inertial size is motion and their resonance (duality) is accelerated motion or force. Inversion of such linear force is decelerated motion and inertion of such duality is motion in opposite direction.

Inversion of such motion is inverse size etc. Inversion of size is rotation and inversion of rotation is linear motion. In 6D curvature—it is size, motion and accelerated motion in all 6 directions (3 positive and 3 negative directions), which is 6x3=18 different fractal curvatures (waves) of one inertion, where difference between two different curvatures is always infinite/singular (light speed) difference. Each successive curvature is further 2x1/2 multiplication/division of inertial space/time duality, which is relatively different density and different physical reality, which is infinity of double/inverse shells without/within inertial shell here and now. Fractal shells (i.e. time and gravity) of inertial space here and now are not physical reality, but only fractal ½ and/or ¼ cross-section of such reality (surface and linear direction). Physical reality of spatial reality of time

(4D) is not on surface but in depth of reality here and now, while physical reality of gravity (5D) is in depth of surface time.

Illustration 16 is only one out of three relatively different fractal 2/3 densities. It is only one out of 3 surface curvatures (and/or 6 in both directions). It is only xy-surface in z-direction of spiraling. Z is linear direction (constant) only in context of xy-duality, while zx in y curvature and zy in x curvature are its double/inverse spiraling curvature. Those three positive/negative curvatures are 1D and 2D (positive/negative) fractal difference, which are linear direction, its acceleration (round curvature) and double acceleration (parabolae). They are size/motion/gravity without and within, which are 6 curvatures that are 4D fractal difference in between (6/4=6x4=24-fold repetition). Xy in z curvature is only one 2/3 fractal of larger 3D xyz, which is 24 such different fractal curvatures. If x and y are space and time, than z is their gravity (equal opposite or negation of xy), which is only 2x1/2 (double/inverse) space/time. Illustration is only one coil, while linear direction of such coiling (z) is only fragment of infinitely larger coil. It is one spiral in context of another, or an infinity of spirals, one within/without the other. Point of intersections of all infinity of such spirals is always inertial here and now. Without such point, it is meaningless description. In 3D context, it is one spiral (xy as per illustration) which spirals around z, where z is jet another spiral which spirals around y—and such zy is yet another spiral which spirals around x.

All 3 of such curvatures are only ½ (singular quantum) of 3D, which is exclusive positive or negative 3/2, where 3rd direction of such duality is neutral reality in between (positive/negative and their neutral particle). Such 3 "particles" are +1,0,-1, or positive/negative infinity of singularity and/or positive/negative singularity of infinity. Such triple double helix is triple density of space, time and gravity (or spatial, massive and gravitational density). Each density is duality of singular and double fractal difference. It is spatial density of time and gravity and equally vice versa. Only here and now is singular density of all directions, while different/distant reality within/without is its fragmentation in context of equal and opposite multiplication and/or vice versa (anti-parallel space/time curvatures). Spatial fragmentation is angular multiplication and angular fragmentation is spatial multiplication. Illustration 16 is only linear direction of 3D double helix. 3D double helix is dense sphere (sphere with content), where size/motion/gravity (density) of all infinity of such spheres is determined exclusively by inertion here and now (and equally vice versa—inertial here and now is inverse reaction of infinity of such spheres without/within). It is inertion of such point in infinite number of fractal curvatures, where each curvature is different perpetuation of same inertion. 3 different directions of double/inverse curvatures of double helix are infinite/singular or light speed difference in between, which is gradual change (time) in between double/inverse instant realities (space/gravity) without and within.

Only surface is inertial here and now (duality). Depth or its inversion is NOT real in its context, unless singular quantum, which is instant inversion of inertion that creates double inertion (change of surface reality). Depth is vertical (angular) direction or infinity of surface singularity, therefore only gradual (fractal) increase in tension. Inverse reality is infinity of different possibilities in depth of singular possibility that is its surface. Such inverse infinity is 2x½ of surface singularity, which is its uncertain/relative reality without/within, which is real only fractally (½)—i.e. only here but not now and/or now but not here. It is empty space (time) of dense mass. Infinity of empty space of singular dense mass here and now is its inversion, which is empty space of no massive content, in direction without, which is double/inverse fractal curvature (direction toward core of solar reality within, which is singular difference and direction toward core of galaxy without, which is double difference). 3rd direction is inversion of empty space into empty time, which is not direction toward singular core within but infinite surface without (expanding universe). Inertion here and now (massive Earth) is resonant with such progressive empty space in exponentially larger periods of uninterrupted passage of time, when not 2 but 3 directions are resonant (not only dual space and time, but also their gravity)—i.e. when Earth, Sun and black hole of galaxy are concurrent (5184-year period). **3D spatial alignment is 2D time and 1D gravitational alignment, where gravity is double/inverse depth of space and time.** Only when Earth's and solar time (which are solar and galactic space) are equal, are dense mass and empty space resonant (quantum change of space/time).

Only galactic eclipse during solstice/equinox is instant synchronicity between Earth, solar reality and galaxy, when they vibrate with same frequency (resonance). There is no interference, unless equal opposites are resonant reality. Earth and solar reality are spatial wave of time and/or time wave of space, which are synchronized (eigenvalue) only on equinox/solstice day—i.e. only 4 days per year (360/4 = 90°). Such 4 days are space/time (Earth/Sun) synchronicity, which is instant reversal of vertical (angular) motion or change of seasons. Unless Earth and Sun are synchronized, they are duality and not singular inertion (quantum) of galactic reality. Earth is real in context of galaxy only as synchronized inversion of solar inertion. Such synchronized duality is singular 3rd direction of solar/galactic duality (physical reality). Each solstice/equinox is only fractal, infinitely small change of galactic reality. Only **260**x10^2=26 000 such fractal changes is accumulation that is equal to singular quantum of galactic reality. 26 000 equinox/solstice days is 6 500 years of uninterrupted passage of time. 26x10^3 is perfect duality (non-inertial reality), while 25920 is double duality (4x1/4 or 2D wave). 5D (3D or 6x1/6 wave) is 20736-year period. 260/360 =0.7222 is infinite non-inertial repetition of duality here and now (spatial Earth) in inertial context of galaxy. 260 solstice/equinox days is 65 years and 10^2x65 = 6500 years. 4/5x10^2=80 and 80x65 or 26000/5=5200 years, which is 20800 days. In 5200 years it is additional 5200x5=26000 days or 72.222 years. 4/5x72.222=57.777 years that is one day created by perihelion precession (infinite repetition of galactic reality in solar context). It is also 1300 days of additional ¼ day, which is 3.6111 years. 72.222x3.6111=260 and 72x3.6=259.2x10^2, which is 3D sequence of such infinite curvature. 72.222/3.6111=20, which is inertial difference. It is instant 20 years discrepancy between non-inertial 6500 and inertial 6480 years period without and/or 20-fold gradual repetition within. 4/5x20=16 years difference between perfect 5200 and actual 5184 years and/or 20 800 and 20736 days (16-fold repetition without). 260-fold curvature is eigenvalue (gravity) between time of rotation and orbital motion. Singular quantum of such resonance repeated in units of 20 is 13-fold curvature (20x13=260 and/or 20x260=5200).

Inertial context of physical reality is its abstraction, which is its double/inverse physical reality. Spatial periodicity is Bode's periodicity, which is 8-fold 2x1/2 multiplication/division of inertial 2x3=6. It is **3x2**=**6**x2=**12**x2=**24**x2=**48**x2=**96**x2 =**192**x2 =**384**, where inertion of inner 2 harmonics is 1 and inertion of outer 6 harmonics is 4, because Earth's 3D in between is their double/inverse inertion (3+1=4—Mercury and 6+1=7—Venus within are 12+4=16—Mars, 24+4= 28-asteroids, etc. without). Inertial difference of inner planets (1D—linear passage of time) and inertion of Earth (3D space) are space/time inertion of outer planets (4D). Earth's 3D is fractal 3/2 in between Venus and Mercury (6+4=10=16-6, which is 6/4=3/2 fractal). Bode's periodicity is exclusive spatial or time (2/3) periodicity, while 2^nx3^4, which is its 4-fold difference is space/time (inclusive) periodicity, which is **81**-162-324-**648**-1296-2592-**5184** within (n=0 to 6) and 10368-20736 without. 5/2 (gravity) of such 3/2 (time) is 1/10 fractal difference, which are 5-fold multiples of ½ difference (405-810-1620-3240-12960-25920-51840).

Difference between such two waves is 3-fold multiplications/divisions (double/inverse fractal difference)—i.e. difference between 648 and 5184 and/or 81 and 648 etc. It is 3D difference in horizontal and 10D difference in vertical (angular) direction or 10/3 =3.333 surface frequency, which is 30-fold inertial curvature (1/3.333=0.3=30/10^2). 30x3.333=10^2 double/inverse (light speed) difference, which is resonance of inertion/inversion that generates double inertion. 3.333 is singular ½ quantum of double/inverse 6.666, which is 15-fold inertial curvature (6.666x15=10^2). 86400 sec within (one day) in context of inertial 3.333 in between is 25920 year without (precessional period), which is boundary (time) without of such time within. Time of one day and time of one precessional period are 3D inertion/inversion—one is space, while another one is its time. 40000 km of Earth's size is 12000 km (non-inertial) diameter in context of inertial 3.333 in between and 12000 km within is 3600 or 10-fold (fractal) multiple of time of one year (360) without. Inertial 12732 km diameter is 12x10^3+6x12x10+12 or 12012 km in context of inertial 720 km difference in between, which is fractal 6-fold double/inverse multiplication/division (720 is 0.06x12000 without and/or 60x12 within, which is double/inverse 10^2 difference). 360x10^2/3.333=10800, which is ½ of 21600 and 108x10^2/3.333 =3240, which is ½ of 6480.

Inertion (quantum) of 5184-year wave is **3^4=81** and inertion of 6480-year wave is **10/2x81**, where 3^4 is inertion here and now (3D space in curvature of its time (4D)). 3^4=81 is double/inverse curvature of double helix, which is 1/81=1.234567../10^2 (1.111^2). 81-fold curvature of fractal ¾ reality is singular quantum of infinity of relative space/time waves. I.e. orbital diameter of 300 mil km is 90° curvature in context of inertial 3.333-fold infinite curvature in between (300/3.333=90), where 1/90=1.111/10^2; 90/3.333=27=3^3 and 27/3.333=8.1 or 1/10 of 81 (1.234567 infinite curvature of double helix. 3-fold 3.333 curvature of 8.1 is 300 (orbital diameter), which is 10^3 fractal difference (multiple) of light speed. Earth and solar reality are double/inverse light speed difference in all 3 directions of inertial 3D. 10^3 is 1000 sec of relative discrepancy between solar and Earth's time, which is 16.666 min or **60**-fold discrepancy between sec and min and/or 0.2777 hrs or **3600**-fold discrepancy between hour and sec, which is **86400**-fold discrepancy between time of one second and time of one day (86.4-fold discrepancy). 81-fold curvature is fractal curvature between double/inverse 9-fold curvature, which is infinite 1.111 repetition. (81=9^2, which is 1.234567.. infinite differentiation between double/inverse infinite 1.111 repetitions). 9 is 3^2, which is fractal 3/2 without (time) of fractal 3x2=6 within (space). 81 is fractal between 90=10x9 without and 72=8x9 (Earth/Sun) within, which is fractal 9^2/(8x10) or fractal **81/80=1.0125** (singular quantum within and/or infinity without).

1/0.9=1.1111 is one strand (space), 1/0.81=1.234567 are two intertwined strands (time), while 1/0.72=1.3888 is 3 intertwined spirals (gravity). 81=80+1 or $\mathbf{2}^3$x10+1, i.e. fractal ¾ (81) without is double/inverse fractal (decimal multiple) of fractal 2/3 (8) within in context of singular inertial difference in between (1—here and now). 2/3 and ¾ are 1D fractal difference (10-fold multiple) or space/time duality (3^4=9^2=81-**1**=8x10 and/or 8+**1**=9=3^2 and 9x8=72). Fractal 81/80 without (1.0125) is fractal 9/8 within (1.125), which is fractal 1/10 difference. 2x1/2 multiples (Bode's periodicity) of 81-fold curvature are infinity of space/time waves, where inertial 6480-year period is 80x81 (3^8-3^4, 9^4-9^2 or 81^2-81, which is singular ½ quantum or time without of inertial space within).

Different space/time waves are parallel spaces in context of singular passage of time and/or parallel passages of time in context of inertial space. They are parallel physical realities, which are equally real, only not here and now. They are relatively different possibilities of inertial here and now, which are real in its context only fractally—i.e. they are here but not now and/or now but not here. Such different possibilities are real only in context of past/future time and/or smaller/larger space (in context of different densities), which is E/m=C^2 or double fractal difference in between. Different spatial realities are inversely different passages of time (time reality of gravity) and different gravitational realities (double difference) are twice different passages of time (spatial reality of gravity). Spatial reality here and now consists of infinite number of 2D cross-sections of different passages of time (different spatial possibilities which are real here and now only as their surfaces). Likewise—space here and now is double/inverse infinity of linear (1D) cross-sections of different gravities (twice different spatial possibilities which are real here and now only as their linear corners between surfaces). Space here and now also consists of 4x1/4 infinity of different antigravities (contraction within of expansion without or dense mass of empty space), which are three times different passages of time (0D cross-sections or spatial possibilities that are real here and now only as corners between 3 different surfaces).

Different physical realities are resonant here and now in 2 (time), 1 (gravity) or 0 directions (inverse space/time). Singular quantum change (5184 years) is synchronicity of two parallel physical realities (2 spaces in context of one time), double quantum change (103680) is simultaneity of 3 different realities (2 passages of time in context of one gravity) and triple quantum change (2073600) is resonance between 4 different realities, etc. Singular quantum change is synchronicity between Earth's and solar physical (space/time) realities, double change is additional synchronicity of galactic reality and triple change is synchronicity of physical reality of universe. Such different physical realities are not real here and now, unless accumulated passage of time is their singular spatial quantum. In context of inertial passage of time, they are only 2D, 1D and/or 0D eigenvalue, which is exponential reduction of their reality. Exponentially larger period of time (20-fold multiples) are exponentially larger synchronicity (synchronicity of 3D with 4D, 5D, 6D etc). Only boundaries or surfaces between different realities (beginnings/ends) are double (overlapping) reality of both—i.e. they are resonant only in singular points of beginning/end of linear passage of time.

Earth's synodic/sidereal space/time difference of 240/360 or double 2/3 duality is also reality of all other planets (stars, galaxies etc), only in different fragments/multiples. Bode's periodicity is frequency of space/time interference (interruptions). It describes consistently spatial periods in time (i.e. planets embedded in solar space, which are periods of mass in empty space), as well as time periods in space (quantum changes, which are periods of energy embedded into constant passage time). Bode's fractal 2/3 periodicity of double helix is abbreviation (1D) of such 24/36 duality. 2/3 periods are only surface of their 3D periods, which are 24/36 duality. 2/3 is 2x1/2 or ½ of 4x1/4 wave, which is interference in 24/36 periods. **24+36=60** is inertion/inversion =double inertion of fractal double helix, where 24=6x4 (i.e. space) and $36=6^2$ (i.e. time) in context of inertial 60=10x6 (i.e. gravity), which is 2.5x24 (fractal 5/2 and/or 24/25), while 36 is 0.6x60 (2.5x0.6=1.5).

Orbital position of Mercury (2/5=0.4, where 0.4 is 1/10 of 4.) is singular boundary within of fractal 2/5 duality (gravity). It is fractal inversion of inertial 4D here and now. In context of fractal 24/36, which is space/time sequence of gravity (2/5), its position is **24/60=0.4**. Position of Venus is ¾ difference, which is **(24+¾x24)/60=42/60=0.7**. Venus is 2nd, while Earth is 3rd harmonic (resonance), which is yet another ¾ increase. Orbital position of Earth is **(42+¾x24)/60=60/60=1** (here and now). Earth is double/inverse refraction point between reality within (solar space) and reality without (galactic space). Inner planets are space/time inversion of double inertial reality of outer planets in context of inertial Earth in between (space of outer planets is time of inner planets and vice versa). Mercury/Venus within are spatial curvature of time, while planets without are time curvature of space. Inner 2 planets are reaction of outer 6 planets and vice versa, in inertial context of Earth, which is 3rd direction or framework of such duality. Inner planets are fractals of inertial 24 (spatial curvature of time), while outer planets are multiples of inertial 36 (time curvature of space) in context of inertial here and now, which is reality of both 24+36=60. 18 is ½ of 36 (saros). Inner planets are negative multiples of 18, while outer planets are positive multiples of 36. **Earth=60, Venus =60-18=42 and Mercury = 60-2x18 or 60-36=24. Venus is only planet, which is fractal ½ and not multiple (whole number) of 36.** Fractal ½ is spatial reality turned inside out (repulsion of attraction). All other planets are space, while Venus is time—hence its inverse (retrograde) motion. Inverse motion is reaction of inverted arrow of time, which is direction from future to past and not from past to future. Venus' spiraling direction is anti-parallel from all other planets—while they spiral away, Venus spirals toward Sun, where away in space is toward in time and equally vice versa.

Venus is only fractal (temporary) member of solar system, because its reality is not in space but in time, which is limitation of its spatial reality to only singular quantum period of time (5184 years). Such 5184-year period of constant passage of time here and now (Earth) is no time at all, but instant spatial reality of Venus. Time of all planets is solar space, while time of Venus is solar time or galactic space. Venus is anti-planet stuck in space for one quantum period of time, which is passage of time only here and now and not in context of double (solar) time that is also Venus' time. Venus' orbital motion is not its inertial passage of time, but its double/inverse time. It is its instant quantum change in context of no time. Venus' inertial passage of time is not solar but galactic orbital. Venus is infinitely eccentric solar orbital and perfect round galactic orbital. 5184 years of its solar orbital, which is passage of time here and now, is only 1-day quantum change of its space/time reality. Venus is captured in inertial spatial sequence of solar system by quantum change of Earth's reality and it is released by its next, equal and opposite change. Venus is time balance of spatial Earth. Every quantum period of constant passage of time here and now is relatively different 2nd planet (Venus). Earth instantly leaps in time at the end of 5184-year period of passage of time because of accumulation of fractal gravitational imbalance caused by such anti-planets.

Gravity is 2D fractal difference, which does not differentiate between space and time, which are 1D fractal difference. It transcends space and time boundaries, because it is double/inverse time of both. Constant passage of time is consequence of interaction between double/inverse gravity (attraction/repulsion) of such spatial and time planets, which are action/reaction of each other in inertial solar context. Venus is time of spatial Earth. Its spatial reality is relative space or passage of time here and now, while its time is gravity (space) here and now. Venus' space/time is negation of space/time here and now, which is boundary of its uninterrupted passage of time. Every quantum period of passage of time is relatively different negation of relatively different affirmation, which is relatively different twin planet (Venus). Period of 5184-years here and now is one day of instant quantum change of such inverse planet, which continues its galactic (time) orbital, after such period. It is equally vice versa—instant quantum change of Earth's reality is its capture in time (relative 4D space), just as Venus is captured in space (3D). One day of quantum change here and now is infinitely long period of passage of time in context of such inverse reality. Earth's fractal angular shift of its orbital motion is infinitely long period of time in which it orbits such inverse (galactic) reality. Relative (4D) space is double/inverse passage of time, which is simultaneously infinitely slower passage of time within and infinitely faster passage of time without (space/time is fractal 3/2, which is double/inverse time of inertial space in between).

1 day of quantum space/time change here and now, which is instant 1296 years in spatial context is 4x1296=5184 years of passage of time in context of its inverse reality (Venus) and equally vice versa, because passage of time of one is instant gravity of another. Brief moment, which is no time here and now, is infinitely long period of time in context of such smaller (micro-world) reality and equally vice versa—brief moment of inverse reality is infinitely long period of time here and now, which is its micro-world. Venus and Earth are on anti-parallel course through in-depth reality of surface space/time One is on course from without to within (contraction), while another is on course from within to without (expansion), where expansion in space is contraction in time and equally vice versa. One quantum period of time, one is micro-world of another, while in the following period; it is vice versa, because of equal and opposite angular shifts in time of such twin planets. No time of one is passage of time of another, which is inverse reality in following period. Time is depth of surface spatial reality and Earth/Venus are time of each other's reality. Time or depth is gradual increase/decrease of size/density, which is motion in direction without/within. Venus's gravity is Earth's time and Earth's time is Venus' gravity in context of passage of time and equally vice versa in context of no time (quantum change).

Venus/Earth are simultaneous reality only in context of 7776 years, which is quantum period of time (5184 years) in between double/inverse quantum changes (2x1296), which is fractal 7776/5184=3/2. Double/inverse quantum changes and passage of time in between here and now are two passages of time without/within quantum change in between in context of inverse Venus's reality, which is 2x5184+1296=11664 years and 11664/7776 is fractal 3/2. Quantum change is brief capture of Earth by its inverse reality in context of inertial passage of time, which is infinitely long period in context of such inverse time, when Earth is 2nd planet (Venus) of such inverse reality and equally vice versa. Venus/Earth are fractal ½ of each other's 2x1/2 space/time duality in context of solar reality, which is their inertial framework (angular ½). They are double/inverse realities of solar inertion. Passage of time is 3D, while quantum change is instant time (4D). 3rd ½ is not real in spatial but only time context. 2x1/2 in period of constant passage of time is fractal 3/2 in context of time.

Fractal 3/2 is space/time reality. It is surface spatial reality in context of passage of time and its negation in angular depth, which is spatial reality of time, when time is not passage but instant space. Double/inverse duality of Earth/Venus is exclusive spatial reality, which is gradual alternation of different planets that are synchronized in solar context on their double/inverse motion through depth of spatial reality (time). Duality is surface of 3D (2/3), which is synchronization of 3 planets in 4D context (3/4), which is instant spatial reality of passage of time. Earth/Venus are double/inverse fractal realities of each other, which is exclusive spatial reality (passage of time). Space/time reality is synchronicity of space and time, when not only space, but also time is instant space. While passage of time is exclusive spatial reality (3D), quantum change is inclusive space/time reality, which is synchronicity of 3rd ½ or simultaneous resonance between 3 fractal realities. It is instant Earth/Venus duality in context of its equal opposite, which is 3rd (neutral or inertial) planet of such double/inverse (positive/negative) reality. Quantum change is instant incorporation of inertion and its inversion into double inertion. It is resonance between two waves, which creates double/inverse wave. Resonance between Earth/Venus fractal realities is singular quantum, when they are not 2x1/2 but only ½ of larger/smaller 2x1/2. In context of quantum change, Earth and Venus are not ½ but ¼ each, while reaction of such resonance is 3rd synchronized planet, which is their equal and opposite ½.

3D (passage of time) is synchronicity of two fractal planets, 4D (quantum change) is synchronicity of 3, 5D (double/inverse quantum change) is synchronicity of 4 planets etc. Infinity of different time possibilities of inertial space are equally real, only not here and now. Quantum change is synchronicity between two different (double/inverse) time possibilities; double/inverse quantum change is synchronicity of additional possibility etc. Quantum change is two time possibilities of inertial space (and/or two spatial possibilities of one inertial passage of time), when reality, which is real only here but not now, is real here and now. Inertial space (here) is infinity of different passages of time. It is infinite number of parallel spatial realities, which are not real now. Synchronicity in context of quantum change is instant reality of such different possibility, parallel with reality here and now. Exponentially larger change is synchronicity of larger number of space/time waves, which is synchronous reality of exponentially larger number of spatial possibilities, which are only empty space in context of inertial passage of time. Earth and Venus are fractal ½ each (duality), while quantum changes are instant additional spatial realities of different planets, which are exclusive time reality in context of passage of time here and now. Earth/Venus are 1D fractal difference (space/time), while 3rd planet, which is real here and now in context of quantum change is 2D fractal difference (gravitational difference). 2D fractal difference in context of passage of time is 1D fractal difference in context of no time (quantum change). Quantum change is no difference between Earth and Venus, which are equal opposite of their 1D fractal difference, which is 3rd ½ of such duality.

It is synchronicity with a planet, which is beyond spatial boundaries in context of passage of time, because it is real only in depth of surface spatial reality. Orbital of such planet is larger than 90° angular curvature in context of passage of time here and now. It is real only in time but not in space (opposite side of a sphere). Its spatial reality in context of passage of time is not now. Only when Earth is synchronized with such different passage of time (quantum change), is its planet real now (orbital less than 90° angular curvature). Because it is different angular curvature, passage of time of such planet is infinitely faster (slower spatial motion, which is faster time motion). Such different passage of time (different spatial possibility of time) is real only in context of quantum change, when time here and now is not time, but space. Infinitely faster passage of time of such planet is same passage of time as here and now in context of quantum change, which is acceleration of passage of time here and now and/or deceleration of passage of time of such planet. While such planet of infinitely faster passage of time is beyond boundary of inertial space here and now in direction without, planet of infinitely slower passage of time is beyond boundary in direction within. Infinitely slow passage of time is within (dense mass), while infinitely fast passage of time is without (empty space). One is instant reality within in direction of dense solar core (Venus) and another is gradual reality without in direction of empty of space (dense core of galaxy), which is real only in quantum periods of time.

Space is only surface of time and time is surface of space. Physical (space/time) reality of different planets are synchronized only spatially and not in time. Their times are fractionally faster/slower. Different mass is synchronized reality only on surface (space), but not in depth (time). Spatial reality without its equal and opposite time is meaningless empty shell without its content. Planets that are real here and now are only fractal (less than light speed) difference between their passages of time, while planets that are not real here and now (infinity of different possibilities) are multiple (quantum) difference, where fractal in time is multiple in space and equally vice versa. Planets here and now are fractal of spatial light speed difference that are multiples of time light speed difference, while planets that are not real here and now, are multiple spatial light speed difference and fractal time light speed difference. Larger spatial distance is smaller time difference and smaller spatial distance is larger time difference (Venus is infinite time difference and singular spatial distance, while 3rd planet that is synchronized during quantum change is smaller time difference, which is infinitely larger spatial distance). Positive spatial distance is negative angular difference and equally vice versa. Only Earth is here and now, which is double/inverse reality of its space/time, while any different/distant mass (empty space in between) is exponential angular distortion of its space/time reality. Micro-world within is exponential horizontal distortion of angular difference.

Mass of universe is distorted through angular rotation of space in time, while mass of micro-world is distorted through linear compression. Micro-world is infinite space in depth that is compressed (instantly) into surface here and now, while universe is (gradual) rotation of such space, which is its equal and opposite decompression in time. Only here and now is space/time (physical) reality. Here and not now or now and not here is exponential fractal distortion, which is increasingly only surface (spatial) reality without its content (time). Different/distant mass is not space/time resonance but only its distortion. Different/distant mass (without or within) in context of inertial here and now is not its reality but its negation, just as Earth's reality is not real in their inertial context. Only space/time reality is Earth, while different/distant mass/energy is exclusive spatial or time reality, which is only ½ of their reality. Increased spatial distance is exponential distortion of inertial E/m here and now, which is decreased mass and/or increased (positive or negative) energy of such different mass. Such different/distant spatial possibilities are increasingly distorted time, therefore increasingly alien environment with increasingly unbalanced mass/energy ratio (i.e. quasars, pulsars etc). Infinitely distant space is infinite distortion of its mass, which is not mass but energy, just as energy within (micro-world) is not energy but mass in context of inertial passage of time here and now. Earth in inverse perspective of such different spatial realities is equally alien and unbalanced (fractal) energy/mass ratio, devoid of its double/inverse time content. Reality without/within is infinitely faster passage in context of infinitely slower passage of time and/or vice versa, which is acceleration/deceleration of their inertial passage of time.

Only Earth is whole reality, which has Moon and/or twin planet—Venus, because only Earth is singular point of refraction between all infinite space/time directions. Because of distortion of reality with inertial here and now, which is one out of infinite number of possibilities, all such different possibilities are not equally real. Point of observation is always in perfect rest (no change of size/motion), while its reality without/within is in constant change. Because of such fractal (relative/uncertain) perspective, only Earth is physical reality (space/time), while its universe/micro-world is its double/inverse exponential distortion. Because of inertial perspective here and now, rest of universe/micro-world is non-inertial reality. Universe/micro-world is exclusive Earth's reality, which is totally unreal in context of any relatively different passage of time. It is reaction of earth as focal point of any reality. Regardless of shifting of fractal space/time boundaries from Earth as center of universe to solar reality and subsequently galaxy and beyond as center of universe, Earth (here and now) remains singular point within of infinitely larger reality without, which is reaction of one-sided shifting of spatial boundaries without equal and opposite shifting of time boundaries. Here and now is always ½-way between all its double/inverse directions. It is ½-way between dense mass and empty space, ½-way between universe without and micro-world within, ½-way between past and future time etc. Expansion of spatial reality is simultaneous contraction of its time and equally vice versa—expansion of time is contraction of space.

Double/inverse expansion/contraction of space/time is change of changeless reality, which is here and now (space/time) in every point and not only here and now. Earth as central point of all its spatial and time directions is infinite relativity without that is infinite uncertainty within. It is double/inverse infinite distortion of physical reality without/within, which is same reality as here and now. Earth as central (inertial) point of spatial and time reality, where only reality without/within changes is not only infinite distortion of distant/different reality, but also infinite distortion of reality here and now, which is not only inertial but also non-inertial point. Every change without/within is double/inverse change here and now and equally vice versa—every change here and now is double/inverse fractal change of universe/micro-world. Every planet (and every different/distant point of empty or dense space or time) is same reality. Every planet (and every star, galaxy, photon or electron) is Earth, which has one Moon and one fractal twin (Venus)—only not here and now.

Such different possibilities, which are larger spatial distance and/or larger angular (time) difference, are synchronized only in exponentially larger quantum periods of passage of time (where angular difference within is spatial distance without and angular difference without is spatial distance within). Universe/micro-world in perspective of inertial here and now is only fractal (exclusive) spatial or time eigenvalue of only point that is space/time eigenvalue in between (Earth). **Fractal synchronicity is synchronicity that is less than light speed difference (increased relativity/uncertainty of reality without/within). Larger than light speed difference is not synchronicity at all (more than uncertain/**

relative, which is negative probability). It is synchronous only in context of quantum changes, which are instant reductions of inertial difference between here and now. Boundary of fractal spatial synchronicity is its double/inverse (future/past) time and boundary of fractal time synchronicity is its double/inverse (larger/smaller) space. There is no space that is real here and now beyond boundary of time and/or time beyond boundary of space, unless in context of quantum change, which is instant double/inverse change of boundaries and their content. Quantum change is change of space/time boundaries within/without, as well as change of inertial here and now, because it is double/inverse light speed change.

All planets are only fraction of light speed difference, while Venus is complete light speed difference (almost complete, because there is always small discrepancy, which is inertion of here and now in context of such duality). Spatial surface of Venus projected into inertial passage of time here and now is not only its orbital/rotational motion—it is double/inverse motion or additional light speed motion. Spinning wheel appears as rotating in opposite direction from actual spinning, because of precessional effect of such motion, which is reaction or equal opposite of such spinning. Such inverse rotation is infinitely slower than inertial spinning, because precessional curvature is space/time (light speed) fractal difference. Double precessional curvature is not motion at all—it is double motion that is inertial size. Because Venus is time of inertial space, it is additional light speed motion in opposite direction of its rotation, superimposed over its spatial surface. It is spinning in opposite direction of retrograde rotation with light speed motion, that is not motion at all but instant size (double motion). Such light speed motion is surface (spatial) reality of Venus. Venus is precessional reality—its reality here and now is not its gradual motion but precessional inverse rotation. It is only inverse spinning of a wheel and not its actual, infinitely faster motion that is space of such inverse rotation.

Small inertial difference between actual (time) and light speed (spatial) motion is time of its retrograde rotation. Spatial reality is multiple of 36 (360 days of a year), while time is multiple of 24 (synodic/sidereal 240 sec difference) and/or vice versa. 36 is $(2x3)^2=6^2= 4x3^2$ or double 3D (curvature), while 24 is flattened double 2D—$6x2^2$. Period of Venus' retrograde rotation is 243 days, which is 10x24+3, which is 240 in context of inertial 3D here and now (just as Earth's orbital is 10x36+5, which is 5/3 or double fractal difference, where 2D is time of 3D space). Synodic/sidereal spatial difference of Earth's motion is time of Venus' rotation, because its spatial reality is constant passage of time here and now. It is 36/24 or fractal 3/2 inversion. **243= 3x81, which is fractal 3D wave between 2x81=162 and 4x81=324.** Venus is synchronized spatial reality exclusively while its relative period of rotation is 243 days (and/or orbital of 224 days). Such relative ratio between Earth and Venus space/time changes with quantum change of Earth's reality. Time difference between Venus' rotation and orbital is 18.3 days, which is reflection of its fractal spatial reality (36/2=18). Earth's orbital motion is $3x10^2+60=360$ (36x10+5), while Venus' orbital motion is $2x10^2 + 24 = 224$ ($24x10-4^2$). Time difference between Earth and Venus' orbital motion is φ (golden mean), which is only possibility (eigenvalue) of such space/time duality (inertion/inversion = double inertion). 360/220=1.6363...(perfect dual ratio) that is inversion of 0.6111 or 0.6111x10=6.111, which is infinite repetitive strand of double helix in context of inertial 6=3x2 (here and now).

All planets, except Venus, are multiples of 36. They are difference of whole numbers, while Venus is only fractal ½ (18) difference. Other planets are space/time difference (2x18=36), while Venus is only time difference. Mars is **36**+60 (Earth)=96/60=1.6 (equal opposite of Mercury), asteroid belt is **3x36**+60=168/60=2.8, Jupiter is **7x36**+60=312/60 =5.12, Saturn is **15x36**+60=600/60=10 and Uranus is **31x36**+60=1176/60=19.6. Each following 2x1/2 multiple is twice previous multiple + inertion here and now (1)—i.e. 2x1 (Mars)+1=3 (asteroids). 2x3+1=7 (Jupiter); 2x7+1=15 (Saturn) and 2x15+1=31 (Uranus). Mercury is—**1**, Venus is—**½**, Mars is **1**, asteroids are **3**, Jupiter **7**, Saturn **15** and Uranus **31**-fold multiples of 36 (2^n periodicity for n=0 to 4 without and n=-1,0 within). Earth is 0 point or singularity of such double/inverse curvature without and within, which is 5/3x36 or (1+2/3)x36=60 in 3rd (inertial) direction (where 5/3=1.666 is inverse $0.6=60/10^2$). There is no irregularity in Bode's periodicity—it is perfect fractal curvature of double helix (3/2). It is singular curvature that determines infinite reality—regardless if its quantum periods (loops) relatively describe space, time or gravity. Mercury/Venus are fragmentation (time) and outer planets are multiples (space). Mercury/Venus are negative multiples of 18 and outer planets are positive multiples of 36 (double multiples without of inverse multiples within). Only Earth is eigenvalue of both (60=24+36 or 240x360=86 400)—space and time. Only Earth is whole number of double/inverse fractals, which are spatial surfaces of time and/or time surfaces of space.

Mercury/Venus within and Mars/Uranus without are action/reaction of inertial here and now in between (Earth), which are double/inverse 1D fractal difference (relativity/uncertainty). They are resonant only in singular point of inertial here and now (Earth). It is exclusive Earth's reality, which is totally unreal from any different vantage point. 3rd direction (inertion) of such dual action/reaction of inner/outer planets (2x1/2 fragmentation/multiplication) is Neptune/Pluto duality, which is gravity without of inertial space/time within. Neptune/Pluto are last 2 out of 10 planets, which is 5th quarter (5/4) or inertion (gravity) of such symmetrical space/time wave. Neptune/Pluto without and Mercury/Venus within are 2x1/2 multiplication/division of inertial 6 planets (Earth/Uranus) in between. They are double/inverse time without and within of inertial 6-fold spatial sequence in between. Neptune/Pluto are spatial expansion (acceleration) without, just as Venus/Mercury are spatial contraction (deceleration) within of linear 3/2 periodicity in between (where expansion in space is contraction in time etc). They are equal and opposite non-Euclidean curvatures of Euclidean curvature of 6 planets in between. They are equal and opposite 2D surfaces of 3D cube in between etc. Orbital position of Neptune is 29.4=1764/60, where 1764=**49x36 (49x36=3/2(31x36+60)).**

Earth is only fractal (½) inertion of Neptune's reality. All 6 planets are its singular quantum (inertion), because 6 planets are singular (space/time), while Neptune is double fractal difference (space/gravity) of Earth's reality. Earth's space/time shell (exclusive solar gravity) ends with Uranus, while Neptune/Pluto are outer gravitational shell, which is overlap between solar and galactic gravitational realities. Venus/Mercury (2) are ¾ fractals within, Earth/Uranus (6) are multiples in between while Neptune/Pluto (2) are inverse $^2/_3$ fractals without. Neptune is fractal ½ reality between second (Uranus) and third shell (Pluto), just as Venus is fractal ½ between first (Mercury) and second (Earth) shells. Neptune is fractal ½ (time) reaction of 2nd shell (Uranus) and Pluto is space/time or double/inverse reaction, which is $31+2^5$=**63x36**+60=2328/60=38.8.

Orbital position of Saturn (10-fold repetition) is fractal inversion of one direction of inertial here and now (¼ of 4D). Uranus' position of imperfect 20-fold repetition (0.4 relative discrepancy) is inversion of 2 (½ of 2D), while Neptune's orbital, which is imperfect 30-fold repetition (0.6 fractal discrepancy), is inversion of 3 out of 4 directions (¾ of 4D). ¾ of 4D are 3 spatial directions of 4D space/time. Neptune is fractal boundary between 3D space and its time without—Pluto is not 3D (space) but 4D (time). Orbital position of Pluto is 40-fold repetition (1.2 angular distortion) or 2x20, where 20-fold repetition is singular quantum of fractal 4/5. Neptune/Pluto is equal and opposite (inversion) of Earth/Uranus (inertion). Uranus is ½-way between Earth and Pluto in context of gravity (just as Saturn is ½-way between Earth/Uranus and Jupiter is ½-way between Earth/Saturn etc). Saturn/Uranus/Neptune/Pluto are 4 inversion points of 4x1/4 wave without, just as Mars/asteroids/Jupiter/Saturn are 4 inversion points of smaller 4x1/4 wave within. Such 4 quantum steps without (inertion/inversion=double inertion) are only 2 quantum steps within (Venus/Mercury), because of fractal 3/2 asymmetry that is expansion/contraction (angular acceleration vs. linear deceleration) without/within. Venus/Mercury are horizontal, while outer planets are angular surfaces of inertial 3D sphere in between.

Orbital position of Pluto is space/time (4D) boundary of inertial here and now. Reality beyond is not real here but only now and/or vice versa. Reality beyond is galactic and not solar spatial reality, which is not real in context of passage of time, but only in context of gravity, which transcends time. **Spatial orbital reality of 10 planets is singular spatial quantum of relative (solar) space, just as 5184-year period is its singular uncertain quantum.** Infinite number of 5184-year periods is such singular spatial quantum, just as infinite number of such spatial quanta is singular time period, because time is spatial density and space is equal and opposite time density (fractal content of empty space/time). In context of all its time spatial here and now expands from singular point to size of entire spatial quantum of 10 planets, just as space here and now is condensed time of all such time periods. There is no further spatial orbital reality (4D) in context of inertial here and now. Any further reality is not space/time but gravitational reality, which is not orbital in context of solar space, but orbital in context of solar time or orbital around core of galaxy, which is infinitely eccentric solar orbital (linear direction). Such galactic (gravitational) reality is 5D or double fractal difference from inertial space here and now (singular fractal difference of its time).

Fractal curvature is infinite, because 2/3 within is ¾ without, ¾ within is 4/5 without, 4/5 within is 5/6 without etc, where within in time is without in space and equally vice versa (anti-parallel double helix). It is infinite double/inverse precessional curvature of inertial here and now. Pluto is 5th quarter, Neptune is 4th third and Uranus is 3rd half etc. 5th 1/6 and further fragmentations are equally real, only not here and now. Fractal 4/5 is boundary of inertial 3/2 (here and now), because it is double fractal difference. It is relative boundary in context of constant passage of time (3D), but not boundary in 4D context, which is not passage but instant time (space), while passage of such instant time is solar time. Pluto is relative boundary in context of Earth's passage of time, but not in context of any other passage of time. In context of inner planets, solar system starts and ends before in space, while in context of inertial passage of time of outer planets, it starts and ends later in space, because spatial reality without/within is expanded/contracted in space and/or vice versa in time. Quantum change of space/time is instant broadening of horizons, when boundary of space is not time but gravity. It is instant change when double/inverse relative space and not only its singular quantum (10 planets) are spatial reality.

In context of quantum change—time **is** spatial reality (and gravity is time). If here and now (passage of time) is fractal $^2/_3$, then context of quantum change (instant time) is ¾. Boundary of $^2/_3$ is 4/5, while boundary of ¾ is 5/6. 5/6 is boundary of spatial reality in context of quantum change of reality here and now, which is additional spatial orbital of 11th planet of solar reality that is not real in context of passage of time, because its angular curvature is larger than 90°, which is infinitely eccentric solar orbital (round galactic orbital). 11th planet is inversion of inertial here and now, which is 90° angular curvature of space/time or total (light speed) relative distortion of reality, which is exclusive Earth's reality. 11th planet is massless particle in context of passage of time here and now, because its gravity is real, while its spatial reality is not. It is wave (time) in context of passage of time here and now and particle in context of no passage of time (which is passage of solar time that is gravity here and now). In context of repetitive double helix (1.111) there are 11 spatial orbitals, just as there are 11 years (time) of solar activity. 11th planet is fractal (½) solar reality, which is ½-way between solar and galactic cores. It is overlap between double/inverse solar/galactic realities, which is reality without of inertial reality of 10 planets within. 11th planet is binary orbital between Sun and core of galaxy. Just as Pluto is spatial boundary, 11th planet is time boundary of solar space/time (which are time and gravitational boundaries of Earth's space).

11th planet is real here and not only now (and/or now and not only here) in context of solar passage of time, which is time of quantum change of Earth's reality. 11th planet is inversion or negation of inertial reality here and now, because time here and now is its instant space and space (gravity) here and now is its time. It is 3rd planet of Earth/Venus duality, which is spatial possibility of time only in context of solar passage of time (quantum change). Passage of time of 11th planet is infinitely faster (light speed difference). Gravity is double fractal difference—it transcends space/time boundary. Although such additional mass is not spatial reality here and now, its gravity is, which is cause of additional gravitational disturbances in Pluto's orbital. Small gravitational disturbance of such inverse reality (Pluto) is infinitely large (double) disturbance, which is not disturbance at all but constant spatial reality here and now. Small gradual attraction in context of Pluto's orbital, which is inversion without of Earth's reality within, is instant large repulsion here and now (equally vice versa). Small inertial disturbance in time is large inverse disturbance in space and equally vice versa.

Spatial disturbance of Pluto's orbital, caused by gravitational reality of 11th planet, is disturbance of time here and now, where time disturbance is not disturbance but constant reality in spatial context. Reaction of Pluto's orbital are small spatial eccentricity, while reaction of Earth's orbital is infinitely large inverse spatial eccentricity, which is spatial reality here and now, as well as its boundary, which is small eccentricity of its passage of time. Disturbances are exponentially smaller with spatial difference (smaller fractal time synchronicity); therefore such infinitely large inverse disturbance in distance is not constant space as here and now, but only its minor (inverse) discrepancy. Spatial reality here and now is minor discrepancy between attraction and repulsion (gravity/antigravity) without and within (spatial and time gravity). Increased spatial distance is decreased time synchronicity, therefore increased asymmetry between gravity/antigravity, which is additional gravitational disturbance of such distant mass. Spatial reality here and now is instant reaction of such double/inverse gravity of 11th planet, which is its twin in time, just as Venus is its twin in space. 11th planet is space of Earth's reality and time of Venus' reality in context of passage of time and vice versa in context of no time (quantum change of reality).

One binary orbital of such planet, which is in relative space is period of 6480 years in context of inertial passage of time here and now, because it is period of synchronicity between two equal opposites that are 1D fractal difference. 6480 years is inertial period of non-inertial 6480x1.111=7200 years, which is ¼ of 28 800 years (where 28800 is round curvature without of a square within, which is 4x6480=25920). 25 920 is only fractal reality within of quantum reality without, which is difference between pentagon and square. It is only relative/uncertain curvature of inertial time of 28800 years. Space is fractal interruption of time and equally vice versa. Consequently—interrupted time period of 25 920 is 28 800 years of uninterrupted time and/or vice versa (and/or 7200 of 6480 years). 0.8x6480=5184 years of passage of time within is 1/0.9x6480=7200 years without (where 7200-5184=2016 or 9.333x216, which is 144 difference from 10x216=21600). 7200-6480 is 720, which is 1/10 or inertial 1D difference of such duality, which is carried into next cycle of time. Interrupted time is not time but space. It is only gradual (constant) change of such 4D spatial reality in context of its time, which is gravity here and now. Consequently, ½ orbital or motion from one to another focus of its eccentricity (its perihelion/aphelion distance, which is distance from core of galaxy and here and now) is period of 3600 years. 3600 year of passage of time here and now is 25 920 light years in spatial context of galaxy (7.2-fold difference, which is inertial reality of such double/inverse reality). As such planet on its eccentric (linear) orbit approaches Earth's orbital, it is an exponentially increased gravitational tension, which concludes in double gravitational inertion and instant quantum change here and now.

Binary orbital is simultaneous—it is synchronous double/inverse orbital around two focal points at the same time. It is not linear but 2D (double) motion or rotation and orbital motion. Surface of Earth is also binary orbital—orbital around double/inverse centers of Earth's and solar gravity. Orbital around center of Earth's gravity is dense mass, while orbital around center of solar gravity is empty space—one is motion in space, while another is motion in time. Orbital/rotation is double/inverse motion, which is not motion but instant size in between, which is their line of symmetry or (eigenvalue). Consequently, such 11th planet in context of passage of time here and now, is not at certain point in its orbital, which is beyond Pluto's orbital—it is everywhere in its orbital at the same time, because time is relative space and/or space of such planet is uncertain time here and now. Binary (2D) orbital of such planet, which is its simultaneous solar and galactic orbital, is its instant spatial surface. Its solar (2D) orbital is its rotation. 11th planet is planet beyond inertial spatial boundary—it is planet that has infinitely expanded and its dense mass within is empty space without inertial Earth. Such planet is 2D fractal difference—it is energy of massive Earth and/or repulsion of its attraction. It is equivalent of solar reality—it is its outer surface, which is not infinitely expanded space without but dense mass within in context of quantum change here and now, when Sun is not equal opposite but only equal of Earth's reality. In context of quantum change, Sun and Earth are not attractive but repulsive reality—i.e. they are planets in their galactic orbital.

11th planet is outer (time) shell of solar reality and its rotation is double solar orbital, while its singular (1D) orbital is galactic orbital. Sun is gravitational center of 11th planet, while Earth is ½ way between its center and its surface. Quantum change is instant expansion (acceleration) of Earth in time and equal and opposite contraction (deceleration) of such 11th planet, when they are not inverse reality of universe and its micro-world (solar/Earth duality), but two equal opposites in context of inertial galaxy. Consequently—in its solar orbital, it does not approach from linear spatial distance beyond Pluto's orbital. It densifies (contracts) from angular (2D) direction, perpendicular on ecliptic just as Earth de-densifies (expands) in such outward time directions. Point of intersection of such double/inverse motion (pulsation) in depth of spatial reality (time) is instant quantum change of reality here and now. It is ½ expansion of Earth and ½ contraction of such additional planet in time in context of inertial gravity (galaxy or 3rd ½ of solar/Earth duality). Its orbital is in time—it contracts in inertial space here and now and materializes in context of spatial solar system, swings around Earth and replaces Venus, which swings in another direction and contracts within. Such binary (2D) orbital without is binary particle of one nucleus within

(Earth/Venus)—i.e. binary motion without is inverse binary size within. Such swing around Earth is instant quantum change of Earth/Venus space/time duality. Period of 6480-years here and now is instant quantum change of reality of Venus (one day), while instant quantum change of Earth's reality (one day) is infinitely long 6480-year period of Venus' inertial reality.

One day of quantum change in context of Earth is infinitely long time in context of Venus' reality, because Venus is micro-world (inverse time within) of Earth's spatial reality, just as Earth/Venus duality is micro-world of such 11th planet (double time without). Venus is only an atom that has instantly expanded in its time (which is duration of 6480 years here and now), just as Earth is only an atom within such 11th planet and one day of its quantum change is no time at all in context of such planet. Venus/Earth/11th planets are 3 space/time shells—gravity, space and time (or mass/space/time). They are 3 inertial curvatures that are double fractal difference in between (Earth). It is rotational light speed within (Venus) and linear light speed without (11th planet), which is double light speed or gravity here and now. Earth is double light speed motion, that is not motion, but instant size (reality of its surface). Venus/Earth/11th planets are inertial here and now in 3 directions that are 90° curvature in between (inertion/inversion = double inertion). Their interaction is triple self-interference of double helix that generates new sequence of reality every 6480 years.

Their interaction is spatial reality of time. Such fractal quantum steps of their space/time synchronicity are not only spatial but also time motion through depth of space (density). It is motion without/within and 6480-years is period when reality of one precedes reality of another. They are 3 different time possibilities of inertial space and/or 3 different spatial possibilities of inertial time. Venus/Earth is dual spatial possibility of time (11th planet), just as 11th planet is dual time possibility of Earth. Likewise—Venus and Earth are two time possibilities of same spatial reality—they are same space in two different times. Venus and Earth are two time mirror images of same space (mirror, which is 11th planet), just as Earth and 11th planet are two spatial mirror images of same time (mirror, which is planet Venus). They are 3 different densities of same reality, which are only 1D and 2D (quantum) fractal synchronicity. 129 000 year period of double/inverse precession is 4D eigenvalue or eigenvalue of 4 different possibilities of inertial here and now, which is yet another—12th planet that is galactic shell. Earth is atom turned inside out; solar reality is Earth turned inside out or an atom turned inside out twice (in two out of 3 directions). Galaxy is solar reality turned inside out or an atom turned inside out three times in all 3 directions, while universe is galaxy turned inside out etc. Singular inversion without is binary nucleus within (or binary rotation without); while double inversion is triple nucleus etc. 9-fold inversion is 10-fold nucleus within. 9-fold or triple 3D inversion is 10 particles (planets) within. 10 planets/electrons within are reaction of 9 space/time shells without in context of inertial here and now (Earth), which is their inertial difference. 10 planets of solar system are reaction of fractal 9/10 reality of universe, while infinite numbers of electrons are reaction of infinite fractal reality (2x1/2 multiplication/division) of double helix.

CHAPTER 16

TIME BUBBLE

Infinite fractal differentiation or total surface randomness is largest symmetry—it is symmetry of 1. Such infinitely asymmetrical surface is reflection of infinite symmetry or self-symmetry that has no comparable opposite. Only two out of infinite number of equal opposites are even or symmetrical reality. Incomparable reality without equal opposite is infinite surface randomness. Infinite randomness is incomparable singularity turned inside out. There is no difference between 1 and infinity of larger/smaller numbers—they are two equal opposites of same inertion. Only nothing is something, just as something is nothing. Only no freedom is infinite freedom of choice and equally vice versa. Only singular non-dimensional point of no size/motion/density etc. is infinite variation without, where all such variations are equal in sum of all their fractal attributes. Only nothing is perfect—something is always diminished perfection, which is inertion or gravity of such "void". Something of nothing is eccentricity (obliquity) of perfect sphere. Such something in context of its nothing is infinite resonance of its 2x1/2 harmonics (reaction).

It is infinite double/inverse precessional 2x1/2 curvature, because it is reaction of nothing; therefore nothing is its inertion. Harmonics (quanta) of such inertion is infinity of fractal 3/2 (double/inverse) Fibonacci sequences. Reaction of singularity is infinity and equally vice versa. Infinity of vibrations is step by step or quantum change of such inertion, which gradually restores the balance of non-inertial "void", which is impossible possibility, because restoration of spatial balance is imbalance in context of time and equally vice versa, which is double/inverse infinity of singular here and now in between. Gradual reconciliation of two equal opposites in their 3rd direction (gravity) is space and time of such reality, which is double/inverse self-interference of inertial here and now. Reality is double/inverse infinity of such self-interferences, of which only one possibility is inertial here and now. An object in motion moves forever through such void, because it is infinite double helix, where every fractal is equal and simultaneous. Regardless how many times such motion is 2x½ multiplied/fragmented, it is still the same motion. Any qualitative or quantitative description of such motion is only relative/uncertain (fractal or comparative) reality of inertial here and now.

Absolute description is only subjective perspective, which is exclusive reality of such inertion. Absolute reality is always duality of equal and opposite polarities (there is up without down etc). Because duality is infinity of singularity, any absolute or constant reality is not absolute but infinitely relative/uncertain. Physical or dual reality is never absolute—it is infinite relativity/uncertainty of singularity, which is its only absolute reality. Absolute reality changes parallel with changes of inertion (here and now). Only inertion is absolute reality of dual action/reaction, which is infinity of different variables. Only 3rd direction in between is absolute description of 2 equal opposites. Singularity, which is inertion of duality, is non-polarized or neutral reality—it is its 3rd direction, and because it has no opposite it is neither physical, nor describable in context of physicality (duality). Π and/or φ, which are irrational sequences of random numbers, are reflections of such impossibility of dual description of singularity. Total randomness of such infinite sequence is description of its singularity and its perfect undivided symmetry. 0 and 1 are two equal opposites—they are same reality in inverse perspectives. Space/time is double/inverse reflection of here and now in between. Inertial here and now (singularity) cannot observe itself but as a mirror image. Infinity of its harmonized (resonant) 2x1/2 vibrations is its infinite mirror image, where angular difference between reflections (harmonics) is always inertial here and now.

Each sequence or each frequency is a whole. **Double helix multiplies through division and divides through multiplication. In curvature of each other such multiplication/division is same (undivided and non-multiplied) inertion.** Double helix is inversion or mirror image of surface physical reality—it is physical reality turned inside out. It is fractal 3/2, where ach fragment is a whole. There is no difference between multiplication and division, because there is no difference between a whole and its fragments (inverse fractal is a whole and equally vice versa, which is double multiplication without of double fragmentation within). Each fragment is not just same abstraction, i.e. they do not share same reality—each fragment **is all reality** there is. Differentiation (1.23456…) is only surface of repetition (1.111..), just as repetition is surface of singular point (double/inverse echo in space and time). Only description of reality is—that it is and is not, therefore it is twice (and it is not, therefore it is 3 times, and is not, …etc). Only irrational description portrays rational reality and rational description portrays irrational reality. Affirmation of negation is negation and negation of affirmation is again negation. Only double negation is affirmation! It means—reality is and is not are both its (double/inverse) negations, and 3rd direction of such dual negation is affirmation. Only two equal opposites—only two simultaneous illusions are real. Space is an illusion of time and time is equal and opposite illusion of space. Neither is exclusive and both are inclusive physical reality. Fibonacci sequencing or double helix is perfect mirror image of such reality.

Abstraction is inertial framework of physical (space/time) reality. Abstraction is its negation, which is infinitely larger reality of singular here and now, which is neither size nor motion, by being both. Physical reality is infinite 2x1/2 echoing of such abstraction, which is infinite sequencing inertion/inversion=double inertion ($C/(1/C)=C^2$). 2x1/2 of fractal 5/3 reality is 1.666/0.**6**–0.**36**/2.777–7.716/ 0.**1296** etc, which is $6x10^{-1}$-6^2x10^{-2}-6^4x10^{-4} It is exponential increase of inversion (6-space/time) in context of equal and opposite decrease of inertion (10-here and now, which is quantum of its inversion). It is acceleration (expansion) of one in context of equal and opposite deceleration (contraction) of another. $6x6^2=6^3=216$ is fractal 2/3 in between double/inverse wholes ($36/36^2$), because 5/ 3-2/3=1 (one is positive/negative eccentricity or discrepancy of perfect curvature of another). Infinity of waves of such fractal are 216x0.111=**24**-fold repetition (galactic or gravitational fractal difference 6x4), 216x0.333 =**72** (solar time difference 360/5 or 36x2, 6x12 etc), 216x0.444=**96** (gravity, 100-4 or 60+36 etc), 216x0.555=**120** (360-240, 3x4x10, 6x20 periods or 40+80 etc), 216x0.666=**144** (12^2 or 4x36 etc), 216x0.777=**168** (1/2 of 336, where 336+24=360) and 216x0.888=**192** (double/inverse gravity). They are fractal space/time waves of inertial 2x3=6, which is Bode's periodicity.

$6^2x6^4=6^6=36x36^2=46656$ (inertion/inversion=double inertion). 46656x0.111 =**5184**, 46656x0.222=**10368**, 46656x0.444= **20736**, 46656x0.555=**25920**, which are time waves or double fractal difference without of singular fractal difference ($216=6^3$) within. Fractal in between 2x3=6 is $2^2=4$ and $4x4^2=4^3=64$ is 2/3 of 96 (gravity). 64/0.111=**576**, which is 2/3 of **864**, 64/0.222=**288**, which is 2/3 of **432**, 64/0.333=**192**, 64/0.444=**144**, which is 2/3 of **216** and 64/0.555=115.2, which is fractal (inversion). 1152x3/2=1728 and 3x1728=**5184**. 64/0.666=**96** and 64/0.777=82.285..., which is infinite fractal, because 6D is boundary of fractal 2/3 (64/0.888=**72**). Space/time is double/inverse fractal 4-6-8 or 2^2-2x3-2^3 sequence. It is infinite 2x1/2 multiplication/division of fractal 2/3 base. Singular 2/3 inertion is infinity of spatial and time waves that are physical reality of universe/micro-world. Abstraction is infinitely larger reality (here and now). Quantum boundary of inertial 6-fold repetitive sequence is $4x6^4$=5184, which is ¼ of $3.333x6^5$ =25920. 28800=1.111x25920 is circle without a triangle (3x8640=72x36x10=25920), while square (4x6480) is its precessional (fractal) curvature, succeeded by pentagon (5x5184) and hexagon (6x4320 etc).

There is infinite number of fractal harmonics between two whole equal opposites and there is infinite number of different passages of time between two spatial realities, just as there is infinite number of different spatial sizes between two times. Only resonant harmonics here and now are real—rest is empty space. Change of time is equal and opposite (inverse) change of space. Through precessional (quantum) changes of time, such "hidden" spatial harmonics are gradually brought to surface and equally vice versa—through spatial changes, unreal time possibilities are brought from its angular depth. There is infinite number of planets in context of solar reality and infinite number of stars in context of galaxy, which are not real, because they are not eigenvalue of inertial here and now. They are real only in context of relatively different quantum time periods (light speed difference). There is spatial reality between two times and there is time between two spatial realities—one is surface of another. It is back and forth infinite fragmentation/multiplication of such fractal duality, which is double/inverse infinite density here and now. It is infinite spatial density of time, which is multitude of different spatial possibilities of one time and/or infinite (inverse) time density of space, which is multitude of different time possibilities of one space.

All possibilities are equally real, only not here and now, which is one such possibility. Only in context of all its time, all spatial possibilities are real, each one only in fractal period of time. Likewise—only in context of all space are all time possibilities real, each one only in fraction of space. Spatial and time density is mass/energy of such empty space, which are its direction without/within—they are size within size (double space) and/or time within time (double motion). Inertial here and now is 2D angular cross-section through such double density—rest is empty space. Only resonant waves/particles (harmonics) of inertial here and now are real—rest is real only in context of infinity of quantum changes of inertion here and now.

Each harmonic of base inertion (i.e. Earth) is increased eccentricity (fractal difference) of perfect curvature, where increase in time is decrease in space and equally vice versa. It is always an additional discrepancy between perfect surface (dual) symmetry and its inertion. Such discrepancy between perfect and imperfect is infinite curvature of irrational (fractal) numbers, such as Π and φ, which are 2D/1D abstraction of particle/wave duality. They are abstraction of only possibility that is real here and now out of infinite number, because they are eigenvalue of inertion/inversion, which is double inertion. They are fractal supersymmetry between reality without (form) and within (content). Irrationality (infinite numerical sequence) is reflection of infinity of time fragments in context of one space and/or infinity of spatial fragments in context of one time. It is infinity of angular surfaces, which are one horizontal cross-section on surface.

Discrepancy is perpetual inertion (momentum) of infinite fractal (precessional) spiraling. It is not discrepancy, but angle of perspective that changes and it changes constantly with passage of time and/or distance. Discrepancy is inertion of surface duality—it is its reality, which changes with passage of time or space (in inverse proportions). Consequence is that any natural (surface) ratio is slight difference from perfect symmetrical ratio. Slight difference is physical reality here and now—not ratio itself, which is only its particular frequency. Duality (Earth) is inertion (singularity) of larger solar/galactic (5/7) duality. Description of one as larger and another one as smaller or one without and another one within is only fractal (relative) description, which has no meaning in context of its inertial 3rd direction. Earth is fractally within and fractally without solar reality (2x1/2). It is fractally larger and fractally smaller, where such fractals are inertion/inversion—one is instant (quantum) reality, while another is long period of constant passage of time. Within/without solar reality is meaningless description in context of galactic (double) inertion, because it is both at the same time. In such 3rd direction—Earth and solar duality are not two equal opposites but singularity, which is fractally within and fractally without such galactic reality (fractally faster and fractally slower etc). Such fractal periods are again instant quantum periods of no time (accelerated time) in context of long periods of uninterrupted passage of time (decelerated passage of time or time bubble).

Φ is supersymmetry (eigenvalue) of inertion/inversion (form/content) in all 3 directions. Its inertial difference of 3D sequence is always 1, which is here and now (φ=1.618, 1/φ=0.618 and φ^2=2.618). It is double fractality or double/inverse spiraling (precessional) curvature of inertial space, which are its time and gravity. It is singular (linear) description of fractal space/time supersymmetry. Φ is numerical abstraction of double/inverse resonant duality in context of inertial here and now. It is double/inverse echo within and without, which is infinite anti-parallel spiral, where within multiplies in context of reality without and without multiplies within through inversion of inertion that creates double inertion in exponential 2x1/2 quantum steps. It is infinite progressive incorporation of two equal opposites into singular anti-parallel direction or infinite double helix, which evolves through reconciliation and incorporation of its fractal (relative) opposite polarities. Φ is numerical abstraction of infinite double/inverse growth in space and time, where any inertial spatial form gradually grows in time, and when such growth is resonant (inertion=inversion and/or space=time), it is instant 2x1/2 multiplication/division, which is incorporation of both opposites—space and time into double inertial space and its growth yet another fractal Fibonacci sequence.

When equal opposites of duality (singular affirmation and infinite negation) resonate in context of each other, it is time of "hatching", when new (double/inverse) singular form breaks through its membrane (surface), which is its periodic quantum change. When inertion within is equal to its space/time boundaries, there is no more room for growth, unless such reality is instantly inverted, which is infinite room in opposite direction, because previous infinity is singularity in double/inverse context. Infinity of fractal choices in context of gradual growth are reduced to singular quantum choice, which is no choice at all, in moment of truth, when all relative/uncertain (subjective) inhibitions of previous sequence fall apart, where falling apart in space is coming together in time and equally vice versa. Every periodic double/inverse synchronicity, which creates double reality (duality), is instant φ. It is moment of destruction of old and birth of new sequence of reality, because spatial destruction is creation in time and equally vice versa (rising of phoenix from ashes of old). Every infinitely long period of time, which is constant and gradual change is only bubble of time, which allows for evolution of inertial reality within. Time is space of evolution and growth within.

Φ is irrational because it is infinite fractal and its every 3D (triple) sequence is one particular inertial reality. Each sequence is one double/inverse fractal in context of its angular (inertial) fractal, which is space/time in context of singular gravity, where space is twice larger choice than time and time is twice larger choice than gravity. It is exclusive surface reality, that such sequences follow each other in time. In context of inertial 3D, all sequences are simultaneous, because time is simultaneous negation of space. Only their periodic synchronicity gives an illusion of continuity of time. Space is inversion of time; therefore all such time periods are different only in context of its inertial space. Space is only fractal of time. In context of time, which is whole number, all time periods are instant space. Each "egg" (space) evolving within its own shell (time) is real only from within. From without—infinity of eggs are simultaneous and instant spatial reality. Φ is singular time number that is infinite sequence of spatial numbers. **2φ=3.236 is synodic/sidereal difference of 236 sec of inertial 3D here and now.** 0.618x2=1.236 is 4D sequence of fractal space (3), time (2) and their gravity (6) in context of inertial here and now (1), which is their double/inverse inertion. 4φ=6.472 is double duality of **64/72** (8^2/(8x9), which is inertial 6480-year period in context of its inertial discrepancy (**2^3**=8=72-64), where 6480/8=**81**x10=**3^4**x10 (fractal 2/3 within, which is fractal ¾ without) and 72/64=9/8 (**$3^2/2^3$**). 2x1.236=2.472 is double **24/72** duality (3x8/3^2x8), where 72/24=3 and 72-24=48, which is 6x8 (fractal 24 is 1D fractal difference of fractal 64, which are space/time in context of inertial 72 in between, which is their gravity). 64x72/24=**192** or 2x96, 2^2x48, 2^3x24, 2^4x12, 2^5 x6 and 2^6x3 (6th harmonic of inertial 3D here and now, which is its spatial boundary, i.e. position of Uranus in Bode's periodicity), while 24x72/64=**27** or 3^3, where **27x192=5184** years of its passage of time (time boundary). **Reality is no coincidence.**

Time shells of inertial Earth are multiples of 4x5=20. 2/3/4D are size/rotation/orbital, 5D is precessional and 6D is double/inverse precessional motion, which are periods of 1 day, 1 year, 6480 and 129 600 years of inertial size here and now. 6D is time boundary of inertial size, which is its gravitational shells beyond (time of time or solar time shells). 7D is 2.5 mil years, 8D is 50 mil, 9D is 1 bil and 10 D is 20 bil years, which is gravitational 3D boundary of inertial here and now (6+4=10=2x5 harmonics of fractal 5/2). 7 through 10 are 4 gravitational shells (change of size/density) here and now, which are 2/3 of 6 solar time shells, because 5/6D in between is overlap between Earth's and solar times. 5/6 are double/inverse Earth's quantum time shells without, which are solar passage of time (fractal rotational/orbital shells), while its precessional and double/inverse precessional motion (quantum reality) are 7/8 shells. 9/10 shells are overlap between solar and galactic passage of time, which are its passage of time (rotation/orbital). Quantum periods of galactic time (solar gravitational shells) are beyond time boundaries here and now. Quantum periods (5/6) are ¼ of passage of time, i.e. 3/4D rotation and orbital motion are 4x90=360° curvatures, while 5/6D are ¼x360=90° curvatures, because inertion in space is inversion in time and equally vice versa (double/inverse symmetry, where 4 in space is ¼ in time and equally vice versa).

7/10D periods are inertion/inversion=double inertion, which are 5-fold curvatures of previous ¼ curvatures—i.e. period of 2.5 is 5x518400 (648000) years, which is singular gravitational quantum (6480x10^2, which is double/inverse time boundary); period of 50 mill years is 5x10.368 mill year periods and period of 1 bill are 5 solar orbitals round core of galaxy (5x20736x10), while singular ¼ quantum of 20-bil years is 5.184 bill years (25 out of 10^2=100 solar orbitals, which is time boundary of time here and now or angular double/inverse density differential between singular size of an atom and infinite size of solar reality). Earth (5 bil.) is ¼ or 90°x10D wave, 1/80 or 4.5°x11D, 1/1600 or 0.225°x12D and 1/128x10^5 of 13D wave. In opposite direction it is 5x9D, 100x8D, 2000x7D, 40 000x6D and 800 000x5D cycles. Larger than 4D cycles (inertion of passage of time) are successive (precessional) surfaces of time bubble here and now (fractal ¾). Time within a bubble is 2x1/2 infinitely inflated, which is temporary discontinuity in curvature of double helix (see illustration 17). Double helix is double/inverse contraction of time without/within, which is equal and opposite spatial expansion. It is inverse expansion of time in direction here and now, where expansion from within is instant surface of dense mass, just as expansion from without is inverse surface of empty space, which are fractal passage of time in context of concave and instant density of no time in context of convex curvature within.

Illustration 17 - Space/time bubble of double helix

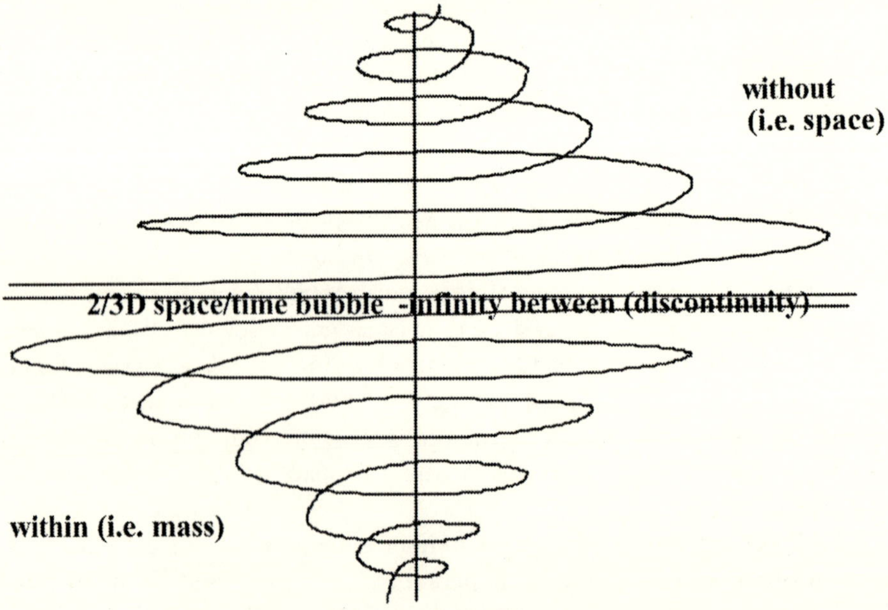

4x5=20-fold wave is surface or inversion of such bubble (4/5=0.80 within, which is 0.81 without in context of inertial 1 in between, which is double/inverse spiraling curvature 1.23456…). Reality within a bubble is inversion of fractal 4/5 without. Reality within is exponential increase (inflation) of linear 4x5 time periodicity without. 5/6 reality is yet another harmonic without. While relative reality without is positive curvature of double helix, reality within a bubble is its fractal inversion—i.e. it is instant uncertain reality. While time without a bubble is not time but instant space, time within a bubble is passage of time, which is exponential deceleration of passage of time and/or acceleration of spatial reality without (infinite universe). Time without the bubble is linear (constant) periodicity of 20-fold repetition, while time within is twice decelerated, through orbital and rotational motion (double fractal difference or 2D time that is 3D space here and now). In context of linear 4x5 constancy, orbital period should be 6480/20=324 years. Orbital motion (wave) is 324 times faster, which is 324 times slower passage of time. 324-fold inflation of time in context of orbital is inversion (fractal difference) of 360 days within (1.111 or 1/10 fractal difference). It is curving of linear time and/or flattening of round space, which is creation of time bubble that is space/time here and now (inflation of time).

1/324 is 0.0030864, which is 86 400 sec (time) of 1 day in context of 3D here and now, as well as its boundary, which is 3x86 400=25 920 years of precessional period. 86 400 sec of rotation within and 25 920 years of precessional motion without are action/reaction of inertial orbital motion in between. They are double/inverse time boundaries of orbital time. Orbital time is singular and rotational time is double fractal difference of inertial passage of time without time bubble. Orbital/rotation are singular and double fractal difference within of relative spatial reality without (singular and double uncertainty within, which are time and gravitational difference). Orbital time is reaction of spatial (solar) reality and rotation is reaction of gravitational (galactic) reality without. It is inflation of solar relative space and galactic relative gravity (twice relative space) into passage of time here and now. Rotational time here and now is twice exponential acceleration/deceleration of double fractal difference (4x1/4), which is double expansion/contraction of inertial (relative) 4/5 wave. Orbital time is 324 or 360/1.111 slower, while rotational time is 324x360x1.111= 129 600 slower passage then time of relative (4/5) space.

20 times faster (slower time) than orbital wave is 324/20=16.2 and/or 360/20=18-years period, which is rotation (18/1.111=16.2). In context of linear time of relative space (periods of 20), 1 rotation should be 18 years (saros). Time of one day (86400 sec) is double/inverse curvature of such linear time of relative

spatial reality of the universe, which is 18x360=6480-fold inflation of time here and now (massive Earth in context of empty space). 6480-years is boundary of reality of such time bubble. 16.2 is ½ eigenvalue (line of symmetry) between 18 and 14.4 (non-inertial and inertial periods of time), which is 1.8-years difference within and without or 3.6 year inertial discrepancy, which is inertial period of 14.4x360=5184 years of uninterrupted passage of time within a bubble (6480 years in context of orbital time is 5184 in context of rotation and/or vice versa in spatial context). Passage of time here and now is inertion between double/inverse times of rotation and orbital motion, which is surface spatial reality here and now. Boundary of such inertial space/time bubble is 5184 years of passage of time between periodic synchronizations with inertial passage of time of relative spatial reality, which is instant inversion of inertion that creates double inertion (burst of bubble that is creation of double/inverse bubble). It is flattening of time curvature of spatial universe without created by equal and opposite spatial curvature of time here and now. It is resonance between round and linear time and/or linear and round spaces, which is reduction of infinite (light speed) fractal difference to no difference.

16.2 years per rotation is 16.2/20=0.81=$3^4/10^2$, which is double/inverse repetition (1.2345...), while 18/20=0.9=$3^2/10$, which is singular repetition (1.111...). 0.81 years is 291.6 days and 0.9 years is 324 days or 32.4 days difference, which another 1.111 repetitive curvature (1.111x1.111=1.23456..). Double helix is inertial background (framework) of inertial duality here and now, which is its inversion (negation) or discontinuity here and now (surface of Earth). Such discontinuity or negation of reality is real only in context of passage of time within a bubble, which is limited 5184-year sequence that is negation of reality here and now. Passage of time is negation of larger reality, which is its only cause. It is only temporary inflation of limited duration, which allows for uninterrupted evolution within a shell, which is instant reconciliation with reality at the end of such period. It is particle in context of 2x1/2 motion (wave), which is equal opposite of such wave. Passage of time is only relative reality of its larger non-relative reality, which is singular time without expanded into infinite time within and infinite space without contracted into singular space of a massive particle within. It is twice inflated time in context of twice-deflated space, which is twice-empty space of double motion without and twice-dense mass within. Twice-empty space without is air of atmosphere (rotation) and empty space (vacuum) of orbital, while twice-dense mass within is water (liquid) on surface and solid mass within (which are fractal 2/3). Solid/liquid/gas are fractal ¾ of 4x1/4 wave in context of their inversion, which is singular ¼ quantum that is vacuum of empty space. ¼ fractal difference (empty space) is time (negation or boundary) of ¾ dense mass within. Empty space is 4th (relative) direction without of inertial 3D of dense mass within, which is its inversion.

Water within and vacuum without are double/inverse reality of inertial air in between, because vacuum is inversion of water, just as water is inverse vacuum, which are double/inverse waters of inertial here and now in between. Empty space of solar reality is inverse water. Dense Earth within (air, water and solid mass) is inversion of such water of empty space, because it is only a bubble of air floating within. Water is base element of double helix. Double helix and water are same fractal 3/2 reality—one permeates another. Double helix of Earth's reality is infinitely densified, while same curvature is infinitely de-densified in opposite direction, creating a bubble of inertion/inversion= double inertion. Empty space is water turned inside out. There is no coincidence that water is 2/3 of solid surface, just as there is no coincidence that day/night ratio fluctuates between 2/3-1/2-1/3 in context of one orbital period (or that wake/sleep period is 2/3 and gestation period of 9 months is ¾ or human lifespan is 25920 days=72 years or number of heartbeats in one day =103680=4x25920, which is 72 per min and/or 6480 in 3/2 hrs etc). It is fractal double helix in infinity of relatively different fractal surfaces. Water/air is 1D fractal difference of perfect duality (1/2), while solid mass in between (or solid mass in between two waters) is fractal 3/2—solid mass is double inertion of inertion/inversion (water/air). Solid mass on surface is only fractal reality of water (whole number). Water is constant surface reality, while solid mass changes in context of different quantum periods of time. Water on surface is twice larger inertion than mass, which is its inversion.

5184-year period is duration of time bubble and 432 000=2/3x648000-year period is duration of gravitational bubble (518400 in between). Gravitational wave is inertion/inversion = double inertion of time wave, just as time is double inertion of spatial wave. Constant inertion is instant inversion and equally vice versa, because one is fractal (2D) of another (3D). Fractal is only surface boundary between two multiples, just as multiple is surface boundary between two fractals. Fractal of multiple and/or multiple of fractal are instant (light speed) reality—they are dawn/sunset between day and night, solstice/equinox between seasons and/or quantum changes between space/time (Earth/Sun) and time/gravity (Sun/galaxy). They are instant surface shells without of gradual changes within, which are its 3D content (passage of time). Such expanded time bubble is exclusive reality of precessional quantum cycle of 6480 years and gravitational bubble is exclusivity of double quantum cycle of 432000 = 2/3x648 000 years. Double/inverse expansion of time/gravity in direction within is reaction of gradual spatial expansion in direction without (expansion of universe is passage of time in reverse motion). Spatial expansion, which is time contraction of solar/ galactic (empty) space, is time expansion here and now (dense Earth) and equally vice versa. Time bubble is inverse Bode's periodicity, which is spatial expansion. Bode's spatial periodicity (expanded spatial periods between mass) mirrors time expansion in opposite direction (toward Earth). Earth is time bubble, solar reality is spatial bubble and galaxy is gravitational bubble, while universe is antigravitational or double/inverse gravitational bubble, which is not only static (spatial), but dynamic (time) expansion. Galaxy is instant (spatial) expansion, while universe is gradual (time) expansion or expansion in double/inverse (space/time) context. Solar/galactic realities are double/inverse (instant) acceleration of spatial size, while universe is additional gradual (fractal) acceleration of its time.

Outside of such quantum spatial or time period (bubble), it is meaningless reality. Spatial bubble of solar and/or time bubble of Earth's inertion are reduced through singular fractal difference, while galactic or gravitational bubble is reduced through double fractal difference. Inverse gravitational bubble of universe is reduced through triple (total) fractal difference, which is double gravitational difference, because it is simultaneous expansion of space in time and time in space (gravity/antigravity). Each successive space/time shell without is additional fractal spatial expansion, while here and now is additional inversion of such expanded (relative) space into passage of time here and now. Each expansion/ contraction is fractal (light speed) difference between singularity and infinity. Each shell is complete inside out inversion. Micro-world is equal opposite of cosmic reality—it is instantly contracted (densified) infinite passage of time, while here and now is its expanded (de-densified) space. In context of inertial micro-world—here and now is not time, but spatial bubble, which is its expanding universe. Here and now is only absolute (whole) refraction point between multiplications and fragmentations without/within, while each shell is only fractal refraction. Here and now is only whole number, while its universe/ micro-world are double/inverse spatial or time fractals (where fractal in space is multiple in time etc). Every multiplication is equal and opposite fragmentation (2x1/2) of only one whole number—here and now.

Reality within surface of Earth (in its depth) is inversion of passage of time in context of empty space, which is infinitely larger time densification of space or instant reversal of time bubble without (exponentially increased density/gravity in depth, which is instant spatial reality of passage of time). It is double/inverse time density of space or spatial bubble within of time bubble without (surface reality). Spatial bubble is twice time bubble and equally vice versa. Singular time bubble is empty space, double/inverse time bubble is dense mass and triple time bubble is its gravity (equally vice versa). Reality within is flattened 3D, where 2 out of 3 spatial directions are compressed into surface density (space/mass inversion). Gravity or density of space is mass, while double density/gravity is density of mass. Singular gravity/density of empty space is here and now (Earth), double density is mass of an atom, and triple density is its nucleus. Double spatial density is singular massive density, which is less than certain space, because it is smaller (denser) than its singular quantum. Atoms are not reality of universe, because its atoms are solar spatial shells. Atoms of micro-world are exclusive reality of inertial density here and now (inside out inversion).

Singularities/infinities are relative points of inversions, which are double inertion. 0 is only point of inversion, which is totally relative/uncertain reality here and now. 0 of physical reality is not 0 but 1/10 fractal difference, which is complete reduction of infinity to singular point. 0 is only reduction of content of a shell and not shell itself, which is double/inverse 0. 0 is not nothing but an empty shell without content (1D fractal difference). 0 is reduction of effect and not its cause (and/or vice versa) to singular quantum of something, which is nothing. Reduction of cause and effect to nothing is double/inverse 0, which is again only an empty shell of no context in context of inertion of such dual action/reaction (double fractal difference). Nothingness (0) is real only in context of surface duality and not in its depth, which is infinity of different something (Balmer—infinity of uncertain subordinates of any relative principal). Each successive 0 within or without is only 1/10 fractal difference of inertial reality in between, which is double/inverse nothing inverted inside out (double/inverse or space/time infinity). Infinite reality here and now is nothing (0) in inverse context, just as infinite inverse reality is nothing here and now.

There is no absolute something, just as there is no absolute nothing. There is only fractal duality in between, which is relative/uncertain reality without/within. Reality is not incoherent or paradoxical—it is its description that is inconsistent, because of contradictory duality, which is affirmation through double negation. There is only fractal (dual) 0 and not absolute 0. Singularity of multiples is infinity of fragments and singularity of fragments is infinity of multiples. Fibonacci sequencing is equally inconsistent in dual terms. It starts with 0+1=1, 1+1=2, 2+1=3, etc. 0 is real only on surface. In its depth (3D), it is refraction point between +/- depths of such surface. It continues on other side as negative (inverse) curvature—i.e. 0-1=-1,-1-1=-2,-2-1=-3 etc. 0 here and now is relative limitation of dual perspective, without any meaning in inertial 3D context, which is not duality of equal opposites but triangulation of action/reaction in its double/inverse inertial context. In context of infinite fractal dimensionality of space/time, it is infinity of relative/uncertain (dual) waves, one within the other, which are its density, of which only resonant harmonics are real here and now. Larger number of resonant waves is larger density, where spatial density is emptiness of time and time density is spatial emptiness.

Fibonacci sequencing is density curvature; where density is ratio between inertion and inversion, therefore double inertion. Singularity/infinity is only surface reflection of 3D inversion—it is sphere in +/-90° curvature. Curvature has 4 points of infinite/singular inversions at 90° intervals, therefore every gradual density curvature or gradual change is real within two such surface realities, where full curvature is 4 surfaces. Consequently—any reality here and now is only ½ of complete curvature, while its fractal opposite is 2. Reality of both is 2+½ or 5/2=2.5—i.e. from perspective of any inertial surface; there are 5 remaining surfaces of a cube. Inertial ½ within (here and now) is fractal 3/2 (space/time) without—i.e. 2-1/2 =3/2=1.5 (where reality of both is 1.5+2.5=4D). In context of double fractal difference—inertial here and now is ¼, while its reality without/within is ¾. Inertial here and now is only meaning of any physical reality, which is infinite repetition of its double/inverse fractals and multiples. When resonant such fractal waves combine into singular wave, which is infinite anti-parallel fractal curvature of double helix. Inertion (singular quantum or negation, which is cause of inverse perpetuity) of any relative/uncertain physical reality is its inversion, which is non-physical reality that is double/inverse physical reality. Only singularity/infinity—i.e. only relative 0 (nothingness) of physical reality is always only here and now.

Instant surface or instant space/time inversion, which is beginning/end of space/time, is no different reality than here and now. **It is instant inversion only in context of equally instant inertion here and now. Such beginning/end or surface of inertial reality is inversion created by inertion.** Apart from here and now—it is not real. It is instant change of space/time only in context of inertial here and now. In context of inertial space/time—it is not space/time, but inertial here and now (point of observation) that changes. In context of double helix, it is always gradual curvature of no change. Infinite/singular fractal space/time (3D) between two surfaces is only space/time bubble created by inertion (gravity) of duality here and now, which is its relative discrepancy or overlap. Beginnings/ends of such bubble in either spatial or time directions are instant (quantum) inversions of inertion. Instant inversion of spatial inertion (constant passage of time) is its uncertain/relative reality within/without. Instant inversion of double inertion (singular gravity, which is not dynamic passage but static changeless inertion) is accelerated (double) relativity/uncertainty. Instant inversion of triple inertion (singular here and now) is twice accelerated relativity/uncertainty, which is accelerated relativity/uncertainty of space and time (and/or constant linear change of gravity and antigravity—i.e. expansion of space in context of contracting mass and equally vice versa). Expansion of universe is reflection of gradual evolution of non-physical inertial 0D point here and now.

Space/time within such relative surfaces is exclusive reality of inertion of such bubble. Duration of such bubble is determined by size and size by duration. Size and duration are determined by gravity/density and equally vice versa. Gravity/density is determined by inertial here and now, just as here and now is determined by gravity/density. Physical reality of such bubble (its space/time, which is double/inverse density) is exclusive function of inertial here and now. Such bubble bursts when spatial (inertial) tension is equal opposite of time (inverse) tension (perfectly symmetrical duality), which is eigenvalue (resonance) between inertion and its physical reality. Bubble expires when curvature of evolution of inertial here and now within is perfect negation of double helix—i.e. when space/time reality created by its inertion is in direct disagreement with itself (self interference). It is a point when inertial surface illusion of reality is in direct conflict with its 3D reality, which is depth of such surface projection. Burst or instant inversion that creates double inertion is correction of such disparity between surface and its depth. In perspective from within, it is instant action/reaction of no delay (inertion) in between, while in perspective from without, it is only perfectly symmetrical gradual progression of no difference. Quantum space/time inversion within is only fractal change of inertion without. Reality within is only one sequence of inertion/inversion= double inertion. It is singular DNA sequence of double helix, which is its reality without.

End of one is instant beginning of another sequence, which is double fractal repetition of the first one—infinite number of times. Each sequence is instant incorporation of two equal opposites into its 3rd direction, which is its direction of growth. Space/time is evolution of inertial here and now, through infinite gradual incorporation of its dual opposites. There is no real difference between organic/inorganic matters—it is same reality in inverse space/time perspectives. Organic in space is inorganic in time and inorganic in space is organic in time. Organic/inorganic is relative/uncertain (fractal) difference, which is exclusive surface reality here and now, without any meaning in its depth. **There is no difference between fractal double helix of space/time and DNA-molecule—they are double/inverse surfaces of same reality**. 4 basic DNA proteins are double/inverse curvature of 4x1/4 wave, which is fractal 5/4 reality or double duality of mass/energy in context of space/time. 10 layers of nucleotide pairs in each turn (loop of double helix) is fractal Bode's periodicity (6-fold space/time curvature in between double/inverse gravity, which is fractal 5/2 of 10 inversion points that are 10 planets and/or 10 electrons of $1^{st}/2^{nd}$ shell).

Number of hydrogen bonds is same fractal 3/2 reality—i.e. A/T is 2-hydrogen bonds (space), while G/C is 3-hydrogen bonds, which is 1D fractal difference (time). 2 and 3 hydrogen bonds are horizontal (spatial) and angular (time) curvatures (2+3=5 in context of 4x1/4 is fractal 5/4). Inertial difference of such curvature (¼) is water, which is its basic building block. DNA in water and permeated by water is fractal 3/2 double helix of space/time. There is no difference. Differentiation is only relative/uncertain surface of 3D reality. Spatial double helix (i.e. DNA) is only ½ of its dual space/time reality and/or only ¼ of its gravitational reality. Spatial double helix is only surface and/or linear projection of its space/time and/or gravitational reality. It is only fractal ½ and/or ¼ of its total reality, which is beyond boundary of its limited spatial perspective. Only ½ of reality (1/2 of 3D sphere) is eigenvalue here and now. Its inverse ½ is beyond spatial boundary and ¼ (depth of sphere) is beyond boundaries of space and time. 1/8 (fractal 8/9) is beyond boundaries of space, time and their gravity, etc. Double helix is infinite fractal curvature and spatial DNA is only its relative/uncertain (½) surface. Spatial (inertial) DNA sequence is only ½ of its time and/or ¼ of its gravity (density). Only instant past time, which is its mass (surface) is real here and now, while another ½ is its future time (depth), which is equally real only not now.

Only surface of 3D is instant (spatial) reality, while its equally real depth (reaction of surface reality), which is its time (negation or boundary) is gradual change, where only change of one is simultaneous equal and opposite change of another. There is no change of inversion (time), without equal and opposite change of inertion (space). Future time is only one possibility, unless changed by equal and opposite spatial change. Quantum change is multiplication/division of only one possibility of future time into two equal and opposite fractal possibilities (2x1/2). Time is only surface (fractal) of space, just as space is only surface of time, which is double/inverse fractal curvature of double helix. Massive DNA here and now is only one surface or only ½ of reality, while another ½ is beyond singular/infinite (relative/uncertain) spatial boundaries, because it is its time. Time is singular/infinite nothing of space (something), which is negatively probable (quantum) reality within of its surface reality without. Inertial reality here and now is always singular ½ quantum, without instant reality of its equal opposite (time of space and/or gravity of space/time). Instant space is ½ of time (past), while future is another ½. Such future time is passage of time only in context of inertial here and now (time bubble), while in context of inertial 3D, it is instant space of double/inverse (relative/uncertain) spatial reality, which is infinitely larger and/or smaller than here and now (light speed difference).

Surface reality is only fractal ½ of any possibility, while its equal opposite is its depth (invisible connections between infinitely small particles of micro-world). Time is passage (duration) only here and now, while in its double/inverse fractal context, it is same instant spatial reality as here and now, which is real only fractally as passage of time. Time is different fractal possibility of only one spatial reality, whose reality is parallel with inertial spatial possibility that is real here and now. Not only particles of micro-world are connected by such invisible (non-local) connections in depth of surface reality, but also massive objects of the universe. Non-local (time) connection is angular connection of surface space, just as spatial connections are non-local (angular) in context of time, which is double/inverse space (light speed difference). Local connection in inertial context is non-local in inverse context. Space of micro-world is time; therefore its non-local depth is space, while non-local depth of universe is time. Difference is only relative—while such non-local connections of the micro-world are infinitely small passage of time (0) in between, in context of universe passage of time between action and its reaction is infinitely long period of time. Time (2D) is double/inverse spatial density (mass/gravity), just as space is double/inverse time density (energy/antigravity). Time is instant density of relative/uncertain space, which is passage of time only within boundary of bubble created by inertion here and now.

Density of one is inverse or no density of another polarity. Dense space is empty time and dense time is empty space. Double/inverse density of space/time is its mass/energy and density of mass/energy is its gravity/antigravity, which is equal opposite (inversion) of space/time. Space/time without and gravity/antigravity within are double/inverse negations of singular affirmation in between (mass/energy). They are convex/concave curvatures without/within flat surface in between. Gravity/antigravity is negative space/time, just as space/time is negative gravity/antigravity in context of inertial E/m in between. Gravity is double/inverse space/time density and space/time is double/inverse gravitational density. Energy is inverse mass and mass is inverse energy. Double energy is double inverse mass, which is empty space. Double or infinitely larger energy (light speed difference) of inertial mass here and now is its gravity/antigravity, which is its empty space/time without and within. **Energy of vacuum or empty space is infinitely larger (double/inverse) energy of dense mass. No energy on surface**

is infinitely larger (accelerated or squared) energy in its depth. Empty space is double/inverse (infinitely larger) mass/energy, because it is mass/energy in context of passage of time, which is double/inverse spatial reality. E/m of instant space is only ½ of E/m in context of passage of time. E/m of dense mass is only ½ reality of past time; while E/m of empty space is its other ½ reality of future time, where past and future is fractal asymmetry (light speed difference in between). E/m of empty space without is square of E/m of dense mass within, because E/m of dense (instant) mass is infinitely accelerated in context of passage of time.

E/m are double/inverse space/time waves, where spatial wave is particle in time and equally vice versa. Particle/wave are double/inverse densities of dense mass or size in context of empty space or its motion. Wave (motion) is inflated instant density of a particle in context of passage of time, which is its empty space. Each double/inverse curvature is eccentricity in between and equally vice versa, where eccentricity is double/inverse particle/wave of such equal and opposite curvatures. It is inertial (3D) gravitational wave of two double/inverse surface space/time waves (one is angular distortion of surface reality of another). Wave in space is time frequency (cycles per sec), while wave in time is spatial frequencies (sec. per cycle). Spatial waves have constant speed, while time waves are accelerated/decelerated (light speed difference), which is instant (double/inverse) reality of time. Such double speed is not constant motion, but instant period or size of no motion. Size is therefore measured not with cycles per unit of time, but unit of time per cycle, which is periodicity (frequency) of such spatial reality. I.e. spatial or radio waves of different frequencies are periodic, infinitely small fractal changes of inertial time, while time waves are infinitely long periods of time between infinitely larger quantum changes of time. Spatial and time waves are equal and opposite fragmentation/multiplication—they are equal and opposite inversion/double inertion of inertial here and now. **Spatial waves are real in time, while time waves are real in space.** Time of spatial and time waves is double/inverse (light speed) difference, which is singular gravity (double fractal difference). Time of time waves is accelerated/decelerated time, while time of spatial waves is constant passage. Time of gravitational waves is twice accelerated/decelerated time, which is twice larger bubble and/or twice larger period of time.

Infinity of spatial (radio) waves is one time wave, and infinity of time waves is one gravitational wave. Time wave/spatial wave = inertion/inversion, which creates double inertion of gravitational reality here and now. Each gravitational wave produces equal and opposite time wave and each time wave produces equal and opposite spatial wave. Schumann resonance is spatial reaction of time wave Earth (such resonance likewise has 6 harmonics, where each successive echo is fractal time reduction). It is 7.8 or 7.5+0.3 Hz (¾), where 78/3 =26 and 75/3=25, which is fractal 26/25 inertial discrepancy (reaction of 24/25). It is $1/5^2$ inertion (singular quantum), which is double gravitational fractal difference (galaxy). 7.5 is ¾ inertial wave between galactic (7)/solar (5) duality, which is spatial reaction of its time wave—**(6480/86400)x10^2=7.5**. Schumann resonance is spatial inertion of double/inverse times, which is infinite repetition. 86400/C=0.288 and 6480/C=0.0216, while 288-216=72 and **216/288=3/4** (288x72=20736 and 288/72=4, therefore 20736/4=5184 etc). 0.288 sec^2/km is 3.4777 km/sec^2, which is 0.02777 km/sec^2 inertial difference (singular quantum) of fractal 7/2=3.5 duality (0.02777 is 5^3 fractal difference of 3.4777, which is inversion of inertial 6^2=36 sec^2/km). 3.47x2.16=3.47/0.463 (rotational velocity) = 7.5 Hz and 3.47x0.463 = φ. Earth's spatial reality (3) pulsate in tune with galaxy (7), while its time (4) pulsate in tune with solar reality (5), which is inertial here and now (6) in between such 4x1/4 wave. Reality is always only here and now in infinite double/inverse fractal echo in space and time.

CHAPTER 17

INVARIANCE

Infinity/singularity is an illusion. It is not boundary of space/time reality but boundary in its understanding. Singularity/infinity points to error in absolute determination of reality, which is duality of two equal opposites. Reality is fractal; therefore ½ relative without and ½ uncertain within, where relativity in time is uncertainty in space and relativity in space is uncertainty in time. Every singular determination is simultaneous infinity of its relatively different negations, because only equal opposites (affirmation and negation) are real—there is not one without another, because something and nothing are ½ of a whole, which are equal fractal realities. Any variance of invariable reality, which has no opposite to compare with, is totally relative/uncertain. Only relative/uncertain (fractal) determination is real. Singularity/infinity is relative/uncertain boundary, without any meaning in context of inertion of such duality, which is double/inverse cause of its singularities/infinities (boundaries). Any subjective perspective (physical or abstract) is at the same time its inverse self-limitation, which is its double/inverse boundaries (negations), because such inertial perspective is real only in context of its equal opposite. Any variance of invariance is fractal duality, which is uncertain/relative reality. There is no singular definition, because reality is comparative. Without equal and opposite surface boundaries, there is no content within and equally vice versa.

Singularity/infinity is beginning/end only in context of totally relative/uncertain (fractal) sequence, while in larger context they are only points of inversion between two fractal sequences. It is only relative beginning/end of equally relative reality here and now. Beginning and equal and opposite end are refraction points between different possibilities, all equally real, only not here and now. They are real only in context of limited inertion (curvature or density) here and now, without any meaning apart from it, because infinity in inertial context is singularity in inverse context. Singular here and now is overlap between double/inverse infinities and equally vice versa (i.e. renormalization). Infinite spatial distance is singular angular (time) distance. Infinite distance is singular difference and infinite difference is singular distance and equally vice versa. Singularity/infinity here and now are only two adjacent points on gradual curvature of double/inverse reality. They are totally relative/uncertain points of inertial here and now, which are its double/inverse negations that are singular affirmation in between. They are double/inverse limitations of unlimited, totally invariant reality, which are exclusive reality of inertial here and now, totally unreal in its larger context. Variance in space is invariance in time and/or vice versa.

Singularity/infinity is not limitation of reality but limitation in understanding, caused by absolute determination of relative (dual) reality. It is consequence of impossible possibility, which is singular (absolute) determination of duality. It is causeless effect or reaction without equal and opposite action. It is unreal reality that does not observe its only law—law of conservancy. It is comparable reality without anything to compare with. It is not infinity or singularity of reality but its infinite/singular distortion, which is equal opposite of inertial here and now. It is boundary of self-imposed self-interference. Renormalization of reality is correction of such double/inverse distortion, because infinity is only a singular number, just as singularity is inverse infinity of numbers in context of fractal ½ in between (here and now). Renormalization is broadening of inertial boundaries, which is introduction of equal and opposite infinity. Only double/inverse infinities are singular point in between, just as only double/inverse singularities are one infinite reality in between. Renormalization is introduction of duality in place of previous singularity (singular infinity), because only duality of equal opposites is physical (comparable) reality. Without anything to compare, one infinity is boundary. With something to compare, such infinity is reduced to singular point of double/inverse reality, which is inversion between quantum and fractal realities. Inversion is singular quantum of inertion and/or vice versa, which are boundaries of each other. Singular quantum is always ½, which is ½ without and/or ½ within in context of inertial ½ in between, which is fractal 3/2 reality or infinite space and infinite time of singular point ½-way in between.

Fractal 3/2 is supersymmetry between symmetry and broken symmetry (asymmetry), because its line of symmetry, which is infinitely larger randomness in depth, breaks symmetry on surface and equally vice versa—symmetry in depth breaks symmetry on surface. Spatial symmetry is broken time symmetry, because flat time is round space and equally vice versa. Line of symmetry between two equal opposites is their overlap, which is duality of both. It is $1.111^2=1.23456$...that is infinite randomness of double/inverse symmetry. Two symmetries is randomness in between, just as two random realities are one symmetry in between. Dual symmetry is exclusivity of two equal opposites, while their line of symmetry is both at the same time. Line of symmetry, which is inertion of surface duality is fractal (light speed) difference of both, which is their negation. Line of symmetry negates symmetry, just as symmetry negates line of symmetry, which is not part of either side, by being both (90° fractal difference). Fractal supersymmetry is reality of both—surface symmetry and its breaking in depth. It is reconciliation between impossible and possible, real and unreal, up and down, wrong and right, by being both and neither. It is duality of double/inverse or relative/uncertain fractal parts in context of anti-parallel double helix, which is singular whole of double/inverse infinities and./or vice versa. It is gravity of space/time, which is double/inverse space and time.

Any gauging is only comparative, relative or dual determination of undeterminable invariance. It is braking of invariable reality into two equal opposites (dual symmetry). Symmetry is broken invariance, just as infinite broken symmetry is singular invariance. Asymmetry is negation of symmetry, which is infinitely larger symmetry in depth. It is not dual, but symmetry of 1, where infinity of fractal parts are all equal opposites of each other. It is invariant reality, where sum of all changes is no change at all. Physical reality is always gauged symmetry on surface, which is broken in its depth, because depth is inertial background of surface reality—it is invariant reality of infinitely differentiated (fractal) reality on surface. Gauging of invariance is infinitely relative/uncertain reality without/within. Only inverse broken symmetry is inertial symmetrical reality and/or vice versa. Only invariant reality in depth is gauged reality on surface, where gauging is always dual (double/inverse) relativity. Gauged reality is comparison between two equal opposites (symmetry), which is real only in context of its inertial framework (here and now), which is its line of symmetry (its negation or inversion).

Gauged (symmetrical) reality is breaking of infinitely larger (invariant) symmetry and equally vice versa. Space/time (physical) reality is negation of inertial here and now and equally vice versa, which is breaking of each other's symmetry.

Reality is simultaneous duality of both equal opposites (½'s). It is simultaneous reality of both—here and now and its space/time reality without and within, which are its double/inverse boundaries. Fractal supersymmetry is invariant reality of no difference between here and now (point of observation) and its universe/micro-world. Universe/micro-world is double/inverse reaction of inertial here and now in between and equally vice versa. There is no difference between point of observation (singular point here and now) and its infinite universe and micro-world without and within. All infinity of different points of here and now (universe/micro-world) are one invariant and amorphous reality. It is total freedom of choice, which relative/uncertain (fractal) form it is given, because infinite choice without is no choice within. It is absolute compliance in which all possibilities are equally real, because sum of all possibilities is no possibility at all. There is no reality without simultaneous, equal and opposite negation, which are its singular/infinite beginning/end, where end of one is beginning of double/inverse sequence and vice versa—beginning of one is end of another, where beginnings/ends are determined entirely by inertial choice here and now, which is ½-way in between. Change of inertion here and now is change of its space/time beginnings and ends and equally vice versa.

Infinity is not perpetuation, but limitation of infinite reality. It is its singular boundary, because infinity without is singularity within. Double/inverse infinities of singular point that is in perfect rest are size and motion (C/(1/C)), which is particle/wave duality (C^2), where one particle in between is duality of double/inverse (space/time) waves without/within and equally vice versa—a wave is dual interaction between double/inverse particles, which are fractal difference in between. Double/inverse (space/time) wave is inertial particle and double/inverse particle is inertial wave. Only self-interference (resonance) is real, which is double/inverse (fractal) reality. It is not particle or a wave—it is particle and a wave, where one is infinity of singularity of another. It is infinite echoing of inertial here and now, where within in space is without in time and vice versa. It is universe that is here and now and here and now that is universe, in context of inverse universe of its micro-world and equally vice versa. Infinity/singularity is point of inversion between within and without, which are its relative boundaries as reaction of total freedom of choice. Space/time between its double/inverse infinities/singularities is same fractal matrix of double/inverse quanta, which are harmonics of inertial here and now, and where gradual curvature of one is instantly interrupted by double/inverse (fractal) curvature of another in points where affirmation (surface reality) and negation (its depth) meet, which is instant inversion of inertion that generates double inertion. It is double/inverse self-interference, which is infinitely larger (denser or accelerated) self-interference.

Boundaries of self-interference or self-interaction of inertial overlap that is real here and now are its singular/infinite points. Particle interferes with its double/inverse wave(s) and wave interferes with its double/inverse particle(s), where surface or 3D wave is field of such particle in between. Self-interaction of a singular particle is infinite, just as its mass is infinite in inverse context, because spatial singularity is time infinity and equally vice versa. Singular particle in context of infinite field is its singular point, which is total uncertainty within and/or relativity without. Field without of singular particle is its infinite self-interaction; just as such particle is infinite mass in inverse context (from within), because singularity in context of convex is infinity in context of concave curvature (singularity within is infinity without). Difficulty is not infinity/singularity, but its one-sided (½) context, because every infinity within is equal and opposite infinity without, which are only two successive whole numbers. Here and now is ½-way (overlap) between them, which are total freedom of choice in description. Infinity is without is reaction of singular point within and singularity within is reaction of infinity without.

Singular point of observation is double/inverse infinity and infinite point of observation is double/inverse infinity. It is double/inverse reduction of infinity of different choices to only one and/or expansion of one into infinity of choices, all equally real, which is invariant reality, because it is infinite choice of no choice. Renormalization is correction of two exclusivities in context of one inclusive reality, because reality that has no cause is not real, and cause of all infinity of different realities is singular here and now. Only double/inverse gauging is one finite result in between, while singular gauging is meaningless distortion, because it is both at the same time. It is incomparable self-gauging, which is meaningless with nothing to compare and infinity/singularity is boundary of such meaningless reality. Only duality of action/reaction in context of their inertion is real—one without the other is only its infinite/singular limitation. Causeless reality is infinite distortion of singularity and/or singular distortion of infinity, while cause and effect is finite result in between. Causeless reality is real only ½-way in between, while angular and/or spatial increase of distance/difference is its exponential negation with boundaries in points where inversion=inertion, because one is singular quantum of infinite reality of another.

Here and now is causeless because it is invariant and indeterminable and because causeless on surface is causative in depth and equally vice versa—causative on surface is causeless in depth, which are two equal opposites of ½ inertion in between. Gauged duality without is undetermined reality here and now in between and equally vice versa. Only both at the same time are real, because one is negation of another. Indeterminable reality here and now is inversion of gauged (space/time) reality—it is its equal opposite, which is double/inverse or infinitely larger fractal reality. It is causeless because it is expansion of reality that is no choice into infinite freedom of choice, where any choice is equally real. It is only causeless surface that is negation of only one cause in its depth, which is cause of all infinity of relatively different causeless realities on surface, which are its totally relative/uncertain fractals. Singularity/infinity are boundaries of such causeless here and now, which is causeless only because it is indeterminable. In context of its singular/infinite boundaries, which are its 90° curvatures, it is negation of such causelessness, because they are its cause and inverse determination in which there is no difference between cause and effect (here and now and its universe/micro-world).

Inertion (particle/wave) interferes with itself only in context of double/inverse inertion (resonance), which is quantum interruptions of gradual reality, where gradual in time is abrupt in space and equally vice versa. Gradual is unreal without equal opposite, which are abrupt beginnings/ends (inversions), where such duality is exclusive double/inverse projection of inertion ½-way in between. Gradual in context of one fractal inertion is abrupt in context of relatively different fractal and equally vice versa. Non-resonant space/time is empty space (nothingness) in context of passage of time, which is instant dense mass in inverse context and equally vice versa—instant dense mass here and now is empty space in context of passage of time of its double/inverse reality without/within. Empty space and/or passage of time are discontinuity of physical (space/time) reality, which is fractal inversion of inertial sequence here and now. Empty space that is passage of time is discontinuity of inertial here and now through its double/inverse singular infinite boundaries that are finite size/motion in between (i.e. Earth). Only whole numbers or quanta (exponential harmonics) of inertial here and now are real. Universe is infinity of exponentially larger spatial quanta in context of exponentially fragmented time, while micro-world is infinity of exponentially larger time quanta in context of exponentially fragmented space. They are double/inverse exclusive spatial or time multiples in context of their inverse fractals of singular space/time quantum here and now (Earth).

Here and now is only quantum that is simultaneously space and time, while its universe is fractal time curvature and micro-world is fractal spatial curvature. Boundaries of inertial here and now are its double/inverse infinite/singular curvatures, which are round Earth without and/or round atom within flat here and now (flat surface in context of linear passage of time). Such double/inverse curvatures are inverse points that are discontinuity of inertial here and now. Discontinuity of inertial space/time is singular fractal difference (passage of time and/or empty space), while inversion of such discontinuity (mass/energy of empty space/time) are double fractal difference, which is renormalization of inertial here and now, which are its different (surface) fractal possibilities. Renormalization in spatial context is different mass/energy of universe, while renormalization in context of time is its quantum change. Renormalization of space/time is both—it is different mass/energy in context of no time of quantum change here and now. Singular fractal difference is space/time difference, which is passage of time of inertial space here and now. It is discontinuity through exponential acceleration/deceleration or densification/de-densification of inertial here and now, which is exponential fractality of its reality. Infinite distance of empty space (passage of time) is only singular distance in context of its inversion, which is instant dense mass. It is temporary expansion of surface (2D) of dense mass into 3D empty space and its linear passage of time, which is its fractal 1D negation (3-1=2).

Infinite empty space is only singular distance in its dense (inverse) context. Interruption in inertial context is gradual uninterrupted reality in inverse context. Singularity/infinity and/or double/inverse density are boundaries of here and now in between. Double fractal difference is renormalization of interrupted reality, because it is inversion of inversion, which is not space/time difference but difference between two space/time sequences, which is gravitational difference (double negation that affirms here and now). It renormalizes inertial spatial reality here and now (surface between dense mass and empty space). Renormalization is harmonization through resonance. Instant mass of universe/micro-world is only ½-renormalized reality (2x1/2 space), without its time context. It is exclusive spatial renormalization, while quantum time change is also its time renormalization (whole quantum—4x1/4), when empty space is not empty and time is not passage but instant dense mass of solar (and/or galactic) reality. Space/time renormalization is different mass in context of no time (light speed difference), which is restores discontinuity of inertial here and now. Quantum change is restoration of interrupted discontinuity of space/time through empty space and/or passage of time. It is instant double/inverse (exponential) broadening of singular/infinite boundaries of inertial here and now, when inertial sequence (Earth) is extended to include entire solar (and/or galactic etc) space/time reality.

Space/time is double/inverse curvature of double helix, which is evolution through infinite incorporation of past and present sequences into its future sequence, which is gradual built-up of its genetic code. It is infinite built-up of past time into instant dense mass within. Evolution of space/time is life. In context of inertial here and now, its double/inverse boundaries (atom/Earth) are boundaries of life. Double/inverse boundaries of empty space and its passage of time discontinue inertial space/time sequence here and now. It discontinues evolutionary double/inverse curvature through time bubble that isolates and preserves evolution of life within its shell. It is relative/uncertain reality of inertial here and now, which is only organic matter of its universe/micro-world, because it is exclusive space/time quantum. Rest is either fractal space or fractal time, which is not whole and undivided quantum of space/time. It is exclusive reality of inertial here and now, which is evolution of one out of infinite number of possibilities, all equally real (but not here and now). Universe/micro-world is life only not in context of inertial passage of time (and/or gravity) here and now. Universe is fractal time and micro-world is fractal space, therefore its life (space/time) is only fractally real here and now. At the same time, here and now is inverse/double fractal reality of its universe/micro-world. Just as uninterrupted life of universe/micro-world is not real here and now, life here and now is unreal in context of any different inertial space/time sequence of universe/micro-world, because their passages of time are only double/inverse fractals.

Universe/micro-world in context of passage of time here and now (linear direction of time) is only surface reality devoid of its depth in context of time curvature. Relative/uncertain reality is distortion of space/time here and now, in which only surface of such different/distant reality is real. The only depth, which is real in context of closed time bubble, is here and now (within its fractal boundaries). There is only one Earth in context of universe, which is its infinite inversion or negation. There is no life in context of spatial universe here and now, because it is its negation through negation of only space/time sequence that is singular space/time quantum. Universe is flat (horizontal) cross-section of life and micro-world is its inverse (angular) compression. Universe/micro-world is infinity of different possibilities of life, all equally real, only not here and now. They are real only in context of its spatial depth, which are relatively different inertial passages of time. Life of universe is real here and now only in renormalized space/time reality (double fractal difference), which restores interrupted space/time reality here and now. Life is real only in context of spatial (instant) reality of time, which is relative spatial reality of universe that is light speed difference of inertial passage of time here and now. Although universe/micro-world is uninterrupted evolution of life, it is not real here and now, because its universe is infinite double/inverse (fractal) distortion and fractal life is not life.

Life either is or is not (quantum) and any fractal surface is not. The only reality that is not fractal but singular quantum is here and now (Earth). Universe/micro-world in perspective of inertial passage of time are its double/inverse fractals, which is negation of life. Such perspective is infinite double/inverse distortion of uninterrupted life that is universe/micro-world. Life here and now is only one out of infinite number of possibilities in evolution of life (space/time), all of them equally real, only not here and now. Life that is real here but not now (micro-world) and/or now, but not here (universe) is exponential fractality of life and only quanta are real. Life in its horizontal and/or angular depth is only 2D cross-section, which is not life, but its fractal negation. Only synchronicity between two relatively different passages of time is renormalization of interrupted space/time (life) here and now. Burst of time bubble is renormalization between uninterrupted evolution of life between Earth and rest of solar reality and burst of gravitational bubble is renormalization of interrupted evolution of life between here and now and the rest of galaxy. Only instant inversion of inertion is double inertion. Only instant space/time inversion is double space/time inertion, which is end/beginning of inertial passage of time here and now. Singular points of beginning/end of linear passage of time in between are instant (spatial) reality of time, when time is not passage but instant spatial density (mass/energy), which is renormalization between mass/energy here and now and its inertial solar (and/or galactic etc.) background.

Life of double/inverse sequences interact only in context of quantum changes, which are light speed difference from inertial passage of time here and now. Empty space of universe is likewise empty of life, because its life is infinitely expanded in its depth and frozen in decelerated context of inertial passage of time here and now. Micro-world is infinitely contracted and accelerated life. Infinite number of life sequences of micro-world is one life sequence here and now, just as infinite number of life sequences here and now is singular life sequence in context of relative (solar) reality and infinite number of life sequences in context of solar reality is one sequence in context of galaxy (double fractal relativity) etc, where infinity of different sequences are all simultaneous, because passage of time is only fractal (½) reality between double/inverse spatial sequences. Passage of time in inertial context is instant empty space without and/or dense mass within in double/inverse context. Only self-interaction or resonance is real and only quantum change is instant interaction between double/inverse space/time sequences—the rest is empty passage of time, in which such excluded possibilities are only fractally real—i.e. only here but not now and/or vice versa. Life is only here and now and here and now is anywhere, because it is any relative/uncertain point of space/time (universe/micro-world). Different/distant life that is not real here and now is different possibilities of life here and now. It is future/past time of inertial life sequence here and now, which are its larger/smaller spatial sequences and/or different densities. Quantum change of reality is eigenvalue with inertial background of life here and now, which is instant quantum leap in its evolution. It is instant leap or transition from one to another relative sequence, in which past sequence of passage of time is added as its instant spatial DNA sequence in context of next sequence of passage of time.

Quantum change is change between double/inverse (relative) uncertain dualities (2x1/2) in context of their singularity (5th quarter) in between. Duality is real only in context of passage of time and quantum change is its negation, just as duality (passage of time) is negation of its singular inertion. Singularity is cause of duality, which is its inversion. Singular inertion that is cause of space/time reality is unreal in context of passage of time. Time bubble is exclusive dual reality. Singular inertion that creates duality of passage of time is real only in singular points that are beginning/ends of passage of time that are its fractal surfaces. Fractal reality (triangle, square, pentagon and hexagon) is reality of a sphere only in corners, which are instant duality between such fractal realities within and/its inverse quantum reality without. Corners between fractal surfaces are only points that are simultaneous with its inertial reality (perfect sphere). Passage of time is duration of exclusive dual reality in context of its inertion, which is instant reversal in points of resonance (corners). They are instant reversal of built-up tension of such relative dual sequence, which is its end and beginning of another, relatively different sequence in evolution of space/time (life).

Every duality is surface, which is double/inverse illusion of 3D reality. Evolution of double helix is instant advance in context of quantum changes and gradual reversal in context of passage of time. Passage of time is only relative/uncertain evolution of dual sequence that is meaningless in its invariant singular context. Duality in context of passage of time is temporary departure from its singular source and infinitely long period of passage of time is not evolutionary but relative reversal of evolution, because longer duration of time bubble is larger separation from its cause, which is increase of dual and decrease of singular evolution in context of its invariant source. Passage of time is horizontal direction of infinitely larger angular evolution, which is real only in quantum leaps. Time is not perpetuity of evolution—it is its limitation. Duality is double/inverse contradiction, which is real only in its inertial context and its boundary is ½, because any larger dual evolution is relative regression in inertial (singular) context. Only interaction with its inertial background in context of quantum change arrests negative curvature and adds inverse momentum. Duality (surface reality in context of passage of time) is evolution only in short term, while in long run; it is its negation (fractal 2x1/2).

Inertial physical adaptation to passage of time is only ½ of double/inverse curvature of double helix. It is only surface of double/inverse evolution in context of instant (accelerated) passage of time. Evolution by means of adaptation to duality of passage of time is inversion of inertial, infinitely larger evolution in context of instant (double/inverse) time. Gradual adaptation to passage of time (space/time duality) is negation of singular, infinitely larger evolution that is inertial background of infinity of relatively different surface realities (passages of time). Adaptation to passage of time is departure from its only source, which is not evolution but devolution. Exclusivity of physical (space/time) adaptation is negation of evolution of point of consciousness that is infinitely larger evolutionary adaptation. Point of consciousness is indeterminable singular point of its space/time reality without and within. It is its inversion or negation; therefore in context of inertial space/time, such point is equal opposite of physical reality. Evolution in context of space/time is devolution in context of its negation, which is indeterminable point here and now. Exclusivity of physical adaptation, without equal and opposite adaptation to its inverse reality, which is its inertial framework, is not evolution at all. Only duality of both is evolutionary sequence. Only physical adaptation in context of passage of time, which is simultaneous with evolution of point of consciousness in context of no time, is infinitely larger (double/inverse) evolution. One without the other is only infinite distortion of evolution.

Physical adaptation and survival has no purpose, unless there is equal and opposite evolution of consciousness, which is not gradual development but quantum leaps in ¼ periods of uninterrupted passage of time. Evolution of consciousness (self-awareness) is double/inverse adaptation, which is infinitely larger tool for survival than exclusive physical (surface) means. Adaptation to passage of time is adaptation to relative/uncertain sequence of reality, which is not adaptation but departure from its singular source, which is double/inverse cause of both—instant space and its inversion, which is passage of time. Adaptation to relative/uncertain sequence of reality is not affirmation, but negation of reality. Only inversion or negation of surface (physical) adaptation is infinitely larger evolutionary step that is also evolution of self-awareness. Self-awareness is infinitely larger adaptation, because it is not adaptation at all, but its reversal. Only reversal of gradual physical adaptation is return to the source of physical reality, which is quantum leap in evolution of self-awareness. Self-awareness is only tool that ensures long-term survival, while physical adaptation to passage of time is only a short time measure, which is negation of its long-term purpose. Exclusive physical adaptation and survival in context of passage of time is meaningless in context of evolution of double helix, which is evolution in self-awareness of inertial point here and now.

Physical adaptation is adaptation to subjective sequence of reality, which is meaningless in inertial context. Exclusivity of physical (relative/uncertain) evolution is its negation (boundary). Only its negation is double/inverse (non-physical) evolution that is twice as large evolution. Only instant reversal and return to singular source, enriched by gradual adaptation in between, is double/inverse evolution. To repent means to accept the truth, because it is only possibility that is evolutionary. To repent is to denounce surface (physical) reality as exclusive and accept infinitely larger (double/inverse) reality as its only source. It is realization of relativity/uncertainty of present dual sequence (passage of time) in context of its double/inverse (inertial) reality, which is instant addition of present to past sequence that generates future fractal sequence. Evolution in context of passage of time is evolution in context of infinite fractality of invariant reality. It is evolution in context of infinitely fragmented invariant reality, which is real only here and now (inertial 5184 year period). Such differentiated physical reality consisting of double/inverse infinity of fractal parts is exclusive relative sequence of reality, which has no meaning in its invariant context. Infinity of fractal parts in context of passage of time is one invariant reality in context of no passage of time (which is accelerated passage of solar and/or galactic reality).

Infinite double/inverse variability is totally relative/uncertain sequence (positive/negative 90° curvature) of one gravity. It is one out of infinite number of possibilities, all equally real (only not here and now). Double/inverse infinity of relatively different space/time (physical) sequences is excluded from such singular (subjective) choice, by its relative passage of time. Double/inverse space/time difference that is physical reality here and now is only one relative sequence of its inertial reality (gravity), which is simultaneous infinity of all such different possibilities. Fractal evolution in context of space/time duality is unreal in context of its inertion (double fractal difference); just as such invariant inertion is unreal in context of physical duality (fractality). Adaptation in context of double/inverse infinite fractality of physical reality here and now (differentiation of different parts in passage of time) is not evolution of invariant inertion. It is only fractal ½ evolution of a whole. Time is negation of any relative spatial sequence; therefore infinite time is negation of all of them, because it is inertial framework of spatial reality, which is its double/inverse space. Evolution in context of passage of time is only spatial evolution (change of form), while evolution in context of no time (quantum change) is evolution in context of time (change of content), which is infinitely larger evolution of self-awareness and not only its physical form.

Without evolution in context of spatial reality of time (quantum change), there is no physical evolution, because it is its framework. Invariance in context of instant time is reality (quantum) of fractal spatial reality in context of passage of time. Physical (dual) and invariant (singular) or quantum evolutions are light speed difference. Invariant evolution is evolution in context of relative spatial reality of universe, which is infinitely larger evolution than local evolution in context of limited time bubble that is Earth in context of 5184 years of passage of time. Evolution in context of relative spatial reality is non-local or double/inverse evolution in horizontal and angular (in-depth) directions, while local evolution here and now is its singular limitation. Local evolution in context of only one, out of infinite number of different possibilities, is not evolution but its limitation in context of infinitely larger evolution of solar and/or galactic relative realities. Local evolution is limitation of unlimited (infinite) space of time to only its singular quantum (curvature) here and now. It is reduction of infinite spatial reality of time to singular sequence of uninterrupted passage of time of 5184 years. Non-local evolution is not gradual but reality of quantum leaps, in which such local evolution is curved in angular direction, which is difference between circular evolution in context of passage of time and double/inverse spiraling evolution of double helix.

Evolution in dual context is random choice. It is one out of infinite number of choices, all equally real. One random and subjective choice precludes infinity of relatively different choices that are totally equal to such inertial choice. Physical evolution is total freedom of choice in context of no choice, which is its invariant (inertial) reality that is simultaneity of all different (relative) choices. In context of invariant (inertial) reality, which is time of space and/or gravity of space/time, there is no difference between infinite numbers of relatively different choices, because they are all only fractal choices of space/time in context of no choice (gravity). Space/time difference is only relative/uncertain difference in context of gravity, which is no difference. Limitation to only one relative choice is limitation of infinite reality to only one of its fractal possibilities, which is limitation of such sequence to a period of passage of time, which is its boundary. Passage of time here and now is reaction of such limitation of reality. Limitation is not reality—it is its equal opposite (negation) and so is evolution in such limited local context. Quantum change of reality here and now is end/beginning of such limitation, which is quantum step in real evolution. It is instant negation of fractality of infinite number of different fractal parts in context of their singular invariant reality.

It is instant negation of physical duality, which is double/inverse illusion of space in context of passage of time, which is affirmation of their invariant (inertial) reality. It is wake-up moment in context of a new down.

Duality is double/inverse complexity of singular and invariant simplicity. In dual context, invariant inertion is double/inverse paradox and equally vice versa—in context of invariance, physical (relative/uncertain) reality is double/inverse contradiction. Only limitation of understanding creates infinity/singularity. Infinity/singularity is negation of dual (fractal) reality in between. It is its negation, which is boundary of invariant reality without/within inertial bubble here and now, which is only its relative/uncertain fractal. Just as singularity/infinity are not real in context of relative sequence here and now (passage of time), so is here and now unreal in context of its singularity/infinity. Singularity/infinity are indicators of relativity of reality here and now, because they are its boundaries. They are perfect indications of total subjectivity of reality here and now, which is exclusive reality, without any meaning in its invariant context. Reality is perspective and perspective is total freedom of choice, where any choice is same choice of no choice. Singularity/infinity are boundaries (inversion) of such subjective choice—they are infinite size of its empty space without and/or singular size of its dense mass within, which are double/inverse light speed difference in between. Quantum change is light speed change that is instant inversion (negation) of such subjective choice, which is inversion of all changes in context of passage of time into singular spatial DNA sequence, which is beginning of yet another random subjective choice of physical reality in context of following period of passage of time.

Duality is real only in context of passage of time. In invariant (inertial) context, it is unreal—it is its empty space between double/inverse densities. Quantum change is reversal of its double/inverse complexity in context of singular simplicity. It is subjective choice, how observation of universe/micro-world is interpreted. In context of duality, such observations are cause of its double/inverse complexity and contradiction. It is not reality that is contradictory, but dual framework of its observation. Reality is paradoxical only in context of relative (subjective) sequence, understood as ultimate. Only in limited context of passage of time is reality at odds with itself, because of double/inverse subjectivity in which it is observed. Relative sequence here and now is double/inverse illusion of invariant reality and understanding of such limited sequence as only possibility and not as one out of infinite number, all equally real, is its double/inverse negation. Negation of reality through subjectivity of relative choice is paradoxical and not the reality itself. Comprehensive understanding of relativity and/or quantum uncertainty is sufficient in realization that subjectivity of inertial point of observation is singular cause of its double/inverse paradoxical interpretation. It is not observation that negates invariance as singular source of space/time duality, but unwillingness to accept its results in objective perspective.

Interpretation in context of relative/uncertain (fractal) subjectivity of only one choice, where there is no choice, because all choices are equally real, is limitation of invariance to double/inverse (relative/uncertain) contradiction. In context of objective interpretation that accounts for subjectivity of inertial choice as only cause of such double/inverse distortion, interpretation is unambiguous. Reality is not a matter of acceptance of any subjective interpretation. It is what it is. It needs no interpretation, because any interpretation is subjective; therefore its double/inverse distortion. Only numerical abstraction and its geometry are unbiased (objective) interpretations—the rest is infinity of fractal theories, all equally real, only not here and now. Fractal theories are exclusive reality that is either here and not now (uncertainty) and/or now, but not here (relativity). Such double/inverse contradictions are only fractal ½ each in context of their invariant inertion that is inclusive (angular) ½ of fractal 3/2 (space/time) reality here and now. Inertial here and now is line of symmetry that breaks surface 2x1/2 symmetry of space and time and equally vice versa, which is double/inverse asymmetry or distortion of invariant reality through inertion of subjective choice here and now.

The only reason, why obvious should not be accepted, is predetermined inertial choice of what reality should be. Only subjective distinction between possible and impossible, real and unreal, something and nothing, limits infinite invariant reality that is not duality of distinction, because it has no opposite. Only reconciliation of affirmative (subjective) choice with its equal and opposite negation is a sequence of double helix, which incorporates both within its structure, because only both are invariant, while their distinction is double/inverse subjectivity. Only double negation (subjective choice and its negation) is affirmation. Incorporation of both in context of invariant double helix is renormalization of continuity that is interrupted by passage of time in context of empty space between double/inverse densities. It is extension of local (physical) evolution in its non-local (solar/galactic etc.) context, where non-local evolution is square of limited local evolution within boundaries of closed shell of inertial passage of time here and now. Only least resistance or least spent energy is evolutionary sequence—rest is its empty space in context of passage of time. Only simplicity of double/inverse complexity is invariant reality. Forcing reality into a shell created by subjectivity of inertial choice here and now is not reality but its negation, where reaction of negation is equal and opposite negation and double negation is affirmation. Subjectivity is circular route that requires a passage of time between double/inverse infinities (negations) that are inertial here and now in context of no time in between. In context of invariant reality—any subjective and circular (dual) evolutionary sequence is the same, because its end result is always inertial invariance of such fractal duality.

Only invariance is evolution, while its fractal and subjective duality in context of passage of time is its temporary arrest. Quantum uncertainty is end result of invariance between double/inverse densities of universe/micro-world, initiated by works of N. Bohr (model of atom derived from classical description of universe). De Broglie's work is equally applicable vice versa, which is inverse invariance between uncertain mass within and its relative empty space (universe) without. Only resonance is real and its singular quantum is ¼, which is instant change (double/inverse waves create new inertial wave in between). Only orbits that have quantized angular momentum are instant change of inertial duality. There is no difference between particle of micro-world (electron) and particle of universe (planet)—they are double/inverse wave/particle duality. There is no reason not to apply results of observation in context of micro-world to universe (De Broglie), just as there is no reason not to apply results of observation of universe to its micro-world (Bohr). If something is not prohibited than it is allowed. ¼ of precessional motion of rotating/orbiting particle (double/inverse waves of a particle and/or singular wave between double/inverse particles) is singular quantum or instant change of relative/uncertain sequence, regardless of its relative size/motion, because infinite repetition within is singular quantum without and equally vice versa.

¼ precessional motion of electron is instant E/m change (quantum) and so is ¼ precessional motion of any orbiting/rotating particle—planet, star, galaxy or universe. Reality is invariance between equal opposites, which is anti-parallel spiral of double helix and each ¼ (90°) is its double/inverse change (resonance), where fractal (gradual) in space is quantum (instant) change in time and/or vice versa. Micro-world is instant infinity of quantum jumps, because its space is instant time, while universe is infinite passage of time between two successive jumps. Micro-world (mass) is instant infinity of all relatively different possibilities, while universe (empty space) is only one possibility at a time, because universe is passage of time and in inverse context of mass, it is instant spatial density. Micro-world is sum of all spatial possibilities of time, while universe (here and now in context of quantum period of passage of time) is only one possibility. Universe is sum of all spatial possibilities of time only in context of all its time, while here and now (5184 years) is only its singular quantum or one out of infinite number of possibilities, all equally real. Micro-world is integrated time (path integrals by Feynman), while here and now in context of passage of time is only one derivate of integrated time of universe (its instant spatial density). Here and now is exclusivity of one out of infinite number of derivations of space/time reality of empty universe, just as space/time of dense mass within is integration of all its possibilities, because mass is instant past time, just as universe is gradual future time, which are 90° angular difference in between. Reality is fractal duality and right angle is its inversion point. Passage of time is boundary of one sequence, which is real only in context of ¼ (90°) curvature.

Quantum change (De Broglie's quantum leap) is singular point between two relatively different linear passages of time. It is simultaneous reality of both (resonance), which is instant space/time duality that creates mass/energy of empty space/time, through inertion/inversion=double inertion (Bode's periods). Resonance between inertial and inverse waves $(d/t)^2$ creates one particle (E/m) in between, just as two particles $(E/m)^2$ create one wave (d/t). Each quantum change of passage of time is singular beginning of dual (physical) reality, because it is affirmation of following sequence through double/inverse (dual) negation of past sequence. Each period of passage of time is one spectral line of solar space/time reality, which is instant infinity of them all. Solar reality is spectrum, while here and now in context of passage of time is one of its spectral lines. Likewise—here and now is spectrum of infinity of different spectral lines in context of its micro-world, which are different relative/uncertain spatial possibilities of passage of time here and now. Each spectral line is double/inverse infinity, where one principal has infinite number of subordinates (Balmer), which are different gravitational possibilities within/without spatial possibilities of time (singular fractal difference in context of double fractal difference, which is double/inverse or light speed difference). Double/inverse infinity of different spectral lines of dense mass of micro-world that are instant here and now, are double/inverse passages of time in between different lines in context of empty (solar) space of universe.

Difference between double/inverse passages of time (spectral lines) is double/inverse density of empty space of universe, which is singular ¼ quantum of time wave (5184) and/or ¼ of gravitational wave (103680). Such double/inverse passages of time are double/inverse boundaries of inertial duality here and now. 5184 years of passage of time is boundary of inertial spatial reality, while 103680 years is boundary of space and time. They are singular quantum or one out of infinite number of different possibilities, all equally real (instant) in context of solar and/or galactic space/time (physical) reality. Reality is invariance of C=d/

t=const (and/or E/m=C^2=const), and any passage of time is only one of infinity of its relatively different fractal surfaces. Passage of time is only eccentricity (overlap or discrepancy) between two perfectly round spatial curvatures (rotation/orbital and or size and its rotation etc) and its limit is De Broglie eigenvalue, which is ¼ of precessional motion of such massive particle. Because every reality is particle/wave duality, where one particle is inertion (discrepancy) between two waves and/or one wave is inertion between two particles, each particle is reality in context of its double/inverse passages of time. Each particle is real only in context of its synodic/sidereal difference. Infinite repetition of such difference is its precessional motion or passage of time, which is fractal ¼ reality, between two relatively different space/time (mass/energy) quanta. Each space/time quanta is one possibility of physical reality (derivate) in context of integrated infinity of such possibilities that are instant spatial density of solar and/or galactic realities (double helix of inertial here and now).

Duality of action/reaction is real only in context of its inertion, which is their line symmetry that breaks such surface symmetry, because small within is large without. Inertion that is small spatial discrepancy between equal opposites is infinitely larger spatial reality—it is spatial reality of time. Time is framework in which space is real, just as gravity is time of time or framework in which time is real. Any local evolution in context of inertial duality is only ½ of its evolution in context of its inertial (non-local) framework. Local evolution is unreal in context of non-local evolution, which is instant infinity of all relatively different local space/time sequences. Any differentiation from invariance of inertial framework of space/time in context of passage of time is exclusive reality of such limited period of time and quantum change is its end. Such surface physical reality, which is exclusive and only possibility in period of 5184 years, does not exist (is not real) beyond its singular beginnings/ends.

Surface reality (space/time) is real only in context of its inertion. Physical reality is action/reaction (2x1/2) of inertial here and now, which is its angular ½. It is only one out of infinite number of relative/uncertain (fractal) reflections of such inertion. Subjectivity of inertial here and now **is** its physical reality. **There is no objective description of reality that does not include its subjective inertion, which is its only cause. Dual description that precludes its inertion (here and now of space/time without/within) is not reality but its double/inverse (relative/uncertain) illusion.** It is reality taken out of its inertial context, which is its double/inverse distortion (with limits of such distortions in singularities/infinities). Only objective description is one that accounts for inertion of here and now in context of its space/time. There is no orbital/rotation without massive particle that is its inertion. There is no space and passage of time without inertial mass in between. Duality of action/reaction is real only in context of its inertion, which is singular point within of double/inverse infinite reality without and/or vice versa. Only infinite repetition of such inertion is its double/inverse space and passage of time. Inertial framework of duality is its fractal line of symmetry that is depth of such surface reality. It is infinitely small surface discrepancy (eccentricity, overlap, friction, broken symmetry etc) between action and its reaction, which is infinitely larger depth (density) of such surface reality. Small within is large without, just as large within is small without. Infinite reality within is singular point of twice larger (squared) reality without.

Smaller discrepancy (inertion) here and now (within) is larger reality without. 0-discrepancy (nothingness) is perfect non-inertion here and now that is infinitely large inertion (gravity/antigravity) of the universe. Distortion of space/time reality through lack of its inertial framework is corrected by renormalization, which is introduction of inertial point here and now in context of singular/infinite duality. Introduction of inertial framework is integral/derivate difference, which is light speed difference that reduces infinity to singularity (finite whole number). Only boundary (singular/infinite limit) of reality is subjectivity of inertial here and now. Here and now is only singularity/infinity of its reality without and within (space/time). Reality is 3D, which is triangulation of action/reaction in context of its inertion, while duality is only one out of infinite number of its fractal surfaces bound by its singularity/infinity (double/inverse difference). Result of attempts to define reality without inertial point of observation, which is its inertial framework, are only double/inverse infinities of fractal and exponentially complex theories. Theory is not definition, but fractal negation/affirmation of reality. It defines reality as fractal surface and at the same time it negates it through created paradox. Fractal (dual) theories are definition (inertion), just as much as they are its negation (inversion).

They are fractal 2x1/2, without its inertial ½, which breaks such surface symmetry. There is no definition of 3D reality in dual (2D) framework and there is no action/reaction that is real without its inertial (subjective) inertion here and now that is line of symmetry in between (eigenvalue). Only objective definition is the one that incorporates subjectivity (relativity) of its inertial here and now in its dual framework, because duality (space/time) is only double/inverse (singular/infinite) repetition of such inertion. Inertion is infinite boundary without of reality within (micro-world), as well as singular point within of infinite reality without, where without in space is within in time and equally vice versa. Only fractal supersymmetry, which is duality in its double/inverse fractal context, is 4x1/4 that is definition of 4D reality (3D space in context of passage of time without, which is instant mass within and/or vice versa). Only fractal supersymmetry is not fractal theory but whole quantum of infinity of different fractal (relative/uncertain) theories, because it is definition of invariance. Supersymmetry is double/inverse symmetry, which is symmetry on surface and at the same time broken symmetry in its depth, where spatial symmetry is broken time symmetry and equally vice versa.

Supersymmetry is fractal (light speed) difference from infinity of its fractal theories, because it recognizes invariant reality here and now as only cause (inertion) of infinity of different space/time (physical) realities, as well as fractality of its dual theories. It is path integral of all its relative/uncertain derivates, which is singular possibility within of infinity of its fractal possibilities without, all equally real. Supersymmetry is invariance, which has no equal opposite (negation or boundary) in context of physical reality (duality). It is reconciliation of all fractal paradoxes and contradictions arising from its fractal theories in its inertial context, which is boundary (singularity/infinity) of infinity of relatively different physical (space/time) realities. It is simultaneous spatial and time curvature, which is total curvature of space/time. Curvature is finite boundary or limitation of infinite reality, which is its negation. Supersymmetry is negation of physical reality in context of its invariant inertion, which is inexpressible here and now. Negation of physical reality is double/inverse or infinitely larger/smaller physical reality, which is its inertial framework. It has no equal opposite in context of physical reality (duality); therefore no dual contradictions arise that are its limits (negation). Supersymmetry is not matter of possibility but certainty. It is not a choice but reduction of infinity of relatively different choices to only one choice, which is not choice at all. It is not a matter of discussion, because it is beyond arguments, but matter of willingness to accept its implications, which is total freedom of choice, because total freedom without is no freedom within.

Surface (dual) reality and/or its dual interpretation are not clarification but obscuration of invariant reality. Collapse of wave function (uncertainty) is not collapse of reality but collapse of its relative/uncertain (subjective) comprehension. It is collapse of surface (dual) reality in its 3D context. It is collapse of inertial passage of time here and now in its inverse context, where time is not passage but instant spatial density. Reality is not convention (Copenhagen) or compromise; it is what it is, regardless of subjective exclusivity of only one out of infinite number of possibilities, all equally real. Reality is inclusiveness of all its fractal (exclusive) possibilities at the same time and its limitation to only one is reduction of passage of time to inertial spatial sequence. It is reduction of time that is exploration of all possibilities to only one possibility, which is not time but instant space (inertion or negation of time). It is reduction of dynamic non-local evolution to static local evolution, which is not evolution at all but circular stagnation. Infinite number of double/inverse possibilities is infinity of fractal surfaces of one reality that is not possibility but certainty, where certainty in space is uncertainty in time and equally vice versa. It is reality that is larger than all its fractal possibilities, because it is simultaneously all its fractal impossibilities as well. Each fractal duality is one possibility in context of all its fractal impossibilities (negations) and/or one impossibility in context of all its fractal possibilities. Supersymmetry is both. Only incorporation of infinity of negative choices of one positive choice (and/or vice versa) in curvature of double helix is evolution of invariant reality of inertial point here and now.

Bias is only origin of surface (dual) reality. Inertion of subjectivity of one relative choice (out of infinite number, all equally real) is singular cause of double/inverse space/time infinity. Transformation of one time possibility into inertial spatial sequence here and now is temporary halt in evolution, which is passage of time of such spatial inertion. It is temporary halt only here and now, because in double/inverse context of invariant inertial framework of such duality, inertial passage of time is not real. Time is inconsequential in context of double/inverse reality of inertial here and now. Passage of time is exclusive reality. In context of invariance, it is uninterrupted evolution. Subjectivity of inertial choice is duality of two equal opposites, without their inertial framework, which is only source of reality of both. Every subjective choice creates duality, which is fractal surface of invariance, because it does not account for inertial background of such temporary duality. Every subjective choice is one relative/uncertain theory of reality, out of infinite number of equally real theories. Duality is double/inverse infinite differentiation of only one invariant reality. Double/inverse differentiation between two equal opposites requires renormalization in order to be real. Differentiation between two equal opposites, without accounting for their inertial (invariant) background is original "sin" of mankind (Adam and Eve parable).

Duality of two equal opposites (i.e. wrong and right, just as much as bellow and above) is double/inverse surface illusion of reality. It is not its comprehension, but obscuration, because reality is not surface exclusivity of equal opposites but their inclusivity in depth of such surface reality, which is its only evolutionary sequence. Duality is subjectivity that obscures reality (3D) and is the only source that creates impression that surface physical reality is ultimate reality (yet another parable of idolatry or giving precedence to temporary surface or materialistic aspect of reality before its invariant source). Final boundary of such dual sequence is total (double/inverse) distortion of reality through complete exclusion of inertial point of observation (here and now) from such duality of action/reaction, which is ultimate self-negation. Physical reality without its inertial here and now, which is its only cause, is ultimate self-denial as well as infinite departure from invariant context of physical reality, which is complete arrest in evolution. Such infinitely distorted reality is corrected only through renormalization, which is inversion of such surface reality in its inertial (invariant) context.

Duality is temporary arrest in evolution and only change of inertial choice, which accommodates such duality in its invariant context, is its renormalization. Renormalization is double/inverse change of inertion, when exclusive duality is equal to its double/inverse (singular/infinite) boundaries. Only inversion of inertion is double inertion. Only inversion of inertial here and now is breaking through singular/infinite boundaries of temporary duality, which is infinitely larger/smaller reality without/within. Renormalization of physical reality is its quantum change, which is breaking through boundaries of such temporary arrest in evolution, caused by inertion of one subjective choice of surface (relative) reality as ultimate. It is instant inversion of such singular spatial possibility of time created by subjectivity of one inertial choice. Quantum change is instant negation of physical reality, which is duality without its inertial context; therefore instant breakthrough in arrested evolution, through negation of subjective choice that created it. It is instant disillusionment of reality of such subjective (dual) choice and revelation of infinite invariant depth of surface physical reality created by such choice. Only acceptance of such invariant reality as inertial framework of all relatively different physical realities is evolutionary, while its denial is self-denial in partaking of continued evolution. Repentance is acceptance of truth that leads to salvation, which is a choice for continued evolution, while obstinacy in accepting self-reality is choice for oblivion. Reality is total freedom of choice, because all choices are same choice of no choice.

CHAPTER 18

FRACTAL SUPERPOSITION

Light speed is constant of inertial space and variable of time, just as gravity is constant of inertial time and variable of space. There is no exclusive duality without its invariant inertial context. Stationary spatial wave is dynamic time wave and equally vice versa (Wheeler-DeWitt). Micro-world within (infinity of fractal spatial waves) is reaction of one spatial wave (universe) without, which is stationary, therefore no wave at all (reduction of infinity of choices to one choice that is not choice at all), just as universe is infinity of time waves of singular time wave that is not wave at all but instant mass of its micro-world. Stationary in space is light speed motion in time, just as light speed motion in space is stationary in time. Any dual exclusivity is only fractal ½ of its inclusive invariance. Reality is superposition of infinite number of fractal possibilities, all equally real only not here and now. Such infinite superposition of relatively different states is reduced to duality (2x1/2) in context of inertial passage of time here and now, which is only one (inertial) possibility out of infinite number, all equally real, only not here and now. Infinity of relatively different dual superpositions (i.e. up/down, left/right, etc) of quantum micro-world is instant (spatial) reality here and now, because micro-world is spatial reality of time, where infinity of different passages of time of such inverse reality is instant spatial reality of inertial here and now.

Universe (solar/galactic reality etc) is likewise, instant superposition of infinity of different passages of time, all equally real, because they are its instant spatial density, while here and now (5184 years of passage of time) is only one of its relative (fractal) possibilities. Invariance is sum (integral) without of infinity of relatively different superpositions within, where each superposition is one space/time or fractal physical reality (derivate). Here and now (passage of time) is one uncertain superposition of states (duality) in context of massive solar reality, which is limited 5184-year sequence. Such fractal superposition is passage of time only here and now, while in solar context, it is instant space. It is totally unreal sequence in context of any relatively different passage of time (different fractal superposition). There is no such reality beyond its 5184-years boundary. 5184 years of passage of time here and now is singular quantum of spatial reality of time, which is singular quantum of past time as instant spatial density in context of future time that is instant inverse density of future empty space. 5184 years is boundary of static spatial wave of universe (Wheeler-DeWitt), which is inversion of such wave into dynamic wave of its micro-world (one spectral line) in context of future static wave that is dynamic wave of its micro-world here and now. 5184 years is inertial discrepancy (overlap) between such static wave without and infinity of inverse dynamic waves within, which is its gravitational inertion in between (massive Earth). It is its line of symmetry that breaks surface symmetry in its depth, because there is no perfect duality without its inertial context, which is small discrepancy here and now that is infinitely larger reality without/within.

Supersymmetry of infinite number of relatively different fractal superpositions of states is invariant inertial context here and now, which is no fractal difference between instant spaces and their passages of time. Fractal theories are its relatively different superpositions that are real only in their exclusive (subjective) dual frameworks. Such fractal theories are not real beyond boundaries of their fractal dualities, because they negate each other (paradox), while they are all equally real in context of their invariant supersymmetry, which is none of them, by being all of them. Supersymmetry is incorporation of all fractal theories, which are limited evolutionary sequences in context of passage of time into singular spatial (instant) sequence, which is singular step of evolution in angular direction. It is overlapping end of evolution in context of one passage of time that is simultaneous beginning in context of another, relatively different passage of time and/or relatively different fractal superposition between space/time, which is 5th quarter between double/inverse sin/cos waves (5/4), where each wave is one fractal sequence of superposition. All fractal (surface) theories meet within singular focal point in their depth, which is point of their supersymmetry. Only in context of such focal point within, fractal surface theories attain their full meaning, because supersymmetry is relatively different 90° (¼) angular curvature of them all.

Relativity, as well as different quantum interpretations are fully comprehensible dualities only in context of their invariant supersymmetry. Without such context, they are exponential curvature of double/inverse distortions (paradoxes) in each other's context. Only in context of invariance (supersymmetry) are implications of transformation theory (Dirac) fully appreciated, because it is an accurate fractal description of space/time reality (2x1/2 particles in dual context, 3x1/3 in 3D and/or infinite fractal multiplication/division in double/inverse invariant context, which are different possibilities, all equally real in context of micro-world, which is instant spatial reality of all different passages of time). It is infinity of relatively different possibilities, where infinite number of possibilities within is only one possibility without (here and now). Infinity of different fractal realities of particles within is no particle at all on surface and equally vice versa (invariance in which all fractal particles are fused into one whole). In other words—reality within is same gradual reality as here and now (inertial spatial sequence in context of passage of time). It is reality that is here but not now, because it is all of time, which is its space (curvature or boundary of all infinite passage of time here and now). It is singular reality (invariance) that is neither particle nor wave, by being both at the same time.

Particle/wave is only dual (double/inverse) distortion of their invariant reality, because their reality is reaction of double/inverse perspective, where space is distorted in time and time in space (contracted space and/or expanded time, as opposed to expanded space and contracted time of universe, which is double/inverse particle/wave duality—i.e. spatial waves of micro-world are time waves of universe and/or vice versa). Mental picture is subjective interpretation that creates double/inverse inertial illusion of only one invariant reality, which is abstraction of all such angular surface distortions (projections). Interpretation (Copenhagen or any other) is not an evolution but its seizure, because it impedes further evolution through infinity of relatively different interpretations, till a point where their infinity inverts into invariant inertial interpretation, which is not just mental picture but reality, because it has no equal opposite in context of physical reality to compare with. Inertial context of something is nothing, which is infinitely larger (double/inverse) something. Likewise—inertial context of any comprehension is its negation. Realization that there is no difference between relative sequences is double/inverse comprehension.

Negation of relative understanding is infinitely larger understanding, because largest (invariant) comprehension is the one that accounts for inertion of point of observation—i.e. one that realizes that any knowledge is subjective (relative) knowledge perpetuated by inertial choice of point of observation. Negation is equal opposite of affirmation. Only when negation and affirmation have same meaning, observed reality is not just mental picture of subjective choice, but invariant reality in between such action/reaction. Only resonance of inertion/ inversion is double inertion; therefore only perfect duality (symmetry) on surface is line of symmetry in between, which is invariant breaking of such symmetry in its depth. Only superposition between affirmation and negation (equal opposites) is neutral (invariant) reality in between. Opposition, which is not symmetrical, is not resonant 2x1/2 here and now, but only distorted angular projection of ½ of whole reality. Only affirmation and equal and opposite negation are full curvature. Only total negation of any relative (subjective) interpretation that has same meaning as its affirmation is whole reality. Only realization that any interpretation and comprehension is relative and totally subjective choice of observer, which is its inertion, and therefore only affirmation that also incorporates its negation, is full comprehension, which is closure of one evolutionary sequence. Larger comprehension than inertial choice is its negation and largest knowledge is no knowledge, because any knowledge in context of passage of time is only temporary (fractal) sequence, which is not real in its double/inverse (spatial) context. Only comprehension that is aware of its fractality (relativity/uncertainty) is larger comprehension.

Q-space (Schroedinger) or 6D quantum space is double/inverse spatial reality of micro-world, which is simultaneous infinity of different possibilities, compared with only one possibility here and now, which is only fractal (exclusive) ½ of such space (3D). Q-space is double/inverse space or spatial reality of time in context of inertial space here and now. It is renormalization of fractal reality here and now and/or inverse renormalization of such inverted spatial reality. 6D space is simultaneous affirmation of 3D in context of its equal and opposite negation, which is 2x3=6D or fractal 2/3 of 3/(1/3)=9D reality (double/inverse 3D). Q-space is relatively different fractal interpretation of same invariant reality described by transformation theory (Dirac) or any other duality. It is recognition of inertion of surface duality here and now (3D), as only source of its inverse reality (2x3=6), which renormalizes its duality. 3D without is double/inverse 3D or fractal 6D within. There is no difference—micro-world is 6D Q-space within, just as universe is 6D Q-time without, which is double/inverse 6D here and now in between (fractal 6x4=24, which is 2/3 of 6^2=36), where 6D time here and now is instant 6D Q-space in inertial (spatial) context of universe. It is fractal 3D without and inverse 3D within in context of inertial here and now, which is non-inertial 7D of its micro-world (7 shells of an atom) and its reaction in context of universe, which are 7 time shells of inertial spatial Earth.

Duality of both is 2x6=12D without/within (universe/micro-world) in context of inertial here and now, which is non-inertial (whole) 13D. Double/inverse (space/time) duality of 12D is 24 in context of its fractal difference (passage of time here and now), which is fractal 24/25 of inertial (invariant) $6^2=36$, etc, which is infinite double/inverse fractal helix curvature. Exclusion of invariant inertion (here and now) from observed reality is its double/inverse obscuration, which is renormalized by its inclusion. Such inclusion is instant double/inverse expansion of infinite/singular boundaries of inertial choice, which is duality in context of its invariance, which is infinitely larger (double/inverse) duality. Duality in invariant context is double duality, which is 4x1/4 wave of inertial 2x1/2 (½ wave). Duality is surface without, invariance is 3D in between and double duality is 4D within. Such double duality is again real only in larger inertial context, which is yet larger invariance in context of proportionally larger period of uninterrupted passage of time, etc, which is infinite anti-parallel spiraling curvature of double helix. All different fractal theories are singular affirmation of invariant reality and fractal negations of relatively different theories (equally correct), which is infinity of paradoxes in between. All different theories are equal description only in context of their supersymmetry, which transforms (fuses) all fractal theories into only one invariant reality of no difference (paradoxes) in between. Only fractal reality is paradoxical in context of another, equally real (correct) fractal reality, while invariant reality is consistent.

Duality is not paradoxical—paradox is that such duality does not include its invariant (inclusive) inertial context. Paradox is that duality is regarded as real without its equal opposite (invariant singularity of its inertial here and now). Paradox is that any subjective choice out of infinite number of equal choices is regarded as only one. Particle/wave duality is not a paradox (double slit experiment) but double/inverse reaction of singular invariance, because particle in space is wave in time and equally vice versa. Physical reality is not space **or** time—it is space **and** time; therefore any singular form is simultaneously particle in inertial and wave in inverse context (and/or vice versa). Ray of light is particle (size) in context of spatial reality and at the same time, it is wave (motion) in context of time. It is not question of dual exclusivity (or) but its singular inclusivity (and). Size (particle) is inertion between double/inverse waves (motions) and motion (wave) is inertion between double/inverse waves, where inertion is fractal (relative or comparative) discrepancy or overlap between two equal opposites. Inertion is only relative (comparative) differential between two equal opposites. Only determination of its size/motion is its comparison with inertial point of observation (here and now, which is its double/inverse inertion). Atom is light speed reduction or spatial curvature of linear direction of time. Time in its context is not linear, infinitely slow passage of infinitely large empty space, but infinitely fast and small rotation, where every spin is one quantum of passage of time. Its passage of time in context of inertial here and now is light speed difference, which is not passage of time, but instant space and it is not motion but instant size (equally vice versa—passage of time here and now is light speed difference, which is universe that expands at the rate of light speed).

Size of an atom (particle) is motion of its electrons (waves). Motion of electrons is in inverse context of passage of time (negative or past direction of time, which is instant spatial density). Angular distance between here and now and surface of atom (shell) is increased spatial uncertainty and/or decreased relativity of time. Relativity of time here and now is in context of an atom reduced to its absolutely spatial reality. Smaller/denser size than an atom is inverse reduction in context of inverse passage of time (space as passage of time). Angular distance between here and now and surface of atom is gradual curvature of linear (infinitely large) passage of time into round (infinitely small) curvature. Distance between surface of atom and surface of its particles (electron/nucleus) is inverse non-Euclidean curvature, which is gradual uncoiling of such round time into linear direction that is opposite vector than passage of time here and now. Surface of atom is line of symmetry that is convex/concave curvature of time without and within. Surface of atom is double/inverse light speed difference between its reality within (orbiting electrons) and without (here and now), which is $E/m=C^2$ (energy within is mass without and energy without is mass within). Surface of atom is curvature of one and surface of its particles is curvature of 2 linear directions here and now (where distinction between spatial and time directions is relative).

In context of an atom, one spatial direction is inverted into inverse passage of time and in context of its electron/nucleus, two instant spatial directions here and now are converted in dual passage of time (instant double/inverse or equal and opposite passages of time). A photon is yet another light speed difference or curvature of 3rd spatial direction here and now into passage of time. Photon is inversion of 3 spatial directions here and now into linear passage of its time and passage of time here and now, which is fractal spatial difference, into its instant space (fractal ¾ inertion/inversion). Atom and its orbiting electrons are double/inverse fractal differences (½ and $^2/_3$) between two spatial quanta that are triple light speed difference or light speed difference in all 3 directions of inertial 3D here and now (¾). It is double C^2 (double mass/energy) and/or 4x1/4 space/time difference (3/4 inertial space in context of fractal ¼ difference, which is its passage of time). Photon is complete 3D inversion of spatial reality here and now; therefore it is complete inversion between particle and a wave (size and motion), while atom and electron are only fractal inversions in between. Photon is 3D particle of time, just as here and now is 3D particle of space. Photon is spatial negation of passage of time here and now. It is infinite passage of time of universe (15 bill years) compressed into singular spatial curvature (boundary). It is infinite limitation of passage of time, while atom and electron are only its fractal limitations (passage of time in context of solar and/or galactic reality). Photon, which is particle of time (singular spatial quantum) in context of passage of time, is double/inverse negation. It is negation of passage of time here and now (invariant reality), which is equal and opposite negation of its reality as a particle (therefore it is wave).

Particle in spatial context is wave in context of time and equally vice versa, which is fractal light speed difference in between (inertial here and now). Duality is not paradox but surface reality of invariant 3D, which is double/inverse duality (4x1/4). Reality is 4 fractal shells without and within in context of inertial here and now (1) in between, which is fractal $2^3/3^2$=8/9 that is 9x8=72-fold fractal multiple (singular quantum) in context of its double/inverse (singular/infinite) boundary, which is its square—72^2=5184 years of passage of time of such inertial spatial sequence. Out of 8 shells, 6 are space (Q-space) and 2 are their fractal double/inverse passages of time. 8 space/time shells are 10 gravitational shells, where additional 2 shells are passage of time of spatial reality of passage of time here and now (2 time shells are spatial boundaries and 2 gravitational shells are double/inverse spatial boundaries or boundaries of time). Space/time is fractal 8/9 and its double/inverse gravity (invariance) is fractal 10/11, which is boundary of physical reality that is 10/9=1.111 fractal difference in between (infinite repetition of inertial here and now in relatively different fractal (decimal) sequences). Micro-world is Q-space in context of inertial passage of time here and now, just as here and now is Q-space in context of passage of time of universe (invariance).

Only surface duality lacking its invariant (inclusive) context is paradoxical. There is no paradox that reality is multiple of light speed differences, because C=const, which is space/time boundary is double/inverse light speed (C^2) in context of double duality (mass/energy), quadruple light speed in context of gravity/antigravity etc. Each fractal 2x1/2 multiplication/division is 2x1/2 fragmentation of invariant light speed, which is constant of any fractal duality. It is infinite variation of invariant reality, because in every dual context, light speed is difference between double/inverse (singular/infinite) curvatures. Light speed is singular fractal difference between instant size of no motion and or infinite motion of no size, where motion in time is size in space, just as spatial infinity is time singularity and/or vice versa. Reality is infinity of light speed differences or infinite multiple of light speeds, where each light speed is one fractal 2x1/2 duality in context of angular ½ (fractal 3/2). Light speed is angular difference between double/inverse inertions, because light speed of space is twice light speed of mass, which is light speed difference in between (passage of time). It is always C=d/t=const (C is invariant context of space/time duality, just as time is invariant context of light speed and spatial reality and/or space is invariance between passage of time and light speed, where spatial light speed is singular time size and/or vice versa).

There is no disparity between quantum reality of micro-world, which is duality of two spatial possibilities of time and universe, which is only one spatial possibility (i.e. question by Einstein why there is only one and not two Moons). There is infinite number of Moons, only not here and now. Two instant spatial possibilities of quantum world (uncertainty) are two time possibilities of universe (relativity), which is infinite number of its fractal spatial possibilities—only not in instant, but gradual passage of time. Moon in its orbit is infinite number of different spatial positions, all equally real, but not in instant time (now). Infinite multiplication within is infinite fragmentation without and/or vice versa. Lunar orbital is infinite fragmentation of one spatial reality in time, which is its infinite multiplication in space (double/inverse repetition or saros). Moon (space) and its orbit (time) are two particles of quantum world. Infinite number of different lunar spatial possibilities is infinity of different quantum periods of passage of time, where such different periods (multiple Moons) are instant (spatial) reality in solar context, because here and now is its quantum micro-world. Dual reality (fractal space/time superposition) is meaningless double/inverse distortion without its invariant context. Dual space of quantum micro-world is reaction of passage of time here and now as its instant spatial reality within, just as passage of time here and now is reaction of spatial reality of relative universe without. Space and passage of time here and now are dual space of relative (solar) space and 4-fold spatial multiple of galactic (twice relative) space, 8-fold spatial possibility in context of universe etc.

EPR (Einstein-Podolsky-Rosen) paradox (in which particle is exclusive finite particle **or** wave) is affirmation of physical (surface) reality as double/inverse illusion (angular distortion) of its invariant (inertial) reality. It is paradox only on surface, while in depth it is perfect description of invariance. Only subjectivity of inertial choice prevents its realization, because reality is not a theory, but is what it is. Reality does not conform to fractal theory; theory ought to conform to reality. A singular particle cannot simultaneously be determined **here and now** as size **and** motion or be particle **and** wave, because it is C=d/t, i.e. duality of size/motion and/or particle/wave is singular invariance (C). Duality in invariant context here and now is simultaneous. Particle is inversion of inertial wave and equally vice versa. It is particle/wave duality exclusively in context of micro-world and not here and now. It is double/inverse duality only in context of light speed difference in between, because particle in spatial context is simultaneously wave in context of passage of time and equally vice versa. It is not exclusivity (or), but inclusivity (and), because here and now is not exclusive space or passage of time—it is both at the same time and so is particle/wave in its context. Duality is fractal double/inverse differentiation of invariant reality. It is only one relative sequence out of infinite number, all equally real sequences only not here and now.

Paradox of such experiment is caused by lack of invariant inertion here and now in context of such duality. It is inverse spatial reality in context of inertial passage of time here and now, which is its double/inverse distortion. EPR experiment is confirmation of paradoxical choice of surface reality as exclusive reality. It needs renormalization by inclusion of inertial here and now as invariant background of such surface duality, which is only its double/inverse angular projection. Paradox on surface is perfect consistency in its dept and equally vice versa—perfect consistency on surface is paradoxical invariant reality. Only duality of affirmation/negation on surface (paradox or contradiction) is invariant reality, which is its source in depth. It is not reality that is paradoxical, but its (mis)understanding, caused by insistence on preconceived ideas of what reality should or should not be, according to particular subjective choice, which is only one out of infinite number, all equally real. A choice that reality can be determined without inertial subjectivity of point of observation, which is its only cause, in its double/inverse context is major stumbling block that arrests any further evolution. Only infinity of surfaces paradoxes is instant inversion of such surface infinity in singular quantum of invariant reality in its depth.

Physical reality is fractal reflection of invariant reality and any evolutionary sequence in context of passage of time is contrary from invariant evolution in context of no time. Advance in context of passage of time is possible only through regression, because regression in context of passage of time is invariant advance, just as advance in passage of time is invariant regression in context of infinitely larger reality that is cause of physical reality (passage of time) here and now. Advance in context of passage of time is only its negation (paradox), while advance that is not negated is regression. Only unlearning of all that is learnt in context of passage of time is an invariant advance, because physical reality is 2x1/2 duality of learning and subsequent unlearning or advance and its subsequent regression and equally vice versa. Learning in context of passage of time is reality that is inverted (vice versa)—it is opposite of its surface appearance, because positive here and now is negative in invariant context and/or vice versa. Only reversal of adaptation through natural selection in context of passage of time is invariant evolution, because any selection is natural—there is no unnatural selection. Every selection is only subjective (relative) choice out of an infinite number, all equally natural. Only realization that choice is relative and not absolute is renormalization (inclusion of inertion in context of surface duality) and instant advance. Only negation of surface advance is advance in its infinitely larger depth.

EPR understood as paradox is affirmation of deep misunderstanding of reality, because particle/wave duality is particle in its double/inverse rotational/orbital motion, where orbit is size of infinitely larger particle in its rotation/orbital, which is infinite double/inverse fractal helix curvature of shell within a shell. Orbit is motion in perspective within and size in perspective without, while rotation is fractal passage of time between such double/inverse particles without/within (empty space/dense mass). If such duality is understood as size or motion (particle or wave) is subjective choice, because size in inertial context is motion in (equal and opposite) inverse context (light speed difference in between). Only surface reality (which is not reality but its double/inverse projection) is exclusive distinction between equal opposites (Adam and Eve parable), while its invariance, which is infinitely larger context of such reality is no distinction between infinity of such equal opposites, all equally real relative (2D) sequences of one invariant reality (3D) that is inertial framework of all of them. Reality is perfect simplicity, which is double/inverse infinite complexity only on its fractal and subjective surface.

Abstraction that does not differentiate between its fractal surface projections is invariant description. Pythagorean theorem—$\mathbf{sin^2+cos^2=1}$, is complete space/time description of reality, while all subsequent fractal theories are only its relative/uncertain derivates. It is description of double/inverse fractal waves that are ¼ fractal difference (90°) in between in context of their invariance here and now (1). Infinite numbers of successive fractal theories on surface are advancement only on surface, while in infinitely larger depth they are gradual regression from its invariance. Pythagorean theorem is triangulation of two equal opposites in context of their inertial invariance (here and now), which is an infinite improvement over its fractal (dual) theories. It incorporates all of them in its singular context. Pythagorean theorem is path integral of all its variables (derivates). Triangle is static (spatial) and its wave function is dynamic (time) illustration of invariant 3D. Any two out of three sides in context of remaining one side of a triangle are space/time or physical duality in its inertial context, which is set of 3 relatively different sequences, all equally real. Each is 3D, while one in context of another is 3^2=9D One in fractal ½ context of another, which is exclusive space or its passage of time is 3x2=6D and all 3 in context of each other are 3^3=27D. 6D is fractal between 3D and 9D (double/inverse reality), just as 27 is fractal in between 9 and its square (double/inverse inertion)—81. 9 is singular and 81 is double fractal difference of inertial 3D here and now, which are space and time of inertial gravity and/or time and gravity of inertial space.

27-fold repetition is boundary of inertion and 81-fold repetition is boundary of its inversion (time). 81^2=6561 is boundary of both—space and time, which is non-inertial 81^2-81=6480 years of passage of time. Non-inertial 27 is 27-3=24 or 6x4-fold sequence of fractal 6/4=3/2, just as non-inertial 81 is 81-9=72 or 9x8-fold sequence of fractal 9/8 (fractal ½ increase of inertial 6 and double increase of inertial 4). 72/24 is 3 (inertion) and 72x24x3=5184, which is 4/5 limitation (curvature) of non-inertial 6480-year time wave. Pythagoras theorem is complete description of infinite 2x1/2 multiplication/division of inertial 3D here and now, which is infinity of its fractal space/times in context of double/inverse spiraling helix. Sin wave is space in time curvature and cos wave is time in spatial curvature. Sin/cos are space/time and their tg/ctg are its mass/energy. Each wave is double/inverse quantum interruption in ¼ period of their precessional (90°) curvature. It is simultaneous relativity without and uncertainty within in context of their invariant 3D in between. Only resonant 4x1/4 points (positive/negative space in context of positive/negative passage of time) of inversion are eigenvalue here and now and rest is its empty space in context of passage of time, because only perfect symmetry on surface (negation/affirmation) is broken asymmetry in its depth, which is one Fibonacci sequence of inertion/inversion=double inertion or duality that creates double duality. $Sin^2+cos^2=1$ is fractal supersymmetry or superposition of double/inverse states in their invariant context. Pythagoras' theorem is definition of a triangle and triangle is 3D, which is inertial context of its fractal space/time dualities.

Distinction between equal opposites is only double/inverse projection of invariance, which is reality of both. Power of distinction (Adam and Eve parable), without its invariant context is not evolution but its temporary regression. Without invariant context, it is not knowledge but its negation. Fermions and bosons are double/inverse reaction of invariant here and now, which is inertion of such duality. Fermion in space is boson in time and equally vice versa—in context of invariant inertion, it is both at the same time ($\sin^2+\cos^2=1$, where wave in time is particle in space and/or vice versa). Likewise—if an atom is sin wave, electron, which is its inversion (affirmation in context of negation), is $\sin/(1/\sin)=\sin^2$ and photon is $\sin^3$, which is exponential 2x1/2 multiplication/division of invariance. Electrons are ½spin, while photons, which are 2x1/2 inertion/inversion = double inertion are 1 spin (Dirac), where ½ spin in space is 1 spin in time and 1 spin in time is ½ spin in space (massive in space is massless in time and equally vice versa, because spatial motion in instant size in time, just as spatial size is motion in time). Electron is spatial spinor and vector in time, while photon is spatial vector and spinor in time (and/or vice versa). Tensor is their inertion in between (invariance), which is angular ¼ within and or 2 without of horizontal 2x1/2=1 in between (double duality of 4x1/4 without/within, which is gravity/antigravity and/or space/time of mass/energy in between).

Scalar is 0D, vector is 1D, tensor is 2D and spinor is 3D, which is exponential 2x1/2 fractal curvature of singular point here and now. Vector is dual scalar, tensor is dual vector and spinor is double/inverse tensor, which is fractal 90° flattening of non-Euclidean curvature in context of equal and opposite curving of Euclidean curvature (4x90=360°). They are fractal particle/waves duality of dense mass within, which are 90°, 180°, 360° and 720° (¼, ½, 1 and 2 spins) that are 9, 18, 36 and 72-fold repetition (curvature) in inverse context of empty space of universe without (1/10 fractal difference). Round instant curvature within (space) is linear repetition without (passage of time) and equally vice versa, which is invariance or renormalization of such double/inverse infinity/singularity in between. Instant fragmentation within (space) is gradual multiplication without (time) and equally vice versa. Only perfect space/time duality (symmetry), which is right angle between double/inverse fractal realities, is invariance, which creates mass/energy, while rest is empty space (nothing). They are 4 fractal sequences of relatively different passages of time (light speed difference) that are real in context of inertial 4D here and now. Only such quanta are resonant with inertial passage of time here and now; therefore they are only possibilities of relatively different spatial density (mass) without (universe) and/or within (micro-world).

Fermions and bosons (half-integer and whole spin) are negation (inversion) of each other's reality, which is one 2x1/2 sequence in context of its double/inverse (invariant) reality that is their angular ¼ and/or 2 spins. Fermions are massive and bosons are massless (energetic) particles, where massive in space is massless in time and equally vice versa. Photon is massless particle of inertial space and/or massive particle of passage of time, while electron is massive in space and massless in time, which is light speed difference in between (difference between infinite/singular size in context of inertial here and now). Such dual determination is only surface (fractal) reality that is real only in context of inertial passage of time here and now. Except here and now, it is totally unreal reality. Atom and electron are fractal 1D and 2D negations (inversion) of inertial 3D spatial reality here and now; therefore they are fractal mass that is once and twice infinitely smaller and denser mass than here and now. Such mass is only fractally real in context of inertial (invariant) passage of time, which is singular and double uncertainty of inertial spatial reality here and now. A photon is 3D spatial inversion or total negation of 3D space here and now. It is complete inversion (negation) of spatial reality; hence its reality is massless. It is time particle or instant time density, which is singular quantum (negation or inversion) of spatial reality here and now (universe without).

Boundary (inversion) of reality of massive particle is its instant empty space without (universe), which is its passage of time. Massive particle is real exclusively as singularity within of its infinite empty space without, where fractal difference (overlap) between such singular/infinite size and their inverse densities is passage of time in between. Boundary of massive particle (fermion) is space, while boundary of its inversion—massless or energetic particle (boson) is time. Massive particle is instant in space (gradual change in time) and massless particle is instant in time (gradual change in space). Massless particle is angular depth (non-local spatial curvature) of surface reality of massive particle (local curvature) and equally vice versa. Particle is instant in space and wave (massless particle) is instant in time. Wave is space/time inversion of a particle and while particle is bounded by its infinite empty space, wave is bounded only by its passage of time, while it is instant in space. Massive particle is not real (probable) beyond boundaries of its infinite space, just as wave is not real by boundaries of its infinite passage of time. Infinite space without of massive particle within, which is its passage of time (gradual change of density) is instant (singular) reality of its wave. Wave is everywhere instantly within empty space of massive particle. Infinite space of massive particle of micro-world is here and now, just as infinite space of here and now is universe. Singular point of observation (surface of Earth) is infinite universe of massive particle of its micro—world, and its negation (inversion) is instant reality of such infinite space, because universe/micro-world are inertial and inverse space/time, which are negations of each other (double negation=affirmation).

Wave is instant spatial reality of infinite universe of a particle (uncertainty—everywhere at same time). Universe of a particle is micro-world here and now (inversion). Its propagation in context of double/inverse spatial reality of universe is its limitation within exclusive passage of time here and now (infinite multiplication without of singular point within in context of passage of time, which is singular inverse fractal of space). Particle (mass) is real only in its double/inverse context, which is its infinite empty space, regardless of its size—micro-world is inverse infinity of inertial infinite universe renormalized through singularity or fractal overlap of surface reality here and now in between. In other words—only motion less than light speed, which is smaller than infinite size of its universe, is reality (possibility) of inertial mass within, regardless if such mass is particle of an atom or an orbiting planet of solar reality (and/or star of galaxy etc). Beyond such infinite boundary, singular massive particle within is not real as particle, but only as a wave, because it is its light speed (quantum) change (impossibility in its double/inverse or more than totally uncertain/relative context). Inertial mass in spatial context is inverse mass or a wave in context of time and equally vice versa. Just as massive particle of universe (i.e. Earth) is bounded by its infinite space (universe), so is wave of its micro-world, because space (universe) of massive Earth is time in context of massive particle of micro-world.

Propagation of waves is exclusive reality in context of passage of time here and now. In context of such massive particle within that is its cause, it is negation of its reality (impossibility). Massive particle is enclosed by its infinite space (universe), which is its instant double/inverse reality (massless wave), while such massless particle (wave) is limited by passage of time. Wave is double/inverse reality of a particle in between. It is its infinitely larger/smaller spatial reality without and within. Spatial reality of a wave is not space but time, which is double (squared) space—i.e. relative space without **and** uncertain space within. Wave is infinitely larger/smaller (twice multiplied) reality of a particle (equally vice versa in context of time). Massive particle is instant in space, while massless particle (wave) is instant in time. Just as massive particle is not real beyond boundaries of its infinite space (here and now), wave is not real beyond boundaries of passage of time here and now (empty space of universe). Reality beyond its boundaries is impossibility or inversion of singular possibility that is real here and now (passage of time that is 5184-year period). Such division into massive/massless particles is exclusive reality of singular quantum of passage of time, while different periods are gradual change (negation in time) of such surface reality.

Massive objects of universe are its massive particles. Their reality is limited by size of empty space in which they are enclosed (which is variable of their density).

Beyond such boundaries, they are massless particles or waves. Triple fractal curvature of universe (mass within and empty space of solar and galactic realities without and/or triple empty space of solar/galactic/universal realities) is triple fractal reduction of mass and equal and opposite increase of their massless (wave) reality, just as inverse triple fractal curvature of atom/electron/photon is gradual decrease of mass and increase of energy (wave) in opposite direction. Mass of photon is totally inverted into wave here and now, while electron/atom in between are its gradual (fractal) changes, just as mass of Earth is reduced to a wave in context of universe, while intermediate solar and galactic realities are its fractal changes. Atom/electron within and/or solar/galactic reality without are double/inverse particle/wave dualities of perfect particle within and wave without (Earth of universe and/or photon of inertial here and now). Space/time (universe/ micro-world) is consequently 6-fold curvature of inertial here and now (6x60=360° of a hexagon, which is Schroedinger Q-space within and/or precessional perihelion advance of universe without). Massive/massless particle are 2x1/ 2 inertion/inversion=double inertion, which are limited by its space and passage of time, which are its double/inverse fractals without/within.

Massive particle of micro-world is wave of spatial universe, just as massive particle of spatial universe is wave in context of its passage of time, which is double/ inverse or relative (solar) spatial reality of universe. Wave is real only in context of passage of time, while mass is real instantly. Wave/particle is duality that is real only in context of its invariance (inertion), which is double/inverse or infinitely larger duality, which is 4x1/4 or double duality. Mass limited by its space and waves limited by time are real (possible) only in their invariant gravitational context, which is double/inverse boundary of both. Massive particle in inertial context (i.e. dense micro-world) is wave in inverse context (i.e. empty space solar reality). In twice-inverted context of empty space of galaxy, it is neither by being both (likewise—massive Earth of solar reality is wave of empty space of galaxy, which is universe of solar reality and neither, by being both, in context of empty space of universe without galaxy). Waves of micro-world are limited by empty space of solar reality, which is inertial passage of time here and now. Solar empty space is boundary of propagation of waves of micro-world. Beyond such infinite boundaries, they are inverted into their negation through 90° angular curvature of space/time (light speed distortion).

Waves are massless particles (time) only in context of relative (solar) reality, while in context of twice-relative galaxy, they are fractal 2 spin, which is 720° curvature and equally vice versa—waves in context of galaxy are fractal ¼ spin in context of micro-world here and now. Each fractal difference is 90° curvature of double helix or light speed difference (difference between singular and infinite sizes). Photon (massless particle) here and now is graviton in context of empty solar space and infinite reduction in context of galaxy. Photon is inversion of 3D space, limited only by passage of time here and now, while graviton is twice massless particle that is limited by neither space nor passage of time. It is instant in space and time (double light speed difference). Gravitational reality is complete 4D inversion of inertial here and now, which is inversion of space **and** time (4x90=360°). Mass is instant in space and energy (wave) is instant in time, which are their boundaries. Gravity is instant in space and time, which is unlimited either by space (size) or by passage of time (motion). Graviton is twice massless or double/inverse wave. Mass/energy are limited by space and time, while their gravity is limited by mass and energy (inversion). Graviton is wave within a wave in context of micro-world and/or wave without a wave in context of universe. It is infinitely smaller wave that is instantly everywhere within massive particle of the micro-world and not its empty space without. Graviton (mass) of gravitational wave is infinitely smaller (light speed difference) size than photon, which is its empty space without.

Photon is boundary of its instant massive reality within, while surface of atom is boundary of its propagation in time (boundary of its wave reality). Propagation of inverse gravitational wave of universe is limited by empty space of solar reality within. Gravitational wave is not real here and now, because it is interrupted by passage of time, which is space between singular size of an atom within and infinite size of solar reality without. There is no difference between strong nuclear force within and electromagnetic force without or electromagnetic force within and gravitational force without—they are 90° space/time curvature or triple inversion of inertion without/within, where each subsequent surface (nucleus/atom/here and now/solar reality etc) is fractal resonance in between two equal opposites (wave without of particle within and/or vice versa). Strong nuclear force is wave of massive graviton within and at the same time it is massive particle of a nucleus. Electromagnetic force is wave of nucleus that is massive atom on surface and massive atom on surface is wave in context of empty space without. Likewise—gravitational force of Earth in context of empty space of universe is electromagnetic force of its dense mass within, which is double/inverse difference.

Gravitational force without massive particle of universe and/or electromagnetic force without nucleus is fractal (decrease proportional with inverse square of distance), therefore limited by passage of time, while force within is multiple (increase proportional with inverse square of mass, which is double/inverse or horizontal and angular distance); therefore limited by instant size (space). Without/within is reality of inertial and inverse (convex and concave) curvatures and increase in context of one is decrease in context of another curvature. Massive reality in context of spatial curvature is wave (energy) in context of time curvature and equally vice versa, which is 4x1/4 reality. Force within (i.e. strong nuclear force) and force without (i.e. electromagnetic force) are double/inverse inertion of invariance. They are action/reaction without/within of invariant inertion in between (surface of mass). Strong nuclear force within nucleus is inversion of electromagnetic force without and equally vice versa (also electromagnetic force within and gravity without). They are action/reaction of each other in context of inertial 10^2=100-fold difference in between (double or gravitational fractal difference). Wave (motion) without is mass (size) within and vice versa. Inverse square of distance without is inverse square of mass within, which is double (spatial and time) distance.

Gravitational wave, which is instant in space and time, is unreal in context of passage of time (time bubble), which is its double/inverse interruption. Wave (energy) is instant only in time, which is limited by size of solar reality. Gravity (double wave, which is double light speed difference) is instant everywhere in galaxy at the same time (no passage of time in between) and propagated at light speed beyond boundaries of galaxy (light speed expansion of universe). Such gravitational wave is disrupted by passage of time here and now and is real only in context of quantum (precessional) changes, when such passage is halted. Gravitational waves are 2x1/2 curvature of time waves (passage of time), which are its double/inverse boundary. Gravitational wave is not real in context of instant spatial reality here and now, but only in context of spatial reality of passage of time. It is exclusive reality of relative space without and/or uncertain space within (smaller than atom). Gravitational wave is infinitely slow 25920-year precessional wave that is real in spatial context only in singular ¼ quantum (where ¼ within is 1/5 without, therefore 6480 years in spatial context and 5184 years in context of passage of time). Graviton of micro-world is likewise infinitely fast double wave (tachyon) that is instant spatial (surface) reality of a photon (massless particle). It is infinitely small wave of micro-world that propagates twice-light speed here and now and/or infinitely large wave of universe that propagates infinitely slow (inverse light speed) in context of passage of time.

Micro-world is gradual (fractal) inversion of time into instant space and photon is its boundary (3D inversion). Reality smaller than photon is twice-spatial reality of passage of time, which is double light speed difference ($E/m=C^2$). Such wave is twice faster than passage of time here and now—i.e. its rate of precession ahead of passage of time here and now is light speed difference, therefore it is instant not only in space but also in time. It is instant reality of all infinite space of galaxy and/or all future passage of time of inertial mass here and now. Gravitational wave is instant in future and past time, while singular wave (energy) is real only here and now (limited by passage of time). Gravitational wave is inversion of singular wave, therefore infinitely smaller and faster wave, where singular point of a wave (time) is expanded into additional linear passage of time. It is wave of a wave or wave of a photon, which is wave of mass here and now. Energy and mass are double/inverse waves of space and time, while gravity is double/inverse wave of mass and energy (i.e. mass/energy is empty space and/or passage of time of gravitational waves). In context of gravitational wave, photon is not massless particle within, but massive particle without (i.e. rotating planet of solar reality). **Gravitational wave is wave of time particle, just as wave (energy) is wave of spatial particle**, which is double fractal difference in between. Reaction of gravitational wave of micro-world within (C^2) is gravitational wave of universe without ($1/C^2$), which is likewise wave of time of inertial mass here and now. It is wave, where passage of time here and now is its singular (¼) spatial quantum. Singular quantum of energy (wave) is instant mass here and now, while singular quantum of gravitational wave is infinite passage of time of such massive particle (¼ precession).

Particle/wave duality is only ½ of 4x1/4 gravitational reality, where space and time are only its double/inverse invariant space and time. Gravity is double duality or duality in which space and passage of time are not equal opposites, but two instant spatial possibilities. Gravitational reality is simultaneous (invariant) reality of relative space without and its uncertain (inversely relative) space within. It is not real in their fractal context, which is passage of time, but only in context of quantum (whole) or invariant space/time reality. Space and time here and now are its double/inverse fractal boundaries (negations), which is interruption of such invariance in context of passage of time of empty space (here and now). It is real (possible) only in context of invariance, without surface differentiation between instant space and its fractal (gradual) passage of time. Consequently, its spatial reality is limited to smaller size than massless particle (photon) and/or passage of time larger than infinite (6480 years). It is C^2 reality, which is beyond boundaries of C (light speed), because it is double fractal difference or difference of space and time, which is whole difference (2x1/2=1), while exclusive surface reality of instant space and/or its passage of time are only duality of two equal and opposites fractal ½ 's. Space and its passage of time are not quantum, but only fractal ½ of gravitational reality, which are its double/inverse infinite distortion (interruption of invariance of physical reality through relative/uncertain passage of time that is exclusive reality here and now). Future (gradual) time is real only in context of passage of time, which is limitation of invariance by inertial here and now, while in context of gravity, it is its instant past time (instant spatial density).

Atom is inversion of one spatial direction, which is fractal ½, electron is inversion of two, which is fractal $^2/_3$ and photon is inversion of 3 directions, which is fractal ¾ reality, where remaining fractal difference is always relative/uncertain passage of time of such inverse fractal (spatial) reality. Graviton or double/inverse wave is fractal 4/5 etc. Universe without is inversion of such fractal reality of micro-world within, where passage of time within is instant space without and equally vice versa—passage of time without is instant space within. Here and now in between is double/inverse discrepancy between such equal and opposite spatial realities without and within, where infinitely small passage of time of infinitely small spatial reality within is infinitely long passage of time of infinitely larger spatial reality without. Small period of time in context of inertion is long period of time in context of inversion, which is double/inverse surface distortion of invariant time here and now. Massive within is massless (wave) without and equally vice versa, which is infinite double/inverse fractal helix of exponentially smaller periods of time within that are reaction of equal and opposite exponentially longer periods of time without. Mass/massless is only relative/uncertain determination, which is real only in context of invariant inertion here and now (singular quantum of passage of time and/or singular quantum of space). In context of double/inverse helix, it is double/inverse infinite curvature of exponentially smaller/larger spatial spheres that are exponentially smaller/larger periods of time in which such spatial spheres are real. Space/time is only singular fractal of such double/inverse infinite curvature.

In context of gravity there is no difference between infinitely large and infinitely small, because they are same. Small and large space is passage of time in between, which is not real in gravitational context, which is singular size between infinitely small and infinitely large. Infinitely large space without and infinitely small space within are only temporary distinction in context of passage of time of inertial density here and now. In context of gravity, which is invariant physical density, such double/inverse infinities are contracted into singular point of space that is double/inverse (fractal) or light speed difference. In context of relative/uncertain spatial reality, which is infinitely larger (4D) spatial reality without and within (passage of time here and now is its instant spatial direction), all the space between infinitely small and infinitely large is its singular quantum (infinite space here and now, which is completed passage of time, is its singular boundary within). Gravitational reality (C^2) is continuity of inertial (invariant) reality here and now (C), interrupted by fractality of passage of time.

Massless photon in context of passage of time here and now is massive particle in context of no passage of time (its instant spatial reality). Because reality is space and time, photon is both. Larger period of time is smaller its massive reality. It is massive only instantly, i.e. in context of 0-passage of time, while any passage of time is instant change of such mass into massless wave. Larger fractality is larger multiplication/division of such singularity in opposite direction. Singular massless photon that mediates between larger massive realities is infinitely fragmented in opposite direction of passage of time (past), which is spatial reality smaller than singular photon (spatial reality of passage of time). Weak nuclear force, which is light speed difference from electromagnetic force, is not mediated by singular photons but its fragmentation into 3 intermediate bosons, which are further fragmented into 8 particles etc. Such smaller particles are inversion of massive reality here and now. They are inverse (negative) mass, which is negation of spatial (positive) mass that is larger than photon (line of symmetry (time) in between). Such smaller massive particles are negative probability or fractal impossibilities (between once and twice impossible) of possible reality that are massive particles larger than photon. Photon is singular time, while smaller particles, which are intermediate bosons of massive reality smaller (or denser) than inertial here and now are its fractals.

Photon is negated space. It is not real in space but only in time. It borrows its spatial reality (mass) from spatial (relative/uncertain) reality of passage of time. Photon is singular (space/time) fractal difference, while smaller fractal particles are double (gravitational or space/space) fractal difference. Smaller fractal particles are not fractal (½) time difference but whole (gravitational) difference; therefore they also borrow their spatial reality from their negation, which is relative/uncertain spatial reality. Time is real in context of space and equally vice versa. Gravity is not real in space, but only in context of spatial reality of time (relative/uncertain reality). Gravity is singular/infinite boundary of finite size of spatial reality in between (renormalization), while fractal time in between is double/inverse (horizontal and angular) distance between such finite and infinite sizes without/within (radius between spheres without/within). Gravitational reality is infinity of different space/time possibilities that are not real here and now, which is empty space (vacuum) in context of passage of time. Photon borrows its spatial reality from passage of time (1D space) and smaller intermediaries borrow it from gravity (0D space), which is nothing at all (empty space) in context of passage of time. Such smaller fractal particles borrow their spatial reality from nothingness of empty space, which is instant spatial reality of time (relative/uncertain space).

Micro-world is inversion of empty space of universe without. Its inverse passage of time is inertial space of universe and inertial passage of time of universe is its inverted spatial reality. Nothingness of micro-world is infinitely large space of universe and equally vice versa. Infinitely small period of time of micro-world is infinitely large period of time in context of universe and equally vice versa, because negation in infinitely small spatial context is equal and opposite affirmation in infinitely large spatial context, etc. Spatial reality of photon is temporary in context of passage of time here and now (from which it borrows its spatial reality), which is its boundary. Photon is real only in context of passage of time, which is limited by ¼ precessional period (5184 years). Photon is eigenvalue between space/time. Smaller intermediaries (bosons) are its double/inverse reality—they are exponentially smaller periods of time in context of micro-world (i.e. action) that are exponentially larger period in context of universe (i.e. reaction). Larger massive reality of such intermediate bosons is shorter period of passage of time in context of micro-world and/or larger period of passage of time in context of universe. Mass is equal opposite of energy and space is equal opposite of time, which is double duality of 4x1/4 wave. Every mass is real only in context of its negation (empty space), which is double fractal difference of inertial (invariant) passage of time in between (fractal ½). Increased mass is decreased period of passage of time in context of uncertain micro-world and increased period passage of time in context of universe, where increased period of time is faster its passage and decreased period is slower passage (invariance). Infinitely small period of time is infinitely slow and infinitely large period of time is infinitely fast passage of time, which is double/inverse distortion of invariant (inertial) passage of time here and now.

Mass smaller than its singular quantum (photon) is exponentially smaller periods of time (slower passage) and mass larger than photon are exponentially larger periods of time (faster passage), where photon is singular quantum (eigenvalue) of space/time (inertial space in context of its singular quantum of passage of time, which is ¼ precession). Mass borrowed from empty space (gravity) is exponentially shorter duration proportional with its size/density in context of micro-world. In context of universe, which is negation (inversion) of micro-world, **larger borrowed mass is exponentially larger period of time. Larger mass of empty space is larger period of passage of time in which such possibility is real**. Its boundaries are infinite period of passage of time, which is singular quantum of gravity, where gravity is spatial reality of galaxy (i.e. 20 bill.

years). Shorter duration of fractal massive particles that are intermediaries (waves) of inverse spatial reality of micro-world is also longer duration of fractal massive particles in context of inertial spatial reality of universe. Earth is atom in context of solar, electron in context of galactic and photon in context of spatial reality of universe. In context of passage of time of spatial universe (relative universe), it is intermediate boson of weak nuclear force. Consequently, such inertial mass is longest duration in context of passage of time of universe and exponentially smaller periods of time in context of smaller fractal spatial spheres, where shortest period is 5184 years in context of solar spatial reality.

Larger period of time is larger fractal negation or spatial change and smaller period of time is proportionally smaller spatial change. Photon is negation of space; therefore it is fractal passage of time, which is only ½ (space/time) change, while smaller particles are additional negation of such time, therefore fractal gravity or double/inverse space within/without inertial space (gravity) here and now. Photon as fractal reality of passage of time is interruption of invariant spatial reality within/without its fractal boundaries. Photon is reduction of infinite 3D here and now to a singular invariant point that is equal opposite of inertial here and now (time of space and/or space of time), while smaller (inverse) particles are its inverse increase, which is Bode's periodicity. Photon is reduction of 3D space to spatial reality of linear (1D) direction of time, which is singular possibility of 3D space ($3^{\circ}=1$). Intermediate bosons of weak force are inverse increase to singular spatial possibility, which is triple or 3D ($3^1=3$). Photon is reduction of 3 (space) out of 4 (space/time) directions to 0, which is only one possibility of inverse duality ($2^{\circ}=1$ and/or $2^1=2$, i.e. it is singular negation or equal opposite of inertial here and now). Intermediate bosons of weak force are increase to 2 inverse possibilities (impossibilities in context of passage of time here and now), which are 3 (spatial) possibilities of duality in context of fractal passage of time ($2^2=4$, which is 3 different spatial possibilities of inertial here and now, which is its 4th or inertial time possibility).

Photon is singular inverse possibility (½) and intermediate bosons are further fragmentation into 3 relatively different inverse possibilities (¾) of inertial passage of time here and now. Intermediate bosons are 3 relatively different directions of inverse passage of time, all equally real, while photon is singular inverse possibility of inertial passage here and now, which is its fractal difference. Further fragmentation of inverse passage of time and/or inverse spatial reality (inverse mass) is $3^2=9$ and/or $2^3=8$, which is fractal 8/9 reality. It is 8 different intermediaries (quantum chromodynamics) in context of inertial here and now, which is their fractal difference or negation (9th particle here and now). Fragmentation/

multiplication of fractal 8/9 is $3^3=27$ in context of inertial $3^1=3$ double fractal difference, which is 6x4=24 different particles of smaller mass and shorter duration. $3^4=81$ in context of inertial $3^2=9$-fold difference is 72 different particles etc, which is infinite double/inverse fractal helix curvature of dense mass within and empty space without in context of inertial passage of time here and now ($3^5-3^3=243-27=216$ and $3^6-3^4=729-81=648$ particles, which is boundary of 6-fold Q-reality). Propagation, which is inverse square of distance without is inverse square of double/inverse distance (mass) within, because propagation in time (fractal) is inversion of instant spatial reality within, which is multiplication within of fragmentation without and/or vice versa. Smaller time of larger mass within is its larger time without.

Infinitely small period of time within is infinitely larger period without and equally vice versa—infinitely small period of time without is infinitely long period within. Reality is double/inverse infinite space/time fractality, whose boundaries are determined only by subjective (inertial) choice here and now. Larger subjectivity or inertion is smaller number of fractal sequences that are real here and now, which is larger limitation of invariance, while reduced subjectivity (inertion) is increased relativity/uncertainty of its surface projection, which is larger number of sequences and therefore more comprehensive (larger) surface duality. Infinite double/inverse fractal curvature is determined exclusively through relative/uncertain perturbations, because only invariance is absolute determination. Larger number of fractal perturbations is larger surface reality and equally vice versa, where limit of such perturbations is subjective choice of inertial point of observation, because any surface (physical) reality is double/inverse limitation of unlimited invariance.

Any surface reality is real only in context of its equal and opposite negation, which is infinity of different choices of inertial choice here and now. Space/time is double/inverse shadow of its invariance (here and now), where such inertial surface reality is real only in context of its equal opposite (negation), which is its reaction. Universe/micro-world is real only in context of its opposite, which is inverse or shadow universe/micro-world, where space is time and time is space (negative direction of passage of time). Shadow spatial reality here and now is only fractal passage of time, which is its gradual negation in context of inertial gravity that is invariant reality of both. Gravity transcends spatial and time boundaries, which are fractal difference (1/10) in between. Gravity in context of inertial reality is instant space, which is passage of time in context of its inversion and equally vice versa. Surface reality here and now is real only in context of its negation, which is its symmetrical mirror image, where time is not passage but instant (inverse) spatial reality. Such double/inverse realities are not real in each other's space/time context (except as passage of time), but only in context of gravity, which is real in context of space/time only during quantum changes. Passage of time is only fractal ½ of singular quantum, which is gravity. Only gravity is reality of space and time. Quantum change is intersection between such two possibilities of invariance, when passage of time is not passage but instant space. It is meeting point of inertial surface reality with its negation, which is instant merging of two different realities that create another, infinitely larger reality, which is equally real only in context of its negation that is another, relatively different passage of time.

Such inverse reality of different possibilities is passage of time of inertial space, where passage of time is not real as spatial possibility, unless singular ¼ quantum (De Broglie). Passage of time is fractal boundary between inertion and its inversion or singular possibility within and infinity of its different impossibilities without (mirror-image). It is boundary between real and unreal, something and nothing, etc. Unreal is not real here and now (passage of time), unless such impossibility is incorporated into inertial possibility, which are points of inversion of 4x1/4 wave and/or points of intersection between sin/cos curvatures (90° or light speed difference in between), which is triangulation of Pythagorean theorem (see illustration 12, 15). Impossibility or negation of inertial reality here and now is not real in context of passage of time, because it is 90° angular (inverse) curvature. Nothingness or negation of something that is real here and now is real only in quantum points, where such linear possibility/impossibility meet. Only resonance is real; therefore impossible possibility or unreal reality is real only when inertion of one is resonant with inverse inertion of another, which is creation of double reality that does not differentiate between previous reality and its negation. Affirmation and negation in inertial context are double negation in inverse context—they are only equal and opposite negations of invariant reality, which is affirmation of both.

Chapter 19

DISCLOSURE

Perturbation, which is subjective (inertial) choice of point of observation, is only limit of unlimited invariance here and now. Not only universe is participatory (Wheeler)—any physical reality that is space in context of passage of time (duality) is participatory, because inertion here and now is always invariance of double/inverse (space/time) reality without and within, regardless of its relative/uncertain (fractal) boundaries. Physical reality that is space in context of passage of time is fractal surface (½) of a whole; therefore only shadow of its deeper reality, which is not real in context of fractal passage of time but only in context of its negation, when such passage is instant space here and now. Space/time (physical) reality is double/inverse infinite echo of singular point here and now in between. Space/time (universe/micro-world) is one with singular point here and now and any fractal differentiation of such point is only temporary and limited reality in space and time. Such limited and temporary space/time (physical) reality is exclusivity of such fractally differentiated point, with no real meaning in context of its invariance, because all future passage of time of such point is its instant past time (mass). Physical reality is fractal 3/2=1.5 time bubble, which is +/- ½ variability in between singularity (invariance=1) and its larger duality (2), and where ½ variability is ½-way between something (1) and nothing (0) that are double/inverse invariant reality.

Lifetime of such fractal duality that is space in context of passage of time is invariably 1.5 of its ½-time (i.e. half-time of radioactive decay), where

1.5x0.96=1.44 (inertial lifetime of non-inertial passage of time and/or vice versa). Lifetime of inertial mass (size/density) here and now is 72x1.5x0.96=103.68, which is 103680 years (10^3 fractal difference, which is difference between 3D space and its linear passage of time), which are 4 precessional periods of 5 different passages of time and/or 5 periods of 4 different passages of time (fractal 5/4 without of fractal ¾ within). Lifetime of inertial massive sequence (gravity) here and now (Earth) is 103680 years of passage of time, while lifetime of its passage of time is ½ x10368=5184 years (5400x0.96, where $5400=2x27x10^2$ and $2700+5400=81x10^2$ or 5/4 of 6480, which is 5/4 of 5184 etc).

Spatial reality is instant, while time is fractal (gradual). Space is simultaneous infinity of all its relatively different fractals (inversions), while passage of time is only one fractal at a time (infinity without is singularity within etc). Different passages of time are not simultaneous (instant), but follow each other. Space is parallel and time is serial sequencing, because infinity of quanta of one is singular quantum of another. There is no different passage of time that is real here and now, unless light speed difference (90° curvature). Passage of time is 4 fractal curvatures within and without, which are eigenvalue (resonance) of inertial passage of time here and now (each is 90° of total of 360° for 4D reality). They are 4 spheres of relatively denser mass within and/or emptier space without (2x4=8). Such inertial reality does not change, unless accumulated difference is singular quantum (light speed). Only light speed change is fractal change of passage of time. Accumulated difference between rotation/orbital motions is light speed difference in 6480 years (¼ precession, which is invariant inertion of such duality). ¼ precession is instant (quantum) change of passage of time (double/inverse changes of rotation/orbital motion), which is light speed change. 4 such periods (precessional cycle) are 4 fractal changes, which is complete 360° curvature (4x90°).

Light speed difference is 4x1/4 change or 90° angular change (curvature) of inertial passage of time here and now. It is ¼ discrepancy between double/inverse waves, which is ¼ precessional motion (regression of one that is equal and opposite advance of another). Angular difference is time difference. Angular difference between different densities of micro-world within and/or universe without is 90°, which is 4 double/inverse fractal spheres within/without in context of inertial sphere in between, which is their line of symmetry (2x4+1=9x10=90°). It is 360° positive and negative curvature without/within in context of inertial 90° here and now in between (2x360=720+90=810 or 81x10). Just as different particles (waves) within are exponential 2x1/2 difference of their spins, which are light speed difference of passage of time (90°), so are massive particles without (particles of empty space of universe, as opposed to particles of dense mass of micro-world). Passage of time here and now (Earth) is light speed difference (90°) from passage of time of solar reality (empty space of Earth's orbital), while passage of time of solar reality is light speed difference of passage of time of galaxy, which is empty space of solar orbital (180° or double light speed difference). Passage of time of universe is light speed difference of galactic passage of time (270° or triple light speed difference), which is fractal ¾ of inertial ¼ here and now (overlap between 4D without and 4D within, which are 2x3D=6D in context of double/inverse 1D in between).

Solar reality and galaxy are 90/180° different passages of time, which are boundaries of linear passage of time and constant (changeless) gravity, which is double/inverse (invariant) time here and now. Rotating/orbiting particle in between solar/galactic reality is a planet, which is mass of double/inverse empty space. Massive planet is inertial discrepancy (overlap) between double/inverse passages of time of solar and galactic realities. It is wave (its double/inverse motion) between double/inverse particles (solar and galactic realities within/without), as well as particle between double/inverse motions (orbital/rotation). Ecliptic (orbital plane) is solar and rotational plane is galactic reality. Solar reality is east/west and galaxy is north/south direction. Angular inclination between ecliptic and rotational plane (polar axe) is space/time reality here and now (Earth). Ecliptic is round 2D curvature, while constant inclination of its polar ace is linear 1D direction, which is time and gravity of inertial space here and now. Space of galaxy is passage of time of solar reality, just as solar space is passage of time here and now, which is angular 90° difference (perfect symmetry). Space and time are horizontal/angular 90° (light speed) difference.

Polar axe is fractal of orbital plane of galaxy. Rotation is passage of time here and now (invariant inclination of polar axe), while its precessional wobble is passage of time of solar reality, which is real here and now only in ¼ quanta. Galaxy is double and solar reality is singular fractal difference. While Sun is singular 0D point of gravity, galaxy is linear (1D) direction, which is angular difference from plane of ecliptic. In context of solar passage of time, Earth rotates around such linear direction, which is its precessional wobble. One precessional period is one rotation around linear center of galactic gravity. Angular difference between polar axe and orbital plane is double/inverse difference between solar and galactic passages of time, which is size (gravity) here and now in between ($m=t^2$). Orbital position of Uranus is eigenvalue between solar/galactic passages of time (double/inverse gravity that is space/time here and now), which is reflected in its inverted rotation, where polar axe faces Sun and its rotation faces galaxy (Uranus is inversion of Earth's space/time reality, and so is Venus in opposite direction—hence their retrograde motions). Uranus' equator is aligned (perpendicular) with gravitational center of galaxy in the same manner that Earth's is not—Earth's equator is parallel, and its poles are perpendicular to such gravitational direction of galaxy (Pleiades), just as Uranus poles are parallel with its orbital plane.

Just as rotation of planet in context of solar reality is east/west (rotation is vertical direction of solar gravity)—rotation in context of galaxy is north/south (rotation that is vertical direction of galactic gravity). There is no change of inertial space/time (passage of time) here and now, unless accumulated difference is perfect dual symmetry, which is ¼ quantum of precessional motion. Galaxy is solar reality turned inside out, where inside in space is outside in time and equally vice versa. Only resonance is real; therefore only perfect symmetry between inertion/inversion creates double inertion. Only perfect symmetry between double/inverse passages of time of solar and galactic reality, which is perfect symmetry between time and gravity in spatial context and/or space and time in gravitational context, is eigenvalue (resonance), which is light speed or 90° (¼) difference. Only perfect symmetry between time (orbital motion) and gravitational angular inclination of polar axe (rotation) is instant change of passage of time here and now, because accumulated difference is light speed (see text bellow). Frequency of such perfect symmetry between solar reality and galaxy is 6480 years, which is solstice/equinox point here and now, because only such points are real in context of solar passage of time.

It is perfect 2x1/2 superposition between solar reality and galaxy in context of inertial space/time here and now (Earth) that repeats (saros) in 6480 (5184) year intervals, when Earth is obscured by solar reality and/or Sun is obscured by Earth's reality from galactic center, which is instant interruption of passage of time here and now (instant acceleration/deceleration). Solar/lunar eclipse is spatial interruption of flow of energy (light) from solar reality, while gravitational eclipse is interruption of invariant gravity here and now, which is negation of inertial passage of time. Solar eclipse affects linear flow of energy, while galactic eclipse affect mass, which is instant change of its double/inverse motion (change of rotation/orbital, which is change of passage of time that is eigenvalue in between). Solar eclipse interrupts flow of photons (waves), while gravitational (galactic) eclipse interrupts flow of gravitational waves, by interpolation of Sun between Earth and galaxy and/or Earth between Sun and galaxy. Solar/lunar eclipse is interpolation of one mass between another mass and its energy (Sun), while galactic eclipse is interpolation of mass between energy and gravity and/or energy between mass and its gravity, which is fractal difference in between (light speed). Perfect conjunction (syzygy) between Sun and galactic core, which is singular point of end/beginning between two different passages of time here and now is winter solstice 2012, when discrepancy between orbital and rotational motion (perihelion/precession) is perfect 15x0.96=14.4 days and/or regression of precessional motion is one season (90 days that are 90° angular or time curvature).

Gradual regression of time in context of precessional motion is not real (resonant), unless singular quantum (¼ of orbital, which is 90 days and/or 1 hour slowing down of rotation, which is 1/24 double/inverse time difference). Only ¼ regression of rotation in its orbital and/or vice versa is light speed (90°) difference, which is quantum change of passage of time here and now. **Winter solstice 2012 is end of passage of time that started 5184 years before and beginning of different (accelerated) passage of time that will last another 5184 years.** Invariant passage of time is singular quantum of relative/uncertain spatial reality and change of its speed is change of surface (space/time) reality, when reality and its negation merge (resonate) into larger reality, which is its following fractal (dual) sequence.

Change of angular rotation (seasons) is 0.5 arcsec per year change of angular inclination, which is 1° per 7200 years (10^2 or double/inverse fractal difference from 1° per 72 years of precessional motion), where 7200=1.111x6480 year. 6480 years is 0.9° and 5184 years is 0.72° change in angular inclination (0.18° discrepancy), which is double/inverse angular reaction (10^2) of accumulated precessional horizontal difference. At the time of conjunction of core of solar reality (Sun) and core of galaxy on winter solstice 2012, Earth's 3D space/time coordinates are: orbital period of 365.**234375** days, angular inclination of polar axe of **23.4375°** and equatorial circumference of 2490**2.34375** miles, which are fractal 1/10 difference in between. All 3 space/time coordinates (space, time and gravity) resonate on that day, which is once in 6480-years occurrence. Because of tidal drag, number of days per orbital period gradually (fractally) increases, which is equal and opposite equatorial expansion (increased obliquity) and increase of eccentricity of its orbital motion. Resonance between all 3 directions is critical point of inversion of gradually accumulated inertion that restores balance (symmetry). It is instant reversal of accumulated difference in context of previous passage of time. Winter solstice is 23.4375° or 0.9375° accumulated difference from perfect 22.5° (¼ quantum). 0.9375 is fractal 24/25 difference of 0.9° that is 6480 years (duality, which is inclusive of its inertial distortion).

0.0375, which is 24/25 discrepancy is difference between 6480 and 6780 years or 300 years relative/uncertain discrepancy, where 6480/300=21.6 and/or

300/6480=0.0463, which is rotational velocity (1/10 fractal difference). Rotational velocity is space/time here and now, which is inertial discrepancy of space/time reality without/within (universe/micro-world). Perfect 0.9° in context of such inertial (in-built) discrepancy is therefore 0.9375°. Perfect 22.5° (¼) inclination is in context of 6480 (5184) years increased through gravity in context of passage of time (tidal drag and/or disparity between rotation/orbital) to 23.4375°, which is fractal 24/25 (gravitational) difference, which is singular gravitational quantum (24x0.9375=22.5 and 25x0.9375 =23.4375). There is no reaction of such fractal accumulation, unless singular 24/25 quantum, which is instant reaction (reversal) of gradual action in context of passage of time. Only quantum, which is 0.9° or 90° in context of 10^2-fractal difference, is light speed difference, which is instant (quantum) change of passage of time. 23.4375x3600 =84375 arcsec and 84375x0.96=81000 or 3^4x10^3 arcsec (full curvature of fractal ¾). Difference between perfect 365.25-days sequence and accumulated difference of 365.234375 days is 0.015625 days per year, which is inversion of 64=2^7 and/or 4^3 (full curvature of inverse ¾). 0.015625 is quantum of fractal ¾ reality and any discrepancy less than quantum is smaller than light speed difference, which is not real here and now. 0.96 of accumulated difference of angular inclination and 64 accumulated difference of its orbital motion are 96/64, which is fractal 3/2 discrepancy between space and time (in context of inertial 10^2 difference within/without).

Winter solstice 2012 is perfect space/time supersymmetry, when space and time are not two equal opposites, but only equals (no fractal difference in between), because inertion in inverse context is equal to inversion in inertial context, which is resonance between two waves that merge to create larger/smaller wave. It is reduction of double/inverse (space/time) surface reality into singular gravity, which is their inertial invariance, which is instant reversal of interruption caused by fractal passage of time. It is instant burst of time bubble, when its solar (and/or galactic) framework is continuity of inertial sequence here and now in context of no passage of time and/or no empty space of solar reality but its dense mass. Fractal 3/2 symmetry is space/time eigenvalue, when time is not passage but singular quantum of relative spatial reality and when unreal is real and impossible possible, because surface reality is unreal in its depth and equally vice versa. It is instant broadening of horizons, when boundary of inertial life here and now is extended beyond its surface boundaries of round Earth, to include entire solar reality (and/or galaxy in context of double/inverse quantum change). Separation is real only in context of passage of time and its singular beginning/end are its instant negation, which is double/inverse infinite extension of life here and now, when Earth is not exclusive singular point of life of infinite universe. Larger quantum change is larger negation of life here and now (Earth) and larger affirmation of life in context of exponentially larger spatial sequences that are larger periods of uninterrupted passages of time, which is instant incorporation of both into singular, infinitely larger sequence.

Circumferential difference of 2.34375 miles from perfect 24900 is 1.625 difference from synodic/sidereal 236 sec discrepancy (fractal 10^2 without/within difference), where $6.25=5^4/10^2$ is difference between passage of time and gravity or quantum of fractal 4/5 (i.e. 5184x6.25=32400 etc). Perfect 24900 miles is 25000 miles or singular ¼ quantum in context of 10^2 discrepancy (inertion). Nothing exists without reason and nothing is real except reality. Coincidence is illusion created by subjectivity of inertial choice, which limits invariance that is double/inverse (dual) causative curvature. Coincidences are reaction of limitation of 3D reality to its surface projection here and now. It is spatial projection without time and/or time without spatial reality. Without non-local (angular) depth of surface (local) duality, physical reality is only coincidence. There is no space or time for any non-causative coincidence in double/inverse fractal helix curvature, which is perfect supersymmetry between surface symmetry and its braking in depth that repeats infinitely many times in relatively different dual sequences. Universe is no coincidence, neither is its micro-world, because it is only possibility of its inertial passage of time in between (surface of Earth), which is exclusive reality in duration of 5184 (6480) years.

Spatial distance of Earth from center of solar reality is 150-mill km, which is fractal 3/2 of its time (inertial discrepancy of 6 mill km, which is its 24/25 variability, is non-inertial distance of 144 mil km). 150-mil km is 500 sec or 8.333 min (fractal 5/6). 3600 sec per hour is frequency of 3600/500 or $6^2/5=7.2$ cycles per hr, which is fractal difference of Schumann frequency here and now, which is 7.2 cycles per sec. 24/25 difference of 500 sec is its inertial discrepancy of 20 sec, which is non-inertial time of 480 sec or 8 min, which is 3600/4800=7.5 frequency of fractal ¾ reality. Orbital position of Mercury is 0.4x7.2=2.88 and/or 0.4x7.5=3, which is double light speed difference. Uranus' orbital is 20x7.2=144 and/or 20x7.5=150-fold difference ($3/2 \times 10^2$), which is space/time inversion, while Pluto's orbital is 40x7.2=288 and/or 300, which is double/inverse or gravitational inversion ($10^2=100$-fold or gravitational difference between Mercury and Pluto). Orbital position of Mercury is $0.4=4\times10^{-1}$ and orbital position of Pluto is 40=4x10AU, which is 1/10 fractal difference within and without or $10/(1/10)=10^2$ (double/inverse) fractal difference of inertial 4D in between (Earth).

Inertial 4D here and now is passage of time that is 1.54321 km/year and/or 0.648 year/km of precessional motion, which is singular ¾ discrepancy of passage of time here and now ($86400x3/4=64.8x10^3$ etc.). Solar system of 10 planets is double fractal difference ($10^2=100$) of inertial 64.8 years that is 4D here and now (6480x4=25920 years of total curvature), where 64.8x1.111=72. 72 years/° of precessional rotation is 72-7.2=64.8 years here and now, which is 64.8-7.2=57.6 years/° of precessional orbital motion (perihelion advance) and $57.6-5.76=51.84x10^2=5184$ years of uninterrupted passage of time. Infinite number of different spatial curvatures (sizes/densities of dense mass within and empty space without, as well as their orbital/rotational motions) is singular possibility, which is double/inverse infinite spatial differentiation of singular passage of time, which is singular inertion of such infinite spatial universe/micro-world in infinity of its different curvatures.

Invariant here and now is singular cause of double/inverse space/time infinity without and within. Universe/micro-world is only double/inverse (serial/parallel) fractal repetition of such inertion, where only resonant harmonics (inertion=inversion) are real (E/m), while rest is empty space of passage of time (vacuum) in between such fractal surfaces. Distance to nearest star is 4 years, which is 4x360x1440 or $1440^2=2073600$ minutes ($1/10^2$ of solar orbital in years) or 2073600/320=6480, where 320 min is 40 AU or 8x40, which is distance of Pluto from center of solar reality. Nearest star is $64.8x10^2$= 6480-fold multiplication of solar gravitational reality, which is distance from its center to Pluto's orbital, just as solar reality within is $64.8x10^2$-fold multiplication of inertial gravitational reality of Earth (here and now). Solar system is 1/10 fractal difference without/within of inertial 4D here and now (0.4 AU position of Mercury within and 40 AU position of Pluto without, which is double/inverse acceleration/deceleration or densification and de-densification without/within), while nearest star is space/time inversion, where instant 4 sec of synodic/sidereal angular difference (240-236=4) here and now is 4 years of linear distance between stars. Solar space is 100-fold repetition of inertial 64.8 here and now, while interstellar space yet another 6480-fold repetition. Moreover—galaxy is 25920 light years radius, which is 25920/4=6480-fold repetition of singular distance between two stars.

$6480=6x10^3+480$ sec of linear distance between Earth and center of solar reality, which is discrepancy or broken symmetry of perfect 6-fold surface curvature and 480= 6x8x10 or 6-fold repetition of 80-fold repetition, which is again 4-fold repetition of 20-fold repetition of fractal 4/5, which is its inertial discrepancy. 6480-fold linear (i.e. local or surface) repetition is 4-fold angular (non-local) repetition and equally vice versa, which is 25920-fold double/inverse repetition that is 4x90=360° curvature, which is 72x360 or 360-72=288 and 288/360=4/5 fractal reality. 72 is 4x18 or fold-fold repetition of singular saros cycle in all 4 directions of inertial 4D here and now, while 57.6=4x14.4 or 4-fold repetition of Sothic cycle, which is its 5/4 or 3.6-fold discrepancy. It is double inverse repetition of saros within (2x1.8=3.6) in context of its 4-fold inertial repetition without (4x18=72), which is fractal ½ of reality that is real here and now, i.e. surface or physical reality that is exclusive reality in context of inertial passage of time. Reality cannot be explained away by coincidence, because coincidence is not reality but its causeless illusion. Conservancy principle is only law of invariant reality and causeless effect or reaction without equal and opposite action is its total distortion, caused by inhibition of inertial choice, which limits (curves) such unimpeded reality into a circular curvature, which is real only in context of saros (repetition), which is fractal passage of time. Time is not evolution but its arrest into repetitive circular curvature, which breaks out of it, only when inertion of such curvature is equal to its inversion (resonance), i.e. when negation of invariant reality through surface physical reality repeated infinitely many times is singular (negative) quantum of its affirmation, which is instant leap in evolution through its past gradual regression, because temporary regression is gradual accumulation of momentum for such instant leap in angular direction of surface reality. Surface evolution is regression of double (invariant) evolution and quantum change is its instant reversal; therefore double negation that is singular affirmation.

Surface reality is double/inverse Doppler effect of inertial and invariant here and now, where space is distortion in time and time is distortion in space. Surface (local) reality is only ½ of its (non-local) depth and equally vice versa. Subjective or participatory universe is double/inverse infinite distortion of invariant here and now that is inertial context of infinity of different surface realities. Winter solstice 2012 is confrontation of such double/inverse distortions, when **hidden variables** of surface reality are brought to surface, which is their instant incorporation (fractal Fibonacci sequence of inversion of inertion that is double inertion). It is instant disclosure that reveals superficiality of physical reality in its invariant (infinitely larger) context. Micro-world (dense mass within) is instant past time and universe without is gradual future time in context of fractal passage of time, where such future time here and now is instant past time in context of larger (invariant) solar reality. Micro-world and universe are **hall of records** of infinite past and future times, and while all records of past times are open in context of passage of time, all records of future time are open only in context of no time. **Winter solstice 2012 is opening of records of future time and closing of all records of past time.** It is turning to new page, which is new reality.

Surface perspective is double/inverse distortion, which is only fractal ½ of space/time reality. Spatial inertion is distorted by its uncertain/relative time and time inertion is distorted by its uncertain/relative space. Historical perspective is also double/inverse surface illusion, because it is distorted by passage of time. It is reconstruction of past time in context of inertial time here and now. Such reconstruction is exponentially distorted with passage of time, because larger time distance is exponentially smaller fractal of such past reality that remains real here and now (on surface). Time is spatial curvature and equally vice versa. Time is angular depth of surface spatial reality. In other words—in context of exponentially larger periods of passage of time, there is exponentially smaller fractal of such past sequence that remains real on surface of inertial here and now. Time curves surface reality of past sequence into its angular depth, which is beyond boundaries or horizons of surface perspective here and now. Past is curved out of horizon of surface perspective through passage of time. Larger distance of past time means exponentially smaller number of aspects of such past time remain fractally real here and now. Passage of time gradually erases all traces of such past sequences, by curving them in angular depth, which is negation of their surface reality here and now. Larger past period is larger fractal proportion of its reality that is curved in its depth (erased from surface); therefore such past is increasingly unreal and/or improbable in context of horizontal perspective here and now.

Increased past time needs increased resolution of observation, which will bring such reality that is hidden within on its surface here and now. Past/future time in context of inertial passage of time here and now is double/inverse distortion through inertion or limits of inertial resolution (distinction) here and now. Consequently, larger distance of past time is inevitably exponentially distorted by surface perspective, which is exclusive reality here and now. Larger distance of time is exponentially larger interruption of reality of such past periods, which is increasingly empty space or blanks of past reality. Infinite past time is infinite emptiness or total void that allows for infinity of different possibilities (total uncertainty of past time, which is infinitely smaller space here and now). It is one possibility here and now smeared all over surface in its depth. Unless interpolated by allowance for its time curvature, past is totally distorted, where unreal past is real, while reality of past periods is negated here and now. **Gaps in knowledge are as real as knowledge itself.** There is no objective reconstruction based only on infinitely fragmented knowledge, without interpolation of its increasingly larger gaps, because only both are reality. One without the other is not reality but its negation.

Surface perspective is reconstruction based only on fractal knowledge of past, without accounting for its increasingly larger gaps in between (non-active DNA is as real as its active parts). It is inertial spatial perspective that does not account for exponentially larger fractality of past reality that is increasingly curved in depth (within), beyond horizons of surface perspective (without). Only reconstruction that merges surface knowledge with its gaps, allows for complete reconstruction that does not deviate from reality. Reconstruction based exclusively on increasingly fractal remains is not reality but its negation through infinite distortion, because such reality is taken out of its time context. It is only round curvature distorted by flat perspective and or flat reality distorted by round perspective, which is its increased fragmentation in context of equal and opposite multiplication of gaps in between. Larger past is increasingly a gap filled by speculations based on inertion of present sequence, which is not causative but non-causative or coincidental curvature. Past cannot be reconstructed by applying subjectivity of inertial perspective here and now, because past is relatively different subjective choice. All spatial remains of past time (something) must be interpreted in context of their equal and opposite void (nothing), because they are only spatial remains taken out of its time context, therefore only fractal ½ of their whole reality. Their spatial reality in context of passage of time here and now is meaningless. Only in context of their (different) time, is such reconstruction reflection of reality and time is invariance, unless singular ¼ quantum. Gradual change of time is not real on surface, unless singular (¼) quantum of 5184 years.

Historical perspective is superficial distortion of past time, because it is determined in subjective framework that is exclusive reality here and now that. Its gaps are increasingly filled with subjective reasoning that is exclusive reality here and now and increasingly meaningless in context of past time. An effect has different causes now than before (or after) and equally vice versa—one cause produces different effects now than before (or after). Motivation that resulted in spatial monuments and/or oral and written tradition was different in past than is now. Applying subjective measure of present period of time to its past periods results in its total distortion. Unequal things have to be measured with unequal measures in order to be equal, because only double negation is affirmation. Equal measurements of unequal things result in inequality (distortion). Largest difference is not superficial surface difference, but its angular depth, which is infinitely larger difference. Angular difference is only small distortion of surface symmetry. Smaller surface discrepancy is larger difference in its depth. Surface ambiguity or contradiction that cannot be filled by inertion of subjective reasoning here and now is infinitely larger reality of such past times. Depth of past time, which is not real here and now is infinitely larger reality than its remaining surface; therefore small error in depth is large error on surface and equally vice versa—large error on surface is only small error in its total reconstruction, because depth is infinitely larger proportion of past time.

Small surface error in measurements does not significantly affect reconstruction of past time, while even small error in inertial framework of such reality is its infinite distortion (misunderstanding). It is of infinitely larger importance to view past time in correct inertial perspective of its own time, untainted by subjectivity of passage of time here and now, than infinitely large surface accuracy tainted by inertial subjectivity, which will misinterpret its results. It is meaningless effort to reconstruct past time through accuracy of surface measurements, unless its inertial framework is just as accurately gauged. In context of exclusivity of superficial surface observations, reality of its inertial context is neglected, because it is not recognized. Any interpretation that does not account for its inertion is distortion. There is difference between past and present (and/or present and future) times, and unless such inertial difference is recognized, any surface reconstruction is distorted. Only in context of inertial difference between present and past times is such reality invariant (no difference), while in context of no difference, such reality is infinite difference (distortion). Inertial framework of past time is infinitely larger reality of past time than its surface remains here and now and without accurate gauging of such inertion through interpolation of gaps caused by gradual angular curvature of time, any subjective reconstruction is only its surface distortion.

Past periods are same invariance as here and now, only in different fractal composition—i.e. difference is only superficial (relative/uncertain). Any subjective differentiation of invariance is double/inverse distortion that is relatively different physical (space/time) duality on its surface. Physical reality is fractal (½) of a whole (invariance or gravity of space/time), where one fractal within is infinity of inverse fractals without ($^2/_3$, ¾, 4/5 etc) and equally vice versa. Here and now is singular quantum of double/inverse infinite fractality that are its space/time (physical reality), because ½ within is 2 (duality) without; ¼ within is 4D without etc. Fractal in space is multiple in time and equally vice versa. Here and now is ½-way in between such equal and opposite multiplication/fragmentation of double helix. Physical reality is relative/uncertain (surface) comparison of invariant (3D) reality in between. Any subjective inertion that differentiate is only surface of invariance. Differentiation is only relative or temporary. It is real only in context of passage of time (time bubble). In non-inertial invariant context, such fractal double/inverse difference is meaningless (unreal). It is consequence of double/inverse illusion that surface differentiation between equal opposites is real (Adam and Eve parable), without simultaneous realization that such inertial subjectivity is limited by its passage of time (5184 years).

Comparison between different time periods is relative reality of invariant here and now. It is real only in context of passage of time, where each subjective difference is duality of equal opposites. Comparison is infinite dual variability of singular context. Any surface difference is equal and opposite difference in angular depth, which is not real on surface, unless singular ¼ quantum of passage of time (spatial curvature of time). Different past periods are inversely different in its depth. Their surface reality here and now is only fractal ½ of their reality. Comparison is differentiation and difference is always double/inverse—it is advance of one in context of equal and opposite regression of another. Any advance on surface is regression in its depth and equally vice versa, where advance in time is regression in space etc (regression of rotation through precession is advance of orbital through inverse perihelion precession). Advance/regression is double/inverse evolution of no real meaning in its invariant depth (larger than ¼ or 90° curvature). Comparison is inertial subjectivity. It is subjective or superficial determination of one period of time as advanced in comparison to another, because advance within is regression without and equally vice versa. It is only different composition of fractal surfaces that does not affect its invariant 3D depth. Here and now is one out of infinite number of possibilities, all equally real, only not here and now. Past periods of time are fractally different possibilities of same inertial here and now and any comparison is only relative (real only in context of its time).

Past periods are relatively less developed only in context of inertial subjectivity here and now. Less developed on surface is more developed in its depth, which is beyond horizontal perspective that does not account for time curvature (relativity). There is no superior or inferior evolutionary sequence—only relatively different permutations of its fractal surfaces. Infinity of different spatial forms here and now is infinity of different fractal perturbations of invariance, and so are different time periods. Different spatial forms are parallel and different time periods are serial evolution, which is double/inverse perturbation of only one reality that is no difference in space or time. Comparison in which duality is not two equal opposites but their inequality is not reality but its distortion. **Two periods of time are two equal opposites**. Perspective in which two periods of time are not equal but unequal opposites is consequence of distortion through subjectivity here and now, which does not accounts for reality of gaps in knowledge of different periods of time. Because gaps are unaccounted for, reality of such past periods is infinitely contracted in surface perspective, which is unbalanced duality of subjective inertion that is real only in context of passage of time (5184 years). There is no two different periods of time, in which one is more or less developed than another. There is only two relatively different periods of time in which some aspects in context of one are more developed than in context of another, where such imbalance is compensated by larger development of relatively different aspects in context of another period.

Law of conservancy is only law and conservancy of invariance is always maintained (no choice). Larger development of one fractal aspect is regression in development of another aspect. There are no two same sequences of Fibonacci curvature, because it is infinite double/inverse variability of invariance. Perspective in which past periods are inferior to present period is distortion of reality, because it does not maintain conservancy. It is only subjectivity of inertial here and now, which is exclusive reality of inertial passage of time. Winter solstice 2012 is instant end of such illusion. There is only relative difference between past and present. Any distinction or preference of one over another is subjectivity of inertial choice, which is not real in context of space/time. It is 3D tainted by inertion of its surface perspective. Only relative distinction between two equal opposites is real. Distinction that does not recognizes their equality (Adam and Eve parable) is not reality but its distortion, because it breaks conservancy principle, which is only temporary real in context of passage of time. Past periods of time are not less or more developed than here and now. They are only relatively different periods, which are equally developed. Present period is lengthy development in surface distinction, while past periods are more developed in angular distinction.

Past periods are increasingly void on surface (here and now). Such void is not negation but affirmation of their reality. Present sequence is not the only sequence—it is only one out of infinite number, equally real sequences, only not here and now. Relative/uncertain boundary between two different sequences is 5184-years of passage of time. Any past sequence beyond such quantum boundary is infinitely large gap here and now, because it is its total negation (inversion) in context of inertial passage of time. Past period that is less than 5184 years is only gradual reduction of its reality, while periods larger than 5184 years are complete (whole) negation. Past period that is less than light speed difference (90° curvature) is only fractal negation, while larger period are complete negation, because reality of curvature larger than 90° is beyond horizons of horizontal perspective, regardless of its resolution—90° curvature is absolute boundary of flat horizontal perspective. Reality of different quantum periods of time in the past in only small surface discrepancy here and now, which is surface reality of such past period reduced to only singular line of horizon (boundary). Present period of passage of time (5184 years) is only one in infinite unbroken chain of relatively different periods, where any unequal distinction between them is unreal (subjective).

Present period did not develop from virtual void of past periods. It is not gradual linear evolution, but curvature of 4 equal periods that are one 360° curvature, where such curvature is only one fractal loop in infinite double/inverse helix curvature. There is no real distinction between 4 last periods of passage of time, which are all equally developed. Any distinction between 4 quadrants of a circle is relative—they are 4 equal quadrants that invert into each other every 90°. Past 20736 years is equality of 4 relatively different periods, all equally developed, where successive period is inversion of past period. Present evolutionary sequence (i.e. civilization) that ends on winter solstice 2012 is last sequence out of 4-fold repetition of equal sequences and/or last out of 4x4=16-fold repetition, which is 4-fold repetition of precessional cycles (4/5 of Milankovich cycle of low eccentricity). Great flood of 5184 years ago was ¼, passing of ice age in northern hemisphere 10368 years ago was ½ quantum change, while event of 2012 is 4-fold quantum multiple (difference between 90°, 180° and 1440° change in 2012). It is last sequence of inertial and double/inverse repetition of uninterrupted repetition of $4/(1/4)=4^2=16$ different sequences. It is singular sequence that is pinnacle of evolution only in flat surface perspective that does not account for its time curvature, which is not reality but its distortion and separation of reality through a singular time bubble from its infinitely larger inertial context that is infinity of self-same sequences.

Past periods of time are equally evolved as here and now. It is only relatively different evolution. They are regressed only through non-recognition of surface void as their reality and small surface discrepancy that cannot be accounted for by inertion here and now. Present civilization is only one sequence of unbroken chain of civilizations that inexorably end with beginning of next sequence. Historical perspective that goes back 5184 years is only temporary limitation (boundary) of infinite depth of history, because quantum change is complete interruption of passage of time, where any history larger than its singular period is curved beyond horizons of linear perspective of time. Surface remains of such past periods are reduced to singular line of symmetry here and now, which breaks surface symmetry in its depth. Reality here and now is real only in context of inconsistencies and paradoxes, which are its negation in context of its past periods. Past periods are real here and now only as unexplained and misunderstood discrepancies that cannot be dealt away by inertial subjectivity of present passage of time. They remain only as ambiguous prophecies that have no roots in superficiality of present surface (dual) sequence.

Difference between historical periods is only relative. Past periods of time are equally developed civilizations that ceased in moment of quantum change, just as present sequence ceases on winter solstice 2012. It is only subjective notion that present sequence is technologically or materially more developed than its past sequences, which is evolutionary regression, because it does not anticipates future events in same measure that past periods did. Past periods are infinitely more advanced periods, because their development was not only superficial but also in-depth development. It was larger development of self-awareness, which is infinitely larger reality than its surface (fractal) physical reality. Only evolution of self-awareness is double, which is invariant evolution, because only such evolution is better adaptation. Surface (material) evolution is exclusive spatial evolution, which begins and ends with its singular quantum changes. It is exclusive reality of inertial 5184-year period, which is totally unreal beyond its boundaries. Its limit is 21.12.2012 and beyond such point, it has no reality. Only evolution of self-awareness is evolution in space and time, which is limitless evolution and only surface evolution that is simultaneously evolution in its depth is unbound and real beyond such cut-off date. Increased self-awareness is only means of adaptation that transcends its quantum boundaries. Exclusive surface evolution is meaningless and its reality in subsequent 5184 years of passage of time is reduced to surface void (nothingness).

Present period is exclusivity of surface projection of in-depth reality, which is paradoxical formulation of two equal opposites without their inertial context. It is duality of surface perception as exclusive reality. Boundaries of such subjective sequence are irreconcilable 2x1/2 without their angular ½. Relativity and uncertainty, as double/inverse fractal surfaces are limits of surface perspective. There is no further advance without their incorporation within singular inertial context. Such fractal supersymmetry that recognizes double/inverse helix as ultimate reality and whose recognition avoided present evolutionary sequence (through inertial subjectivity) was a norm of past periods of time. Transience of surface reality in context of passage of time as uncertain/relative vibration of infinitely larger self-awareness was not an end but beginning of past evolutionary sequence. That recognition that evaded present sequence was known throughout infinite number of past periods. Inherited prophecies that have no meaning in context of exclusive surface reality are remainders of that fundamental understanding of past periods of time. Prophecies of past periods are reflection of their intellectual propensity and relative achievements, which are surface void in flat historical perspective here and now. Prophecy of today is science of yesterday. It is infinitely larger adaptation that is able to predict future events with larger accuracy than present period, by an uninhibited approach.

Past period of time is global civilization; same as here and now, in which fundamental tenets of reality were known throughout, just as were known infinite number of periods before. Mayan prophecy, with end date of 21.12.2012, is only last vestige of past periods, which was common understanding throughout at beginning of present sequence of passage of time. Same prophecy is incorporated in Gregorian calendar, which is inherited division of time from ancient Sumer, passages of Bible (Genesis, Book of Daniel), megalithic structures scattered across the globe and in innumerable written and oral myths of larger or smaller magnitudes. Fractality of space/time reality is core of ancient Egyptian, Vedic and Tao philosophies and religious teachings, as well as numerous minor streams of thought that pervaded nations of Earth, before their inhibition by subjectivity inherited from Middle Ages of European and Middle East thought development. Meaning and message they convey is synchronized. Great Pyramid, Mesoamerican pyramids, ziggurats and all host of megalithic monuments are constant remainder of past times and message they convey, which is unambiguous without inhibition by subjective impediments. Spirals, swastikas, pentagons, hexagons, 5 or 6-pointed stars, ankh, cross, wheels, etc.—infinity of different symbolic inscriptions of past times resonate only one meaning—fractal supersymmetry of double helix, which announces superficiality of physical reality in context of passage of time and inertion of subjectivity as only boundary of invariant reality.

Present sequence of passage of time started with inherited understanding of fractality of surface space/time reality, which was gradually suppressed by pervading inhibitions of inertial subjectivity. Egyptian, Sumerian, Vedic and Chinese civilizations at the beginning of present passage of time are not self-starting development but inheritance from past, infinitely more developed periods of time that were able to predict future events with more accuracy than is possible here and now through superficial understanding based on fractality of infinite number of contradictory theories. Period of time that was intimate with ultimate meaning of reality cannot be less, but only more developed than here and now, which is its remote regression. Only cause of such temporary regression in evolution of self-awareness is limitation of unlimited by inertion of subjective choice as only possibility. Superior evolutionary development of past periods of time is unreal here and now, because of its distortion by passage of time. Prophecies, inherited by such times are negation of impossibility that past was more developed than present time. Same prophecies are innate outcome of physical theories here and now as well, which is irrefutable confirmation of past time as superior beginning of an ignominious end.

Disillusionment of subjective choice as only real option is due at long ago appointed date, which arrives on winter solstice 2012, as prophesized by past and present. Winter solstice 2012 is closure of one inertial sequence, when present is returned to its past and/or past advanced to its present, through 4 subsequential ¼ periods of a circle (4x5184=20736 years), which is only $^2/_3$ of larger 6x1/6 double/inverse sequence (6^2=36) of fractal double helix. It is 6x5184 =31104, which is 0.96x32400 and where 31104=36x864 (6^2-fold repetition of inertial time of 86400 sec, which is double 10^2 fractal difference), 216x144 (6^3-fold repetition, which is fractal 3/2; therefore inertial time of 1440 min in context of singular 1/10 fractal difference) and/or 1296x24 (6^4-fold repetition of inertial time of 24 hrs, which is no fractal difference here and now, because 864/36=24, while 1296/24=54, which is 10^2 fractal difference of 5400-years that is ¼ of 21600-year precessional cycle and/or 5400-216=5184 non-inertial time (double/inverse boundary of passage of time). Division of time in current use is encoded with same prophecy as Mayan time count and/or prophetic passages in Book of Daniel that describe passage of time as 4x1/4 periods of precession, where each 90° curvature is quantum change of reality, which is instant end of one and beginning of another period in context of fractal 3/2 space/time reality (time, time and a half time or time, two times and half time, which is 2x1/2=1 etc). 31104x60=1866240, which is 6x60=360° curvature that is 2073600/1.111, which is compete cycle of time (gravitational quantum).

Quantum event of winter solstice 2012 is return of past, which is future of present reality. It is opening of hall of records of future time (universe), which are mirror image of hall of records of past times (massive Earth), which is instant discontinuity of passage of time and its inertial subjectivity here and now. It is instant continuation between past and future times in context of inertial present passage of time in between, which is contracted into singular point between such double/inverse curvatures of a wave. Discontinuity of fractal passage of time is spatial continuity, when empty space of universe (sky), without its passage of time, which is its constant force of repulsion, instantly contracts into instant dense mass of solar reality. It is inversion of nothingness of vacuum into dense mass, which is continuation of temporarily separated singular life sequence (Earth) throughout solar system (and/or galaxy in context of double quantum change, which is beginning/end of 129600-year sequence), inclusive of its 11th planet, which is inversion (negation) of time as passage and its confirmation as singular spatial quantum. It is moment of resonance between two relatively different spatial possibilities of time without/within (relative/uncertain space), which creates different fractal possibility here and now.

It is closure of one and beginning of another sequence of infinite fractality of double helix (snake, dragon, flying serpent etc. in ancient myths), which is infinite repetitive ritual of multiplication/division of self-aware reality here and now. Self-awareness not only permeates every form of physical reality—it is undivided within each of infinite number of its fractal forms, because double helix multiplies through division and divides through multiplication. More fractal divisions within are more fractal multiplications without and larger subtractions from its invariance are larger its additions and growth in time. An idea grows and multiplies, only through its division and subtraction. As spiraling of double helix approaches same point in time from its opposite polarity, forgotten fractality of physical reality is again remembered, because it is not only change of fractal space/time reality but also equal and opposite change of self-aware inertion here and now, which is its only source. Winter solstice 2012 is resonance of past and future times in context of timeless self-aware inertion in between, which is invariance of any fractal duality. It is resonance of self-awareness here and now with double/inverse inertial reality of universe without and micro-world within (self-interference), when they are not separate double/inverse infinity but singular one.

GLOSSARY

Since fractality exceeds inertial boundaries of any theory, it's essential that all used terminology have much broader meaning than accepted through present definitions. Terminology that is used is not limited by present definitions, although it's not in contradiction with it. In context of fractality, same terms are enhanced with non-local depth of surface interpretation, by allowing that relatively different interpretations of one reality meet in singular context. Since reality is one in all its forms, different terms are only relatively different interpretations of same thing. There is no real difference between different terms but only relative (comparative) difference. Fractal or quantum, inertial, symmetrical, eigenvalue or resonance, synchronicity, gravity, relativity, uncertainty etc mean different things in different contexts, but underlying definition of all is the same. When knowledge exceeds its boundaries, the only choice is to expand meaning of different terms as well. One reality of all things cannot be explained by use of limited definitions and interpretations that separate a whole (quantum) in infinity of relatively different fractals. Explanation of larger truth requires equally enlarged interpretation of various terms. An attempt to clarify that expanded interpretation of different terms is listed bellow.

Acceleration (deceleration)—any exponential increase/decrease, regardless of its fractal form. 1D fractal difference from constant (linear or 1D change) and 2D difference from no change (singular or 0D change). Acceleration is twice inverted reality of no change and/or once inverted reality of linear change; therefore it is 2D or surface change. It is accelerated space in time and/or accelerated time in space, which are double/inverse acceleration or accelerated space and time in context of invariant gravity (where positive acceleration is inverse or negative deceleration). Accelerated/decelerated duality is double duality or fractal ¾, which is 3D space in context of passage of time (and/or vice versa). Accelerated space is double/inverse space and accelerated time is double/inverse time, which are mass/energy of empty space/time (and/or gravity/antigravity of mass/energy). Acceleration/deceleration is synonymous with horizontal and/or angular densification/de-densification, which is 1D fractal difference or space/time difference, while double acceleration is gravitational or 2D fractal difference, which is densification in horizontal and angular direction. Acceleration in inertial context is linear direction (constant) in double/inverse context and equally vice versa.

Anti-parallel—parallel directions with opposite orientations (parallel vectors inversely orientated), where any change of one is equal and opposite change of another. Any anti-parallel change is dual (positive/negative or double/inverse) change with sum that is no change at all (0) and difference that is 1-(-1)=2. Any change of anti-parallel curvature is double change of no change, which is dual (space/time) surface or physical reality of invariant reality. See also—double/inverse curvature.

Asymmetry—equal opposite of duality. Any reality that is not exclusivity of two equal opposites. Symmetry of infinite number of relatively different surface parts, which are only relatively different fractal composites of one (invariant) reality in its depth. Infinite number of fractal symmetries. See also—broken symmetry.

Boundary—synonym of horizon. Point of refraction between two different dimensionalities, which is angular 90° curvature of one of its direction. Boundary of 3D is surface, boundary of surface is linear direction and boundary of linear direction is singular point. Surface, linear direction and singular point are 2D, 1D and 0D boundaries of inertial 3D etc (3D is likewise boundary of 4D). Reduction of spatial reality is equal and opposite increase of its time, therefore 2D space (surface) is 4D time, 1D space (linear direction) is 5D time and 0D space is 6D time and equally vice versa, where time is passage of time only here and now (3D), which is boundary of 4D that is spatial reality of time (relative/uncertain space without/within). Boundary is point of inversion between infinity in direction without and singularity in direction within (and/or vice versa). Any boundary is horizon or sky of its 3D within. It is form, which is—1D fractal difference from its content. It limits singular point within (center of a sphere), which is not real beyond its infinite boundaries without (surface of a sphere). Boundary is surface between inversion without and inertial content within.

Broken symmetry—Discrepancy of perfect symmetry, where symmetry can be linear, accelerated or any exponential or inverse symmetry. 3D or physical reality is broken symmetry, which is temporary duration (time) of inertial space and/or temporary space of inertial passage of time. Broken symmetry is space/time of invariance and/or invariance of space/time. Only fracture of symmetry is real, i.e. only both symmetry and its simultaneous breach in angular direction are space/time, because time breaks spatial symmetry and space breaks time symmetry, while gravity breaks space/time symmetry. Broken symmetry is 3rd direction or inertial context of surface symmetry. Symmetry and its fracture are double/inverse negation of each other, which is fractal difference in between.

Balmer series—pattern of spectral lines, where each principal spectral line is sub-divided in number of subordinate spectral lines (different spectra), which are its fractal 2x1/2 multiplication/division. Each consecutive spectrum is created on same principles in which white light refracts into infinite number of rainbow colors.

Causality—fundamental principle that there is no reaction without equal and opposite action and equally vice versa. It is singular principle of reality and any coincidental or accidental explanation is deviation from its only tenet. Coincidental explanation is always an effect without a cause, which is meaningless; therefore no explanation at all. Reality is infinite causative curvature, where effect is inversion of its cause and at the same time it is inertion or cause of yet another effect. Reality (something) is always an effect of its negation (nothing), which is inversion of inertion.

Collapse—infinite/singular boundary, light speed or fractal (quantum) difference, threshold of 90° curvature. Complete negation (inversion) of inertial here and now and/or vice versa, because of total distortion of invariant reality, which is double/inverse negation of causality. Space collapses in all its time (90° time curvature) and time collapses in all its space (90° spatial curvature). Space and time collapse in context of 180° or ½ curvature (gravity), while entire 4D collapses in 360° (excluding inertion of such reality, which is always additional fractal difference or perpetuity of fractal curvature).

Constant—invariant framework of double/inverse variability, i.e. C=d/t etc. Constant is singular inertial framework of double/inverse or dual variables. Constant is fractal difference (larger/smaller dimensionality) from its dual variables.

There is difference between constant and its variables only in dual context. In 3D context, which is triangulation, such constant is 3rd variable of inertial 4D context, which is its constant etc. First instance is fractal 2/3 or two variables of constant 3D, while second instance is fractal ¾ or 3 variables of constant 4D etc, which is infinite fractal curvature.

De Broglie eigenvalue—property of electron, which changes its quantum state in leaps determined by ¼ precessional motion of its orbit.

Densification—exponential increase/decrease in double/inverse context. Exponential increase in inertial (quantum) context is exponential decrease in inverse (fractal) context and/or vice versa. Dense space is empty time and dense time is empty space—see acceleration

Derivate—surface duality of inertial 3D, 3D of 4D, 4D of 5D etc. It is reduction of larger invariance through one (or several) of its directions into its fractal parts. 3D is triple derivate of space in time and/or triple derivate of time in space, which is 6-fold fractal curvature. 3D is 6 relatively different derivates (dualities), while 4D is 12 different 3D derivates; 5D is 20 different 4D derivates etc. Fractal 2/3 is 6, fractal 3/4D is 6x12=72 different 2D derivates or 72-fold fractal curvature, while 4/5D is 20x12=240-fold curvature and 3/5D is 6x240=1440-fold curvature. Fractal 2/5 is 72x240=17280, which is 2/3 of 25920-fold curvature (25920-fold repetition of relatively different fractal surfaces or 25 920 relatively different years, which are duality of orbital/rotation in context of one precessional curvature).

Double duality—horizontal and angular (perpendicular) duality, which is 4x1/4 reality of sin wave and/or 4 quadrants of a circle. It is double/inverse space/time curvatures, which are mass/energy of empty space/time and/or gravity/antigravity of mass/energy. Double duality is fractal ¾ that is 3D space or 3 instant directions in context of 4D space, which is remaining 1D fractal passage of time. Last direction is always smaller than singular quantum, because it is inertial context of remaining instant directions (larger than quantum). It is always passage of time of inertial space, which is instant space only in context of all its time.

Double helix—double/inverse or anti-parallel spiraling curvature, which are spatial curvature of time and/or time curvature of space, best known in molecular biology (DNA). See double/inverse and anti-parallel curvatures.

Double/inverse curvature—curvature in which any change in one is equal and opposite change in another direction. Any change of inertial here and now is cause of double or dual effects, which are negations of each other. Double/inverse is fractal curvature, because change in one direction is exponential change or square (2) of inertion in between (1), while change in opposite direction is its inversion (-1). 2-1=1D fractal difference or 90° curvature between realities that are without and within. 1D fractal difference means that within is inversion of without—space without is time within and time without is space within, which is 2-(-1)=3D in between. 3D in context of 1D fractal difference is surface space/time duality (3-1=2). Duality in between singularity (1D) and 3D is space in context of passage of time, which are linear gravity within (instant spatial contraction of mass) and/or 3D antigravity without (gradual spatial expansion of empty universe). Double/inverse is fractal 3/2 curvature of inertial here and now, which is 2x3=6-fold fractal curvature (6 points of fractal inversions or 6 relatively different surfaces of a cube, all equally real). 6-fold fractal curvature is 3D space in curvature of double/inverse time (time without and time within, which are relative and uncertain space of universe/micro-world); therefore fractal ½ of 3x4=12-fold fractal curvature. Etc.

Duality—2D reality. Exclusivity of two equal opposites without their inertial framework. One of relative/uncertain surfaces of a cube. Its reality is limited by its boundaries, which are double/inverse or positive/negative (spatial and time) inversions. It is duality of two (double/inverse) variables, without their constant framework. Duality is singular/infinite distortion of invariant 3D, which is real beyond its boundaries only through renormalization. Duality is double/inverse or 2x1/2 reality.

Eccentricity (and/or obliquity)—discrepancy and/or asymmetry of perfect curvature. Curvature is 4x1/4 or double duality and its eccentricity is inertial discrepancy of both. Also discrepancy or inertion between two perfect curvatures without/within. Eccentricity is inertion of curvature—it is fractal proportion of particle in context of perfect wave and/or wave within a wave. Eccentricity is singular quantum or boundary of surface curvature, which is its infinite repetition. Eccentricity is time of spatial curvature and/or space of time curvature, which is angular curvature or distortion of surface curvature. It is perfect curvature in angular direction, where perfect surface curvature is its eccentricity. Eccentric in space is round in time and equally vice versa.

Eigenvalue—line of symmetry between equal opposites. Resonance between inertion/inversion (space/time). Point of refraction between inertion/inversion, which is infinite boundary of one (without) and singular boundary (within) of another. It is singular spatial quantum of infinitely long passage of time or singular quantum of relative reality without, which is infinite uncertainty within. Only perfect surface symmetry is inversion of inertion that creates double inertion, because only perfect surface symmetry is non-inertial reality, which is instant surmounting of in-built inertion or instant quantum leap. It is boundary of fractal passage of time.

Equivalence—symmetry, causality, invariance etc

Everett's many worlds—quantum interpretation of infinity of different space/time realities, all simultaneously real (but not here and now, which is singular eigenvalue between such double/inverse infinities).

Exclusivity—synonymous with duality or space that is real exclusively in context of its passage of time, which is its equal and opposite negation. It is even-numbered fractal of odd-numbered quantum, which is inclusivity of both equal opposites in inertial or invariant context. Expansion of Pauli's exclusion principle—i.e. one fermion per one quantum state, which is infinite passage of time or singular time quantum of such fermion (space).

Fission—fractal 2x1/2 fragmentation/multiplication of invariance, which is exponentially less dense mass/energy (gravity/antigravity). Inversion of fusion, which is time of fused spaces and/or space of fused times. See fusion.

Force—invariance of space/time, which is double/inverse curvature. It is overlap or differential between positive/negative non-Euclidean curvatures that is Euclidean (flat) curvature in between. Fractal 1D difference, which is tension (inertion) between space and its fractal passage of time (also between space/time and its gravity, which is double/inverse or infinitely smaller/larger force, which is force within force)

Fractality—self-same repetition of one reality in space and time, where repetition in time is inversely proportional with repetition in space. Fractality is equal opposite of a quantum—it's less than singular quantum, which is space of time and/or time of space. Since space and time are inversely proportional—fractal space is quantized time and vice versa—fractal time is quantized space. Fractality and quantization are equal opposites. Quantization is multiplication, while fractalization is fragmentation of a quantum (inertial here and now). Difference between fractal and quantum is relative, because what's fractal in space is quantum in time etc. Fractality is double/inverse (space/time) fragmentation of singular quantum. Because infinite number of spatial fractals is singular quantum of time and infinite number of time fractals is singular spatial quantum, there is no clear distinction between quantum and fractal. Quantum is inverse fractal and fractal is inverse quantum. Quantum of passage of time is repetitive curvature of 5184 different fractals (1 year is fractal time of precessional time, which is singular quantum etc). Passage of time is fractal spatial reality, just as space is fractal passage of time. Relative or 4D space, smaller than its singular quantum (5184-fold repetition) is not space but passage of time; therefore fractal spatial reality. Likewise—time smaller than singular quantum is not passage but instant uncertain space of micro-world. Space in context of passage of time is duality and duality is always fractal, because of fractal difference between realities without/within (dense mass/empty universe). Fractal in context of one is quantum in context of another, which is double fractal and/or quantum reality, where quantization without (universe) is inversely proportional with quantization within (micro-world). Mass is quantized space (fractal time), while empty space of universe is quantized passage of time (fractal space). Quantized space is fractal time and quantized time is fractal space (quantized space/time is fractal gravity and fractal gravity is quantized space/time). Infinity of inverse fractals is one inertial quantum and infinity of inverse quanta is one inertial fractal. Gradual quantum change is instant fractal change in opposite direction, just as gradual fractal change is instant quantum change in opposite direction. Fractal reality is reality of passage of time, which is 1D discrepancy between double/inverse curvatures without/within. It is reaction of equal and opposite action delayed by passage of time in between, which is their line of symmetry (eigenvalue).

Fractal difference—dimensional difference or difference between instant (spatial) and gradual (time) change. It is ¼ or 90° of a curvature, which is linear light speed difference (light speed difference in one out of 4 directions of double duality). In 5D context, light speed or fractal difference (passage of time) is 1/5 or 72°; in 6D, it is 1/6 or 60° and in 2D it is 180° curvature. In fractal context, such difference is not fractal but multiple—i.e. in context of fractal ½, it is 2x360=720° (spin 2); in context of fractal ¼, it is 1440° (spin 4), in context of fractal 1/72; it is 25920 (72-fold spin) etc, where fractal in time is multiple (quanta) in space and equally vice versa. Fractal difference between double/inverse (relative/uncertain) spaces is passage of time in between and fractal difference between passages of time without/within is space in between. Multiple of time is fractal of space and/or vice versa.

Fractal sequence—one out of 6 surfaces of 3D, one out of 12 3D's of singular 4D, one orbital out of 25920, one out of 360 rotations, 1 out of 10 planets of solar system, one electron of an atom etc. It is one sequence out of infinite number of sequences, all equally real. It is one relative and/or uncertain possibility of infinite number of equal possibilities. Fractal sequence is reality less than quantum; therefore reality that is real only in context of passage of time. Fractal sequence is spatial quantum of fractal passage of time and/or time quantum of fractal space, which are infinite number of repetitive smaller curvatures within 90° of its double/inverse curvature. Fractal sequence is 1D fractal difference from its quantum boundaries.

Fusion—blending of two equal opposites into singular invariance (i.e. space and time into mass and/or energy, as well as mass and energy into gravity etc.). Fractal fusion is fractional blending of lighter into heavier mass etc, which is 4-fold fractal curvature in 4D context and/or triple in 3D context etc. Fractal fusion is ¼ of a whole or 90° of a curvature. In 6D context; it is 6-fold curvature or 7 shells of an atom (inclusive of its inertion), which are exponential 2x1/2 fusion without and/or fission within.

Gauge symmetry—dual symmetry in context of invariance. Double/inverse variability of a constant.

Gravity—inertial context (invariance) of space/time duality. It is double/inverse (2D) fractal difference of its space and time, which are only 1D fractal difference in between, It is infinitely smaller and/or larger (uncertain/relative) space of inertial space here and now in between and their correspondingly (infinitely) faster/slower passages of time. Space/time is duality or 1D fractal difference in between, where 1D is only ½ quantum of duality. Quantum of duality, which is physical (space and time) duality, is 2D fractal difference, which is again only ½ quantum of double duality (4D). Gravity is infinitely smaller and/or larger spatial reality, beyond singular/infinite boundaries here and now. It is continuation of fractal space/time (physical) reality interrupted by passage of time, which fragments/

multiplies (densifies/de-densifies) invariant reality. It is continuity of inertial space/time reality here and now beyond its boundaries. Gravity/antigravity of space/time is renormalization of double/inverse infinite reality here and now, which expands/contracts 1D fractal boundary, which is exclusive passage of time, to include such passage as instant space in context of double/inverse (infinitely larger/smaller) passage of time. Gravitational reality of Earth is its solar reality without and/or atom within in context of no passage of time here and now, which is renormalization of such double/inverse realities. Gravity is non-local or action at distance, because it is not only instant space but also instant time.

Half-life—½ or gravitational quantum. One 2x1/2 fractal sequence, which is difference between invariance (1), ½, ¼, 1/8 etc (which is 2, 4, 8-fold curvature or repetition in opposite direction). Each half-life cycle is one fractal sequence, which is infinitely larger period of infinitely slower passage of time (and/or infinitely larger and emptier space of infinitely smaller and denser mass).

Hidden variables—angular depth of surface duality. Non-locality of local reality. Reality beyond its singular/infinite boundaries, which is spatial reality of time and/or time reality of space. It is gravity or double fractal difference of space/time reality on surface (here and now), which is negatively probable (more than uncertain/relative). Space/time reality that is beyond boundaries of space and time or spatial reality that is beyond its boundaries (time).

Horizon—see boundary

Inclusivity—3D or any odd-numbered space/time reality, which is inclusion of inertial context (gravity) in its double/inverse (space/time) distortion. It is action/reaction that accounts for its inertial framework, i.e. it incorporates its singular/infinite boundaries. See renormalization, exclusivity and triangulation.

Infinity—inversion of singularity. 90° curvature or light speed difference. Boundary of singularity, where infinite space is singular time and equally vice versa. Infinite number of inertial fractals is singular inverse quantum etc. It is point of inversion between double/inverse or space/time inertion. It is beginning/end of passage of time and/or spatial size that is real here and now. Infinity/singularity are inertial framework or invariance of space/time duality within/without its boundaries.

Inertion—a broader term that's not only inertia in classical interpretation but absolutely any resistance to change, regardless of relativity of particular form. It's only reality there is—it's resonance or overlap between equal opposites, which is space of time and/or time of space. It's line of symmetry between equal opposites that is at the same time a quantity that introduces their asymmetry. It's supersymmetry between negation and affirmation, randomness and symmetry, positive and negative etc. It's singular context of any duality, which is their discrepancy, invariance, superposition, harmonization etc. It's a quantum of inversely proportional fractal space and time. Duality is real only as long as it's within its inertial framework, because it's all there is—infinite repetition of harmonized inertial points in space and time—the rest is nothingness of empty space and/or passage of time. It is time delay between spatial action and reaction and/or spatial delay between action and reaction in time (i.e. non-local interaction). It is line of symmetry or eigenvalue, therefore gravity of space/time, where spatial attraction is time repulsion and equally vice versa. Inertion is non-differentiated singularity of dual space and time. It is discrepancy or deviation from perfect symmetry of action/reaction, which is its infinitely larger inertial context.

Inertial difference—see fractal difference, eccentricity

Inversion—inverse inertion; inertion turned inside out. Equal opposite or negation of inertion. Inversion of spatial inertion is time inertion and equally vice versa. Inversion is reaction of inertion—it is time of space and/or space of time (and/or gravity of both). Inversion is singular/infinite boundary without/within inertion in between—it is also space and time of inertial gravity. Distance/difference between inertion/inversion is space in context of passage of time, which are double/inverse boundaries of invariant reality here and now.

Instant—spatial reality in context of no time

Invariance—non-differentiated reality. Singularity of duality (space and time). Inertial context, which is double/inverse cause of its space and time.

Local—exclusivity of surface reality without its angular depth, which is space without its time and/or space/time without its gravity.

Many worlds—see Everett

Mass/energy—any double duality (4x1/4), which is double/inverse density of empty space/time; therefore singularity or finite round curvature within infinite flat reality, which is its inertial context (boundary) and/or infinity or linear direction of infinite angular round curvature. Mass/energy is double space/time, which is accelerated/decelerated or densified/de-densified inertial space/time here and now (in between). Double/inverse space/time is mass/energy (singular density) and double/inverse mass/energy is double density (angular/horizontal density) of gravity/antigravity.

Milankovich cycle—empirical time cycles of +/-100000 years of low eccentricity, followed by +/-20000 years of high eccentricity of Earth's orbital, which are 5 periods of 20736 years followed by 6th inertial period or total of 5x25920=129600-5184=124416 years (fractal 5/6 of perihelion precessional advance, i.e. passage of time limited by precession of orbital motion).

Momentum—angular inertion or force. Gravitational or invariant reality of space and time (conservancy principle). See inertion.

Non-local—angular in-depth curvature of surface reality. Negation or inversion of local reality. Space of time and/or time of space (gravity of space/time).

Participatory universe—space/time reality as reaction of inertial here and now, which is exclusive reality in context of passage of time. Invariant reality distorted by double/inverse inertion, which are space and time of inertial gravity.

Perturbation—synonymous with fractality. Perturbation of surface duality is its small discrepancy (eccentricity). Perturbations are infinite fractal curvature of discrepancy within (and/or without) discrepancy, where small spatial discrepancy within is large time discrepancy or large passage of time without and equally vice versa. Perturbation is inversion or fractured symmetry of inertial surface symmetry. Each successive sequence within and/or without is infinitely smaller spatial discrepancy (light speed difference), therefore larger number of exclusive spatial perturbations has exponentially smaller surface effect. At the same time, smaller spatial reality within is larger period of time without; therefore such exponentially smaller spatial perturbations are reaction of exponentially larger time perturbation (and exponentially smaller space/time perturbations are exponentially larger gravitational perturbations). Perturbation is double/inverse angular (non-local) depth of surface (local) reality. Double/inverse or space/time perturbations are exponential increase of relativity/uncertainty in directions without (space) and within (time). Perturbation is inertial context or infinitely larger/smaller reality without/within inertial surface (space/time) here and now. It is double/inverse exponential fractal increase of inertial here and now. Perturbative process is renormalization of inertial reality here and now beyond its singular/infinite (double/inverse) boundaries, which is zigzag pattern of double/inverse exponential curvature (negative discrepancy of positive discrepancy and/or vice versa).

Principal—inertial surface duality (space/time) or singular starting point of fractal 2x1/2 sequencing. See Balmer

Quantum—inverse fractal (see). Singular quantum of infinitely large reality is inversion of inertion. Infinite reality is infinite multiplication in one direction, while its singular quantum in opposite direction is equal and opposite infinite fragmentation. Singular quantum is infinity of fractals, just as singular fractal in opposite direction is infinity of quanta. Quantum/fractal are double/inverse inertion of inertial here and now in between. Quantum/fractal are singular/infinite boundary or 90° (¼) angular curvature in double/inverse (without/within) depth of inertial surface in between (space/time). They are spatial boundary of time in context of equal and opposite time boundary of space. Reality beyond quantum/ fractal boundaries is beyond boundaries of space and time, which is their gravitational reality (inversely probable space/time). Gravitational reality is ½ or 2x90=180° of inertial 4D curvature in which exclusive space and time are fused into singular mass and/or energy (inverse mass of empty space).

Quantum leap—beginning/end of passage of time (fractal/quantum boundary) in context of perfect double/inverse symmetry that instantly surmounts inertion of surface duality. Instant expansion/contraction of space/time (infinite/singular) boundaries. Instant change of size/motion (fractal density). See De Broglie.

Renormalization—reduction of infinity to singularity or infinite flat reality to finite round curvature, which is inversion of inertion that creates double inertion or turning of inertial reality inside out. Renormalization is inclusion of inertial framework (boundaries) into context of dual action/reaction, which is singular overlap between double/inverse infinity.

Resonance—perfect surface symmetry, which is equality between inertion and its inversion or perfect non-inertial reality in between. Double/inverse curvatures, which are 90° angular difference (sin/cos waves), gradually (passage of time) reduce such difference, because of inertial discrepancy in between (i.e. particle of a wave or inertial size/mass in context of motion/energy etc). Inertial discrepancy is fractal or relative difference (overlap) that is cause of such double/inverse waves. Singular curvature in between such double/inverse curvatures is their precessional curvature (tangent), which is real only in context of passage of time. Reduction of 90° difference (positive or negative reduction, i.e. regression or advance) is their resonance, when such two equal and opposite waves resonate. Resonant waves incorporate each other creating a new wave, which is their sum/difference. Resonant points of a wave are 90° inversion points, which are ¼ periods of their precessional motion. One wave is 4 different inversion points that are 4 periods of precession in which spatial and time waves combine to produce relatively different gravitational wave in between. Full 360° curvature is passage of one wave (i.e. cos) through complete length of another wave (i.e. sin), which is one precessional period of orbital/rotational duality and where only cause of infinity of subsequent fractal waves (curvatures) is singular size (mass) here and now. Precessional wave is singular effect of orbital/rotational duality (cause), which is again an effect of massive particle in between (cause). Precessional wave is only ½ of duality, which are two equal and opposite precessional motions (fractal 90° in between) or two infinitely larger sin/cos waves that are precession and its negation (double/inverse precession). Such double/inverse precessional waves are yet another larger precession in between, which is infinite double/inverse precessional (fractal) curvature of double helix, where points of resonance that create larger wave are exponential 90° (¼) curvatures between equal opposites.

Singularity—inversion (negation) of infinity, which are double/inverse (space/time) boundaries within and without of inertial here and now in between.

Shell—Surface that is duality of double/inverse curvatures. Surface created by horizontal convex and angular concave curvatures. Shell is duality of double/inverse or two non-Euclidean curvatures differentiated for 90°. It is instant reality of both—spatial and time waves, which is spatial reality (size) of precessional motion. Shell is invariance of double/inverse or space/time waves (eigenvalue between spatial and time curvatures). Shell is gravity of space/time—it is singular space (size) in context of infinite time (motion), as well as infinite space in context of singular time. Shell is gravitational renormalization of double/inverse infinite/singular space and time. It is surface (boundary) of infinite number of different fractal space/time dualities. Shell is gravity (3D) of space/time (double/inverse duality), which is their invariance.

Space/time—Any double/inverse duality or any action/reaction. It is fractal duality of inertial 3D. Instant space and its passage of time are negation of each other; therefore double negation that creates singular affirmation (gravity). Space/time are surface of in-depth invariance, which are double/inverse (relative/uncertain) fractal distortion and not the reality.

Spatial curvature—finite round curvature of infinite linear time. Negation or singularity of infinite time. It is fractal density of time, which is double/inverse time. Double/inverse times without and within are singular (inertial) space in between. Two inverse spatial curvatures (times) are one time curvature (space) in between and equally vice versa. Surface of a sphere (i.e. space) is duality of two equal and opposite round curvatures of time in space (spatial curvatures)

Spin—Spatial or time round curvature. Infinite time in singular spatial curvature and/or infinite space in singular time curvature (i.e. round size and/or round motion, which is singular limitation of infinite flat reality). Spin is 4x1/4 or 360° curvature, which is time of space and/or space of time. Spin in space is wave in time and/or vice versa.

Sphere—see shell

Subjectivity—fractal determination of invariant reality. Dualistic expression of inertial 3D in which depth that is only cause of surface space/time duality is overlooked. Relative/uncertain reality understood as absolute. Any subjective determination is double/inverse distortion and not the reality. Subjectivity is perspective in which only one, out of infinite number of equal sequences is regarded as exclusive reality. Subjectivity is double/inverse negation of reality that confirms reality as equal opposite (depth) of surface duality. Any distinction in which one is given preference over its equal opposite (negation) is not reality but its distortion. Reality understood as something without nothing as equal and opposite something is relative/uncertain (fractal) distortion or subjective perspective of reality that is limited by boundaries of its passage of time (¼ precession).

Symmetry—Duality of two equal opposites, negated by line of symmetry, which is their overlap or discrepancy (broken symmetry). Only perfect symmetry that is infinitely thin line of symmetry is non-inertial duality, which is instant inversion of inertion. Only double/inverse symmetrical infinity is singularity in between. See exclusivity.

Subordinate—fractal perturbation in depth of surface duality, which is its exponential 2x1/2 multiplication/division. See Balmer

Supersymmetry—duality of symmetry and asymmetry (broken symmetry). Continuation of spatial reality interrupted by its fractal passage of time and/or continuation of passage of time interrupted by its fractal spatial reality in between. Symmetry between relative/uncertain or double/inverse curvatures. Symmetry that renormalizes broken symmetry, which is infinitely larger (double/inverse) symmetry. Symmetry between local and non-local or spatial and time (fractal) symmetries. Supersymmetry is whole symmetry (quantum) that incorporates two equal and opposite fractal symmetries. It reconciles surface paradoxes and inconsistencies of infinite number of relatively/uncertainly different fractal dualities, because it accounts for their inertial context as their singular (invariant) cause/effect. Supersymmetry is inertion or gravity of infinite number of different space/time realities, all equally real.

Synchronicity—see resonance

Syzygy—conjunction or opposition between two massive objects (particles) in context of inertial here and now (Earth).

Time curvature-Inversion or equal opposite (negation) of spatial curvature. It is finite round curvature of infinite linear space. Time curvature is singular quantum of spatial curvature, which is its infinite double/inverse repetition without and within and equally vice versa.

Triangulation—Pythagoras's theorem. Duality in its inertial context, which is constant (invariance) of its double/inverse variability. Triangulated reality is triple constant, each of them double/inverse variability; therefore totality of 6-fold variable curvature (curvature that is 6 relatively different fractal parts). Double triangulation is 6^2=36 and triple triangulation is 6^3=216 relatively different fractal or surface parts of 3D invariance. Invariant reality is total curvature only in context of all its fractal surfaces; therefore 6-fold (36 and/or 216 etc) repetition of its singular inertial curvature (singular quantum of infinitely fragmented reality).

Each fractal part is one possibility or one out of infinite number of results of exponentially triangulated equation, all equally real, just as duality is equality of two (double/inverse or positive/negative) results. Triangulation is fractal 3/2 duality, just as duality is 2/3 of triangulated reality. Triangulation of duality and/or dual triangulation is infinite double/inverse fractal 3/2 curvature of double helix.

Variable—negation (inversion) of constant, which is fractal distinction between two equal opposites, because constant in space is variable in time and equally vice versa. Variable of inertial reality is constant of inverse reality (1D fractal difference). Variable space is constant time and variable time is constant space, which is double variability in context of equal and opposite double constancy. Twice variable reality is variable space and time of constant gravity and/or vice versa. Variability of a constant and/or constancy of variable is double/inverse infinite fractal curvature, where each additional variability generated as eigenvalue between constant/variable, is space of time and/or time of space; therefore additional fractal curvature of space/time.

Virtuality—Angular depth of surface reality. Non-local reality beyond singular/infinite boundaries of local reality (inversely or negatively probable on surface)

Wave—fractal oscillations of space in time and/or time in space, which are double/inverse or spatial and time waves. Quantum interruptions of gradual (fractal) reality. Larger spatial wave within is smaller time wave without and vice versa—smaller spatial wave is larger time wave. Surface wave is double duality or 4x1/4 sin/cos curvature, while 3D wave is additional angular oscillation (rotation). 3D wave is eigenvalue between sin and cos waves, therefore their tangent in between.

Wave/particle duality—space/time duality of invariant gravity (fractal size in motion and/or size of motion)

0-595-29125-2

Printed in the United Kingdom
by Lightning Source UK Ltd.
109882UKS00001B/52

9 780595 291250